7th International Conference on DV-Xα Method (ICDM 2019)

The Advances-Related Experiments and Theories on Material Science

IOP Conference Series: Materials Science and Engineering
Volume 835

Semarang, Indonesia
2 - 4 September 2019

Editor:

Robbi Rahim

ISBN: 978-1-7138-3611-7
ISSN: 1757-8981

Printed from e-media with permission by:

Curran Associates, Inc.
57 Morehouse Lane
Red Hook, NY 12571

CURRAN ASSOCIATES INC. proceedings .com

Some format issues inherent in the e-media version may also appear in this print version.

Institute of Physics
Dirac House, Temple Back
Bristol BS1 6BE UK

Phone: 44 1 17 929 7481
Fax: 44 1 17 920 0979

techtracking@iop.org

Additional copies of this publication are available from:

Curran Associates, Inc.
57 Morehouse Lane
Red Hook, NY 12571 USA
Phone: 845-758-0400
Fax: 845-758-2633
Email: curran@proceedings.com
Web: www.proceedings.com

TABLE OF CONTENTS

IOP Conf. Series: Materials Science and Engineering **835** (2020) 011001 doi:10.1088/1757-899X/835/1/011001

Preface

It is an honour for Universitas PGRI Semarang (UPGRIS-Indonesia) to be the host of the 7th International Conference on DV-Xα Method: The Advances-Related Experiments and Theories on Material Science (ICDM 2019). We are pleased to have cooperation with DV-Xα Society to hold this international event. As we know, the previous conferences related to this topic had been conducted six times: once in Hungary, three times in South Korea, two times in Japan and supported by some institutions as the hosts: The Institute of Nuclear Research of the Hungarian Academy of Sciences (ATOMKI-Hungary), Seoul National University (South Korea), Kunsan National University (South Korea), National Institute of Physical and Chemical Research (RIKEN-Japan), Korean Basic Science Institute (KBSI-South Korea), and University of Hyogo (Japan).

ICDM 2019 was an international event aimed at presenting contemporary research and accommodating discussion for scientists, scholars, engineers, students, research institutes, and industries in the field of advanced materials, science and information technology. It was successfully held on 2 - 4 September 2019 in Semarang, Indonesia. This event was organized by Universitas PGRI Semarang (UPGRIS - Indonesia) in cooperation with DV-Xa Society (Japan), Kwansei Gakuin University (Japan), Kobe University (Japan), Universitas 17 Agustus 1945 Semarang (Indonesia), Universitas PGRI Madiun (Indonesia) and under the coordination of Centre of Science and Technology UPGRIS (Indonesia).

ICDM 2019 was attended by 110 participants, and a total of 60 papers were presented and discussed. The papers were authored by 30 students and 34 lecturers. Categories of presentation sessions were divided into two sessions; lecture and poster sessions. There were 4 keynote speakers, 18 invited speakers, 20 general speakers who delivered the talks in lecture sessions and 18 poster presenters. Conference presenters were from various countries, 1 presenter from the United States, 1 presenter from Thailand, 1 presenter from South Korea, 1 presenter from Estonia, 25 presenters from Japan, 1 presenter from Ukraine and 30 presenters from Indonesia.

We are proud and grateful for the interest you expressed in the discussion of this theme. It provoked discussion among participants over the best ways to implement DV-Xα Method in advances experiments on material sciences. Throughout the process, hopefully we could contribute answers or alternative solutions for the demand of the progressing industry in regards to advanced materials to support the development of many countries.

In particular, we would like to express my sincere gratitude to the keynote and invited speakers for sharing their expertise with us as well as all participants for the excellent presentations and active discussions. I can conclude that the purpose of the conference has been accomplished. Last but not the least, we are thankful to IOP MSE for producing the proceeding.

Editors:
Robbi Rahim

Chairman of the BICED 2019:
Mega Novita, Ph.D

IOP Conf. Series: Materials Science and Engineering **835** (2020) 011001 doi:10.1088/1757-899X/835/1/011001

IOP Conf. Series: Materials Science and Engineering **835** (2020) 011001 doi:10.1088/1757-899X/835/1/011001

IOP Conf. Series: Materials Science and Engineering **835** (2020) 011002 doi:10.1088/1757-899X/835/1/011002

Peer review statement

All papers published in this volume of *IOP Conference Series: Materials Science and Engineering* have been peer reviewed through processes administered by the proceedings Editors. Reviews were conducted by expert referees to the professional and scientific standards expected of a proceedings journal published by IOP Publishing.

IOP Conf. Series: Materials Science and Engineering **835** (2020) 012001 doi:10.1088/1757-899X/835/1/012001

Chemical forms of rhodium ion in pure water and nitric acid solution studied using ultraviolet-visible spectroscopy and first-principles calculations

Shinta Watanabe[1], Toshikazu Sato[1], Miki Harigai[2], Yusuke Inaba[2], Kenji Takeshita[2], Jun Onoe[1*]

[1]Department of Energy Science and Engineering, Nagoya University, Furo-cho, Chikusa-ku, Nagoya, Aichi 464-8603, Japan
[2] Laboratory for Advanced Nuclear Energy, Tokyo Institute of Technology, 2-12-1 O-okayama, Meguro-ku, Tokyo 152-8550, Japan

j-onoe@energy.nagoya-u.ac.jp

Abstract. We have investigated chemical forms of rhodium (Rh) ion in nitric acid solution by using spectroscopic and first-principles methods, in order to develop the disposal of high-level radioactive nuclear liquid wastes (HLLW). The electronic structure of Rh complexes in both pure water and nitric acid solutions was investigated using ultraviolet-visible (UV-vis) absorption spectroscopy. Comparison with many-electron calculations indicates that Rh ion forms an octahedral bidentate complex coordinated with three nitrate ions, $[Rh(NO_3)_3]$, in both pure water and nitric acid solutions. This is consistent with thermodynamic prediction. Since the overall feature of UV-vis spectra of the Rh complex remained unchanged in a range of 0–3 M nitric acid solutions, the structure of the Rh complex was found to be stable in the range. In addition, since no significant changes in UV-vis spectra of Rh complex were observed under γ-ray radiation, the Rh complex, $[Rh(NO_3)_3]$, can be expected to be stable in actual HLLW.

1. Introduction

High-level radioactive liquid wastes (HLLW in 2M nitric acid solution), which are obtained after recovery of uranium and plutonium from the spent nuclear fuels for reuse of new fuels, are vitrified at a reprocessing plant. In the vitrification processes, platinum-group metals [PGMs: ruthenium (Ru), rhodium (Rh) and palladium (Pd)] and molybdenum (Mo) cause serious problems: PGMs are accumulated on the sidewall of the glass melter, whereas Mo tends to form so-called "yellow phase" in vitrified objects (VOs). These issues result in not only the degradation of VOs but also an increase in cost and disposal space by flushing the glass melter. To solve these serious issues, we have proposed a process for efficiently simaltaneous removal of PGMs and Mo from HLLW prior to being introduced into the glass melter, using metal hexacyanoferrates (HCF) as a sorbent [1]. In this process, chemical forms of PGMs and Mo in nitric acid solution are one of the important information for designing a high-performance sorbent, because the stability of those metal ion complexes in the solution is expected to affect the sorption rate of HCFs against each ion. We have already examined the chemical form of Mo and Pd ions in nitric acid solution [2-3]. In the present study, we report on the chemical form of Rh ion in nitric acid solution.

The different results on the chemical form of Rh complex in nitric acid solution have been reported so far. For examples, Samuels *et al.* concluded from theoretical and spectroscopic results that a bidentate octahedral structure, $[Rh(NO_3)_3]$, is the most stable chemical form of the Rh ion in nitric acid solution [4], whereas Vasilchenko *et al.* concluded from the results of ^{103}Rh- and ^{15}N-NMR and Raman spectroscopy that monodentate octahedral structures, trans-$[Rh(NO_3)_4(H_2O)_2]^-$ and $[Rh(NO_3)_6]^{3-}$, are the stable chemical forms in nitric acid solution [5]. Although they pointed that the Rh concentration and the starting materials are able to affect the chemical form of Rh complex in nitric acid solution, they did not examine the influence of γ-ray radiation on the chemical form of Rh complex in nitric acid solution yet. In this study, we examined the thermodynamic stability and electronic structure of Rh complex in pure water and nitric acid solution with and without γ-ray radiation, using UV-vis spectroscopy and first-principles calculations in order to expect the chemical form of Rh ion in actual HLLW.

2. Methods

2.1. *Experimental methods*

Rh complex in nitric acid solution were prepared by dissolving the $Rh(NO_3)_3 \cdot xH_2O$ (RARE METALLIC Co., LTD., 3 N) powder into 0–3 M nitric acid solutions. The concentration of Rh ion in the solutions were adjusted to be 5×10^{-3} M. UV-vis absorption spectra of these samples were measured using JASCO V-630. The absorption signal of nitric acid was subtracted from the nitric acid solution in the same concentration as the background.

γ-ray irradiation to the samples (Rh in 1.5 M HNO_3 solution) for a total dose of 0.5 and 1.0 MGy was performed using a ^{60}Co radiation source with a dose rate of 3.44 kGy/h. This dose rate includes an absorbed dose by air, and the total dose was controlled by the irradiation time. Thereafter, UV-vis absorption spectra of the γ-ray irradiated samples were also obtained in a similar manner to those of the samples without irradiation.

2.2. *Computational methods*

The geometry optimization of $[Rh(NO_3)_n(H2O)_{6-n}]^{3-n}$ (monodentate) and $[Rh(NO_3)_n(H_2O)_{6-2n}]^{3-n}$ (bidentate) complexes were performed using Gaussian09 based on density-functional theory (DFT) [6]. The change in the Gibbs's free energy (ΔG) for anation reactions at 298.15 K, 1 atom were also estimated using the optimized structures. All DFT calculations were performed with B3LYP/def2-TZVP level [7, 8]. The effective core potential was used for the Rh atom [9], whereas all electrons were explicitly considered for hydrogen, nitrogen and oxygen atoms. The polarizable continuum model was employed to consider the solvent effects of water by using the integral equation formalism variant [10].

Theoretical absorption spectra arising from the 4d-4d transitions of the optimized Rh complex were obtained using the relativistic configuration-interaction (CI) method (DVME) [11-12]. Since the Rh^{3+} ion has the $4d^6$ electronic configuration, we considered the 6 electrons in 10 molecular spinors as the active space of CI calculations. One-electron approximation calculations were performed with the relativistic formalism in the framework of local density approximation proposed by Vosko, Wilk and Nusair [13]. For the many-electron calculations, we adopted the Dirac-Coulomb-Brite Hamiltonian. The calculated oscillator strength was convoluted with 0.3 eV full-width-at half-maximum Gaussian functions.

3. Results and Discussion

Figure 1 shows the ΔGs of the anation reactions for (a) mono and (b) bidentate Rh complexes. The anation reactions for the mono and bidentate Rh complexes proceed by the following equations (1) and (2), respectively.

$$\text{Monodentate: } [Rh(NO_3)_n(H_2O)_{6-n}]^{3-n} + NO_3^- \rightarrow [Rh(NO_3)_{n+1}(H_2O)_{5-n}]^{2-n} + H_2O \quad (1)$$

$$\text{Bidentate: } [Rh(NO_3)_n(H_2O)_{6-2n}]^{3-n} + NO_3^- \rightarrow [Rh(NO_3)_{n+1}(H_2O)_{4-2n}]^{2-n} + H_2O \quad (2)$$

The results of the monodantate complex system indicate that trans-$[Rh(NO_3)_4(H_2O)_2]^-$ complex is the most stable among ten chemical species. In addition, cis-$[Rh(NO_3)_4(H_2O)_2]^-$, $[Rh(NO_3)_5(H_2O)_1]^{2-}$, and $[Rh(NO_3)_6]^{3-}$ complexes are also thermodynamically stable chemical species. The energy difference between these complexes (cis-$[Rh(NO_3)_4(H_2O)_2]^-$, $[Rh(NO_3)_5(H_2O)_1]^{2-}$, and $[Rh(NO_3)_6]^{3-}$) and the most stable complex (trans-$[Rh(NO_3)_4(H_2O)_2]^-$) was 4.40, 0.87 and 15.96 kJ/mol, respectively. As shown in Fig. 1(a), the anation reaction proceeds for monodentate coordination complexes and the complex with $n > 4$ is considered to form stably. This result is consistent with the previous report described above [4].

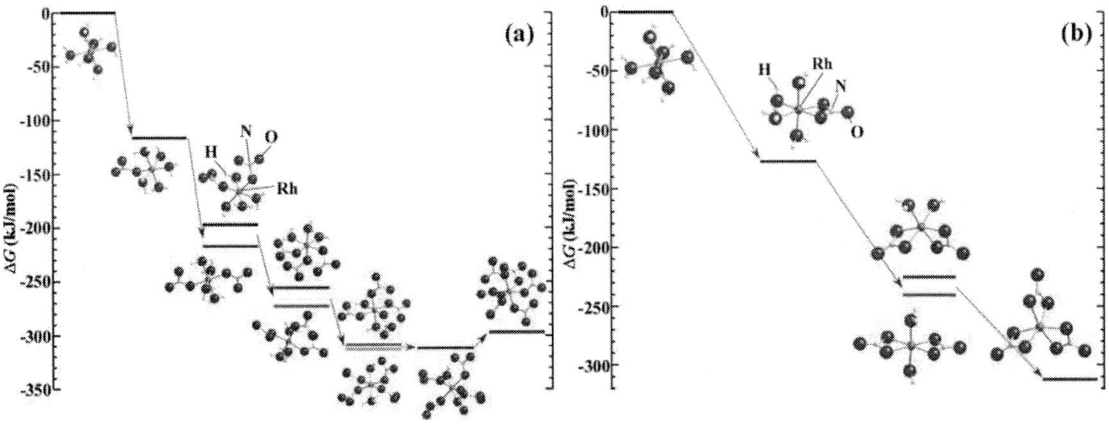

Figure 1. Δ*G*s of the anation reactions for mono (a) and bidentate (b) Rh complexes

On the other hand, the results of the bidentate complex system indicate that $[Rh(NO_3)_3]$ was the most stable complex among five chemical species. Since the energy difference when compared to other species was relatively large by more than 70 kJ/mol, it is considered that this chemical specie was the dominant component in the bidentate system. In other words, the anation reactions of Rh complex proceed smoothly for the bidentate system.

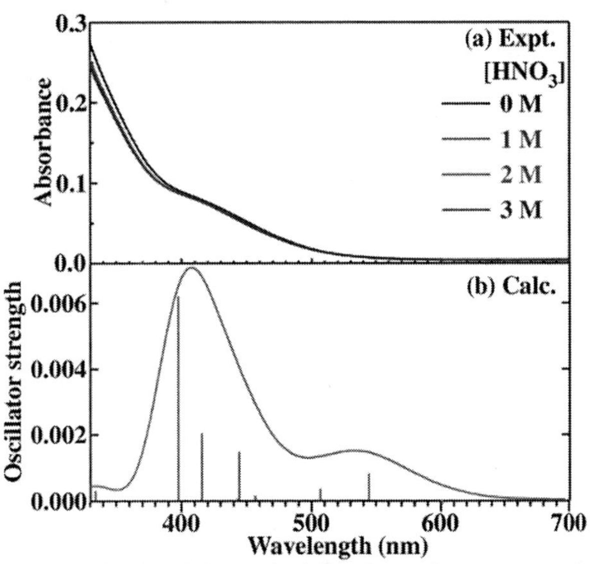

Figure 2. Experimental (a) and theoretical (b) absorption spectra of Rh complexes

Figure 2 shows the UV-vis absorption spectra (top) of Rh complexes in pure water and 0–3 M nitric acid solutions along with theoretical spectrum (bottom) calculated by the CI method. Although the absorption peaks originating from ligand-metal charge transfer (LMCT) and ligand π–π^* transitions were observed around 202 nm for Rh complexes [5], the strong absorption due to nitric acid itself was overlapped in this wavelength region, and thus we could not distinguished between them. Therefore, we focused on the absorption originating from the 4d-4d transitions of Rh ion in a wavelength region longer than 400 nm.

The experimental UV-vis spectral features of Rh ion in 0–3 M nitric acid solutions seemed to be almost same and thus were independent of nitric acid concentration. This indicates that the chemical form of Rh complexes in 0–3 M nitric acid solutions were the same. In addition, for $Rh(NO_3)_3{\cdot}xH_2O$ dissolved into pure water, the coordination number of nitrate ion does not increase. Accordingly, the Rh complex in pure water was coordinated with three or less NO_3^- ions, and in a similar manner, the Rh complex was assumed to be coordinated with three or less NO_3^- ions in 1–3 M nitric acid solution. Figure 2 (b) shows theoretical absorption spectrum of $[Rh(NO_3)_3]$ complex to confirm the assumption. The $[Rh(NO_3)_3]$ complex has the D_3 symmetry with a slightly distorted octahedral field. The ground state (GS) is $^1A_1$ with a low spin state, which corresponds to the $(t_{2g})^6(e_g)^0$ electronic configuration in the O_h symmetry notation.

As shown in Fig. 2, the theoretical absorption spectrum well reproduced the experimental one in the wavelength region longer than 400 nm. Thus, the chemical species of Rh complex in pure water and nitric acid solution can be assigned to $[Rh(NO_3)_3]$. Since the strong absorption band appearing in less 400 nm wavelength is attributed to the LMCT or ligand π–π* transitions [4], and it was not considered in the present many-electron calculations. The absorption band in the range 400–450 nm is due to the transitions from $^1A_1$ (GS) to $^1A_2$ and ^1E excited states, which corresponds to the transition from $(t_{2g})^6(e_g)^0$ to $(t_{2g})^5(e_g)^1$ electronic configuration. These absorption bands are assigned to the 4d-4d transitions of Rh^{3+} ion, and their intensities were relatively week when compared to those of the LMCT or ligand π–π* transitions. Since the previous study focused on the strong absorptions such as LMCT and ligand π–π* transitions [4], these weak absorptions may not have been observed.

Although the present results were consistent with those reported by Samuels *et al*, the theoretical results of the ΔGs suggest that the monodentate coordination complexes could exist in the strong nitric acid solution, which depends on the Rh metal concentration or the initial materials.

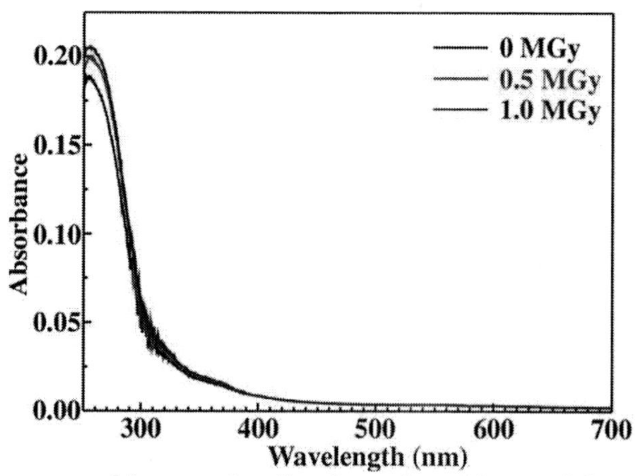

Figure 3. The absorption spectra of the γ-ray irradiated samples with a total dose of 0 (black), 0.5 (red) and 1.0 (blue) MGy.

Considering the actual HLLW, it is important to examine whether γ-ray radiation affects the chemical form of the Rh complex or not in nitric acid solution. We next investigated the influence of the absorption spectra of Rh complex in 1.5 M nitric acid solution by γ-ray irradiation. Figure 3 shows the absorption spectra of the γ-ray irradiated samples with a total dose of 0 (black), 0.5 (red) and 1.0 (blue) MGy. The overall features of these spectra were the almost same. This indicates that there were no changes in the chemical form of the Rh complex against the γ-ray radiation. In other words, $[Rh(NO_3)_3]$ complex structure exhibits a radiation resistant, and is considered to be dominant chemical species even in the actual HLLW.

4. Conclusion

We have investigated the chemical forms of Rh ion in pure water and nitric acid solution, using UV-vis spectroscopy and first-principles caclulations, in order to develop the disposal of HLLW.

$[Rh(NO_3)_3]$ complex was the most stable among the five bidentate Rh chemical species thermodynamically, whereas trans-$[Rh(NO_3)_4(H_2O)_2]^-$ complex was the most stable among ten monodentate ones. In addition, in case of the monodentate system, cis-$[Rh(NO_3)_4(H_2O)_2]^-$, $[Rh(NO_3)_5(H_2O)_1]^{2-}$, and $[Rh(NO_3)_6]^{3-}$ complexes were also stable chemical species.

Comparison between UV-vis spectra and many-electron calculations indicated that Rh forms a octahedral bidentate complex coordinated with three nitrate ions, $[Rh(NO_3)_3]$, in both pure water and nitric acid solution. However, the monodentate coordination complexes still existed in the strong nitric acid solution, which depended on the Rh concentration or the initial materials.

We further examined the influence of γ-ray radiation to the Rh complex, using UV-vis spectroscopy, and revealed that the Rh complex, $[Rh(NO_3)_3]$, exhibits a radiation resistant and is considered to be the dominant chemical species even in the actual HLLW.

Acknowledgement

We thank to Radioisotope Research Center, Tokyo Institute of Technology for γ-ray irradiation to the present samples. The present work was financially supported by "The R&D program for advanced nuclear power systems" of MEXT (Ministry of Education, Culture, Sports, Science and Technology), the government of Japan.

References

[1] K. Takeshita, Y. Inaba, T. Takahashi, J. Onoe, H. Narita, "H. Development of separation process of PGMs and Mo from high-level liquid waste for the stable operation of vitrification process," Proceedings of Global 2015, pp. 20-24, 2015.

[2] S. Watanabe, T. Sato, T. Yoshida, M. Nakaya, M. Yoshino, T. Nagasaki, Y. Inaba, K. Takeshita, and J. Onoe, "Spectroscopic and first-principles calculation studies of the chemical forms of palladium ion in nitric acid solution for development of disposal of high-level radioactive nuclear wastes," AIP Advances, vol. 8, 045221, 2018.

[3] S. Watanabe, S. Sato, M. Nakaya, T. Yoshida, and J. Onoe, "Chemical forms of molybdenum ion in nitric acid solution studied using liquid-phase X-ray absorption fine structure, Ultraviolet-Visible absorption spectroscopy and first-principles calculations," Chem. Phys. Lett., vol. 723 pp. 79-81, 2019.

[4] A. C. Samuels, C. A. Boele, K. T. Bennett, S. B. Clark, N. A. Wall, and A. E. Clark, "Integrated Computational and Experimental Protocol for Understanding Rh(III) Speciation in Hydrochloric and Nitric Acid Solutions," Inorg. Chem. vol. 53, pp. 12315-12322, 2014.

[5] D. Vasilchenko, S. Vorob'eva, S. Tkachev, I. Baidina, A. Belyaev, S. Korenev, L. Solovyov, and A. Vasiliev, "Rhodium(III) Speciation in Concentrated Nitric Acid Solutions," Eur. J. Inorg. Chem. vol. 2016, iss. 23, pp. 3822-3828, 2016.

[6] M. J. Frisch *et al.*, Gaussian 09, Revision D.01, Gaussian, Inc., Wallingford CT, 2013.

[7] A. D. Becke, "Density-functional thermochemistry. III. The role of exact exchange," J. Chem. Phys., vol. 98 pp. 5648– 5652, 1993.

[8] F. Weigend, and R. Ahlrichs, "Balanced basis sets of split valence, triple zeta valence and quadruple zeta valence quality for H to Rn: Design and assessment of accuracy," Phys. Chem. Chem. Phys. vol. 7, pp. 3297–3305, 2005.

[9] D. Andrae, U. Häußermann, M. Dolg, H. Stoll, H., Preuß, "Energy-adjusted ab initio

pseudopotentials for the second and third row transition elements," Theor. Chim. Acta, vol. 77, pp. 123-141, 1990.

[10] J. Tomasi, B. Mennucci, and R. Cammi, "Quantum mechanical continuum solvation models," Chem. Rev. 2005, Vol. 105, pp. 2999-3093, 2005.

[11] K. Ogasawara, T. Iwata, Y. Koyama, T. Ishii, I. Tanaka, and H. Adachi, "Relativistic cluster calculation of ligand-field multiplet effects on cation $L_{2,3}$ x-ray-absorption edges of $SrTiO_3$, NiO, and CaF_2," Phys. Rev. B vol. 64, pp. 115413, 2001.

[12] S. Watanabe, K. Ogasawara, M. Yoshino, T. Nagasaki, "First-principles and experimental analysis of f^n-$f^{n-1}d^1$ absorption spectra and multiplet energy levels of Pr^{3+}, Nd^{3+}, and U^{3+} in $LiYF_4$," Phys. Rev. B vol. 81, pp. 125128, 2010.

[13] S. H. Vosko, L. Wilk and M. Nusair, "Accurate spin-dependent electron liquid correlation energies for local spin density calculations: a critical analysis," Can. J. Phys., vol. 58, pp. 1200–1211, 1980.

Supramolecular Rare Sugar

Tomohiko Ishii,1 Tatsuya Senoo,1 Hiroaki Taguchi,1 Kouta Sogo,1 Bui Quoc Huy,1 Norito Taniguchi,1 Syuichi Katsumoto,1 Natsumi Nagayama,1 and Genta Sakane2

[1]Program in Advanced Materials Science, Department of Engineering and Design, Faculty of Engineering and Design, Kagawa University, 2217-20 Hayashi-cho, Takamatsu, Kagawa 761-0396, JAPAN.
[2]Department of Chemistry, Faculty of Science, Okayama University of Science, 1-1, Ridaicho, Kita-ku, Okayama, Okayama 700-0005, JAPAN

tishii@eng.kagawa-u.ac.jp

Abstract. We introduce rare sugars including their derivatives and supramolecular rare sugars, that have been actively researched at Kagawa University, Japan. Although the rare sugars are special sugars which hardly exist in nature, we has succeeded in mass synthesis of the rare sugars by utilizing biological enzyme isomerization reaction. In addition, the rare sugars have various functions such as blood sugar level suppressing function, cancer cell growth suppressing function, and antibacterial action. In addition, since rare sugars have a large number of hydroxyl groups, they have a possibility for using as devices for expressing various functions by employing them as ligands of transition metal complexes. In recent years, it is successful that we have synthesized the suplamolecular rare sugars (SRSs) from the different types of rare sugars, that is not as simple mixtures but as single crystals having supralattice structure, that can be freely controlled optical rotation. It has been found that the crystal structures of these SRSs mostly follow the Wallach rule, but do not satisfy the Wallach rule especially in the case of D,L- psicose. Therefore, we have investigated the single crystal X-ray structural analyses of SRSs, obtaining the detailed crystal structure data, and analyzed the intermolecular interaction between their sugar molecules in the crystal by means of the DV-Xα molecular orbital calculation. According to our detailed analysis of the research, calculating the intermolecular interaction revealed that the stability of the intermolecular interaction in the crystal can not be explained only by following the simple Wallach rule. Specifically, for example in the case of D,L-psicose, the total energy in a crystal can be stabilized by aligning the polarization vectors of the molecules, and as a result, we have clarified that the SRSs crystal structure can be stabilized, even if it does not follow the simple Wallach rule.

1. Introduction

There are two types of monosaccharides, which are classified into natural sugars that exist in large quantities in nature, and rare sugars that rarely exist in nature. Typical examples of the natural sugars are D-glucose and D- fructose and so on. On the other hand, typical examples of the rare sugars are D-psicose and D-allose, and there are many other types of rare sugars. Since the rare sugars rarely exist in nature, little research has been done up to now. The price of the rare sugar as a reagent is extremely expensive. For example, the price of the natural sugar beta-D-glucose is about 0.001 USD, that is equal to about 14 Indonesian Rupiah per a gram. On the other hand, the price of the rare sugar beta-D-allose is about 1,000 USD, that is equal to about 14 million Indonesian Rupiah per a gram. Therefore, the price of the rare sugar is about 1 million times more expensive than that of the natural sugar. But the difference of the molecular structure between them is quite small, as shown in Figure 1.

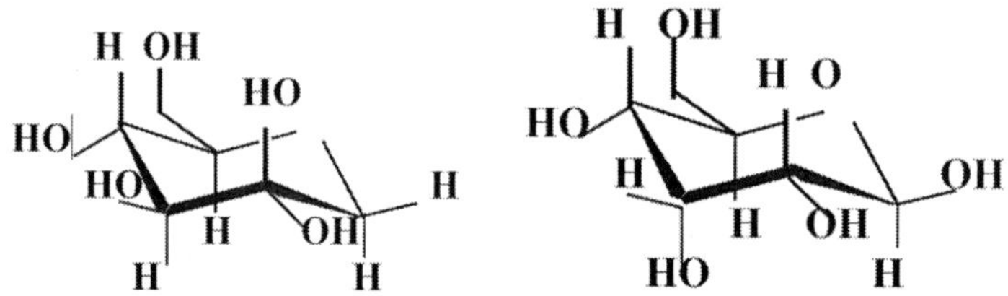

Figure 1. Comparison of the molecular structures between the natural sugar beta-D-glucose (left) and the rare sugar beta-D-Allose (right).

In Kagawa University, Japan, professor Ken Izumori has found the enzyme named "D-tagatose-3-epimeraze (DTE)". By using the enzime DTE, it is very easy to obtain the rare sugar by enzymatic isomerization of the natural sugar. As a result, the strategic way for synthesizing the rare sugar has been presented as a map named "Izumoring", as shown in Figure 2.

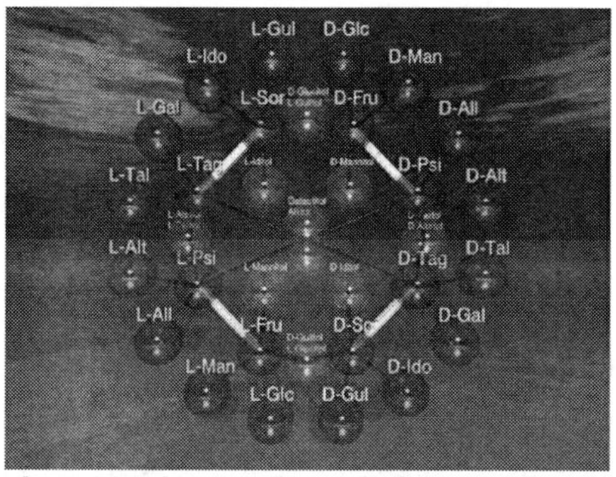

Figure 2. Izumoring, that is the strategic map through the new way and new path for producing the rare sugars.

The purpose of this study is to newly synthesize the supramolecular rare sugars composed of multiple kinds of rare sugars located on the strategic Izumoring map, as shown in Figure 3.

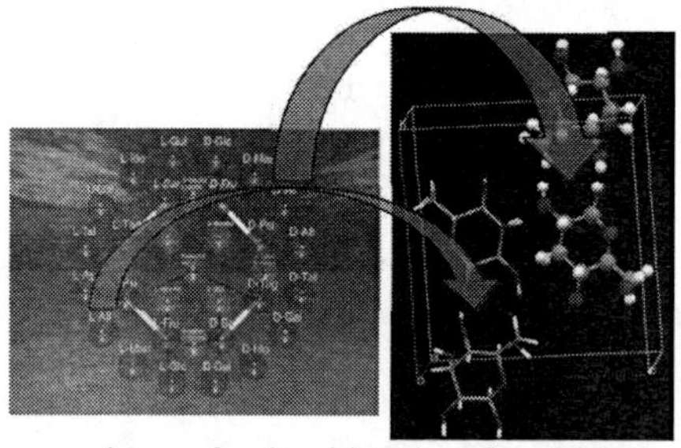

Figure 3. Supramolecular rare sugar is synthesized by the plural kinds of rare sugar molecules located on the Izumoring map, which is used as a parts list of the building blocks of the supramolecule

Each sugar molecule has a specific optical rotation phenomenon. Therefore, if a supralattice structure can be created by freely combining multiple types of the rare sugar molecules having the different kinds of optical rotations, the supramolecular rare sugar can be expected for the application to an optical device that can freely be controlled of the optical rotation.

IOP Conf. Series: Materials Science and Engineering **835** (2020) 012002 doi:10.1088/1757-899X/835/1/012002

2. Experimental and Result

We have already succeed to synthesize a several kinds of the supramolecular rare sugars, for example, D-psicose : D-glucose = 1 : 9, D-psicose : D-allose = 1 : 1, and D-psicose : D-fructoce = 1 : 3. Unfortunately, these three kinds of the suplamolecular rare sugars are obtained only as a powder crystal with a low quality of the crystallinity. On the other hand, we have also succeed to synthesize the single crystals with a high quality of the crystallinity as a D,L-racemic supramolecule, as following,
D-fructose : L-fructose = 1 : 1 (P-1),
D-psicose : L-psicose = 1 : 1 ($Pna2_1$),
D-sorbose : L-sorbose = 1 : 1 ($P2_1/c$),
D-allose : L-allose = 1 : 1 ($P2_1/c$),
D-arabinose : L-arabinose = 1 : 1 ($P2_1/c$).

Scheme. Supralattice structure of the supramolecular rare sugar D,L-allose.

3. Discussion

The results of the single-crystal X-ray structural analysis of the chiral monosaccharides D-psicose, D-allose and D-fructose are summarized in Table 1.

Table 1. Results of the X-ray single crystal structural analyses of chiral D-psicose, D-allose and D-frucose, as well as the corresponding racemic supramolecular rare sugars D,L-psicose, D,L- allose and D,L-fructose.

	D-psicose	**D,L-psicose**	**D-allose**	**D,L-allose**	**D-fructose**	**D,L-fructose**
Formula	$C_6H_{12}O_6$	$C_6H_{12}O_6$	$C_6H_{12}O_6$	$C_6H_{12}O_6$	$C_6H_{12}O_6$	$C_6H_{12}O_6$
Crystal System	orthorhombic	orthorhombic	orthorhombic	monoclinic	orthorhombic	triclinic
Space Group	$P2_12_12_1$	$Pna2_1$	$P2_12_12_1$	$P2_1/c$	$P2_12_12_1$	P-1
a [Å]	7.727	11.263	4.918	4.98211	9.191	5.43124
b [Å]	8.672	5.355	11.925	12.5624	10.046	7.2727
c [Å]	11.123	12.654	12.805	11.8156	8.095	10.1342
alpha [°]	90	90	90	90	90	69.120

The space groups of the chiral compounds were all orthorhombic $P2_12_12_1$ and $Z = 4$. The unit cell volume was about 750 Å3, and there are no significant difference among these three types of chiral sugar molecules. In addition, the results of the single-crystal X-ray structural analysis of the supramolecular rare sugars composed of racemic D,L-enantiomers of these chiral compounds are also summarized in the Table 1. Among these three supramolecular rare sugars, the space groups, symmetry, and their *Z*-values are varied from each other. In the supramolecular rare sugar D,L-psicose, the space group remained an orthorhombic, but the symmetry was changed into $Pna2_1$. This symmetry means that the racemic D,L-enantiomers are existed in the unit cell. On the other hand in the case of the supramolecular rare sugar D,L-allose, the space group was monoclinic and the symmetry was reduced. The space group is $P2_1/c$, which means that the mirror symmetry exists along to the *c*-axis direction. As a result, the D-allose and L-allose coexist in the unit cell. Furthermore, in the case of the supramolecular rare sugar D,L-fructose, the space group was reduced to be triclinic, and the symmetry to be P-1, which is having an inversion center in the unit cell. In the supramolecular rare sugar D,L-fructose, it was found that the *Z*-value is to be 2, the unit cell volume was halved, and $V = 366$ Å3. In all those chiral and racemic forms, the chemical composition per one sugar molecule is $C_6H_{12}O_6$, and the molecular weight is equal. Therefore, if the *Z*-value is the same, it means that the larger the unit cell volume, the smaller the crystal density. Comparing the unit cell volume between the chiral and supramolecular rare sugars in the case of psicose, it can be seen that the increases from 745 Å3 in the chiral one to 763 Å3 in the supramolecular rare sugar. In contrast, in the case of the allose, it can be seen that the cell volume is reduced from 751 Å3 in the chiral one to 739 Å3 of supramolecular rare sugar. Furthermore, even in the case of fructose, it can be seen that the cell volume is decreased from 747 Å3 of the chiral one to supramolecular rare sugar 732 Å3 (= 366.088 Å3 × 2). In general, the racemic crystals with optical isomers are denser and more stably packed than chiral crystals, according to the Wallach's rule. The changing of the density after racemization of these materials are summarized in Table 2.

Table 2. Changing of the crystal density before (chiral) and after (racemic) racemization.

	racemic (g/cm^3)	chiral (g/cm^3)	difference
psicose	1.568	1.600	−0.032
fructose	1.634	1.606	0.028
allose	1.618	1.593	0.025
arabinose	1.655	1.627	0.028
fucose	1.520	1.486	0.034
mannitol	1.504	1.484	0.020

In almost of all materials, the density of the supramolecular racemic rare sugar is larger that that of the chiral one. So, the tendency only in the case of the psicose is unusual. Such an unusual trend has been come from the existence of the intramolecular hydrogen bonding between two OH groups among the 3,5-diaxial structural situation. The hydrogen bonding network in the case of chiral D-psicose, the homochiral bonding network such as 5(D) -> 3(D) -> 5(D) -> 3(D)… can be observed among the D-psicose molecules. On the other hand, the hydrogen bonding network is weakened by the heterochiral intermolecular interaction such as 5(D) -> 3(D) -> 1(L) -> 5(D) -> 3(D) -> 1(L) ->… between two enantiomer sugar molecules.

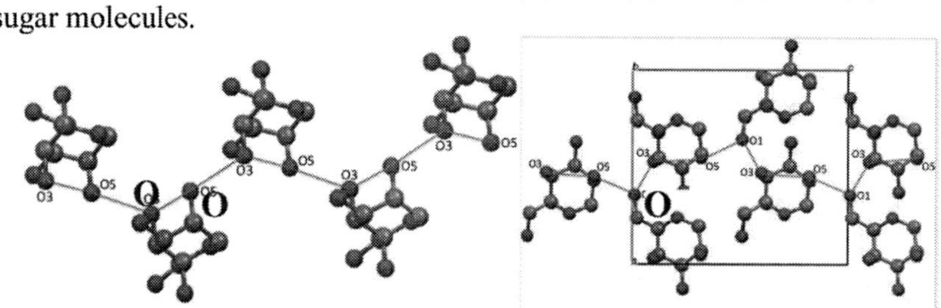

Figure 4. Hydrogen bonding networks in chiral D-psicose (left) and supramolecular rare sugar D,L psicose (right).

To discuss the crystal stability and their packing, we must discuss not only the hydrogen bonding but also the direction of the molecular packing in the crystal. Concerning to the molecular orientation in the chiral D-psicose, the 6 membered ring is stacked along to the *c*-axis in a unit cell. On the other hand in the case of the supramolecular rare sugar D,L-psicose, the hexagonal molecular packing can be included among the heterochiral molecules. As a result, the stacking of the 6 membered ring are directly along not to the *c*-axis but to the b-axis. According to the calculation of the molecular orientation and the dipole moment of these materials, it is revealed that there is quite large dipole moment between the intramolecular hydrogen bonding. In the case of the chiral D-psicose, all these dipole moments have been cancelled into zero due to the antiparallel situation in the $P2_12_12_1$ space group.

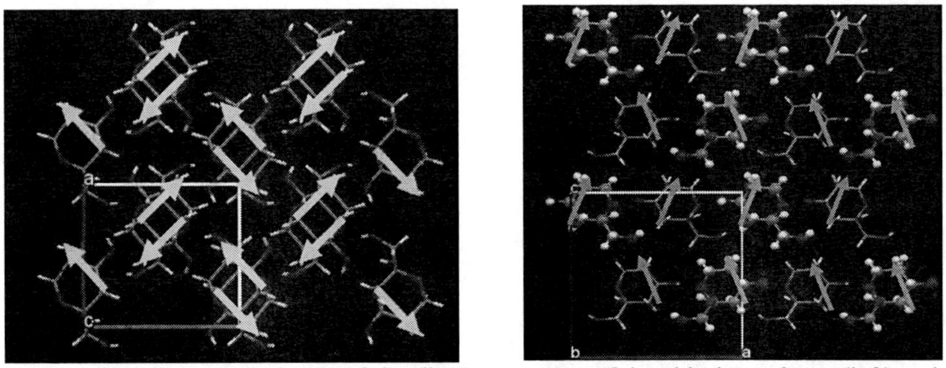

Figure 5. Molecular orientations and the dipole moments of the chiral D-psicose (left) and the supramolecular rare sular D,L-psicose (right).

On the other hand in the supramolecular rare sugar D,L-psicose, the dipole moment has not be cancelled because of the modified space group $Pna2_1$. As a result, there is quite large amount of the total dipole moment can be obtained in the crystal of the supramolecular rare sugar D,L-psicose. The hydrogen bonding among these heterochiral molecules are weakened by the large dipole moment.

4. Conclusion

We can succeed to obtain the new category of the rare sugar "supramolecular rare sugar" in many kinds of the combinations between plural rare sugar molecules located on the strategy map "Izumoring". Change of the unit cell volume in a single crystal of the supramolecular rare sugar can be explained according to the Wallach's rule, except only in the case of the D,L-psicoes. The specificity of the psicose have been realized by the intramolecular hydrogen bonding and the dipole moment observed in the single crystal of the supramolecurar rare sugar D,L- psicose.

Acknowledgement

The authors thank to Profs. Kazuhiro Fukada, Akihide Yoshihara, Kenji Morimoto, Goro Takata, Ken Izumori, Hiromi Yoshida, and Shigehiro Kamitori in Kagawa University for giving us a fruitful advices. This research is supported by The Third Medium-Term Goals and Plans Fund of Kagawa University Research Promotion Program 2019 (KURPP).

References

[1] T. B. Granström, G. Takata, M. Tokuda, and K. Izumori, "Izumoring: A novel and complete strategy for bioproduction of rare sugars", J. Biosci. Bioengineer., vol. 97, no. 2, pp. 89-94, 2004.

[2] S. Kamitori, H. Yoshida, M. Yamada, Y. Ohyama, G. Takata, K. Izumori, M. Yamaji, and **T. Ishii**, "X-ray Structures of L-Rhamnose Isomerase from Pseudomonas stutzeri in Metal-binding Form and Metal-free Form", Proc. Third Symposium Int'l. Soc. Rare Sugars, pp. 293-296, 2006.\

[3] Y. Watanabe, H. Yoshida, K. Takeda, **T. Ishii**, and S. Kamitori, "beta-D-Altrose", Acta Cryst., vol. E65, pp. o280, 2009.

[4] H. Yoshida, M. Yamaji, **T. Ishii**, K. Izumori, and S. Kamitori, "Catalytic Reaction Mechanism of Pseudomonas stutzeri L-Rhamnose Isomerase Deduced from X-ray Structures", Fed. Euro. Biochem. Soc. J., vol. 277, pp. 1045-1057, 2010.

[5] K. Fukada, **T. Ishii**, K. Tanaka, M. Yamaji, Y. Yamaoka, K. Kobashi, and K. Izumori, "Crystal Structure, Solubility, and Mutarotation of the Rare Monosaccharide D-Psicose", Bull. Chem. Soc. Jpn., vol. 83, no. 10, pp. 1193-1197, 2010.

[6] S. Kamitori, A. Ueda, Y. Tahara, H. Yoshida, **T. Ishii**, and J. Uenishi, "Crystal Structures of Rare

Disaccharides, alpha-D-glucopyranosyl beta-D-psicofuranoside, and alpha-D-galactopyranosyl beta-D-psicofuranoside", Carbohydrate Res., vol. 346, no. 9, pp. 1182-1185, 2011.

[7] **T. Ishii**, S. Ohga, K. Fukada, K. Morimoto, and G. Sakane, "beta-D-Gulose", Acta Cryst., vol. E70, no. 5, pp. o569, 2014.

[8] Y. Kitai, K. Hayashi, M. Otsuka, H. Nishiwaki, T. Senoo, **T. Ishii**, G. Sakane, M. Sugiura, and H. Tamura, "New Sesquiterpene Lactone Dimer, Uvedafolin, Extracted from Eight Yacon Leaf Varieties (Smallanthus sonchifolius): Cytotoxicity in HeLa, HL-60, and Murine B16-F10 Melanoma Cell Lines", J. Agric. Food Chem., vol. 63, pp. 10856-10861, 2015.

[9] H. Yoshida, A. Yoshihara, **T. Ishii**, K. Izumori, and S. Kamitori, "X-Ray Structures of the Pseudomonas Cichorii D-Tagatose 3-Epimerase Mutant form C66S Recognizing Deoxy Sugars as Substrates", Appl. Microbiol. Biotechnol., vol. 100, no. 24, pp. 10403-10415, 2016.

[10] A. Yoshihara, **T. Ishii**, T. Kamakura, H. Taguchi, and K. Fukada, "Crystal Structure of 6-deoxy-α- L-Psicofuranose", Acta Cryst., vol. E71, no. 12, pp. o993-o994, 2015.

[11] **T. Ishii**, T. Senoo, A. Yoshihara, K. Fukada, and G. Sakane, "Crystal Structure of beta-D,L-Fructose", Acta Cryst., vol. E71, no. 10, pp. o719-o720, 2015.

[12] **T. Ishii**, G. Sakane, A. Yoshihara, K. Fukada, and T. Senoo, "Crystal Structure of beta-D,L-Psicose", Acta Cryst., vol. E71, no. 5, pp. o289-290, 2015.

[13] T. Kozakai, K. Fukada, R. Kuwatori, **T. Ishii**, T. Senoo, and K. Izumori, "Aqueous Phase Behavior of the Rare Monosaccharide D-Allose and X-ray Crystallographic Analysis of D-Allose Dihydrate", Bull. Chem. Soc. Jpn., vol. 88, no. 3, pp. 465-470, 2015.

[14] **T. Ishii**, T. Senoo, T. Kozakai, K. Fukada, and G. Sakane, "Crystal Structure of beta-D,L-Allose", Acta Cryst., vol. E71, no. 2, o139, 2015.

[15] H. Taguchi, K. Sogo, **T. Ishii**, A. Yoshihara, and K. Fukada, "α-D,L-Sorbose", IUCrData, vol. 3, no. 1, pp. x180114, 2018.

X-ray Absorption Spectroscopy and Its Simulation for Some Disordered Oxide Systems

Deok-Yong Cho*

IPIT & Department of Physics, Jeonbuk National University, Jeonju 54896, Korea

zax@jbnu.ac.kr

Abstract. The local structural properties of representative amorphous oxides, ZnO, HfO_2, ZrO_2, and $In_2Ga_2ZnO_7$ were investigated by using X-ray absorption spectroscopy and an ab-initio theoretical calculation (FEFF). The X-ray absorption near-edge structures of the amorphous systems were compared with those of their crystalline counterparts, and were interpreted in terms of weakened structural orders (in the case of ZnO, HfO_2, and $In_2Ga_2ZnO_7$) or fundamentally different local structures (ZrO_2; monoclinic-like local structure favoured in amorphous phase in contrast to the tetragonal structure in crystalline film).

1. Introduction

X-ray absorption spectroscopy (XAS) is a nondestructive way to probe the chemistry and structural properties of materials of interest by shining x-rays with various photon energies near the absorption edge energies and measuring the absorption coefficients [1]. Since the core level photoionization process itself involves the creation of localized holes on the inner orbitals very near to the nucleus, the XAS spectrum essentially reflects the short-range local structures near the photon-absorbing atom.

Amorphous materials have no long-range orders so that structural characterization using diffraction techniques is not feasible. However, the structural orders can survive in a short range no matter how significant the randomness in the atomic positions is. Thence XAS has a strong advantage in providing structural information in a highly disordered system. In particular, the X-ray absorption near-edge structures (XANES) reflect the geometry of the local coordination. Therefore, one can find clues on the local structural evolution in amorphous phases in comparison to the crystalline phases. This study demonstrates the methodology of structure identifications in amorphous oxide systems with examples of four kinds of oxide systems, that are close to the industrial application; HfO_2, ZrO_2, ZnO, and InGaZnO. HfO_2 and ZrO_2 are the representative high dielectric constant materials which can be used as the gate oxides in metal-oxide-semiconductor field-effect transistors. Besides, they also show ferroelectric behaviors. On the other hand, ZnO and InGaZnO are transparent semiconductors, which can be used as the electrodes or semiconducting channels in optoelectronic devices. All these oxides are now fabricated in industry level, and controlling the structural orders or defects is now very important. The XANES of amorphous systems are compared with those of the crystalline counterparts, and the features are interpreted being aided by an ab-initio theoretical simulation, FEFF. Although not as much accurate as density functional theory-based calculations (like DV-Xα), FEFF can provide fair insights on how the local structure can influence the electronic structure and the orbital configuration.

IOP Conf. Series: Materials Science and Engineering **835** (2020) 012003 doi:10.1088/1757-899X/835/1/012003

2. Methods

2.1. X-ray Absorption Spectroscopy

All the XAS measurements were performed at hard X-ray beamlines in Pohang Light Source in Korea. The samples were prepared by thin-film techniques of either rf sputtering (ZnO and $In_2Ga_2ZnO_7$) or atomic layer deposition (HfO_2 and ZrO_2), except for commercial polycrystalline $In_2Ga_2ZnO_7$ powder (Toshima Manufacturing Inc.). For thin film samples (ZnO, HfO_2, ZrO_2, and $In_2Ga_2ZnO_7$ film), fluorescence yield mode was used, while for powder sample ($In_2Ga_2ZnO_7$), transmission yield mode was used. The photon energies were calibrated before recording the experimental spectra.

2.2. XANES Simulation

The XANES simulation was performed by using an ab-initio real-space multiple scattering code, FEFF [2,3]. For the spectra of ZnO, ZrO_2, and $In_2Ga_2ZnO_7$, FEFF8 was employed, while for the spectra of HfO_2, an older version, FEFF6 was used. The contribution of each of the virtual electron scatterings in the X-ray absorption process can be included or excluded in the simulated spectrum. One convenient way is to consider all the multiple scattering paths up to the infinite number of "legs" in the scattering paths (for instance, 2 for Zn→O→Zn and 3 for Zn→O→O'→Zn) within a given radius of scattering range with respect to the photon-absorbing atom. This is often called full multiple scattering (FMS) mode. Generally, the FMS radius (R_{FMS}) represents the degree of structural order so that a higher (lower) value of R_{FMS} is for the more ordered (disordered) system. Alternatively, the number of legs can be restricted to a fixed number. For example, the signals from the single backscatterings only can be collected by confining the number of legs to be 2. In this work, the amorphousness was mimicked overall either by reducing the R_{FMS} value or confining the number of legs to be 2.

3. Results and discussion

3.1. ZnO

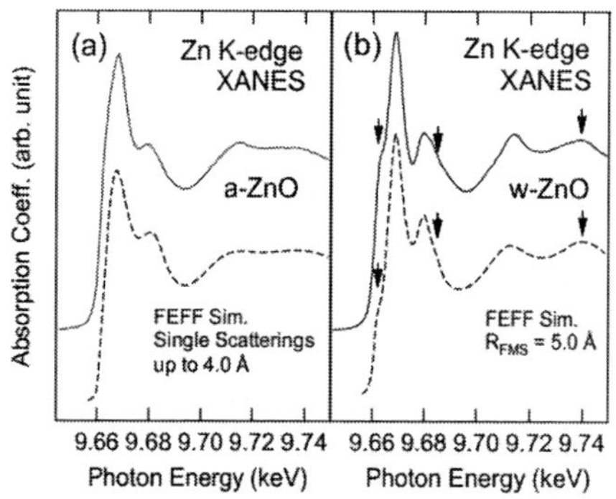

Figure 1. Zn K-edge XANES of (a) amorphous ZnO and (b) crystalline (wurtzite) ZnO, and their FEFF simulations [4].

Figure 1 shows the Zn K-edge XANES of (a) amorphous and (b) crystalline ZnO films taken from Ref. [4]. The ZnO films were prepared by sputtering on Si wafer [4]; the as-grown film with 2 nm thickness was amorphous (a-ZnO), while the film with 50 nm thickness after annealing was crystalline having wurtzite structure (w-ZnO; $P6_3/mc$). FEFF simulation was done assuming the wurtzite structure and the results were attached in the respective figures [4]. For w-ZnO, full multiple scattering up to R_{FMS} = 5.0 Å was considered, while for a-ZnO, single scatterings within half path length of 4.0 Å (i.e. Zn→O→Zn and Zn→Zn'→Zn only) were considered. The simulation well reproduced the major features (highlighted by arrows) in the experimental data. This suggests the preferential loss of multiple scattering events due to structural disorders in amorphous ZnO. Thus, it can be told that the amorphousness would suppress the virtual electron hopping and degrade the electrical property of ZnO for use as a transparent semiconductor [4].

3.2. HfO_2

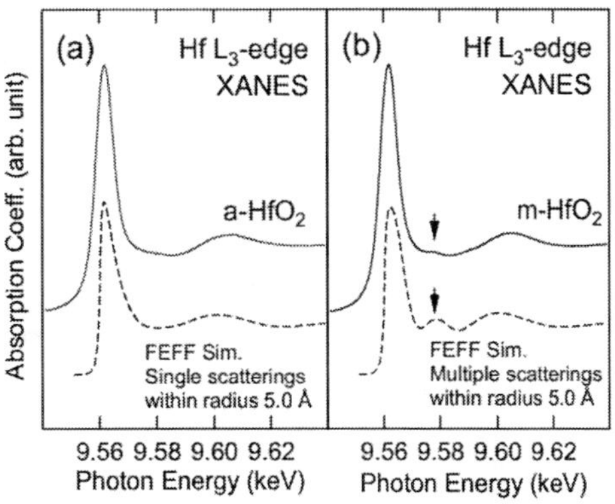

Figure 2. Hf L_3-edge XANES of (a) amorphous HfO_2 and (b) crystalline (monoclinic) HfO_2, and their FEFF simulations [5].

Figure 2 shows the Hf L_3-edge XANES of (a) amorphous and (b) crystalline HfO_2 films and their simulated spectra, taken from Ref. [5]. The HfO_2 films were prepared by atomic layer deposition on Si wafer [5]; the as-grown film with 10 nm thickness was amorphous (a-HfO_2), while the film after annealing was crystalline having a monoclinic crystal structure (m-HfO_2; $P2_1/c$). FEFF simulation was done assuming the monoclinic structure. For m-HfO_2, multiple scatterings with the number of legs less than 8 in a radius of 5.0 Å were considered, while for a-HfO_2, only the single scatterings such as Hf→O→Hf and Hf→Hf'→Hf (the number of legs is 2) are considered within the same range. The simulation well reproduced the major features (highlighted by the arrow) in the experimental data in Fig. 2. This suggests the preferential loss of multiple scattering events in amorphous HfO_2, similar to the case of ZnO in Fig. 1.

3.3. ZrO_2

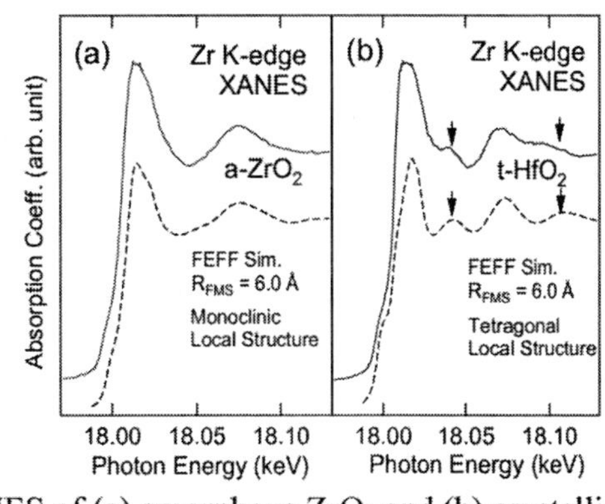

Figure 3. Zr K-edge XANES of (a) amorphous ZrO_2 and (b) crystalline (tetragonal) ZrO_2, and their FEFF simulations [6].

Figure 3 shows the Zr K-edge XANES of (a) amorphous and (b) crystalline ZrO_2 films with their simulated spectra taken from Ref. [6]. The ZrO_2 films with 8 nm thickness were prepared by atomic layer deposition on Si wafer [6]. The as-grown film was amorphous (a-ZrO_2), while the annealed film became crystalline having a tetragonal crystal structure (t-ZrO_2; $P4_2/nmc$). FEFF simulation was done with R_{FMS} = 6.0 Å for both a-ZrO_2 and t-ZrO_2. The experimental spectrum of t-ZrO_2 was reproduced very well by the theoretical spectra based on the tetragonal local structure (see the arrows in the figure), whereas that of a-ZrO_2 was not. Meanwhile, the spectrum of a-ZrO_2 was very similar to that of ZrO_2 powder having a monoclinic structure and was well reproduced by the simulation based on the monoclinic-like local structure model as shown in Fig. 3(a). This suggests the amorphousness in ZrO_2 hinders the metastability of the tetragonal structure in ZrO_2 so that a monoclinic-like local structure is favoured. Thus, the benefit of the high dielectric constant in tetragonal ZrO_2 can be suppressed in amorphous ZrO_2.

3.4. $In_2Ga_2ZnO_7$

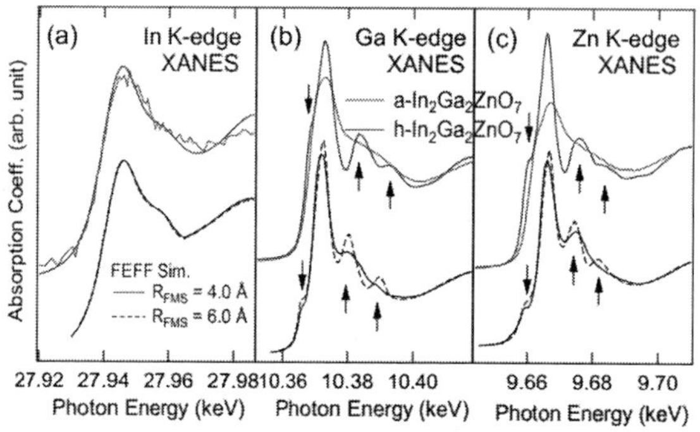

Figure 4. (a) In K-edge, (b) Ga K-edge, and (c) Zn K-edge XANES of amorphous $In_2Ga_2ZnO_7$ [7] and polycrystalline (hexagonal) $In_2Ga_2ZnO_7$ with their FEFF simulations [8].

Figure 4 shows the XANES of a transparent semiconductor, $In_2Ga_2ZnO_7$ at (a) In K-edge, (b) Ga K-edge, and (c) Zn K-edge, and their simulated spectra, taken from Refs. [7,8]. The polycrystalline $In_2Ga_2ZnO_7$ powder had a hexagonal crystal structure (h-$In_2Ga_2ZnO_7$; $P6_3/mmc$), while the sputtered $In_2Ga_2ZnO_7$ film on glass (150 nm-thick) was amorphous (a-$In_2Ga_2ZnO_7$). The spectra of the polycrystalline sample were well reproduced by the full multiple scattering simulations with R_{FMS} = 6.0 Å, while those of a-$In_2Ga_2ZnO_7$ could be done with much-reduced values of R_{FMS}, e.g. less than 4.0 Å as shown in the figure. This indicates that the structural disorders in the film were significant in spite of the relatively high thickness (150 nm). Maintaining the amorphous state even at high thickness is a useful property for flexible devices.

4. Conclusion

The X-ray absorption near-edge structure analyses aided by theoretical simulations are powerful in studying the local structural properties of highly disordered materials. Not only the evolution in the degree of structural disorders (amorphous ZnO, HfO_2, or $In_2Ga_2ZnO_7$) but also the local structural evolution (amorphous ZrO_2) can be scrutinized in comparison with their crystalline counterparts.

Acknowledgement

This work was supported by grants from the National Research Foundation of Korea (NRF-2018R1D1A1B07043427).

References

[1] "X-Ray Absorption", edited by D. C. Koningsberger and R. Prins (Wiley, New York, 1988).
[2] J. J. Rehr, "Recent Developments in Multiple-Scattering Calculations of XAFS and XANES," Jpn. J. Appl. Phys., Suppl. vol. 32-2, pp. 8-12, 1993.
[3] A. L. Ankudinov, B. Ravel, J. J. Rehr, and S. D. Conradson, "Real-space multiple-scattering calculation and interpretation of x-ray-absorption near-edge structure," Phys. Rev. B, vol. 58, no. 12, pp. 7565-7576, 1998.
[4] D.-Y. Cho, J. H. Kim, and C. S. Hwang, "Electron hopping interactions in amorphous ZnO films probed by x-ray absorption near edge structure analysis," Appl. Phys. Lett., vol 98, no. 22, pp. 222108, 2011.
[5] D.-Y. Cho, T. J. Park, K. D. Na, J. H. Kim, and C. S. Hwang, "Structural disorders in an amorphous HfO_2 film probed by x-ray absorption fine structure analysis," Phys. Rev. B, vol 78,

no. 13, pp. 132102, 2008.

[6] D.-Y. Cho, H.-S. Jung, J. H. Kim, C. S. Hwang, "Monocliniclike local atomic structure in amorphous ZrO_2 thin film," Appl. Phys. Lett., vol. 97, no. 14, pp. 141905, 2010.

[7] D.-Y. Cho, J. Song, K. D. Na, C. S. Hwang, J. H. Jeong, J. K. Jeong, and Y.-G. Mo, "Local structure and conduction mechanism in amorphous In–Ga–Zn–O films," Appl. Phys. Lett. vol. 94, no. 11, pp. 112112, 2009.

[8] D.-Y. Cho, J. Song, Y. C. Shin, C. S. Hwang, W. S. Choi, and J. K. Jeong, "Influence of High Temperature Postdeposition Annealing on the Atomic Configuration in Amorphous In–Ga–Zn–O Films," Electrochem. Solid-State Lett., vol. 12, no. 6, H208-H210, 2009.

IOP Conf. Series: Materials Science and Engineering **835** (2020) 012004 doi:10.1088/1757-899X/835/1/012004

Study on the higher-order structure and hydrogen bonding of biodegradable polymer by low-frequency vibrational spectroscopy

Dian Marlina[1,2], Harumi Sato[1]

[1]Graduate School of Human Development and Environment, Kobe University, Tsurukabuto Nada-ku, Kobe 657–8501, Japan
[2]Faculty of Pharmacy, Universitas Setia Budi,
Jl. Letjend Sutoyo Mojosongo Solo, Central Java 57127, Indonesia

157d841d@stu.kobe-u.ac.jp

Abstract. Biodegradable polymer is preferred material to replace the conventional petroleum based polymer materials to protect the nature and conserve oil supplies. Information related to higher-order structure and hydrogen bonding among polymers can be observed in the low-frequency region (3.3–333 cm^{-1}) using terahertz (THz) spectroscopy. THz spectra measurement of temperature-dependent investigation of poly(3-hydroxybutyrate) (PHB) homopolymer and polymer blend of PHB/glycol chitosan with the blend ratios of (80/20) and (60/40) have been performed to investigate the change in higher-order structure and intermolecular hydrogen bonding formed by polymer blends of PHB and glycol chitosan. The THz spectra revealed that the peak at 82 cm^{-1} does not collapse even in the high temperature indicating the important role of hydrogen bonding interaction between poly(3-hydroxybutyrate) (PHB) and glycol chitosan in maintaining the higher-order structure of the PHB in its blending system.

1. Introduction

Polymer have been an important part of our daily life to be easier and better. They can be found almost in all sorts applications from the car we drive to the television we watch, from the food we eat to the refrigerators they are store in, from the contact lens we use to the medicine we take. However, most of them, are made of conventional fuel-based polymers such as polyethylene (PE) and polypropylene (PP) which are not biodegradable. They cannot be decomposed in natural environment causing many environmental problems and harmful for living organisms. Therefore, biodegradable polymer is preferred as a promising material to replace petroleum-based polymer used in contact with living organisms [1,2]. It is characterized by biocompatibility, biodegradability, non-toxicity, and adsorptive ability which is suitable for cosmetics, pharmaceutical, food industries, and biomedical engineering applications [3].Among the biodegradable polymers, chitosan is abundant in nature. It is a polysaccharide produced by deacetylation of chitin and is extensively used in drug delivery applications [4,5]. However, chitosan is difficult to use industrially because it has high glass transition temperature (T_g) and high crystallinity. Therefore, it is necessary to improve their physical properties by blending or copolymerizing with other polymers. In this study, we modified glycol chitosan, a water soluble chitosan derivate with a highly demand poly(3-hydroxybutyrate) (PHB). The PHB/glycol chitosan polymer blend will enhance its degradability [6].

The number of conventional methods, X-ray diffraction, infrared (IR) spectroscopy, UV-Vis

IOP Conf. Series: Materials Science and Engineering **835** (2020) 012004 doi:10.1088/1757-899X/835/1/012004

spectroscopy, mass spectroscopy (MS), and nuclear magnetic resonance spectroscopy (NMR) has been well introduced to understand the structures and properties of biodegradable polymers. [6-8]. However, there is newly developing spectroscopic method in low-frequency vibrational region which still under explored [9,10]. The region of low-frequency vibrational spectroscopy or widely known as terahertz (THz) lies between wavenumber range from 3.3–333 cm^{-1}. Vibrational bands of molecules in this region arise from both inter and intramolecular interaction of the molecules. Therefore, the spectra are sensitive to the higher-order structure and hydrogen bonding which plays an important role in stabilizing polymer structure [11].

2. Methods

PHB (Mn=1000), glycol chitosan with purity ≥ 60% and degree of deacetylation 91.6% [11] were purchased from Polysciences, Inc. and Sigma-Aldrich Co., respectively. They were used as received without any pre-treatments. The solvent of 1,1,1,3,3,3-Hexafluoro-2-propanol (HFIP) was received from Wako Pure Chemical Industries, Ltd. Film samples of PHB homopolymer and PHB/glycol chitosan with blending ratios of (80/20) and (60/40) were prepared by dissolved-casting method. Firstly, the powder of PHB and glycol chitosan were dissolved in HFIP at 60°C based on the calculated weight percentages. After that, the mixture of the blends were discharged into aluminium dishes for evaporation process at 20°C for 12h. The obtained film samples were then moved into vacuum oven and continued the evaporation at 70°C for 12h in order to optimize the crystallization process.

THz spectra of the blend films were measured by using a terahertz time-domain spectrometer (THz-TDS), TAS7400TS from ADVANTEST Co. The spectra were obtained in the transmission mode in the 120–60cm^{-1}region with 1024 scans and 0.25cm^{-1} spectral resolution. Temperature-dependent THz measurements were performed by attaching temperature control stage LINKAM 10002L in the THz-TDS spectrometer with the heating rate of 5 °C/min. The spectra were measured from 30°C to melting temperature in every 5°C.

3. Results and discussion

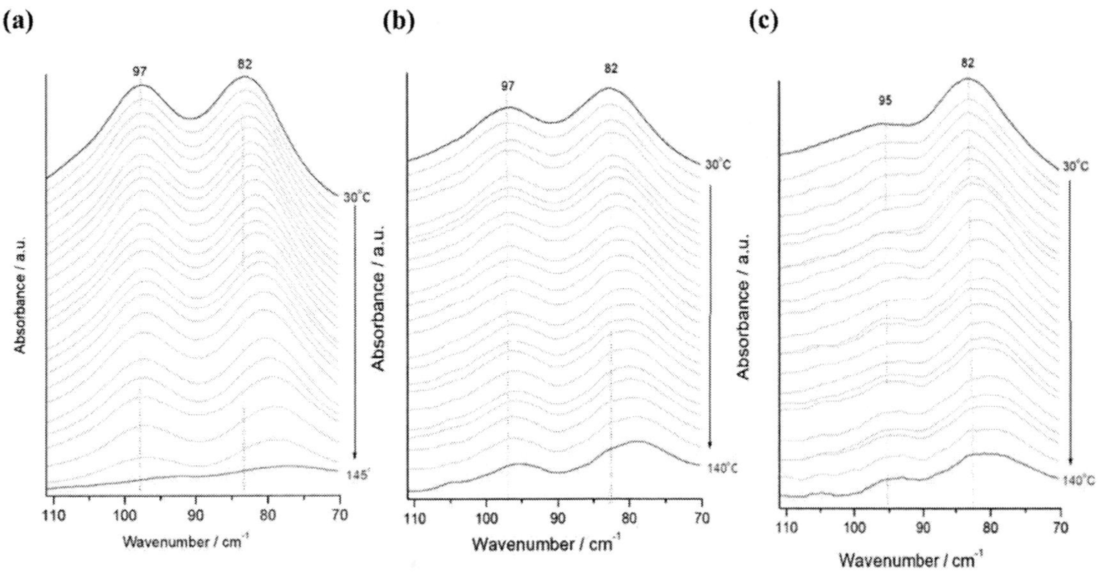

Figure 1 THz spectra of temperature dependent investigation of **(a)** PHB homopolymer **(b)** PHB/glycol chitosan (80/20) and **(c)** PHB/glycol chitosan (60/40).

Figure 1 shows the THz spectra of PHB homopolymer and PHB/glycol chitosan for the blend ratios of (80/20) and (60/40) in the region of 110–70 cm^{-1} obtained during heating process from 30°C to 145°C for PHB homopolymer and from 30°C to 140°C for the PHB/glycol chitosan. The spectra at the lowest temperature (30°C) reveal two clear peaks at 97 cm^{-1} and 82 cm^{-1} for PHB homopolymer and PHB/glycol chitosan (80/20) and the peaks at 95 cm^{-1} and 82 cm^{-1} for PHB/glycol chitosan (60/40). Those peaks of homopolymer PHB are related to the PHB-crystalline form and has been identified as spring like motion of helical structure due to the vibrational motion of CH_2+CH_3 out plane for the one at 97 cm^{-1}, while the other one at 82 cm^{-1} was assigned to hydrogen bonding interaction of PHB due to the vibration of C=O and CH_3 out plane [11].

In the 30°C, PHB homopolymer shows the sharpest and highest absorbance of the peaks at 97 cm^{-1} and 82 cm^{-1} compare than other samples. It indicates that PHB homopolymer forms the higher-order structure and the presence of glycol chitosan can transform PHB into less-order structure.

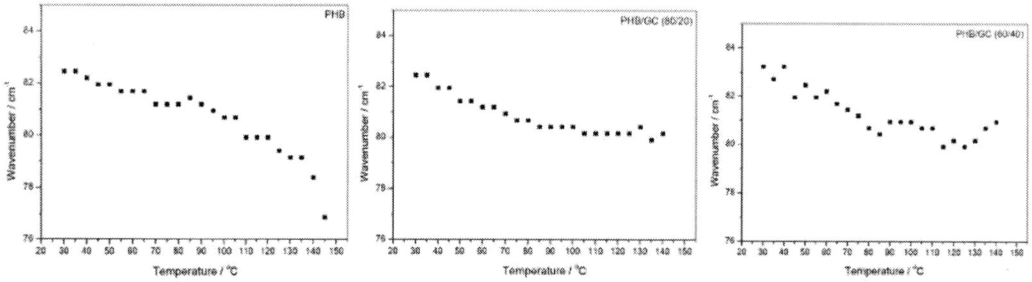

(a) (b) (c)

Figure 2 Plots of the peak position at about 82 cm^{-1} versus temperature of **(a)** PHB homopolymer **(b)** PHB/glycol chitosan (80/20) and **(c)** PHB/glycol chitosan (60/40).

The peak at 95 cm^{-1} of the PHB/glycol chitosan (60/40) has the lowest absorbance and it shows big absorbance ratio with the one at 82 cm^{-1}. Since the peak at 95 cm^{-1} is related to the helical structure of PHB, the result indicating that PHB inside the PHB/glycol chitosan (60/40) system has the shortest helical structure. In this case, further investigation need be performed by wide angle X-ray scattering (SAXS) method to observe the lamella or helical structure thickness of PHB in the blend samples.

The film of PHB homopolymer and PHB/glycol chitosan blends were measured during heating process every 5°C in order to understand the spectral behaviour. In Figure 1 (a), it is clear that the absorbance of PHB crystalline peak decrease during the heating process and finally disappear at 145°C, in this stage PHB homopolymer is in molten stage or amorphous phase. Different with PHB/glycol chitosan (80/20) and (60/40), the peaks are still can be observed in the highest temperature even with the small absorbance. In another word, PHB still has the small helical structure and a very few hydrogen bonding. This indicates that glycol chitosan is able to inhibit transformation process of PHB from the crystalline state to the amorphous state.

Peak related to hydrogen bonding of PHB at 82 cm^{-1} shifts to higher wavenumber as temperature increases for all of the samples. This shifts mean that the hydrogen bonding of PHB weakens by heating process. Plots of the shifts are displayed in the Figure 2 (a), (b), and (c) for PHB homopolymer, PHB/glycol chitosan (80/20), and (60,40), respectively. In the low temperature from 30°C to 80°C, the 82 cm^{-1} peak of PHB homopolymer shifts gradually and the dramatically from 85°C to 145°C. In the case of PHB/glycol chitosan (80/20), the 82 cm^{-1} peak shifts gradually from 30°C to 80°C and it remains in almost same position from 85°C to 140°C. PHB/glycol chitosan (60/40) almost shows the same behaviour of PHB/glycol chitosan (80/20), the 82 cm^{-1} gradually shift in the low temperature and then remains in almost same position in the higher temperature. However, there is some difficulties to specify the peak positions due to extremely low absorbance.

IOP Conf. Series: Materials Science and Engineering **835** (2020) 012004 doi:10.1088/1757-899X/835/1/012004

4. Conclusion

PHB homopolymer and PHB/glycol chitosan with the blend-ratio of (80/20) and (60/40) have been investigated by THz spectroscopy to monitor the transformation in the higher-order structure and hydrogen bonding during heating process. THz spectra of PHB/glycol chitosan (80/20) and (60/40) have the same peaks as PHB homopolymer appear at 97 cm^{-1} and 82 cm^{-1} with slightly different position for the ratio at (60/40). The small peaks absorbance and the big peak ratios of the blend samples indicating that the presence of glycol chitosan in PHB homopolymer can transform the higher-order structure of PHB to the less-order. The hydrogen bonding of PHB homopolymer collapses in the high temperature yet it does not happen in the blend samples. Glycol chitosan can maintain the hydrogen bonding interaction of PHB even in its melting temperature.

References

[1] L.M. Rios, C. Moore, and P.R. Jones, "Persistent organic pollutants carried by synthetic polymers in the ocean environment," Mar. Pollut. Bull., vol. 54, no. 8, pp. 1230–1237, 2007.

[2] S. Sorrel, J. Speirs, R. Bentley, A. Brandt, R. Miller, "Global oil depletion: A review of the evidence," Energy Policy, vol. 38, pp. 5290–5295, 2010.

[3] Y. Shchipunov, "Bionanocomposites: Green sustainable materials for the near future," Pure App. Chem., vol. 84, no. 12, pp. 2579–2607, 2012.

[4] O. Felt, A. Carrel, P. Baehni, P. Buri, R. Gurny, "Chitosan as tear subtitute: a wetting agent endowed with arimicrobial efficacy," Journal of Ocular Pharmacology and Therapeutics, vol. 16, pp. 261–270, 2000.

[5] S. A. Agnihotri, N. N. Mallikarjuna, T. M. Amibhavi, "Recent advances on chitosan-based micro- and nanoparticles in drug delivery," Journal of Controlled Release, vol. 100, pp. 5–28, 2004.

[6] H.Y. Nam, S.M. Kwon, H. Chung, S-Y. Lee, S-H. Kwon, H. Jeon, Y. Kim, J.H. Park, J. Kim, S. Her, Y-K. Oh, I.C. Kwon, K. Kim, S.Y. Jeong, "Cellular uptake mechanism and intracellular fate of hydrophobically modified glycol chitosan nanoparticles," J. of Controlled Release, vol. 135, pp. 259–267, 2009.

[7] H.Sato, R. Murakami, A. Padermshoke, F. Hirose, K. Senda, I. Noda, Y. Ozaki, "Infrared spectroscopppy studies of CH···O hydrogen bonding and thermal behavior by biodegradable poly(hydroxyalkanoate)," Macromolecules, vol. 37, pp. 7203–7213, 2004.

[8] H. Sato, K. Mori, R. Murakami, Y. Ando, I. Takahashi, J. Zhang, H. Terauchi, F. Hirose, K. Senda, k. Tashiro, I. Noda, Y. Ozaki, "Crystal and lamella structure and C–H···O=C hydrogen bonding of poly(3-hydroxyalkanoate) studied by X-ray diffraction and infrared spectroscopy," Macromolecules, vol. 39, pp. 1525–1531, 2006.

[9] D. Marlina, H. Sato, H. Hoshina, Y. Ozaki, "Intermolecular interactions of poly(3-hydroxybutyrate-co-3-hydroxyvalerate) (P(HB-co-HV)) with PHB-type crystal structure and PHV-type crystal structure studied by low-frequency Raman and terahertz spectroscopy," Polymer, vol. 135, pp. 331–337, 2018.

[10] D. Marlina, H. Hoshina, Y. Ozaki, H. Sato, "Crystallization and crystalline dynamics of poly(3-hydroxybutyrate) / poly(4-vinylphenol) polymer blends studied by low-frequency vibrational spectroscopy," Polymer, 2019, DOI: 10.1016/j.polymer.2019.121790.

[11] S. Yamamoto, Y. Morisawa, H. Sato, H. Hoshina, Y. Ozaki, "Quantum mechanical interpretation of intermolecular vibrational modes of crystalline poly-(R)-3-hydroxybutyrate observed in low-frequency Raman and terahertz spectra," J. Phys. Chem. B, vol. 117, pp. 2180–2187, 2013.

IOP Conf. Series: Materials Science and Engineering **835** (2020) 012005 doi:10.1088/1757-899X/835/1/012005

Systematic First-Principles Calculations of Charge Transfer Transitions of Transition Metal Ions (Sc^{3+}, Ti^{3+}, V^{3+}, Cr^{3+}, Mn^{3+}, Fe^{3+}) in α-Al_2O_3 with Structural Optimization

Shota Takemura[1], Mega Novita[2], Kazuyoshi Ogasawara[1]

[1] School of Science and Technology, Kwansei Gakuin University, 2-1 Gakuen Sanda, Hyogo 669-1337 Japan

[2] Faculty of Engineering and Informatics, Universitas PGRI Semarang, Jl. Sidodadi-Timur No.24 Semarang, Central Java 50232, Indonesia

s.takemura@kwansei.ac.jp

Abstract. In order to calculate the LMCT energies for various trivalent TM ions in α-Al_2O_3 using realistic model clusters, we performed structural optimization using the CASTEP code and constructed the optimized TMO_6^{9-} and $TMAl_{13}O_6^{30+}$ clusters (TM = Sc^{3+}, Ti^{3+}, V^{3+}, Cr^{3+}, Mn^{3+}, Fe^{3+}). We obtained the LMCT energies of TM ions by first-principles many-electron calculations using the DVME method. It was found that the differences between the LMCT energies of the optimized clusters and those of the relaxed clusters based on the crystal radii originate from the difference in the multiplet splitting of the LMCT states. The results also indicated that the LMCT energies of transition metal ions in α-Al_2O_3 can be calculated using small clusters composed of the central TM ion and the first-neighbor oxygen ions.

1. Introduction

The position of the impurity level within the band gap is very important in optical materials. It can be estimated by calculating the charge transfer transition energy from the valence band to the impurity levels (Ligand to Metal Charge Transfer: LMCT). Recently, we successfully reproduced the LMCT energies and their experimental trend of transition metal (TM) ions in α-Al_2O_3 using the discrete variational multi-electron (DVME) method [1] considering the configuration-dependent correction

(CDC) and the lattice relaxation (LR) based on the Shannon's crystal radii [2]. We also clarified that the LMCT energy is very sensitive to the bond length. In this work, in order to calculate the LMCT energies for various trivalent TM ions in α-Al_2O_3 using more realistic model clusters, we performed structural optimization using the Cambridge Serial Total Energy Package (CASTEP) code [3-5] and constructed the optimized TMO_6^{9-} and $TMAl_{13}O_6^{30+}$ clusters (TM = Sc^{3+}, Ti^{3+}, V^{3+}, Cr^{3+}, Mn^{3+}, Fe^{3+}). We obtained the LMCT energies of TM ions in α-Al_2O_3 by first-principles many-electron calculations using the DVME method for both the relaxed clusters based on the Shannon's crystal radii and the optimized cluster based on the CASTEP, and compared the results.

2. Computational Method

We constructed a 7-atom cluster with C_3 symmetry consisting of the central aluminum ion and the first-neighbor oxygen ions, and a 20-atom cluster in which the second-neighbor aluminum ions were added, based on the crystal data of α-Al_2O_3 (#ICSD 73724). [6] Then we substituted various trivalent TM (Sc, Ti, V, Cr, Mn, Fe) ions for the central aluminum ion. The effective Madelung potential was considered by setting point charges on the atomic sites around the cluster. The lattice relaxation was considered in two ways. One is the method to estimate the relaxed bond lengths based on the Shannon's crystal radii [7, 8]. The other is a method based on the structural optimization using the CASTEP code. The optimized lattice constants of α-Al_2O_3 by the CASTEP calculation are a = 4.806 Å and c = 13.119 Å. In this calculation, the Vanderbilt ultrasoft pseudopotentials [9] were used and the Generalized Gradient Approximation (GGA) exchange-correlation potential proposed by Perdew, Burke, and Ernzerhof (PBE) were considered [10]. The procedure employed here is similar to those used in Ref. 11.

In this work, we suppose that the LMCT occurs only from the first-neighbor ions to the TM ion. The CI calculation of the LMCT energies requires to consider the ground configuration and the excited configuration after LMCT. The number of 2p orbitals is 18 because there are 6 oxygen ions around the TM ion. Therefore we considered $3d^n2p^{36}$ configuration as the ground configurations and $3d^{n+1}2p^{35}$ configuration as the excited configuration. The $3d^n2p^{36}$ configuration includes the excited states of the d-d transition. Then, the LMCT energy was obtained by the energy difference between the lowest multiplet energy within the $3d^n2p^{36}$ configuration and the lowest multiplet energy within the $3d^{n+1}2p^{35}$ configuration. In calculations using the CI approach, multiplet energies are generally overestimated because of the finite number of Slater determinants. However, this overestimation can be corrected nonempirically by introducing the configuration-dependent correction (CDC) in which a correction term estimated from one-electron calculation is added to the diagonal elements of the many-electron Hamiltonian [1].

IOP Conf. Series: Materials Science and Engineering **835** (2020) 012005 doi:10.1088/1757-899X/835/1/012005

3. Results and discussion

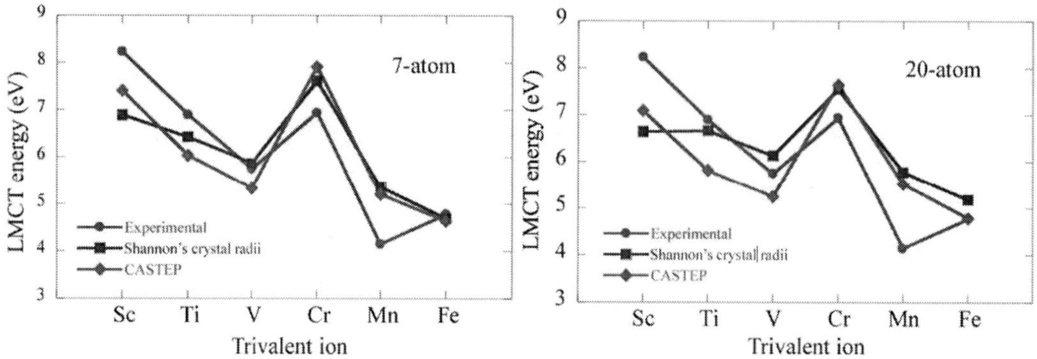

Fig. 1. LMCT energy diagram of trivalent transition metal ions in α-Al_2O_3 calculated by the DVME method with CDC together with the experimental data [12,13]. Left figure shows the LMCT energies using 7-atom clusters while right figure shows the LMCT energies using 20-atom clusters.

The LMCT energy diagram calculated by the DVME method with CDC together with the experimental data [12,13] is shown in Fig. 1. The trend from Sc to V was reproduced better by calculations using the optimized clusters. Although the differences in the LMCT energies of Cr, Mn, and Fe between the relaxed clusters based on the crystal radii and the optimized clusters based on the CASTEP are small, they are relatively large for Sc, Ti, and V. These energy differences are due to the difference in the average TM-O bond lengths between the cluster models of the two approaches. The differences in the average TM-O bond lengths of Sc, Ti, and V are 0.58 Å, 0.16 Å and 0.11 Å, respectively, while those of Cr, Mn, and Fe are less than 0.1 Å.

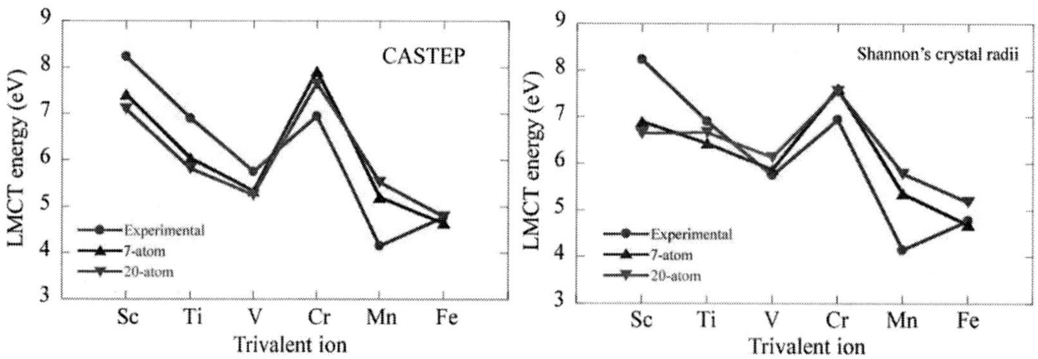

Fig. 2. Cluster size dependence of the LMCT energies between 7-atom cluster and 20-atom cluster. Left figure shows the LMCT energies calculated using the optimized clusters based on the CASTEP while right figure shows the LMCT energies calculated using the relaxed clusters based on the Shannon's crystal radii.

Fig. 2 shows the cluster size dependence of the LMCT energy diagrams where the results of 7-atom clusters and those of 20-atom clusters are compared. The CDC were also considered in these calculations. The differences between the LMCT energies of the 7-atom cluster and those of the 20-atom cluster are small in both the calculation using the optimized clusters based on the CASTEP and that using the relaxed clusters based on the Shannon's crystal radii. The reason can be explained by the difference in the barycenters of the $3d^n2p^{36}$ and $3d^{n+1}2p^{35}$ configurations and the multiplet splitting of the LMCT state. The MO energy diagram of VO_6^{9-} and $VAl_{13}O_6^{30+}$ calculated by the DV-Xα method is shown in Fig. 3.

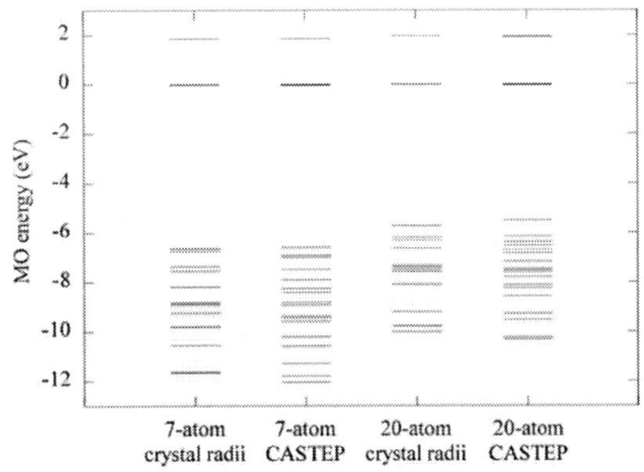

Fig. 3. MO energy diagram of VO_6^{9-} and $VAl_{13}O_6^{30+}$ calculated by the DV-Xα method. Blue and red lines are V 3d levels and oxygen 2p levels, respectively.

In this diagram the highest energy of the t_{2g} orbital of V^{3+} is set at 0 eV. The energy differences between t_{2g} level and the highest oxygen 2p level of the 20-atom clusters are smaller than those of the 7-atom clusters. When the energy differences between t_{2g} level and the highest oxygen 2p level are large, the difference of the barycenters of the $3d^n2p^{36}$ and $3d^{n+1}2p^{35}$ configurations is large. Then the LMCT energy is expected to be larger. On the other hand, the splittings of the oxygen 2p levels of the 20-atom clusters are smaller than those of the 7-atom clusters. When the splitting of the oxygen 2p levels is large, the multiplet splitting of the LMCT state is large. Then the LMCT energy is expected to be smaller. The results indicated that the LMCT energies of the 7-atom and the 20-atom clusters are almost the same due to the cancelation of these two effects. As a result, the LMCT energies of transition metal ions in α-Al_2O_3 can be calculated using small clusters composed of the central TM ion and the first-neighbor oxygen ions.

4. Conclusion

We performed structural optimization using the CASTEP code and constructed the optimized clusters. We obtained the LMCT energies of TM ions in α-Al_2O_3 by first-principles many-electron calculations using the DVME method. The decreasing trend from Sc to V was improved in the results based on the CASTEP compared to the results based on the Shannon's crystal radii. On the other hand, the energy differences between the 7-atom clusters and the 20-atom clusters are small. These results indicated that the LMCT energies of transition metal ions in α-Al_2O_3 can be calculated by calculations using small clusters composed of the central TM ion and the first-neighbor oxygen ions.

References

[1] K. Ogasawara, T. Ishii, I. Tanaka and H. Adachi, Phys. Rev. B **61** (2000) 143.

[2] S. Takemura and K. Ogasawara, Opt. Mater.: X **1**, (2019) 100005.

[3] M. C. Payne, M. P. Teter, D. C. Allan, T. A. Arias and J. D. Joannopoulos, Rev. Mod. Phys., **64**, (1992) 1045.

[4] M. D. Segall, P. J. D. Lindan, M. I. J. Probert, C. J. Pickard, P. J. Hasnip, S. J. Clark and M. C. Payne, J. Phys. Condens. Matter **14**, (2002) 2717.

[5] S. J. Clark, M. D. Segall, C. J. Pickard, P. J. Hasnip, M. I. J. Probert, K. Refson and M. C. Payne, Z. Kristallogr. **220**, (2005) 567.

[6] E.N. Maslen, V.A. Strel'tsov, N.R. Strel'tsova, N. Ishizawa and Y. satow, Acta Crystallogr. B **49** (1993) 973.

[7] S. Watanabe, T. Ishii, K. Fujimura and K. Ogasawara, J. Solid State Chem. **179** (2006) 2438.

[8] R.D. Shannon, Acta Crystallogr. A **32** (1976) 751.

[9] D. Vanderbilt, Phys. Rev. B, **41**, (1990) 7892.

[10] J.P. Perdew, K. Burke and M. Ernzerhof, Phys. Rev. Lett., **77**, (1996) 3865.

[11] M. Novita, H. Nagoshi, A. Sudo and K. Ogasawara, IOP Conf. Ser.: Mat. Sci. Eng. **299**, (2018) 012060.

[12] B.R. Namozov, R.I. Zakharchenya and M.P. Korobkov, Phys. Solid State **40** (1998) 599.

[13] H.H. Tippins, Phys. Rev. B **1** (1970) 126.

IOP Conf. Series: Materials Science and Engineering **835** (2020) 012007 doi:10.1088/1757-899X/835/1/012007

Optimization of first-principles calculation conditions for multiplet energies of Fe^{3+} and Co^{3+} in α-Al_2O_3

Judo Fusamoto, Kazuyoshi Ogasawara

School of Science and Technology, Kwansei Gakuin University, 2-1 Gakuen Sanda, Hyogo 669-1337 Japan

Abstract. Corundum can contain various transition metals as impurities, for example, ruby. Ruby is used as solid state laser. The prediction of the electronic state of these transition metals is very important for the development of novel optical materials. The CI calculation used finite Slater determinants, the calculated mutiplet energies is overestimated generally. In this study, we optimized the calculation multiplet energy of transition metals in α-Al_2O_3 by considering some corrections.

1. Introduction

The prediction of multiplet energies of d^n ions such as transition metal ions in crystals is important for the development of novel optical materials. Although the mutiplet energies calculated by Configuration-Interaction (CI) calculations are generally overestimated, the overestimation can be corrected by considering the Configuration-Dependent Correction (CDC) and the Correlation Correction (CC). However, for d^5 and d^6 ions in Al_2O_3, the conditions of CDC-CC such as the considered transition processes have not been established. In this study, in order to optimize the conditions of CDC-CC, we performed first-principles calculations of the multiplet energies of Fe^{3+} and Co^{3+} in α-Al_2O_3 considering CDC-CC based on various transition processes.

2. Methods

2.1 DV-Xα and DVME

We used Discrete Variational Xα method (DV-Xα) [1] and Discrete Variational Multi-Electron (DVME) method [2] in this study. These methods are way to perform First-Principles Calculation. The wave function of the molecular orbit and the orbital energies are acquired by DV-Xα method. The multiplet energies are acquired by DVME method. The number of sampling points for numerical integration was 100000 for central transition metal ion, 10000 for the first proximity oxygen ions, and 1000 for others

2.2 Corrections

In DVME method, the obtained multiplet energies are generally overestimated because of considering the finite Slater determinants. Therefore, we performed some corrections to calculate the mutiplet energies more precisely.

First, we performed Configuration Dependent Correction (CDC) [2]. CDC is a method to correct the center of gravity of each electronic configuration by using the orbital energies obtained by DV-Xα method.

Second, we performed Correlation Correction (CC) [2]. CC is a method to consider the underestimated electron correlation effect by spin-flip energies obtained by spin-unrestricted DV-Xα method.

Finally, we performed lattice relaxation [3][4]. Lattice relaxation is a method to consider the change of bond distance between the ligand and the central atom in replacing the central atom.

$$d_{relax}=d_0 \times \frac{r(\mathrm{M}^{3+})+r(\mathrm{O}^{2-})}{r(\mathrm{Al}^{3+})+r(\mathrm{O}^{2-})}, \qquad (1)$$

We can take account for many spin inversion transitions in CC, so we considered various spin inversion transition in CC and investigated optimal correction in the calculation of multiplet energy of Fe^{3+} and Co^{3+}.

2.3Crystal constracture and model cluster

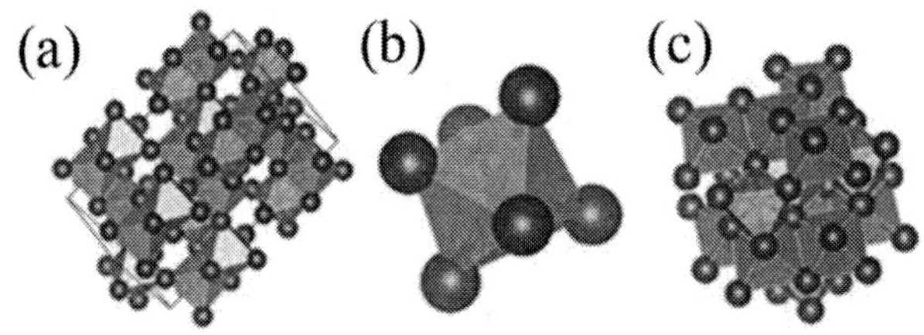

Figure1. The crystal constracture of α-Al_2O_3 (a), The 7-atom cluster of α-Al_2O_3 (b) and 63-atom cluster of α-Al_2O_3 (c)

We showed the crystal constructure of α-Al_2O_3 [5]. We made 7-atom cluster and 63-atom cluster which have Al^{3+} ion in center and O^{2-} ion as ligands. We replaced Al^{3+} with Fe^{3+} or Co^{3+} in calculation.

3. Results

We performed both calculation with lattice relaxation and calculation without lattice relaxation. We considered various spin inversion energies in CC respectively.

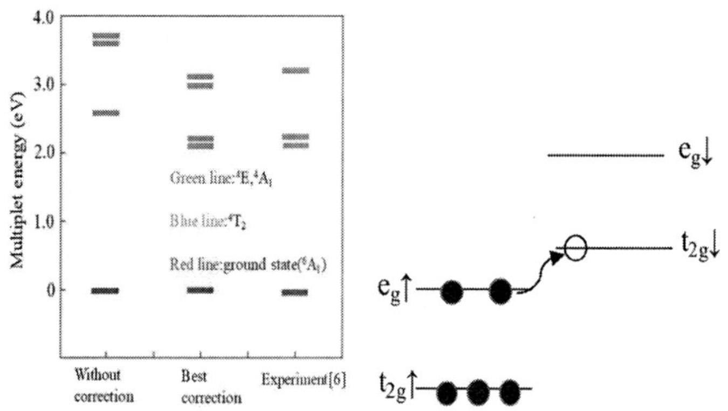

Figure2. The best result about multiplet energy of Fe^{3+} in the 7-atom cluster and the transition considered in CC

in CC without lattice relaxation. We also considered CDC and spin inversion transition energies of t_{2g}up→t_{2g}down in high spin, $e_g$up→$e_g$down in high spin, $e_g$up→t_{2g}down in high spin and t_{2g}up→$e_g$down in low spin in CC with lattice relaxation. Finally, we also performed calculation

IOP Conf. Series: Materials Science and Engineering **835** (2020) 012007 doi:10.1088/1757-899X/835/1/012007

without CC and CDC. As a result, the calculation considered CDC and spin inversion energy of $e_g$up→t_{2g}down in high spin in CC without lattice relaxation is the best result. The mean square error of the best result is minimum in all calculation. In the case of multiplet energy of Fe^{3+} in the 7-atom cluster, the calculated multiplet energies decreased by considering the transition showed at Figure2 in CC.

In the calculation of 63-atom cluster for Fe^{3+}, we considered CDC and spin inversion transition energies of t_{2g}up→t_{2g}down in high spin, $e_g$up→$e_g$down in high spin, t_{2g}up→$e_g$down in high spin, $e_g$up→t_{2g}down in high spin and t_{2g}up→$e_g$down in low spin in CC without lattice relaxation. We also considered CDC and spin inversion transition energies of t_{2g}up→t_{2g}down in high spin, $e_g$up→$e_g$down in high spin, t_{2g}up→$e_g$down in high spin, $e_g$up→t_{2g}down in high spin and t_{2g}up→$e_g$down in low spin in CC with lattice relaxation. Finally, we also performed calculation without CC and CDC. As a result, the calculation considered only lattice relaxation is the best result. The mean square error of the best result is minimum in all calculation. In the case of multiplet energy of Fe^{3+} in the 63-atom cluster, the calculated multiplet splitting decreased by considering only lattice relaxation.

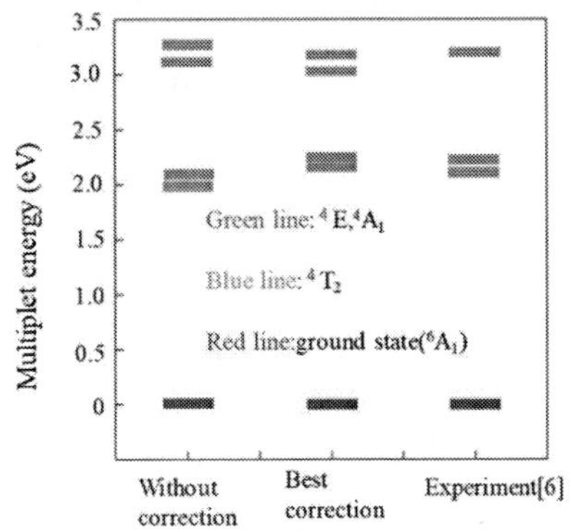

Figure3. The best result about multiplet energy of Fe^{3+} in the 63-atom cluster

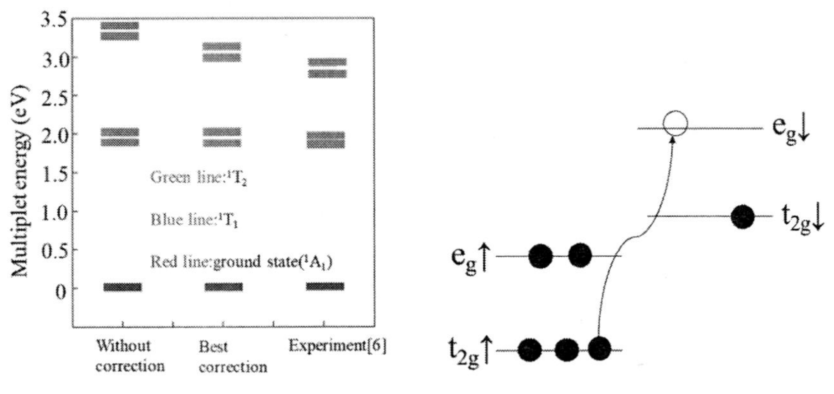

Figure4. The best result about multiplet energy of Co^{3+} in the 7-atom cluster and the transition considered in CC

In the calculation of 7-atom cluster for Co^{3+}, we considered CDC and spin inversion transition energies of t_{2g}up→t_{2g}down in high spin, $e_g$up→$e_g$down in high spin, t_{2g}up→$e_g$down in high spin and $e_g$up→t_{2g}down in high spin in CC without lattice relaxation. We also considered CDC and spin inversion transition energies of t_{2g}up→t_{2g}down in high spin, $e_g$up→$e_g$down in high spin and $e_g$up→t_{2g}down in high spin in CC with lattice relaxation. Finally, we also performed calculation without CC and CDC. As a result, the calculation considered CDC and spin inversion energy of t_{2g}up→$e_g$down in high spin in CC with lattice relaxation is the best result. The mean square error of the best result is minimum in all calculation. In the case of multiplet energy of Co3+ in the 7-atom cluster, the calculated multiplet splitting decreased by considering lattice relaxation and the transition showed at Figure4 in CC.

Finally, in the calculation of 63-atom cluster for Co^{3+}, we considered CDC and spin inversion transition energies of t_{2g}up→t_{2g}down in high spin, $e_g$up→$e_g$down in high spin, t_{2g}up→$e_g$down in high spin and $e_g$up→t_{2g}down in high spin in CC without lattice relaxation. We also considered CDC and spin inversion transition energies of t_{2g}up→t_{2g}down in high spin, $e_g$up→$e_g$down in high spin and $e_g$up→t_{2g}down in high spin in CC with lattice relaxation. Finally, we also performed calculation without CC and CDC. As a result, the calculation considered CDC and spin inversion energy of t_{2g}up→$e_g$down in high spin in CC with lattice relaxation is the best result. The mean square error of the best result is minimum in all calculation. In the case of multiplet energy of Co^{3+} in the 63-atom cluster, the calculated multiplet splitting decreased by considering lattice relaxation and the transition showed at Figure5 in CC.

4. Conclusion

We could optimize the conditions of CC and CDC for Fe^{3+} and Co^{3+} in α-Al_2O_3, and successfully reproduced the experimental multiplet energies. By consideration of CC and CDC, the repulsion between electrons was suppressed and the overestimation of the multiplet splittings was reduced.

References

[1] H.Adachi, M.Tsukada, C.Satoko, J.Phys. Soc. Jpn., **45**, 875 (1978).
[2] K. Ogasawara, T. Iwata, Y. Koyama, and T. Ishii, Phys. Rev. B **61** 143 (2000).
[3] S. Watanabe, T Ishii, K Fujimura, K. Ogasawara, J. Solid State Chem. **179**, 2438 (2006).
[4] R. D. Shannon, Acta Cryst., **A23**, 751, (1976).
[5] H. Sawada, Mater. Res. Bull., **29**, 127 (1994).
[6] Donald S. McClure, J. Chem. Phys., **36**, 2757 (1962).

Water Quality in the Maintenance of *Oreochromis niloticus* through Environmentally Friendly Biofloc Technology

Endah Rita Sulistya Dewi[1*], Ary Susatyo Nugroho[1], Maria Ulfah[1]
Departement of Mathematics, Sciences and Information Technology, Universitas PGRI Semarang, Indonesia

*endahrita@upgris.ac.id

Abstract. This Population growth that continues to increase demands continuous availability of food. Aquaculture is one of the sectors of food production which has the highest growth rate in the world. Facing this opportunity, aquaculture is faced with several challenges related to limited natural resources such as water and land, as well as waste water from aquaculture. Furthermore, the application of best aquaculture practices in aquaculture product certification requires environmentally friendly aquaculture practices. *Oreochromis niloticus tilapia* is a type of freshwater consumption fish and is now a popular pet fish in freshwater ponds in Indonesia. This study aims to analyze water quality in the maintenance of Tilapia through environmentally friendly Biofloc technology. This research method was designed in the form of Research and Development. The results showed that the water quality was proper for pond operation using minimal water exchange, the development of dense microbial populations through the management of microbial populations could control the concentration of ammonia in the water. Bacteria in ponds form biofloc, producing microbial proteins that make it possible to recycle unused feed protein, so as to minimize pollution.
Keywords: Biofloc, Oreochromis niloticus, Environment Friendly

1. Introduction

Aquaculture is a form of maintenance and breeding of various kinds of aquatic animals or plants that use water as its main component. Common activities include fish cultivation, shrimp cultivation, oyster cultivation, seaweed cultivation (algae).
Aquaculture is an activity to produce aquatic biota (organism) in a controlled environment in order to get profit. In the next two decades, it is estimated that the demand for fish or non-fish will double due to the increase in world population. No doubt, aquaculture is a very reliable industry. It is estimated that in 2020, aquaculture will produce more than 68 million tons, reaching 40% of global seafood production [1].

Tilapia (*Oreochromis niloticus*) is a cheap source of animal protein for human consumption. Fish farming can be done in ponds or rearing tanks. In the process of growth, protein metabolism by aquatic organisms generally produces ammonia as a result of excretion. At the same time, the protein in feces and inedible feed will be broken down by bacteria into the same product, thus the more intensive a cultivation activity will be followed by the higher concentration of nitrogen compounds especially ammonia in water [2-3].

The high organic waste from artificial feed (pellets) and feces resulting from intensive fish maintenance will cause buildup and sedimentation at the bottom of the maintenance water media, so a decomposition process is needed. If not decomposed the maintenance media will be decomposed anaerobically by anaerobic bacteria then form toxic gases such as sulfuric acid, nitrite, and ammonia and have a negative impact on the metabolism of aquatic organisms to death. To reduce organic waste and will be discharged into public waters, water quality management is needed so that the maintenance media remain in good condition. One of the efforts is a biological approach by utilizing bacterial activity to accelerate the process of decomposition of organic waste.

Along with the development of technology through a biological approach, biofloc technology has been applied to maintain the quality of aquaculture. Biofloc technology is a technology to use both heterotrophic and autotrophic bacteria that can convert organic waste intensively into flock-shaped microorganisms, which can then be utilized by fish as a food source [4-5]. Within the floc there are several forming organisms such as bacteria, plankton, fungi, algae, and suspended particles that affect the structure and nutrient content of biofloc, but the bacterial community is the most dominant microorganism in floc formation in biofloc [6].

Biofloc technology or active suspension technique, uses constant aeration to enable the process of aerobic decomposition and keep bacterial flocks in suspension [7]. In this system heterotrophic bacteria that grow with high density function as bioreactors that control water quality especially N concentrations as well as a source of protein for the organisms that are kept. Furthermore according to [6], states that the formation of biofloc by heterotrophic bacteria generally aims to increase nutrient utilization, avoid environmental stress and predation.

This research is important to assess the efficiency of water use, ammonia content in water, the use of nutrients as natural food and the reduction of waste pollution in tilapia fish farming. It is expected that perpetrators of aquaculture will continue to prioritize the use of science and technology in the management of sustainable aquaculture businesses.

2. Methods

This research is a research development (R & D) . The development procedure uses the Borg and Gall model. Biofloc technology application is also carried out in this activity, namely making biofloc pools, preparing water media, growing floc, feeding management and water quality management.

Figure 1. Biofloc Ponds

The application of biofloc technology is carried out in a round tarpaulin pond with a diameter of 150 cm, height, 100 cm. Maintenance water media is 1.5 m^3 and fish density is 300 fish/ m^3. At the beginning before the fish were stocked, probiotics were given, namely EM4 floc forming at a dose of 5 cc/m^3. The addition of probiotic bacteria was carried out along with the accumulation of feed given in the maintenance media for 42 maintenance days and the length of the fish was measured at the beginning and the end of maintenance. During maintenance measured water quality including ammonia, temperature, DO, and pH was carried out on days 0, 21 and 42.

3. Results and discussion

From the results of the study water quality data obtained in the measurement period are as follows:
Table 1. Data on Biofloc Pond Water Quality in Tilapia Maintenance

Parameters	Day 0	Day 21	Day 42
Ammonia	0.015 mg / L	0.02 mg / L	0.013 mg / L
Temperature	26^0C	28^0C	28^0C
DO	5.2 mg / L	8.4 mg / L	8.57 mg / L

pH	6	6	7

Next, the ammonia content for 42 days, with sampling on day 0, day 21 and day 42, is presented in Figure 2.

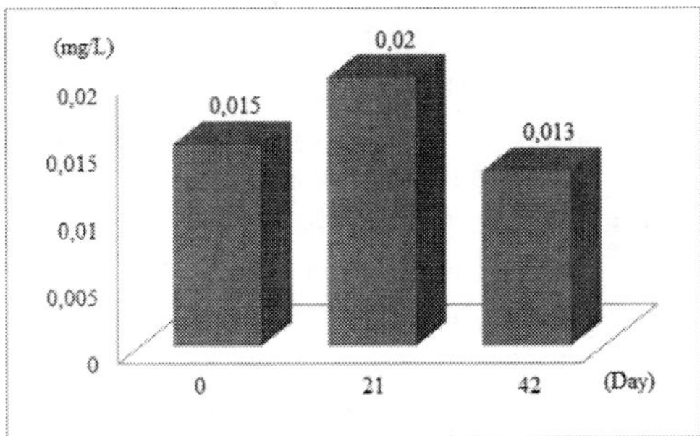

Figure 2. Ammonia Content in Biofloc Ponds

From this graph, it can be seen the average ammonia content of 0.016 mg / L, which means that the ammonia content in the pond is still below the quality standard. This condition can occur due to bacteria that can break down waste from fish, namely the Nitrosomonas bacteria, which convert Ammonia into Nitrites, Nitrites are then converted into Nitrates by Nitrobacter bacteria. According to [8] requirements, the maximum limit of ammonia levels for fish farming activities is <0.02 mg / L.

By [9] asserted that decomposition of ammonia by microbes can be done by deamination process. Deamination is the process of dismantling proteins into amino acids, then amino acids are broken down into ammonia and some other substances by the bacteria producing urease enzymes, with this enzyme urea can be broken down into ammonium carbonate, while ammonium carbonate is easily broken down into ammonia, carbon dioxide and water, also confirmed by [10] that the urease enzyme is the only catalyst enzyme in hydrolyzing urea. [11] added that the genus Bacillus is a microbe from a class of bacteria that is able to synthesize urease well in waters.

The need for good water quality in intensive fish maintenance requires an environmentally friendly technology so that the low organic matter in the maintenance media and the low waste that is discharged into public waters, in this case biofloc technology can provide a solution.

Broadly speaking, intensive maintenance of Tilapia with biofloc technology is more effective than without biofloc technology. And some supporting parameters such as temperature, DO, and pH are still within the maintenance threshold of tilapia. Provision of probiotics such as EM4 (Effective Microorganisms 4) has an important meaning in maintaining water quality by the process of breaking down waste materials into more useful organic matter. According to the opinion of [12] which states EM4 contains fermented and synthetic microorganisms consisting of Lactic Acid bacteria (*Lactobacillus sp*), Photosentetic Bacteria (*Rhodopseudomonas sp*), *Actinomycetes sp*, *Streptomyces sp* and Yeast (yeast)) and cellulose decomposers.

[13] said that EM (Effective Microorganisms) consists of photosynthetic bacteria, lactic acid bacteria, actinomicetes, yeast and fermented fungi. Photosynthetic bacteria form beneficial substances that produce amino acids, nucleic acids and bioactive substances derived from harmful gases and function to bind nitrogen from the air. Lactic acid bacteria function to ferment organic material into lactic acid, accelerate the overhaul of organic matter, lignin and cellulose, and suppress pathogens with lactic acid produced. Actinomicetes produce anti-microbial substances from amino acids produced by photosynthetic bacteria. Yeast produces antibiotic substances, produces enzymes and hormones, yeast secretion becomes a substrate for effective microorganisms of lactic acid bacteria actinomycetes. Fermentation fungi are also able to break down organic matter quickly which produce anti-microbial ester alcohol, and eliminate the foul odor.

These microbes also play a role in floc formation. The formation of flocks in the pool can be seen with the characteristics of the pool does not smell, the color of the flocks in the pool is green which indicates in the pool more phytoplankton is formed.

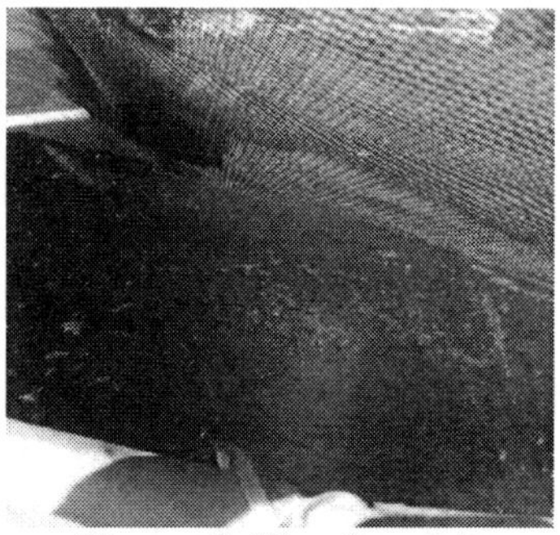

Figure 3. Floc formation

Concentrations of temperature, DO, and pH in each treatment ranged from 26-28°C; 5.2-8.57 mg / L; and 6-7. The three parameters show the results in the maintenance media are within reasonable limits for the growth of tilapia. Temperature requirements according to [8] 25-32°C as well as (DO) dissolved oxygen is within the proper limits for the growth of tilapia, according to the requirements of [8] SNI, DO is a minimum of 3 mg / L tilapia can still grow well at dissolved oxygen 5.2-8.57 mg / L. The pH concentration of water in the pond is within a reasonable limit for the growth of tilapia, ranging from 6-7 while [8] requirements are 6.5- 8.5.

Ammonia's toxicity to aquatic animals is very dependent on pH, temperature and salinity. At high pH the NH3 levels will rise, and vice versa. When ammonia levels are high, the ability of fish to excrete ammonia is reduced. This causes an increase in ammonia levels in the blood and body tissues. This will increase blood pH levels and have a detrimental effect on enzyme sharing reactions and membrane stability. Negative effects include damage to the gills, reducing the capacity of blood to carry oxygen and histological damage to red blood cells [14]. From the results of observations of Ammonia (NH3) in ponds during the study containing ammonia with an average concentration of 0.016 mg / L. This proves that the role of flocks in ponds is able to maintain ammonia concentration during the tilapia aquaculture process. This condition is in accordance with the statement of [15] if the level of ammonia in the waters contained in an amount that is too high is greater than 1 mg / L can not be suspected of pollution.

4. Conclusion

1. Ammonia levels in biofloc ponds contain ammonia with an average concentration of 0.016 mg/ L, still below the quality standard.
2. Water quality in biofloc ponds is declared feasible during maintenance of Tilapia.
3. Provision of EM4 forms biofloc and produces microbial proteins that play a role in recycling unused feed proteins, so as to minimize pollution.

Acknowledgement
Thanks to Direktorat Riset dan Pengabdian Masyarakat, Direktorat Jendral Penguatan Riset dan Pengembangan, Kementerian Riset, Teknologi dan Pendidikan tahun 2019, who have provided this research grant.

References

[1] Pangkey, H. "Pengembangan Akuakultur di Wilayah Kota Tidore Kepulauan". Jurnal Perikanan dan Kelautan. IV (2): 27-34. 2008.
[2] Avnimelech, Y. "Sustainable land-based aquaculture-rational utilization of water, land and feed". Mediterranian aquaculture journal (41):45-55. 2008
[3] Avnimelech, Y., and Kochba M. "Evaluation of Nitrogen Uptake and Excretion by Tilapia in biofloc Tanks", Using15N tracing Aquaculture 287: 163-168.2009
[4] de Schryver P, Verstraete W. "Nitrogen Removal from Aquaculture Pond Water by Heterotrophic Nitrogen Assimilation in Lab-Scale Sequencing Batch Reactors". Bioresource Technology. 100(3): 1162-1167. 2009. http://doi.org/d24f5t

[5] Avnimelech Y. "Biofloc Technology" - a Practical Guide Book, 2nd edition. United States (US): The World Aquaculture Society. 2012
[6] de Schryver P, Crab R, Defoirdt T, Boon N, Verstraete W. "The basics of bioflocs technology: The added value for aquaculture". Aquaculture. 277(34): 125-137. 2008. http://doi.org/d9wr3
[7] Azim, M.E and D.C Little. "The Biofloc Tecnology (BFT) in Indoor Tanks: Water Quality, Biofloc Composition, and Growth and Welfare of Nile Tilapia (Oreochromis niloticus)". Aquaculture, 176:227-235. 2008.
[8] Badan Standardisasi Nasional. SNI 7550 : 2009, "Produksi ikan Nila ((Oreochromis niloticus Bleeker) kelas pembesaran di kolam air tenang. Author. 2009.
[9] Waluyo L. "Mikrobiologi Umum". Malang (ID): Universitas Muhammadiyah Malang Press. 2007.
[10] Gianfreda L, Bollag JM. "Influence of natural and anthropogenic factor on enzyme activity in Soil". Soil Biochemistry. 9: 123-176. 1996
[11] George M, Cyriac N, Nair A, Hatha AAM. " Diversity of Bacillus and Actinomycetes in the water and sediment samples from Kumarakom region of Vembanadu lake". Indian Journal of Geo-Marine Sciences. 40(3): 430-437. 2011
[12] Javaid, A. & Bajwa, R. "Field evaluation of effective microorganisms (EM) application for growth, nodulation, and nutrition of mung bean". Turkish Journal of Agriculture and Forestry, 35(4), pp.443–452. 2011
[13] Irmak, S. & Cil, A. "The Effects of Microbial Fertilizer Application on Yield and Some Yield Elements of Peanut in Cukurova Region in Turkey". Research Journal of Pharmaceutical, Biological and Chemical Sciences, 2(1), pp.880–888. 2011.
[14] Boyd CE. " Water Quality in Warm Water Fish Ponds". Fourth Printing. Auburn. University Agricultural Experiment Station, Alabama, USA.359 p.1988
[15] Irawan. "Faktor-Faktor Penting dalam Proses Pembesaran Ikan di fasilitas Nursery dan Pembesaran". 2009. Diambil dari www.sith.ipb.ac.id

IOP Conf. Series: Materials Science and Engineering **835** (2020) 012009 doi:10.1088/1757-899X/835/1/012009

Prediction of $4f^2$-$4f^1 5d^1$ transition energy of Pr^{3+} in fluorides based on first-principles calculations and machine learning

Hayato Obata, Shota Takemura, and Kazuyoshi Ogasawara

School of Science and Technology, Kwansei Gakuin University, 2-1 Gakuen Sanda, Hyogo 669-1337 Japan

Abstract. The $4f^2$-$4f^1 5d^1$ transition energy of Pr^{3+} in fluorides are utilized for various optical materials such as solid-state lasers, phosphors, and scintillators. Therefore, it is important to predict such energies of unknown materials for theoretical design of novel optical materials. In this study, we tried to predict the $4f^2$-$4f^1 5d1$ transition energies of Pr^{3+} in fluorides based on first-principles calculations and machine learning. The first-principles calculations were performed based on the relativistic discrete variational multi-electron (DVME) method using the model clusters composed of the central Pr^{3+} and the anions closer than the nearest cation. Although the calculated $4f^2$-$4f^1 5d^1$ transition energies of Pr^{3+} in fluorides showed a relatively good correlation with the experimental ones, the calculated values tend to be overestimated by ca. 2 eV. In order to improve the accuracy of the prediction, we used the calculated transition energies as an attribute for machine learning. As a result, the regression formula to predict the $4f^2$-$4f^1 5d^1$ transition energy of Pr^{3+} in fluorides has been derived by machine learning using the calculated $4f^2$-$4f^1 5d^1$ transition energy as well as some other electronic and structural parameters as the attributes. The accuracy of the prediction was significantly improved compared to the simple first-principles calculations.

1. Introduction

The $4f^2$-$4f^1 5d^1$ transition energy of Pr^{3+} in fluorides are utilized for various optical materials such as solid-state lasers, phosphors, and scintillators. Therefore, it is important to predict such energies of unknown materials for theoretical design of novel optical materials. Although we calculated the $4f^2$-$4f^1 5d^1$ absorption energy by relatively simple first-principles calculations, the calculated values are overestimated by ca. 2 eV compared to the experimental values as shown in Figure 1. In order to solve this problem, we used machine learning as a simple correction method. Therefore, the purpose of this study is the following two. For Pr^{3+} in fluorides, regarding the calculated $4f^2$-$4f^1 5d^1$ absorption energy as one of the attributes, create a predictive model of $4f^2$- $4f^1 5d^1$ absorption energy based on machine learning considering electronic and structural parameters as attributes, and regarding the calculated $4f^2$-$4f^1 5d^1$ absorption energy as one of the attributes, create a predictive model of $4f^1 5d^1$-$4f^2$ emission energy based on machine learning considering electronic and structural parameters as attributes. Table 1 shows the fluorides calculated in this work.

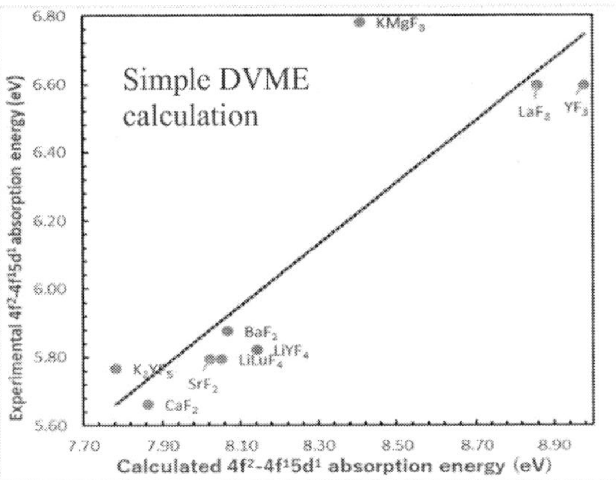

Figure 1. Correlation between calculated and experimental $4f^2$-$4f^15d^1$ absorption energy of Pr^{3+} in fluorides

Table 1. Fluorides calculated in this work

Crystal	LaF_3	YF_3	BaF_2	$LiYF_4$	SrF_2	K_2YF_5	$LiLuF_4$	CaF_2	$KMgF_3$
Cation site	La	Y	Ba	Y	Sr	Y	Lu	Ca	K
Cluster Model	PrF_{13}^{10-}	PrF_{11}^{8-}	PrF_8^{5-}	PrF_8^{5-}	PrF_8^{5-}	PrF_7^{4-}	PrF_{10}^{7-}	PrF_8^{5-}	PrF_{12}^{9-}
Reference	[1]	[2]	[3]	[4]	[5]	[6]	[7]	[8]	[9]

2. Methods

2.1. *Relativistic DV-Xα Method*

3 based on the crystal structure data [4-11] [12] and (CI) dominant dominantDV-Xα method is based on LCAO approximation and Density Functional Theory(DFT). This method enables one to obtain molecular orbital energies and molecular orbital wave function by solving one-electron wave equation

One-electron wave equation (relativistic)

$$h(\boldsymbol{r})(\boldsymbol{r}) = (\boldsymbol{r}) \quad (1)$$

One-electron Dirac Hamiltonian

$$\boldsymbol{h} = \boldsymbol{cp} + - + \boldsymbol{d} + - \quad (2)$$

2.2. *Relativistic DVME Method*

DVME method based on configuration interaction method enables one to obtain multiplet energy. Many-electron wave functions are written as linear combination of the Slater determinants.

Many-electron wave equation (relativistic)

$H = E$ (3)
$=$ (4)
$(,\cdots,) =$ (5)
$H = +$ (6)

2.3 Machine learning

We used WEKA (software) for machine learning. It developed in the University of Waikato. In this work, we used to linear regression (Single layer perceptron) model. It was used for the creation of the predictive model of $4f^2$-$4f^15d^1$ transition energy based on the electronic and structural parameters. We considered the following 6 attributes, calculated $4f^2$-$4f^15d^1$ absorption energy, net charge of Pr^{3+}, bond order between Pr and F, barycenter of 5d levels relative to the lowest 4f level, crystal field splitting of 5d levels, average bond length.

3. Results and Discussion

3.1 Predictive model of $4f^2$-$4f^1 5d^1$ absorption energy based on electronic and structural parameters

Figure 2 shows the correlation between the simple DVME calculated and experimental $4f^2$-$4f^1 5d^1$ absorption energy. The correlation coefficient was 0.8669. Figure 3 shows the correlation between the predicted and experimental $4f^2$-$4f^1 5d^1$ absorption energy. In this model, $4f^2$-$4f^1 5d^1$ absorption energy can be predicted from calculated $4f^2$-$4f^1 5d^1$ absorption energy and net charge. The correlation coefficient was 0.9796. Correlation was significantly improved by considering net charge.

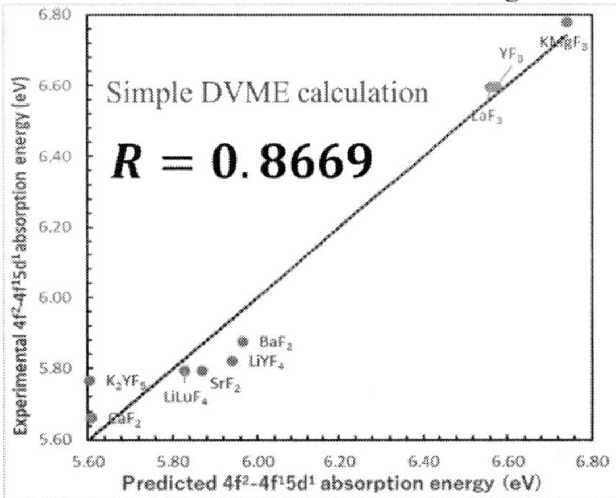

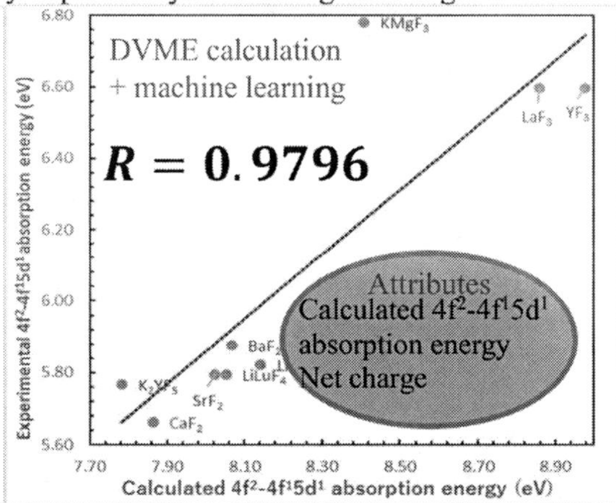

Next, we use cross-validation (leave-one-out-method) as an evaluation of the prediction model of $4f^2$-$4f^1 5d^1$ absorption energy. Cross validation is as follows: Exclude one sample from the data set and the remaining data are used for training. The validation is performed for the left-out sample. This process is repeated for all samples. The correlation coefficient (leave one out) was 0.8778. This predictive model has low generalization ability. Therefore, we added another attribute. Figure 4 shows the correlation between the predicted and experimental $4f^2$-$4f^1 5d^1$ absorption energy. In this model, $4f^2$-$4f^1 5d^1$ absorption energy can be predicted from calculated $4f^2$-$4f^1 5d^1$ absorption energy, net charge and crystal field splitting of 5d levels. The correlation coefficient was 0.9842 and correlation coefficient (leave one out) was 0.9290. This predictive model has high generalization ability. The prediction formula is shown below Predictive model based on machine learning using DVME calculations

$$E = 1.4233E^{calc} - 2.3206Q + 0.0761C - 1.2148 \quad (7)$$

E^{calc}: calculated $4f^2$-$4f^1 5d^1$ absorption energy
Q: Net charge of Pr^{3+}
C: Crystal field splitting of 5d levels

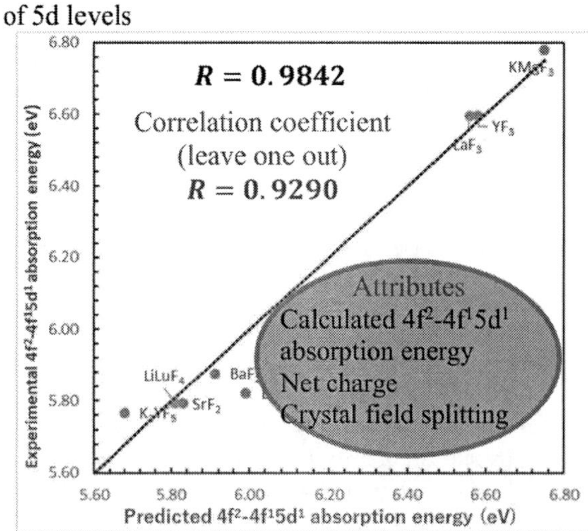

Figure 4. Correlation between predicted and experimental $4f^2$-$4f^1 5d^1$ absorption energy

Table 2. predicted values of $4f^2$-$4f^15d^1$ absorption energy based on machine learning model

Crystal	Prediction (eV)	Experimental (eV)	Difference (eV)	Reference
LaF_3	6.56	6.6	-0.04	[10]
YF_3	6.58	6.6	-0.02	[10]
BaF_2	5.91	5.88	0.03	[10]
$LiYF_4$	5.99	5.82	0.17	[10]
SrF_2	5.83	5.79	0.04	[10]
K_2YF_5	5.68	5.77	-0.09	[10]
$LiLuF_4$	5.81	5.79	0.02	[10]
CaF_2	5.58	5.66	-0.08	[10]
$KMgF_3$	6.75	6.78	-0.03	[11]
EuF_2	6.13			Only prediction

3.2 Predictive model of $4f^15d^1$-$4f^2$ emission energy based on electronic and structural parameters

Figure 5 shows the correlation between the simple DVME calculated and experimental $4f^15d^1$-$4f^2$ emission energy. The correlation coefficient was 0.3559. Figure 6 shows the correlation between the predicted and experimental $4f^15d^1$-$4f^2$ emission energy. In this model, $4f^15d^1$-$4f^2$ emission energy can be predicted from calculated $4f^2$-$4f^15d^1$ absorption energy and barycentre of 5d levels to lowest 4f level. The correlation coefficient was 1.000 and correlation coefficient (leave one out) was 0.9999. This predictive model has high generalization ability.

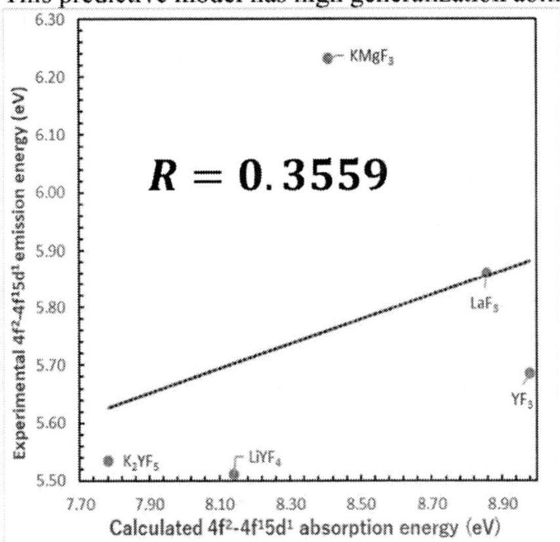

Figure 5. Correlation between calculated and experimental $4f^15d^1$-$4f^2$ emission energy

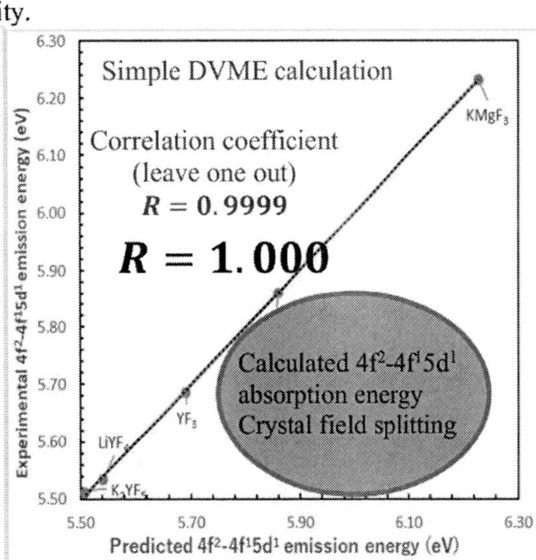

Figure 2. Correlation between predicted and experimental $4f^15d^1$-$4f^2$ emission energy

Correlation was significantly improved by considering barycentre of 5d levels. The prediction formula is shown below.

Predictive model based on machine learning using DVME calculations

$$E = 0.3083E^{calc} - 0.5912E^{5d} + 6.1551$$

E^{calc}: calculated $4f^2$-$4f^15d^1$ absorption energy

E^{5d}: crystal field splitting of 5d levels

Table 3. predicted values of $4f^15d^1$-$4f^2$ emission energy based on machine learning model

Crystal	Prediction (eV)	Experimental (eV)	Difference (eV)	Reference
LaF_3	5.86	5.86	0	[12]
YF_3	5.69	5.69	0	[12]
BaF_2	5.92			Only prediction
$LiYF_4$	5.51	5.51	0	[13]
SrF_2	5.76			Only prediction
K_2YF_5	5.54	5.54	0	[10]
$LiLuF_4$	5.61			Only prediction
CaF_2	5.54			Only prediction
$KMgF_3$	6.23	6.23	0	[12]
EuF_2	5.78			Only prediction

Emisson energy can be predicted from calculated absorption energy.

4. Conclusion

The following predictive models of Pr^{3+} in fluorides were created based on machine learning. Predictive model of $4f^2$-$4f^15d^1$ absorption energy using the calculated $4f^2$-4f15d1 absorption energy and the net charge and the crystal field splitting of 5d levels by first-principles calculations as the attributes and predictive model of $4f^15d^1$-$4f^2$ emission energy using the calculated $4f^2$-$4f^15d^1$ absorption energy and the barycenter of 5d levels relative to the lowest 4f level by first-principles calculations as the attributes.

Acknowledgement

would like to express my sincere thanks to Prof. Ogasawara for his advice and guidance in conducting this research. I would like to thank Mr. Shota Takemura for his helpful advice.

.

References

[1] A. Zaikin et al, Inorganic Chemistry.**5**,1466-1468(1966)
[2] A. K. Cheetham et al, Series A Physical and Inorganic Chemistry.**28**,55-60(1974)
[3] J. M. Leger et al, Physical Review B Condensed Matter and Materials Physics.**53**,13247-13256(1995)
[4] A. V. Goryunov et al, Zhurnal Neorganicheskoi Khimii.**37**,276-279(1992)
[5] J. B. Forsyth et al, Section A Foundations of Crystallography.**45**,244-247(1989)
[6] Yu. A. Kharitonov et al, Kristallografiya.**28**,1031-1032(1983)
[7] A. Grzechnik et al, Journal of Physics Condensed Matter.**17**,763-770(2005)
[8] V. A. Strel'tsov et al, Kristallografiya.**33**,90-97(1988)
[9] F. Demartin et al, Physics and Chemistry of Minerals.**41**,403-407(2014)
[10] P. Dorenbos, Journal of Luminescence.**91**,91-106(2000)
[11] N. J. M. Le Masson et al, Journal of Luminescence.**101**,175-183(2003)
[12] A. M. Srivastava, Journal of Luminescence.**169**,445-449(169)
[13] A. Zych et al, Journal of applied physics.**112**,013536(2012)

Lattice Relaxation Effects on the Multiplet Energies of Ruby Under Pressure using One-Electron Calculations

Mega Novita[1], Nur Cholifah[2], Kazuyoshi Ogasawara[3]

[1]Faculty of Engineering and Informatics, Universitas PGRI Semarang, Jl. Sidodadi-Timur No.24 Semarang, Central Java 50232, Indonesia
[2]Lembaga Penelitian dan Pengabdian Kepada Masyarakat, Universitas PGRI Semarang, Jl. Lontar No.1 Semarang, Central Java 50232, IndonesiaLembaga Penelitian dan Pengabdian Kepada Masyarakat, Universitas PGRI Semarang, Jl. Lontar No.1 Semarang, Central Java 50232, Indonesia
[3]School of Science and Technology, Kwansei Gakuin University, 2-1 Gakuen Sanda, Hyogo 669-1337 Japan

novita@upgris.ac.id

Abstract. Up to recently, it has been difficult to calculate the multiplet energies of compounds using one-electron approach. Since it only considers one electron and one nuclei, the interaction among the electrons are neglected. Previously we have successfully estimated the $^2E \rightarrow {}^4A_2$ (R-line) levels of ruby at 0 pressure using one-electron approach based on one-electron calculation using the first-principles Discrete Variational-Xα (DV-Xα) method. Here we want to perform similar investigation, not only for 0 pressure but up to 110 GPa. We estimated the lattice relaxation effect due to the Cr^{3+} substitution and due to the applied pressure by using two different methods i.e., Shannon's crystal radii and geometry optimizations. Two different types of model cluster consisting of 7 and 63 atoms will be used. The ^2E level is estimated by the barycenter of t_{2g}^3 configuration which consists of 4 multiplets.

1. Introduction

Up to recently, it has been difficult to calculate the multiplet energies of compounds using one-electron approach. Since it only considers one electron and one nuclei, the interaction among the electrons are neglected. Previously we have successfully carried out a comparative study on the transition energies from ^2E to $^4A_2$ (R-line) and from $^4A_2$ to $^4T_2$ (U-band) of α-Al_2O_3: Cr^{3+}, or the so-called ruby, at zero pressure based on one-electron [1] and many-electron first-principles calculations [2]. The estimations were performed without relating to any observed parameter. In Ref. 1, we introduced the improved formula to calculate the R-line energy. Based on Ohnishi's work [3], the R-line energy was estimated by the barycenter of t_{2g}^3 configuration which consists of 3 multiplets i.e., ^2E, $^2T_1$, and $^2T_2$ states. Whereas the U-band energy were simply estimated to be equal to the crystal field splitting, $10Dq$. However, according to the ligand filed theory (LFT), the t_{2g}^3 configuration may consist of 4 multiplet states; not only ^2E, $^2T_1$, and $^2T_2$ states but also $^4A_2$ state [4, 5]. On the other hand, in the case of U-band energies and $10Dq$, they have the distinctly different estimations. In the other papers, we also reported an investigation on the effect of lattice relaxation on the bond length and the molecular orbital energy

[6,7]. The effect of lattice relaxation was estimated by applying pressure for ca. 0-110 GPa. It was performed by the first-principles band-structure calculations using the Cambridge Serial Total Energy Package (CASTEP) code [6]. The results show that when we applied pressure on ruby, the bond length decreased rapidly. On the other hand, the molecular orbital energy was estimated from the one-electron calculation using the first-principles Discrete Variational-Xα (DV-Xα) method [7]. The results show that both t_{2g} and e_g levels of $3d^3$ orbital increased. The crystal field splitting ($10Dq$) especially increased as the pressure increased.

In this paper, we want to perform similar investigation, not only for 0 pressure but up to 110 GPa. This work aims to find out the consistency of our previous results in Ref. 1 for ruby under pressure. Here we also compared the lattice relaxation effect estimated using Shannon's crystal radii and geometry optimizations. Two different types of model cluster consisting of 7 and 63 atoms were used.

2. Computational Method

The model clusters used in this work is similar with those used in Ref. 1. We constructed two sizes of model clusters i.e., 7- and 63-atom. These model clusters were treated without and with considering lattice relaxation effect. The model clusters including lattice relaxation effect were estimated based on Shannon's crystal radii method [8, 9] and geometry optimization method using the Cambridge Serial Total Energy Package (CASTEP) code in the Material Studio 6.0 [10-12]. Those model clusters were pressed from 0 to 110 GPa. The procedure of these steps were explained in Ref. 6. In order to estimate the multiplet energy levels such as R-line and U-band energies of ruby using the one-electron method, here we used DV-Xα method similar with those used in Ref. 1. Figure 1 illustrates the estimation of the energy levels of ruby. In the $3d^3$ configuration, there are several multiplets such as $^4A_2$, ^2E, $^2T_1$, $^2T_2$, and $^4T_2$ states. In the term of Racah parameter, the energies of $^4A_2$, ^2E, and $^4T_2$ can be described as $-15B$, $-6B+3C$, and $-15B+10Dq$, respectively [13]. The R-line energy was estimated by $\varepsilon(2\mathrm{E}) = 0.9 \times \Delta E_R$, where $\Delta E_R = \frac{10}{3}(3B + C)$. On the other hand, the U-band energy ($\varepsilon(4\mathrm{T2})$) was simply calculated as $\Delta E_U = 10Dq$.

3. Results and discussion

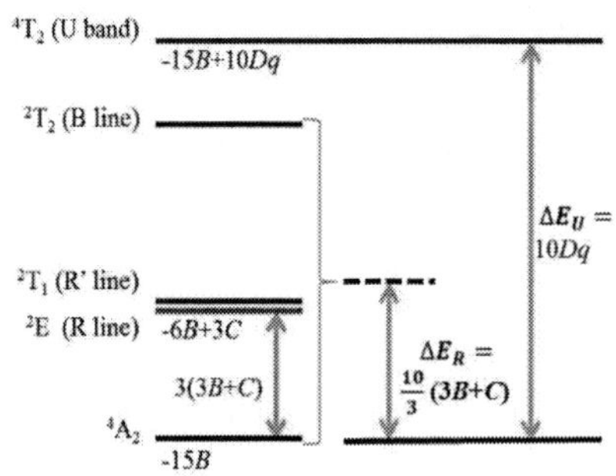

Figure 1. Schematic illustration on the estimation of R-line and U-band energy of ruby based on one-electron calculations.

Figure 1 shows the estimated R-line and U-band energies of ruby under pressure. Figure 1(a), 1(b) and 1(c) are obtained from the 7-atom model clusters. On the other hand, Figure 1(d), 1(e) and 1(f) are obtained from the 63-atom model clusters. The top panel of Figure 1 (a and d) shows the estimated multiplet energies of ruby under pressure without considering any lattice relaxation effect. The middle panel of Figure 1 (b and e) shows the estimated multiplet energies of ruby under pressure with considering lattice relaxation effect based on Shannon's crystal radii. Whereas the lowest panel of Figure 1 (c and f) shows the estimated multiplet energies of ruby under pressure with considering lattice relaxation effect based on geometry optimization using CASTEP code. The R-line and U-band energies are shown by circle (●) and square (■) symbols. The theoretical results are shown by coloured symbols connected with solid line, while the experimental data obtained from Ref. 14-17 are shown by black symbols without any line.

The estimated R-line energies obtained from all model clusters decreased as the pressure increased. It is also shown that by calculations using 7-atom model clusters, the R-line energies were overestimated. However, the agreement has been improved excellently by using larger clusters, 63-atom model clusters. Unlike the R-line energies which were found to slightly decrease along with the increasing pressure, the U-band energies were increased. The estimated U-band energies agree with the experimental data only by the model cluster without considering the lattice relaxation effect, both 7- an d63-atom model clusters. When the lattice relaxation effect was considered, the U-band energies seem to be underestimated. Compared to the results obtained from the model clusters with considering lattice relaxation based on Shannon's crystal radii, the results obtained from the model clusters with considering lattice relaxation based on geometry optimization using CASTEP code give better agreement. The increasing of U-band energy, which is actually $10Dq$ in this work, is due to the decreasing of Cr-O bond length as the result of applying pressure.

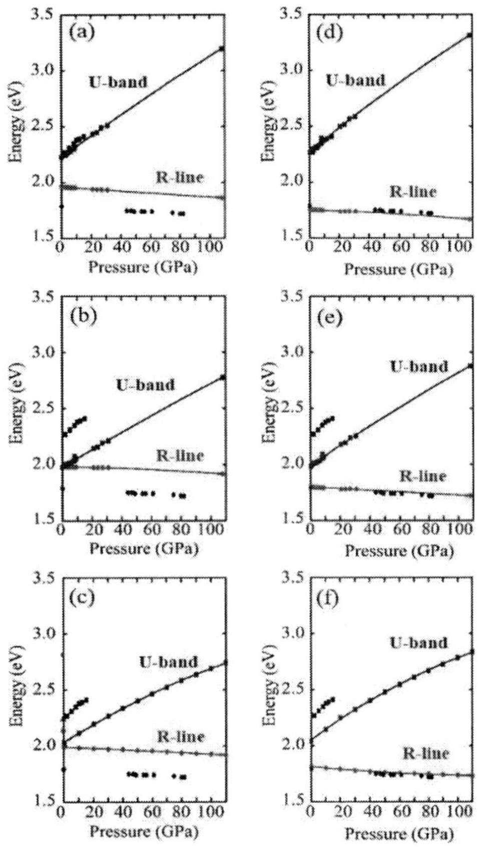

Figure 2. The estimated R-line and U-band energies based on (a) 7-atom model clusters without considering lattice relaxation effect (b) 7-atom model clusters with considering lattice relaxation effect based on Shannon's crystal radii (c) 7-atom model clusters without considering lattice relaxation effect based on geometry optimization using CASTEP code (d) 63-atom model clusters without considering lattice relaxation effect (e) 63-atom model clusters with considering lattice relaxation effect based on Shannon's crystal radii and (d) 63-atom model clusters without considering lattice relaxation effect based on geometry optimization using CASTEP code. The black symbols are the experimental data obtained from Ref. 14-17.

4. Conclusion

We have successfully calculated the multiplet energy levels of ruby bond length under pressure without referring to any experimental parameter using the one-electron first-principles calculations, DV-Xα method. The lattice relaxation effects were calculated by two different methods *i.e.*, the first-principles band-structure calculations using Shannon's crystal radii method and CASTEP method. The results show that the R-line energies decrease as the pressure increase. On the other hand, the U-band energies increase as the pressure increase. The investigation on the cluster size dependence shows that the larger size of cluster improves the accuracy of R-line energies. On the other hand, the increasing of U-band energy, which is actually $10Dq$ in this work, is due to the decreasing of Cr-O bond length as the result of applying pressure. Although the calculation based on 7-atom model clusters gives better agreement with experiment, the underestimation on the barycenter of t_{2g}^3 configuration cannot be neglected. This result also implies that the size of model cluster is crucial.

Acknowledgement

The authors gratefully acknowledge the support from Direktorat Riset dan Pengabdian Masyarakat, Direktorat Jenderal Penguatan Riset dan Pengembangan, Kementrian Riset, Teknologi, dan Pendidikan Tinggi, Indonesia.

References

[1]. Novita, M., Wibowo, S., Nada, N. Q., & Ogasawara, K. (2019, July). Comparative Study on R-line and U-band Energies of Ruby Estimated from One-Electron and Many-Electron First-Principles Approaches. In Journal of Physics: Conference Series (Vol. 1179, No. 1, p. 012104). IOP Publishing.
[2]. Novita, M., & Ogasawara, K. (2012). Comparative Study of Absorption Spectra of V2+, Cr3+, and Mn4+ in α-Al2O3 Based on First-Principles Configuration–Interaction Calculations. Journal of the Physical Society of Japan, 81(10), 104709.
[3]. S. Ohnishi and S. Sugano, Theoretical Studies of High-pressure Effects on Optical Properties of Ruby, Jpn. J. Appl. Phys. 21 (1982) L309-311
[4]. Y. Tanabe and S. Sugano, On the absorption spectra of complex ions. I. J. Phys. Soc. Jpn. 9 (1954) 753-766.
[5]. Y. Tanabe and S. Sugano, On the absorption spectra of complex ions. II. J. Phys. Soc. Jpn. 9 (1954) 766-799.
[6]. Novita, M., Cholifah, N., & Ogasawara, K. (2019, December). Non-empirical study on pressure dependence of ruby bond length. In Journal of Physics: Conference Series (Vol. 1402, No. 6, p. 066005). IOP Publishing.
[7]. M Novita, N Cholifah, and K Ogasawara 2020 ISMEE IOP Conference Series. In Press
[8]. Shannon R D and Prewitt C T 1969 Acta Cryst. B 25 925
[9]. Shannon R D 1976 Acta Cryst. A 23 751
[10]. Payne MC,Teter MP, Allan D C, Arias T A and Joannopoulos J D 1992 Rev. Mod. Phys. 64 1045
[11]. Segall M D, Lindan P J D, Probert M I J, Pickard C J, Hasnip P J, Clark S J and Payne M C 2002 J. Phys. Condens. Matter 14 2717
[12]. Clark S J, Segall M D, Pickard C J, Hasnip P J, Probert M I J, Refson K and Payne M C 2005 Z. Kristallogr. 220 567
[13]. J. C. Slater: Quantum Theory of Molecules and Solids, Vol. 4 of Quantum Theory of Molecules and Solids (McGraw-Hill, New York, United States, 1974).
[14]. J. H. Eggert, K. A. Goettel, and I. F. Silvera, Ruby at high pressure. I. Optical line shifts to 156 GPa. Phys. Rev. B 40 (1989) 5724-5732.
[15]. J. H. Eggert, K. A. Goettel, and I. F. Silvera, Ruby at high pressure. II. Fluorescence lifetime of the R line to 130 GPa. Phys. Rev. B 40 (1989) 5733-5738.
[16]. J. H. Eggert, K. A. Goettel, and I. F. Silvera, Ruby at high pressure. III. A pumping scheme for the R line up to 230 GPa. Phys. Rev. B 44 (1991) 7202-7208.

[17]. D. R. Stephens and H. G. Drickamer, Effect of pressure on the spectrum of ruby, J. Chem. Phys. 35 (1961) 427.

IOP Conf. Series: Materials Science and Engineering **835** (2020) 012011 doi:10.1088/1757-899X/835/1/012011

Optical properties of Co^{3+} doped in α-Al_2O_3 with Considering Lattice Relaxation Effect

Mega Novita[1], Duwi Nuvitalia[2], Nur Cholifah[3], Kazuyoshi Ogasawara[4]

[1]Faculty of Engineering and Informatics, Universitas PGRI Semarang, Jl. Sidodadi-Timur No.24 Semarang, Central Java 50232, Indonesia
[2]Faculty of Mathematics, Natural Science and Information Technology Eduation, Universitas PGRI Semarang, Jl. Sidodadi-Timur No.24 Semarang, Central Java 50232, Indonesia
[3]Lembaga Penelitian dan Pengabdian Kepada Masyarakat, Universitas PGRI Semarang, Jl. Lontar No.1 Semarang, Central Java 50232, Indonesia
[4]School of Science and Technology, Kwansei Gakuin University, 2-1 Gakuen Sanda, Hyogo 669-1337 Japan

novita@upgris.ac.id

Abstract. Search of new luminescence materials to improve the current white-ight emitting diode (LED) is still ongoing. Several investigations such as the ion dependence, the host crystal dependence has been done previously. Since we want to get a complete picture of ion-host combination dependence, we investigated α-Al_2O_3: TM^{3+}. In this work we estimated the optical properties of α-Al_2O_3 doped with Co^{3+}. We constructed model clusters consisting of 7 atoms. The lattice relaxation effects due to the Co^{3+} substitution were calculated using Shannon's crystal radii method and geometry optimizations in the Cambridge Serial Total Energy Package (CASTEP) method. The one-electron Discrete Variational-Xα (DV-Xα) method was used to estimate the molecular orbital energies, while the many-electron Discrete Variational Multielectron (DVME) method was used to estimate the d-d absorption spectra. Since Co^{3+} belongs to $3d^6$ configuration, there are two possible spin states *i.e.*, high-spin (HS) and low-spin (LS) states.

1. Introduction

Up to recently, transition metal (TM) ions doped compounds have attracted great interest due to a wide variety of applications such as phosphors used in white light emitting diode (LED). However, search of new luminescence materials to improve the current white-LED is still ongoing. In our previous studies, we performed several investigations such as the ion dependence and the host crystal dependence. In the case of ion dependence, we found out that the U- and Y-band increase in the order of V^{2+}, Cr^{3+} and Mn^{4+} [1-3]. On the other hand, in the case of host crystal dependence, we found out that when the Mn-F bond length decreased, the U- and Y-band energies increased whereas the R-line energy decreased [4,5]. Nevertheless, similar study for α-Al_2O_3: TM^{3+} ions *i.e.*, Ti^{3+}. V^{3+}, Cr^{3+}, Mn^{3+}, Fe^{3+}, and Co^{3+} need to be conducted to understand the complete picture of TM ions in the octahedral system. Therefore, we have reported a study on the optical properties of α-Al_2O_3: V^{3+} and Mn^{3+} using similar method [6,7]. In the case of α-Al_2O_3: V^{3+}, since V^{3+} belongs to $3d^2$ configuration, there are several transitions which occur from $^3T_{1a}$ ground state to $^1T_2$, ^1E, $^1A_1$, $^3T_2$, and $^3T_{1b}$ states. The transition energies from $^3T_{1a}$ to the singlet states are observed as weak and sharp lines, while the transition energies from $^3T_{1a}$ to the triplet states are observed as strong and broad bands. Our results point out the importance of lattice-relaxation consideration to produce the optical properties of α-Al_2O_3: V^{3+} accurately. On the other hand, in the case ofα-Al_2O_3: Mn^{3+}, since Mn^{3+} belongs to $3d^4$ configuration, there are two possible spin states *i.e*., high-spin (HS) and low-spin (LS) states. In the HS state, the spin-allowed transition is occurred from the ^5E ground state to $^5T_2$ states ($^5E \rightarrow {}^5T_2$). On the other hand, in the LS state, the spin-allowed transition was occurred from the $^3T_1$ ground state to ^3E state ($^3T_1 \rightarrow {}^3E$). Nevertheless, we limit our work in the LS state.

In this work, we want to do similar investigation for α-Al_2O_3: Co^{3+}. The lattice relaxation effects due to the Co^{3+} substitution were calculated using Shannon's crystal radii method and geometry optimizations in the Cambridge Serial Total Energy Package (CASTEP) method. The one-electron Discrete Variational-Xα (DV-Xα) method was used to estimate the molecular orbital energies, while the many-electron Discrete Variational Multielectron (DVME) method was used to estimate the d-d absorption spectra.

Table 1. The Co-O bond length of α-Al_2O_3: Co^{3+} calculated using different computational methods.

Computational Method	d_1 (Å)	d_2 (Å)	Ratio d_1 (%)	Ratio d_2 (%)
No-relax [1]	1.855	1.971	-	-
Shannon's crystal radii	1.864	1.981	100.5	100.5
CASTEP	1.952	1.965	105.2	99.7

2. Methods

In this work, we used 7- atoms model clusters constructed from α-Al_2O_3 crystal structure [8], CoO_6^{9-}. The lattice relaxation effect was estimated by Shannon's crystal radii [9,10] and geometry optimization using CASTEP code [11-13]. The detailed procedures were explained in Ref. 14. Table 1 shows the Co-O bond length of α-Al_2O_3: Co^{3+} calculated using different computational methods. There are two different bond lengths; the shorter bonds indicated by (d_1) and the longer bonds indicated by (d_2). The lattice relaxation ratios described as the ratio between the bond length after relaxation process and the original bonds were shown in the right columns. It shows that d_1 increases when model clusters with considering lattice relaxation effect calculated by both Shannon's crystal radii and the geometry optimization using CASTEP code were used. On the other hand, d_2 increased when we used the model clusters with considering lattice relaxation effect calculated by Shannon's crystal radii, and then decreased when we used the model clusters with considering lattice relaxation effect calculated by the geometry optimization using CASTEP code.

In order to calculate the molecular orbital (MO) energies of α-Al_2O_3: Co^{3+}, We used the one–electron calculations DV-Xα method. The detailed procedure is described in Ref. 15. In this case, only one electron and nuclei are considered, while the other electrons are treated just as potentials. On the other hand, the many-electron calculations DVME method was utilized to calculate the absorption spectra of α-Al_2O_3: Co^{3+}. The detailed procedure is described in Ref. 16. Here, four electrons occupying the impurity levels are treated explicitly. Both above mentioned methods are first-principles calculations which performed without referring to any experimental parameter. The calculations carried in this work were performed under low spin (LS) state.

3. Results and discussion

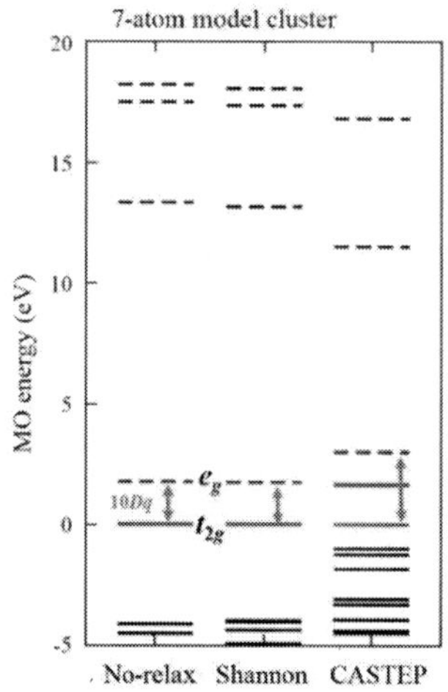

Figure 1. The molecular orbital (MO) energies of α-Al_2O_3: Co^{3+} estimated without and with lattice relaxation effect.

Figure 1 shows the molecular orbital of α-Al_2O_3: Mn^{3+} calculated using different computational methods. The 7-atom model cluster (CoO_6^{9-}) were used in the calculations. Two different types of lattice relaxation estimation i.e., Shannon's crystal radii method and CASTEP method are compared. The conduction band and the valence band are indicated by dashed and solid lines, respectively. The impurity levels consist of t_{2g} and e_g which their difference energy is called as crystal field splitting ($10Dq$). In this case, the lowest impurity levels were set to zero. The results show that the $10Dq$ were found to be 1.78, 1.75, and 1.92 eV calculated using no-relax, Shannon's crystal radii, and CASTEP methods, respectively. As we can see, the crystal field splitting calculated based on the model cluster with considering lattice relaxation effect using CASTEP code has the highest $10Dq$. It might be understood since the t_{2g} and e_g levels split largely.

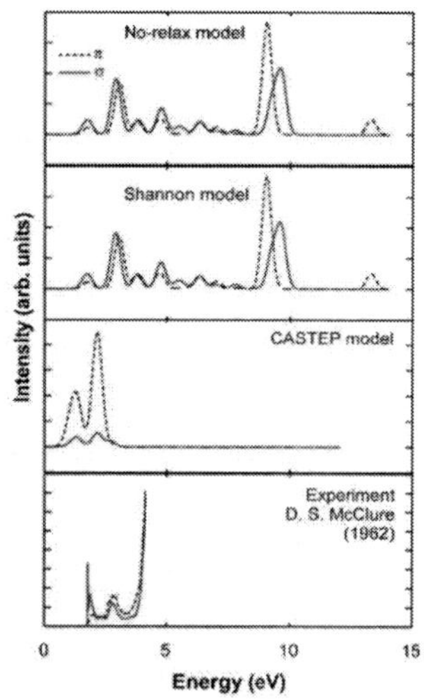

Figure 2. The absorption spectra of α-Al_2O_3: Co^{3+} estimated without and with considering lattice relaxation effect. The observed spectra were obtained from Ref. 17.

Figure 2 shows the theoretical absorption spectra of α-Al_2O_3: Co^{3+}. Several computational methods considering lattice relaxation effect such as Shannon's crystal radii and CASTEP methods. The solid red line indicates the σ spectrum, $E \perp c$. On the other hand, the dashed blue line indicates the π spectrum, $E \parallel c$. The experimental absorption spectra of α-Al_2O_3: Co^{3+} at HS state obtained from Ref. 17 are shown at the lowest panel. In the experimental spectra, two broad peaks were observed either in the σ or in the π spectrum. The first peak appears at ca. 2 eV, while the second peak appears at ca. 2.8 eV.

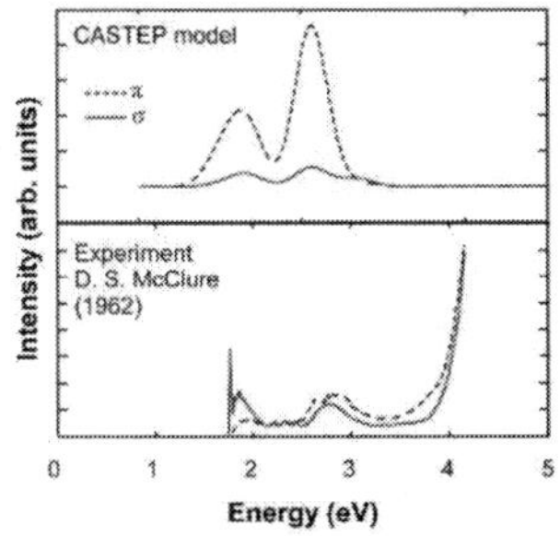

Figure 3. The absorption spectra of α-Al_2O_3: Co^{3+} estimated without and with lattice relaxation effect. The observed spectra were obtained from Ref. 17.

Our results show that the calculated absorption spectra obtained using the model clusters without considering lattice relaxation effect and with considering lattice relaxation effect estimated based on Shannon's crystal radii were identical. The tendency of the absorption spectra was changed significantly when we used the model cluster with considering lattice relaxation effect estimated based on the geometry optimization using CASTEP code. Figure 3 magnifies the absorption spectra of α-Al_2O_3: Co^{3+} estimated with considering lattice relaxation effect calculated by the geometry optimization using CASTEP code. The peaks positions agree with the observed absorption spectra reported from Ref. 17.

4. Conclusion

In this work, the optical properties of α-Al_2O_3: Co^{3+} have been successfully estimated by first-principles calculations DV-Xα and DVME method. The lattice relaxation effect was investigated thoroughly by comparing two different methods using Shannon's crystal radii method and first-principles band structure calculations CASTEP. The estimated absorption spectra shows that the results based on 7-atom model cluster with considering lattice relaxation effect using CASTEP improves the agreement with the experimental data.

Acknowledgement

The authors gratefully acknowledge the support from Direktorat Riset dan Pengabdian Masyarakat, Direktorat Jenderal Penguatan Riset dan Pengembangan, Kementrian Riset, Teknologi, dan Pendidikan Tinggi, Indonesia.

References

[1] M. Novita and K. Ogasawara, "Comparative Study of Absorption Spectra of V 2 þ , Cr 3 þ , and Mn 4 þ in -Al 2 O 3 Based on First-Principles Configuration – Interaction Calculations," J. Phys. Soc. Japan, vol. 81, pp. 1–7, 2012.

[2] M. Novita and K. Ogasawara, "Study on Multiplet Energies of V 2 + , Cr 3 + , and Mn 4 + in MgO Host Crystal Based on First-Principles Calculations with Consideration of Lattice Relaxation," vol. 124707, pp. 1–6, 2014.

[3] M. Novita, H. Yoshidab, and K. Ogasawara, "Investigation of Ion Dependence of Electronic Structure for 3 d 3 Ions in Mg 2 TiO 4 Based on First-Principles Calculations M. Novita," ECS Trans., vol. 50, no. 41, pp. 9–17, 2013.

[4] M. Novita and K. Ogasawara, "Comparative Study of Multiplet Structures of Mn 4 þ in K 2 SiF 6 , K 2 GeF 6 , and K 2 TiF 6 Based on First-Principles Configuration – Interaction

Calculations," Jpn. J. Appl. Phys., vol. 51, pp. 1–6, 2012.
[5] M. Novita, T. Honma, B. Hong, and A. Ohishi, "Study of multiplet structures of Mn 4 þ activated in fl uoride crystals," J. Lumin., vol. 169, pp. 594–600, 2016.
[6] Sawada H 1994 *Mat. Res. Bull.* **29** 127
[7] Shannon R D and Prewitt C T 1969 *Acta Cryst.* B **25** 925
[8] Shannon R D 1976 *Acta Cryst.* A **23** 751
[9] Payne M C, Teter M P, Allan D C, Arias T A and Joannopoulos J D 1992 *Rev. Mod. Phys.* **64** 1045
[10] Segall M D, Lindan P J D, Probert M I J, Pickard C J, Hasnip P J, Clark S J and Payne M C 2002 *J. Phys. Condens. Matter* **14** 2717
[11] Clark S J, Segall M D, Pickard C J, Hasnip P J, Probert M I J, Refson K and Payne M C 2005 *Z. Kristallogr.* **220** 567
[12] M. Novita, H. Nagoshi, A. Sudo, and K. Ogasawara, "Ab-initio study on the absorption spectrum of color change sapphire based on first-principles calculations with considering lattice relaxation- effect," IOP Conf. Ser. Mater. Sci. Eng., vol. 299, no. 012060, 2017.
[13] Novita, M., Nuvitalia, D., Cholifah, N., & Ogasawara, K. (2019, December). Effect of trivalent manganese substitution in α-Al2O3 crystal on the absorption spectra based on first-principles calculations. In Journal of Physics: Conference Series (Vol. 1402, No. 6, p. 066004). IOP Publishing.
[14] M Novita, D Nuvitalia, N Cholifah, and K Ogasawara 2020 ISMEE IOP Conference Series. In Press
[15] Adachi H, Tsukada M and Satoko C 1978 *J. Phys. Soc. Jpn.* **45** 875
[16] Ogasawara K, Ishii T, Tanaka I and Adachi H 2000 Phys. Rev. B **61** 143
[17] McClure D S 1962 *J. Chem. Phys.* **36** 2757

IOP Conf. Series: Materials Science and Engineering **835** (2020) 012012 doi:10.1088/1757-899X/835/1/012012

Effects of Transition Metal Doping to Cerium-Oxides in Thermal Reduction Reaction

Takaki Nishimura,[1] Yuki Tsutsumi,[1] Tomohiko Ishii,[1] Tatsuya Kodama,[2] and Genta Sakane[3]

[1] Department of Advanced Materials Science, Graduate School of Engineering, Kagawa University, 2217-20 Hayashi-cho, Takamatsu, Kagawa 761-0396, Japan

[2] Faculty of Engineering, Niigata University, 8050 Ikarashi 2 nocho, Nishi-ku, Niigata, 950-2181, Japan

[3] Department of Chemistry, Faculty of Science, Okayama University of Science, 1-1 Ridaicho, Kita-ku, Okayama 700-0005, Japan

E-mail : s18g576@stu.kagawa-u.ac.jp

Abstract. Considering our lives, novel energy production is so required, and many scientists have been investigated this problem from a lot of views. One of the ways for the new energy productions is the solar power thermochemical production by an oxidation-reduction reaction of the ceramics CeO_2 (ceria). In this research, the ceria is used for obtaining an H_2 gas in vapor reaction in high-temperature thermal equipment which has a heat production system consisting by focused sun light. Nowadays technology has been also developed to keep a human living, not to harm the Earth environment. From these points, it is realized that how important the H_2 gas is for our lives, so we are trying to improve the thermochemical H_2 gas production by means of a ceria as a catalyst. Our purpose in this study is, to point out what effects occur in doped ceria's thermal reduction (TR) reaction by using DV-X**α** method. Comparing the models between the pure ceria and the Mn doped ceria, we have realized the difference by the reaction. In this appearance, one of the considerations is a view from p-DOSs (partial-Density of States), in which there are different O-2p DOSs by different crystal structures. From this view, we considered that the ceria structure would become more stable to keep the stable cyclic thermal reaction in Mn doped ceria.

1. Introduction

1.1. How to solve the energy problem and get H_2 gas?

We face a big problem. It is an energy problem and we have to sustain energy production even using the limited rare resources such as petroleum and natural gas. Considering our lives, a novel energy production is so required, and many scientists have been investigated this problem from lots of views. One of the ways for the new energy productions is the solar power thermochemical production by an oxidation-reduction reaction of the ceramics CeO_2 (ceria). This thermal oxidation-reduction reacting mechanism is shown in figure 1, there is two-step reaction. The first step is the thermal reduction reaction called TR reaction, the oxygen atoms are reduced from CeO_2 in 1500 °C. The second step is the water decomposition reaction, which is also called WD reaction. As its name, the vapored water is decomposed to the H_2 gas around 1000 °C, it will be decomposed to the O_2 gas completely after the TR

IOP Conf. Series: Materials Science and Engineering **835** (2020) 012012 doi:10.1088/1757-899X/835/1/012012

reaction. The redox cerium reaction is the cycle reaction consisted of two step-reactions, the continuous reaction makes it possible to efficiently produce H_2 gas.

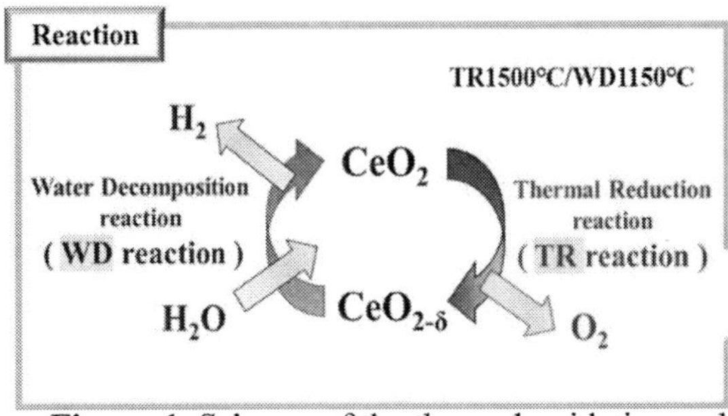

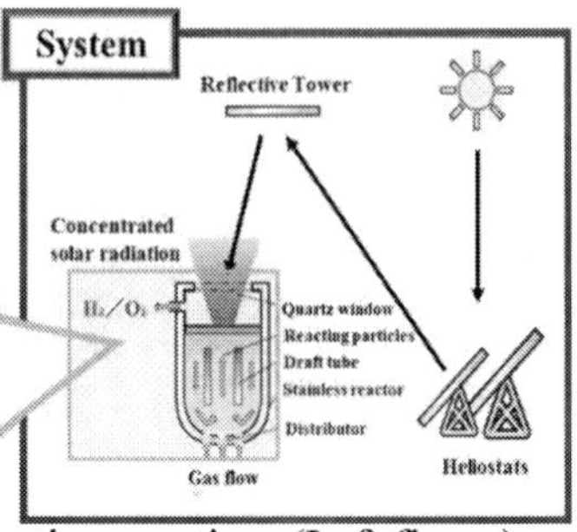

Figure 1. Scheme of the thermal oxidation-reduction reaction. (Left figure)
Figure 2. Floor plan of solar reactor concept for a circulating reaction measure. (Right figure)

In this research, the ceria is used for obtaining the H_2 gas in high-temperature thermal equipment which has heat production systems consisting of focused sunlight. The system is shown in figure 2.

1.2. The advantages of the H_2 gas utility.
The advantages of the producing H_2 gas system are mainly mentioned at three points, which are listed below.

(i) The first point is that it is very suitable for producing H_2 gas in an environment where abundant sunlight can be obtained (called as the sunbelt zone). In this matter, if the solar light can be supplied stably, the thermal redox reaction can be continued. More H_2 gas production is expected well in such areas where have the amount of sunlight irradiation.

(ii) The second point is that H_2 gas can be stored and transported. It is applicable for supplying H_2 gas to countries with poor energy resources such as Japan. Moreover, it has the advantage of being able to store sustainable "solar energy" which can be poured from the sun to the earth as the form of "hydrogen energy". It could be the hope of the energy problem solution in the world.

(iii) The third point is the fact that H_2 gas can be an important resource to support our lives. When H_2 gas is used as energy, it is possible to suppress the emission of harmful substances and produce more electric power energy in the thermal power generation.

From these points, it is realized that how important the H_2 gas is for our lives, we are trying to improve the thermochemical H_2 gas production by means of a ceria as a catalyst. "Hydrogen production using the cyclic thermochemical reaction of cerium oxide" is one of the researches conducted by Kodama *et al.* in Niigata University.

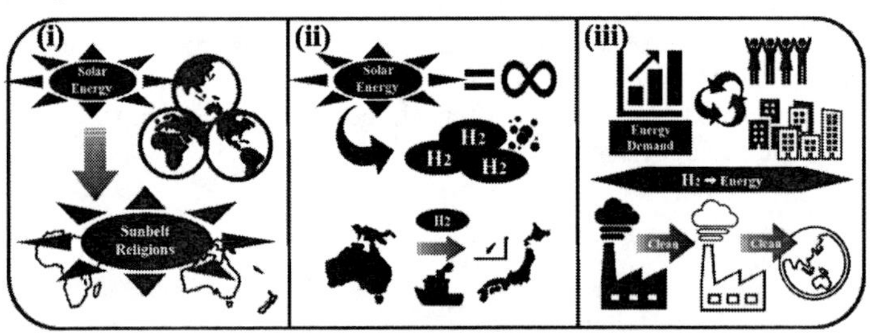

Figure 3. Scheme of the image in each advantage of the H_2 gas utility.

2. The purpose of this study

Our purpose in this study is to point out what effects occur in doped ceria's TR reaction by using Discrete Variational-Xα molecular orbital calculation. The cycle reaction used the manganese (Mn) doped ceria has been reported the high efficiency in the experimental by Kodama et al., and much higher efficiency is also required. We have investigated to solve the mechanism of the ceria's cycle reaction to get higher efficiency.

IOP Conf. Series: Materials Science and Engineering **835** (2020) 012012 doi:10.1088/1757-899X/835/1/012012

It has been also pointed out that there are some extra crystal structures in the CeO_2 after TR reaction from the views of XRD peak patterns. However, it has been confirmed that there is no extra crystal structure in the Mn-doped ceria. This difference is the key point of whether keeping the cycle reaction or not. As the results of our simulation, it was suggested that the crystal of Ce_2O_3 is formed after TR reaction, the crystal structure of the about 30 % ceria has been changed from Fm-3m (#225, CeO_2) to Ia-3 (#206, Ce_2O_3) space groups, respectively. So, the amount of the cerium oxide available for cycle reaction is reduced, which may well reduce the reaction efficiency. In other words, by suppressing the collapse of the crystal, it is possible to explain the efficiency improvement of the cycle reaction, and further efficiency improvement can be aimed at by considering the factor of the peak.

3. Calculation models

We prepared 6 types of the calculation models, the center atom is the metal atoms (Ce or Mn) and oxygen atoms coordinated around the center metal. It was divided for some groups by the types of pure cerium or Mn-doped and CeO_2 or Ce_2O_3. And the 6 prepared calculation models, in which Coulomb potential (Madelung potential) of atoms surrounding the central metal and the oxygen atoms are taken into considerations. However, considering the environment of oxygen atoms around the central metal, there are one kind of environment for CeO_2 and two kinds of environment for Ce_2O_3. For this reason, a total of three models were prepared from the difference of coordination environments, the experimental facts were reproduced by changing the central metals (Ce or Mn) in order to consider the doping effect on ceria. Since cerium atoms with different coordination environments exist in the Ce_2O_3 crystal, they are named as Ce1 and Ce2, respectively, and the Ce_2O_3 crystal structure has cerium atoms consisted of Ce1 : Ce2 = 3 : 1.

4. Considerations

Comparing the models between the pure ceria and the Mn-doped ceria, we have realized the difference by the TR reaction (after TR reaction : CeO_2 100 % changes into CeO_2 70 % and Ce_2O_3 30 %), which are mentioned below.

4.1 The states of electrons.

At first, We want to introduce how the electrons states are in each model to explain well why the Ce_2O_3 structure is formed in pure cerium after TR reaction, also why the Mn-doped model does not have the Ce_2O_3 structure.

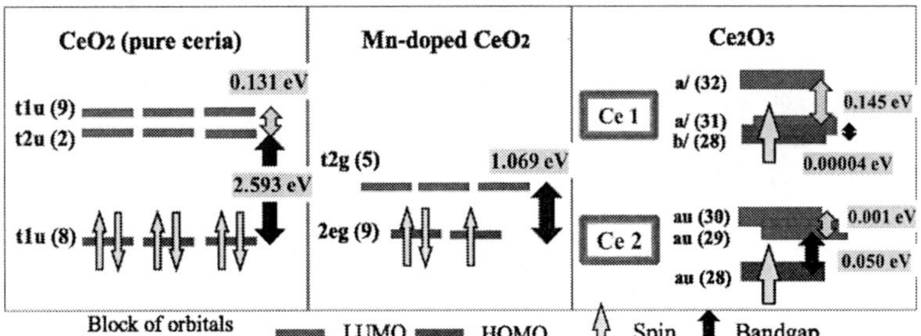

Figure 4. Scheme of the electronic state of each crystal structure.

4.1.1 In the CeO_2 structure.

First of explanations, explaining the pure CeO2 structure. The electrons state (CeO2) is shown in figure 4 (left). The energy gap is 2.593 eV from LUMO (Lowest Unoccupied Molecular Orbital) to HOMO (Highest Occupied Molecular Orbital) levels. The electrons in the HOMO levels are fully occupied on the three-generation orbitals, it would be occupied for LUMO levels beyond for the 2.593 eV gap when adding more electrons into the CeO_2 structure. This means that the CeO_2 structure is reduced by oxygen defects, electrons caused by oxygen defects would not be easily captured in the HOMO levels of the CeO_2 structure.

IOP Conf. Series: Materials Science and Engineering **835** (2020) 012012 doi:10.1088/1757-899X/835/1/012012

4.1.2 In the Mn-doped CeO_2 structure.
On the other hand, the electrons state of Mn-doped CeO_2 is shown in figure. 4 (middle). The energy gap is 1.069 eV from LUMO to HOMO levels, the value of the bandgap is lower than the pure CeO_2 model. The electrons in the LUMO levels are not fully occupied on the two-generation orbitals, it would be occupied for LUMO levels beyond 0 eV when adding more electrons. Even if adding more electrons to the Mn-doped structure at the fully occupied on the HOMO levels, the electrons move to the LUMO levels beyond for the 1.069 eV gap. Compared between the Mn-doped CeO_2 structure and pure CeO_2, electrons caused by oxygen deficiency would be captured easily in hybrid orbitals than the CeO_2 structure. From the viewpoint of electron acceptability of ceria, we have studied the behaviour of electrons provided by oxygen defects and we have found that electrons are more acceptable in the Mn-doped ceria.

4.1.3 In the Ce_2O_3 structure.
We are also to explain the Ce_2O_3 structure. The electrons states of Ce1 and C2 in Ce_2O_3 are shown in figure. 4 (right) and it is one of the models after TR reaction. The energy gap of each structure is 0.00004 eV in Ce1 model and, 0.050 eV in Ce2 model. The electrons in the LUMO levels are not fully occupied in an orbital in each both models, it appears the characteristics like the Mn-doped CeO_2 model. Moreover, there is almost no energy difference, which shows it is so easy to be occupied in LUMO levels too.
From the views of electron's behaviour in the orbitals, the CeO_2 crystal structure is difficult to receive electrons, which makes it difficult to reduce the CeO_2. So, it is sufficiently considered that Ce_2O_3 is formed along with oxygen defects due to the TR reaction and electrons are supplied into the Ce_2O_3 structure. Since Mn-doped CeO_2 is regarded as a pseudo-Ce_2O_3 when Mn is doped into CeO_2, it is concluded that the need to newly form Ce_2O_3 by the TR reaction is eliminated as the amount of Mn-doped increases.

4.2 Density of states in oxygen 2p orbitals.
Figure 5 shows the p-DOS (partial-Density of States), in which there are different O-2p DOS figures by each different crystal structures. In figure. 5, the upper three figures and lower three figures are the results of the p-DOSs in the cases of pure cerias (upper) and the Mn-doped cerias (lower), respectively. And left column, and center and right columns are the results in the case before (left) and after (center and right) the TR reaction. It is mentioned in the above column that there are different coordination environments to the Ce atom (Ce1and Ce2) though they are belonged to the same crystal structures.

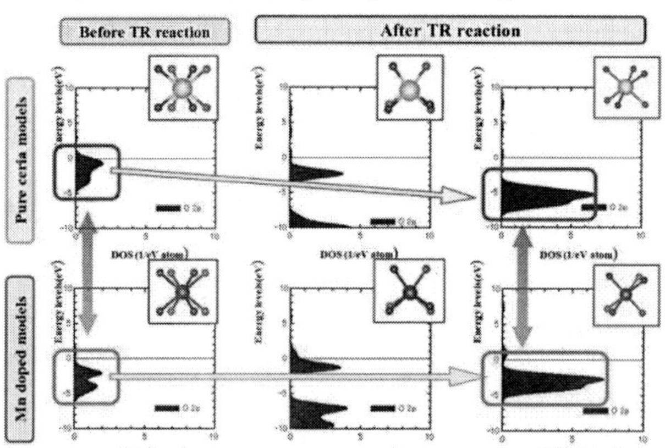

Figure 5. Six characteristic graphs show the O-2p DOS, in which each graph's upper right indicate the state of oxygen atoms coordination. The vertical axis is the energy levels (eV), the horizontal axis shows what ratio is : DOS (1eV/ atom). The left figures belong the crystal structure (#225), both center and right crystal structure (#206). Bottom figure's models are Mn-doped but upper figures are not doped.

The 0 eV on the vertical axis in figure. 5 means that it is the HOMO levels in each crystal structure. The DOS peaks in the O-2p orbitals is lower along with changing the CeO_2 structure in the pure ceria model, but in the Mn-doped model, it can be seen that the energy level itself of the DOS peaks at relatively high energy levels are not changed so much.
It is considered that O-2p DOS peaks is observed in the vicinity of the energy level below lower the HOMO level means that the stability of the oxygen atom becomes higher. This indicates that, in pure ceria crystals, the stability of oxygen atoms due to the structural change (CeO_2 to Ce_2O_3) is enhanced as the result of the TR reaction. And with the changed Ce_2O_3 crystal structure, oxygen atoms would have a lower elimination reaction of oxygen. On the other hand, in the Mn-doped CeO_2, the overall DOS peaks shift downward compared to the pure CeO_2 model. In the initial state of Mn doping (Mn-doped CeO_2), there is a tendency that the oxygen atoms around the Mn-doped ceria are stable and there is a tendency not to break its crystal structure along with oxygen defects more than the pure ceria. Even if the structure is changed by the TR reaction from this state (Mn-doped CeO_2), the DOS peaks shift is not observed, so that the stability of oxygen is lower than that of pure ceria, and the necessity of the structure change by the TR reaction would be lower in Mn-doped ceria, too. From the view of O-2p DOS, it can also be explained that the formation of Ce_2O_3 is difficult when Mn is doped and no peak is observed after the TR reaction.

5. Conclusion

Based on the above conclusion, the TR reaction requires an environment in which electrons are easy to donate, and the environment is such that the bandgap is small and the electrons on the LUMO are not full. The CeO_2 achieves this by changing its own crystal system, but it can also be created by artificial manipulation (Mn doping). From the view of O-2p DOS, it can be explained that the formation of Ce_2O_3 is difficult when Mn is doped and no peak is observed after the TR reaction. We concluded that the reasons for the XRD peak difference between the pure ceria and Mn-doped ceria are given for us by these theories.

Reference

[1] Steffen Gríeshammer, 2017 *J. Phys. Chem.* C **121** 15078-15084.
[2] Nobuyuki Gokon, Toshinori Sudab, Tatsuya Kodama 2015 *Thermochimica Acta* **617** 179-190.
[3] Nobuyuki Gokon, Toshinori Sudab, Tatsuya Kodama 2015 *Energy* **90** 1280-1289.
[4] Roger Jocot, Ren Mor , Ronald Michalsky, Aldo Steinfeld, and Greta R. Patzke, J. Mater. 2017 *Chem.* A **5** 19901.

IOP Conf. Series: Materials Science and Engineering **835** (2020) 012013 doi:10.1088/1757-899X/835/1/012013

Study of LCAO-MO calculation by using completely numerical basis functions

Katsumi Nakagawa

3-1-1-201 Hiyoshi-Honcho, Kohoku-ku, Yokohama, Kanagawa 223-0062 Japan

nakagawa.katsumi@nifty.com

Abstract. Choice of basis functions is decisive in LCAO-MO method to calculate accurately MOs of a molecule under complicated electric/magnetic potential. For such situation, calculating basis functions analytically is generally difficult. To obtain them, the new numerical calculating method based on matrix operation and the new sample point system based on FCC lattice were proposed. Further LCAO-MO calculation using basis functions calculated by this method was applied for small molecules under inner electric potential and showed reasonable results. Finally, two challenges to apply this method for arbitrary magnetic potential are clarified.

1. Introduction

The author has studied calculation of NMR spectrum.[1],[2] In his study, he experienced that calculating nuclear spin-spin coupling(J) accurately is more difficult than the case of chemical shift(δ).

For δ, uniform magnetic field H is applied, H induces current in a molecule and the current causes anti-magnetic field, which shifts the resonance peak. Here the current can be calculated by using MOs of the molecule.[3] In this paper basis function is substantially atomic orbital, so the term "AO" will be used instead of "basis function". For LCAO-MO calculation under uniform H, analytically accurate AO, often called "London atomic orbital", is known.[4] It is very effective to obtain accurate MOs but works only for uniform H. On the other hand, J is caused by interaction between nuclear magnetic moments of neighboring atoms via MOs. J causes splitting of the resonance peak. H caused by nuclear magnetic moments is generally very complicated and analytically accurate AO is unknown. The author imputes poorer accuracy of J to poorer quality of AO under complicated H.

To obtain high quality AO χ, solve numerically for example Hartree-Fock-Slater equation including potential $\phi(r)$ for each atom. In the case that $\phi(r)$ is electrical potential, it is like eq.(1).

$$\left\{\frac{1}{2}v^2 - ZA + \int r - \frac{\rho(r2)}{r12}dv^2 - 3\alpha\left[\frac{3}{4\pi}\rho\sigma(r)\frac{1}{3}\right] + \phi(r)\right\}\chi = \epsilon\chi \quad (\sigma = \alpha, \beta)\} \tag{1}$$

This is the similar as conventional **DV-Xα** method.[5] But we can't use spherical harmonic functions as χ because $\phi(r)$ can creates extreme spherically asymmetric circumstances. We need new strategy.

2. Completely numerical AO

In quantum mechanics, wave functions can be obtained by two ways, differential equation and matrix operation. The former is popular but the latter is more powerful in this case. In matrix operation, physical quantity P is represented by matrix $[P]$ and wavefunction ψ is represented by vector $[\psi]$. $[P]$ act on $[\psi]$ like eq.(2), where $\psi(r_i)$ is value of ψ at sample point r_i in 3D space.

$$[P][\psi] = \begin{bmatrix} P_{11} & P_{12} & \cdots & P_{1n} & \psi(r1) \\ P_{21} & P_{22} & \cdots & P_{2n} & \psi(r2) \\ \cdots & \cdots & \cdots & \cdots & \cdots \\ P_{n1} & P_{n2} & \cdots & P_{nn} & \psi(rn) \end{bmatrix} \quad (2)$$

A new sample point system for AO calculation was developed. Firstly, arrange sample points (= **SP**) like FCC lattice points and let's call them imaginary SPs, each of which occupies the same volume or weight ω_0 and has equivalent 12 neighboring points except peripheral area of the system. Designate one of imaginary SP as the origin and express coordinate of imaginary SP as (x, y, z). This system is very convenient for numerical calculation but its uniform point distribution is inappropriate for AOs. Concentrate SPs toward the origin to calculate precisely around the nucleus. After this transformation, let's call them real SPs. and express coordinate of real SP as (X, Y, Z). Although the direction of each SP around the origin is preserved, the distance from the origin $r = \sqrt{x^2} + y^2 + z^2$ is modified to $R = \sqrt{X^2 + Y^2 + Z^2}$. If the probability of appearance of real SP at R is assumed to be like eq.(3)

$$P(R) = R^{-\lambda} \ (0 < \lambda < 3), \quad (3)$$

R can be related to r by eq.(4), because imaginary SPs exist uniformly.

$$4\pi R^2 P(R) dR = 4\pi r^2 dr \quad (4)$$

Therefore

$$\frac{11}{-1C} \lambda\, R = C3 - \lambda_r C \ or\ r = C3R (C = 1 - 3) \quad (5)$$

where a parameter λ seems to be appropriate around 2.0.

In matrix operation, each term in eq.(1) has to be replaced by a matrix. Matrix form of Laplacian ∇^2 was obtained as follows. Poisson equation (6) is solved like eq.(7).

$$\nabla^2 f(r) = g(r) \quad (6)$$

$$f(r) = -1 \int \frac{g(r2)}{4\pi(r12)} dx \quad (7)$$

Eq.(7) is transformed to numerical integration and represented as matrix operation like eq.(8).

$$[f] = -1 \sum g(rj) w = -1 [R][w][g] \quad (8)$$

where r_{ij} is the distance between point r_i and point r_j and ω_i is weight of real SP r_i. $[g]$ is solved like eq.(9) and matrix form of ∇^2 is found to be eq.(10) by comparing eq.(9) with eq.(6).[6]

$$[g] = -4\pi\{[R][\omega]\}^{-1}[f] = -4\pi[\omega]^{-1}[R]^{-1}[f] \quad (9)$$

$$[\nabla^2] = -4\pi[\omega]^{-1}[R]^{-1} \quad (10)$$

Eq.(10) is the expression for real SPs, but it is modified like eq. (11) for imaginary SPs.

$$-\frac{1}{2} - \frac{1}{1}[v] = -4\pi[w2][R][w2] \quad (11)$$

Diagonal elements of $[R]$ were set to be 0 to avoid infinity. To include contribution from diagonal element, correcting term should be added to kinetic term in eq.(1) like eq.(12).

$$-1-\frac{1}{2}-[v2]+2(4\pi \times 9)3\,[\omega 3] = 2\pi\,[w2][R][w2]+2{,}41798793[w3] \quad (12)$$

Eq.(12) is easier to handle, because it is Hermitian operator. But obtained $\psi(r_i)$ by using eq.(12) includes factor of $\sqrt{\omega_i}$. It has to be corrected to $\psi(r_i)/\sqrt{\omega_i}$ in real 3D space, where we set the molecule. Other terms in the left side of eq.(1) are diagonal matrices, whose diagonal element is the value of each potential at each SP. So every matrices in eq.(1) can be written down easily and eq.(1) becomes simple numerical eigen-value problem, which can be solved by ordinary mathematical routine, even if potential $\phi(r)$ has very complicated form. Therefore, we can calculate easily accurate AOs under arbitrary electric potential.

Table 1 shows energies of lower 9 orbitals of **Ar atom** calculated by this method. AO#3-#5 and #7#9 are perfectly degenerated.

AO	Energy(a.u.)
#1	-114.681094
#2	-11.136150
#3	-8.760035
#4	-8.760035
#5	-8.760035
#6	-1.045049
#7	-0.498346
#8	-0.498346
#9	-0.498346

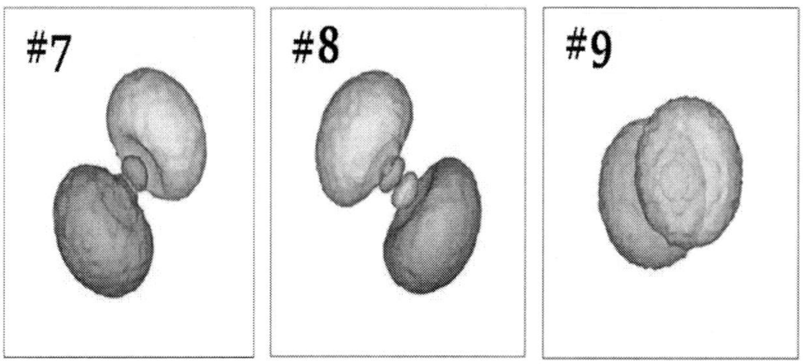

Fig. 1 Pictures of AO#7-9 of Ar atm in Table 1 illustrated by VESTA

Fig. 1 shows pictures of AO#7-#9 by VESTA[7]. They seem to be 3p orbitals. Although spherical harmonics functions weren't used explicitly, they appeared spontaneously.

3. LCAO-MO calculation by using new AOs

For LCAO-MO calculation, Fock Matrix F_{rs} is necessary. It corresponds to an interaction between AO χ_r and χ_s. As F is an operator in the left side of eq.(1), F_{rs} is integrated numerically like (13).

$$F_{rs} = \int \chi * F\chi\, dv \rightarrow \sum \chi * (r)F(r)\chi(r)wrs \quad (13)$$

Eq (13)should be integrated at all SPs of all atoms. Two problems appeared and were solved. Firstly, atom A and atom B have their proper SP systems constructed independently. SPs of both atoms overlap in the molecule and their weights ω have to be unified. It was done in the same way as conventional **DV-Xα** method[5] like eq.(14)

$$1/\omega = 1/\omega_A + 1/\omega_B \quad (14)$$

where ω_A is weight when only atom A exists and ω_B is weight when only atom B exists. Secondly, if χ belongs to atom A, value of χ is given only at SP of atom A. Values of χ at SP of atom B were interpolated by using values of χ at surrounding SPs of atom A. Interpolation was executed as follows. Let r_B be any SP of atom B. Find the SP of atom A nearest to r_B. Designate it r_{A0} and set r_{A0} to be the origin. Expand χ to the second order around r_{A0} like (15).

$$\chi_{xx}\chi_{yy}{}_{zz} = \chi \sim \chi(r\quad) + \chi\ x + \chi\ y + \chi\ z + x^2 + y^2 + {}^{\chi} \quad z^2 + \chi\ \ yz + \chi\ \ zx + \chi xy \qquad (15)$$

Select SPs of atom A around r_{A0} and designate them r_{Ai} $(i = 1,2,\cdots,n)$. By substituting $r_{Ai} = (x_i, y_i, z_i)$ into eq.(15), simultaneous equations (16) are obtained.

As number of unknown coefficients is 9, they can be uniquely solved if condition number $n = 9$ or can be estimated by the method of least squares if $n > 9$. Even if $n < 9$, they can be reasonably estimated by using Moore-Penrose pseudo-inverse matrix.[8] As a lattice point of FCC, each point has 12 equivalent SPs neighboring around it in the space of imaginary SPs, like

$$(a,0,a),(0,a,a),(-a,0,a),(0,-a,a),(a,0,-a),(0,a,-a),(-a,0,-a),(0,-a,-a),(a,a,0),(-a,a,0),(-a,-a,0),(a,-a,0), \qquad (16)$$

where a is a half of lattice constant. If all 12 SPs are used, or $n = 12$, the method of least squares is applicable. But the author used only 8 SPs from the top of (17) and apply Moore-Penrose pseudo-inverse matrix for 9 coefficients. It's because using small number of SPs is advantageous in peripheral area of SP system. **Fig.2** shows example of interpolation by this method for a N2 molecule which is composed with atom A and atom B. Values of 2s orbitals of atom A at SPs of atom A(= real values) is shown by solid curve as a function of distance from atom A nucleus and its interpolated values for SPs of atom B is shown by open circles. They seem to be interpolated but not perfectly. In this method, AOs and MOs are calculated alternatively by SCF method. After one step of AO-SCF and Fock Matrix calculation, one step of MO-SCF proceeds. Then information of orbital populations analyzed from MOs flows to the next AO-SCF calculation and adjusts each AO's occupation number. This is the same process as conventional **DV-Xα** method.[5]. In AO-SCF calculation, eigen-value problem is heavy, because the dimension of its matrix (= n) is number of SPs of a atom and time for eigen-value problem calculation is proportional to n^3 . In MO-SCF process, n is small but Fock Matrix calculation is rather heavy. The cost of AO- SCF is roughly proportional to the number of atoms (= m) and that of MO-SCF is proportional to m^2. As Fock matrix between distant atoms can be neglected in large molecule, total cost of calculation in this method may be $O(m \sim m^2)$.

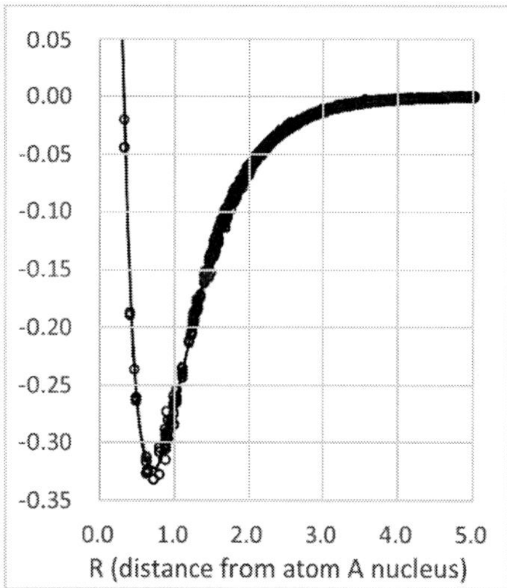

Fig. 2 Interpolation of 2s orbital of N atom at SPs of another N atom in N_2 molecule.

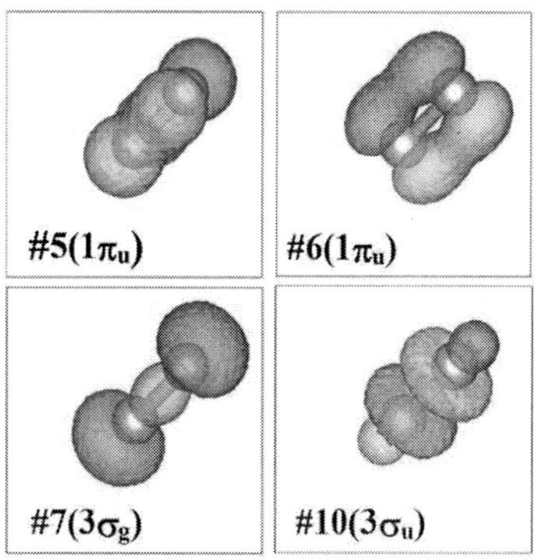

Fig. 3 Pictures of typical MOs of N_2 molecule in Table 2.

4. Examples of MOs calculated by this method

Table 2 Energies of MOs N_2 molecules and their symmetry

MO	Energy(a.u.)	
#1	-14.1297	$1\sigma_g$
#2	-14.1292	$1\sigma_u$
#3	-1.0443	$2\sigma_g$
#4	-0.4022	$2\sigma_u$
#5	-0.3936	$1\pi_u$
#6	-0.3916	
#7	-0.2627	$3\sigma_g$
#8	0.0102	$1\pi_g$
#9	0.0123	
#10	0.7252	$3\sigma_u$

one of σ or π orbitals. MO#5-#6 and MO#8-#9 are degenerated, but their accuracy are not so perfect as the case of AOs of Ar atom. (See **Table 1**) Relatively poor accuracy of interpolation causes it. **Fig.3** shows pictures of typical AOs in **Table 2** illustrated by VESTA. **Fig.4** shows total energy (atomic unit) of N2 molecule as a function of inter-nuclear distance R (relative unit). The curve has the minimum around R=1.0, which means experimental equilibrium distance.

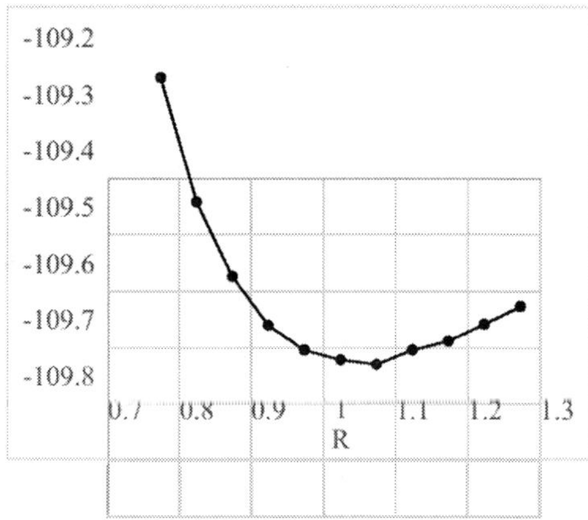

Fig.4 Total energy of N_2 molecule as a function of inter-nuclear distance R

Table 3 shows energies of lower

MO	Energy(a.u.)	
#1	-19.0250	A_1
#2	-18.7845	--
#3	-18.7838	--
#4	-1.2039	A_1
#5	-0.9330	B_1
#6	-0.5906	A_1
#7	-0.4680	B_2
#8	-0.4246	B_1
#9	-0.4210	A_1
#10	-0.2238	A_2
#11	-0.1839	B_1
#12	-0.1594	A_1
#13	-0.1162	B_2
#14	0.1297	A_1
#15	0.2268	B_1

15 orbitals of **O3 molecule** calculated by this method. This molecule has C2v symmetry and **Table 4** is its irreducible represen- tation table. Fig.5 is pictures of typical MOs. Each of MOs except MO#2,#3 corresponds to one of representations (A1,A2,B1,B2). MO#2 and #3 are just 1s oribitals of atom ***O2*** or atom ***O3*** in **Fig.5**. MO#7 corresponds to 3-centered- bond composed with 3 2p orbitals.

Table 4 Irreducible representation table of C_{2v}

C_{2v}	E	C_2	$\sigma_v(xz)$	$\sigma_v'(yz)$
A_1	1	1	1	1
A_2	1	1	−1	−1
B_1	1	−1	1	−1
B_2	1	−1	−1	1

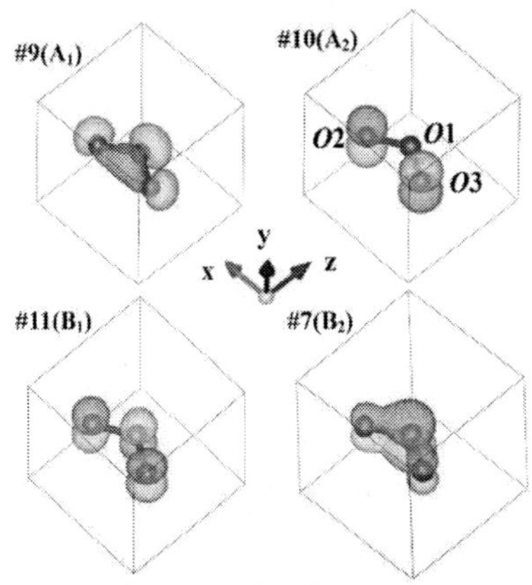

Fig.5 Pictures of typical MOs of O_3

Table 5 shows energies of lower 8 orbitals of**NH3 molecule** calculated by this method. MO#3,#4 and MO#7,#8 are degenerated. This molecule has C3v symmetry and degenerated representation **E.** ψa and ψb in eq.(18) are MOs corresponding to **E** composed by 1s of ***H1***, ***H2***, ***H3*** in **Fig.6**. MO#3,#4 in **Fig.6** show same characteristic as ψa and ψb but include component of 2p of N atom slightly.

Table 5 Energies of MOs of NH_3 molecule calculated by this method.

MO	Energy(a.u.)	
#1	-13.9865	A_1
#2	-0.7106	A_1
#3	-0.3631	E
#4	-0.3628	
#5	-0.1269	A_1
#6	0.2197	A_1
#7	0.2881	E
#8	0.2900	

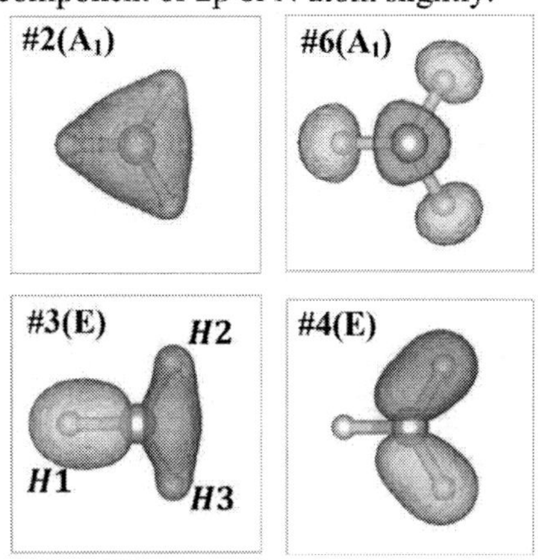

Fig. 6 Pictures of typical MOs of NH3 molecules

$$\psi a = \frac{1}{\sqrt{6}}(2x1 - x2 - x3)$$

$$\psi b = \frac{1}{\sqrt{6}}(x2 - x3) \quad (17)$$

IOP Conf. Series: Materials Science and Engineering **835** (2020) 012013 doi:10.1088/1757-899X/835/1/012013

5. Inner potential and its effect

Table 6 Ionization charge (Z-Q) of atoms in O_3 molecule (Fig.5) and in NH_3 molecule (Fig.6).

O_3	O1	O2	O3
Z	8.000	8.000	8.000
Q	7.646	8.182	8.171
Z-Q	0.354	-0.182	-0.171

NH_3	N	H1	H2	H3
Z	8.000	1.000	1.000	1.000
Q	7.502	0.830	0.834	0.834
Z-Q	-0.502	0.170	0.166	0.166

The sum of orbital populations of the same atom is called as an atomic population (Q). The difference be- tween atomic number (Z) and Q is regarded as ionization of the atom in the point charge approximation. As examples, **table 6** shows Z, Q, Z-Q of 3 3 molecules. Z-Q will form electric potential $\phi(r)$ and $\phi(r)$ will act on electrons in other atoms. Let's call the sum of $\phi(r)$ inner (electric) potential $\phi_i(r)$. $\phi_i(r)$ will reproduce real potential in each atom for AO calculation and completely numerical AOs will be able to reflect such potential. To see its effect, electric dipole moment $\vec{\mu}_e$ was calculated according to eq.(19).

$$\vec{\mu}_e = \sum Z_A \vec{R}_A - \int \sum \psi^* r \psi_i \, dv \qquad i \qquad (18)$$

where Z_A is atomic number of atom A and coordination vector of a nucleus of atom A.

Table 7 Effect of ϕ_i on dipole moments of CO, O_3, NH_3 molecules

		Dipole moment $\|\mu_e\|$ (Debye)	
	ϕ_i	Calculation	experimental
CO	without	0.2990	0.1098
	with	0.3020	
O_3	without	0.5948	0.5337
	with	0.3938	
NH_3	without	1.7763	1.4718
	with	1.3613	

Table 7 shows calculated $|\mu_e|$ of **CO, O3, NH3** molecules and experimental ones for comparison. In case of **O3, NH3**, ϕ_i seems to reduce $|\mu_e|$. In these calculations, only 1s, 2s and 2p orbitals were used. ϕ_i may distort them and shorten substantial distance between ionization charges. But corrected $|\mu_e|$s do not necessarily get closer to experimental ones. This may be the result of insufficient accuracy of $\phi_i(r)$ due to rough approximation of ionized charges.

6. Summary and the next challenges

a. In AOs by the new method, spherical harmonic functions were not used explicitly but appeared spontaneously in spherically symmetric circumstance.
b. MOs by the new method showed symmetries agreeing with the point group theory.
c. Inner electric potential distorts AOs, and modify electron distributions reasonably.

But to include magnetic vector potential A(r) to kinetic energy term (20), there remain two challenges.

$$\frac{1}{2}(vi2 - \alpha A(r)) = -\frac{1}{2}v^2 a2i[v.A(r) - A(r).v] + \frac{a^2}{2}A(r)^2 \qquad (19)$$

d. As gradient $\boldsymbol{\nabla}$ appears alone, $\boldsymbol{\nabla}$ has to be represented as a matrix operator.
e. As imaginary number $\boldsymbol{i}$ appears, operation has to treat complex wave functions.

Acknowledgement
The author would be grateful to Prof. Mizuno of Osaka University for his kind explanation of VESTA[7] to illustrate wave functions clearly in 3D space.

References

[1] K. Nakagawae,Bulletin of the Society for DV-Xα **28**,1&2,70(2015) (in Japanese)
[2] K. Nakagawae,Bulletin of the Society for DV-Xα **30**,1&2,24(2017) (in Japanese)
[3] C.P.Slichter,Principle of Magnetic Resonance(3rd ed.),Springer,(1990)
[4] T.Helgaker,P.Jørgensen,J.Olsen,Molecular Electronic-Structure Theory,Wiley,(2002)
[5] M.Iwasawa,H.Adachi,"Electronic Structure Calculation by DV-Xα method",Sankyo Shuppan, (1996) (in Japanese)
[6] K. Nakagawae,Bulletin of the Society for DV-Xα **19**,1&2,62(2006) (in Japanese)
[7] https://jp-minerals.org/vesta/jp/
[8] H.Togawa,"Numerical Calculation of matrices",Ohm-sha,(1971) (in Japanese)

IOP Conf. Series: Materials Science and Engineering **835** (2020) 012014 doi:10.1088/1757-899X/835/1/012014

The magnetism change by Jahn-Teller distortionin octahedral hexa-coordinate complex

Takeyoshi Oguma,[1] Dan Tsuneda,[1] Yuki Tsutsumi,[1] Yoshiki Fujikawa,[1] Momoe Nakano,[1]Tomohiko Ishii,[1] Genta Sakane,[2] and Kazuyoshi Ogasaware,[3]

[1] Department of Advanced Materials Science, Graduate Schoolof Engineering, Kagawa University, JAPAN
[2] Department of Chemistry, Faculty of Science, Okayama University of Science , JAPAN
[3] Department of Chemistry, School of Science and Technology, Kwansei Gakuin University of Science, JAPAN

s18g553@stu.kagawa-u.ac.jp

Abstract. Metal complexes exhibit significant physical properties such as electrical conductivity, magnetism and optical properties. They can be controlled by changing the ligand field splitting of the d-orbital of the central metal ion. We have investigated the reversible spin crossover (SCO) phenomenon by changing the ligand field splitting in six-coordinated metal complexes ML_6. In this study, we focused on the change of the spin states by uniaxial Jahn-Teller distortion in the case of an octahedral hexa-coordinated metal complexes ML_6, involving two types of different ligand molecules. In addition, we have used DV-Xα and DVME methods in order to calculate the electronic states. In order to reproduce the effect of the Jahn-Teller distortion in the calculation, a uniaxial anisotropic distortion has been applied into the octahedral hexa-coordinate metal complex. The purpose of this research is to study the mechanism and phenomenon of the spin state changing by discussing the calculated electronic states of the spin crossover phenomenon.

1. Introduction

Metal complexes show significant physical properties such as conductivity, magnetism and optical properties. These can be controlled by changing the size of the energy splitting of the five d orbitals of the central metal ion. These are clarified both by the ligand field theory and the crystal field theory. Currently, the research is under way to develop the spin crossover (SCO) phenomenon in which the magnetism of the molecule changes reversibly by changing the ligand field splitting by the energy of the external field such as heat, magnetic field, physical and chemical pressures and a light. We have clarified the possibility of the SCO from the structural change and structural distortion of the metal complex molecules.

In this study, we considered the change of the spin state of the metal complex in Jahn-Teller distortion in order to obtain the novel SCO phenomenon in the octahedral hexa-coordinate metal complex, where the SCO examples have been reported most in this field.

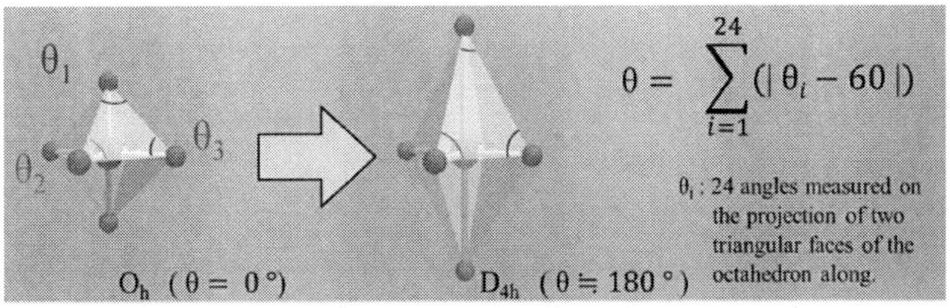

Figure 1. Determination of the distortion parameters θ in hexa-coordination compounds.

2. Methods

2.1. Calculation models

As a method used for the calculation, the electronic state calculation was carried out using DV-Xα method and DVME method, which is one of an ab initio molecular orbital method. The DV-Xα calculation is performed by using the data from the position information of the type of the central metal and a ligand molecule. The DVME calculation can be obtained the spin state, that is, possible to check for the magunitude of the magnetism.

The calculated octahedral hexa-coordinated cluster model [Fe^{3+} $(H_2O)_nCl_{6-n}$](n = 0~6) is consists of an Fe^{3+} metal ion having five d-orbitals (d^5) and two kinds of ligands (H2O, Cl). In addition, these cluster models were created based on the data of an X-ray crystal structure analysis reported in the Cambridge Structural Database (CSD) (Table 1).

Table 1. [Fe^{3+} $(H_2O)_nCl_{6-n}$] metal complex structure models in CSD

	Metal Complex				
	Metal	Ligand (1)	Ligand (2)	M-L Bond Length (Å)....(1)	M-L Bond Length (Å)....(2)
①	Fe^{3+}		$(H_2O)_6$		1.98
②	Fe^{3+}	Cl	$(H_2O)_5$	2.02	2.22
③	Fe^{3+}	Cl_2	$(H_2O)_4$	2.27	2.05
④	Fe^{3+}	Cl_3	$(H_2O)_3$	2.33	2.06
⑤	Fe^{3+}	Cl_4	$(H_2O)_2$	2.36	2.07
⑥	Fe^{3+}	Cl_5	(H_2O)	2.36	2.16
⑦	Fe^{3+}	Cl_6		2.39	

We created the 18 kinds of the cluster models in the 7 kinds of different chemical formula of the octahedral hexa-coordination complexes described in Table1.

Table 2. 18 types of calculation model octahedral hexa-coordination complex.

		$[Fe^{3+}(H_2O)_nCl_{6-n}]$
n = 0	①	$[Fe^{3+}(H_2O)_0Cl_6]$
n = 1	②	a - $[Fe^{3+}(H_2O)_1Cl_5]$
	③	e - $[Fe^{3+}(H_2O)_1Cl_5]$
n = 2	④	trans - aa - $[Fe^{3+}(H_2O)_2Cl_4]$
	⑤	cis - ae - $[Fe^{3+}(H_2O)_2Cl_4]$
	⑥	trans - ae - $[Fe^{3+}(H_2O)_2Cl_4]$
	⑦	cis - ee - $[Fe^{3+}(H_2O)_2Cl_4]$
n = 3	⑧	mer - aae - $[Fe^{3+}(H_2O)_3Cl_3]$
	⑨	fac - $[Fe^{3+}(H_2O)_3Cl_3]$
	⑩	mer - aee - $[Fe^{3+}(H_2O)_3Cl_3]$
	⑪	mer - eee - $[Fe^{3+}(H_2O)_3Cl_3]$
n=4	⑫	cis - ee - $[Fe^{3+}(H_2O)_4Cl_2]$
	⑬	trans - ae - $[Fe^{3+}(H_2O)_4Cl_2]$
	⑭	cis - ae - $[Fe^{3+}(H_2O)_4Cl_2]$
	⑮	trans - aa - $[Fe^{3+}(H_2O)_4Cl_2]$
n = 5	⑯	e - $[Fe^{3+}(H_2O)_5Cl_1]$
	⑰	a - $[Fe^{3+}(H_2O)_5Cl_1]$
n = 6	⑱	$[Fe^{3+}(H_2O)_6Cl_0]$

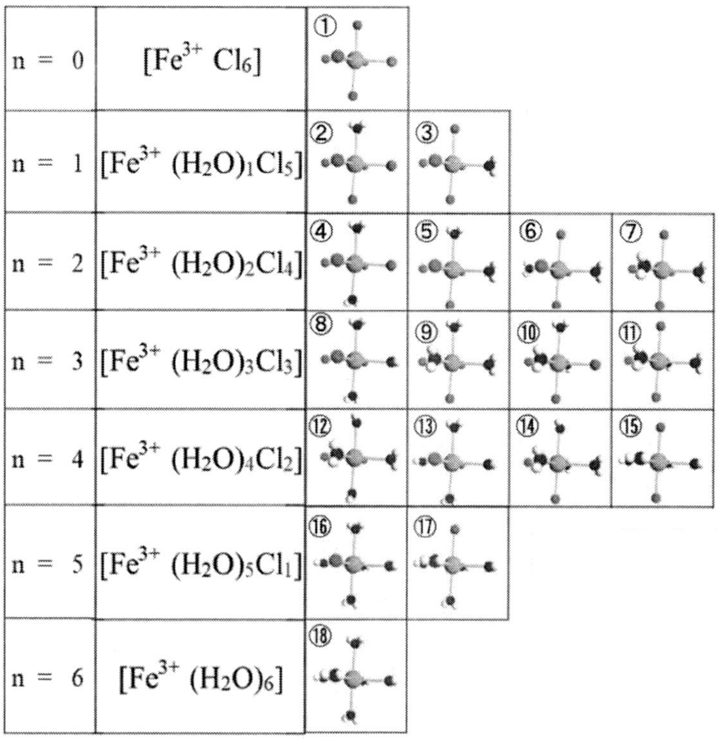

Figure 2. 18 types of calculation model octahedral hexa-coordination complex.

IOP Conf. Series: Materials Science and Engineering **835** (2020) 012014 doi:10.1088/1757-899X/835/1/012014

To reproduce the Jahn-Tellar distortion of the cluster models, we made the eleven kinds of models with uniaxial anisotropy having the ± 0.5 Å difference of the interatomic bond lengths, as shown in Figure 4.

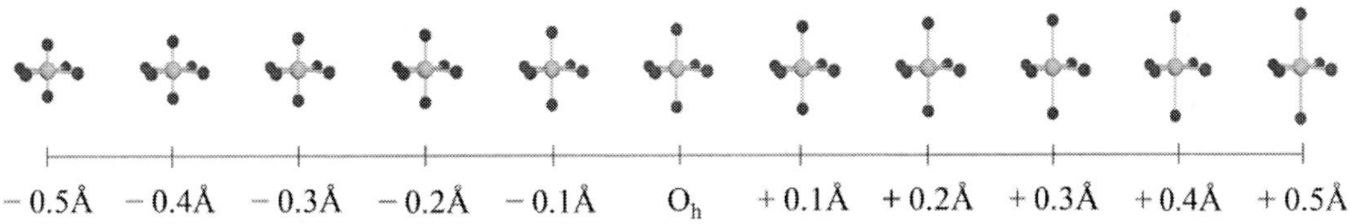

Figure 3. Jahn - Teller distortion (shift of M-L bond length (Å))

3. Results and discussion

In this study, the SCO phenomenon was confirmed with 10 types of metal complexes among the prepared 18 kinds of octahedral hexa-coordination complexes. The Figure 5 indicates that the energy diagrams of these 18 cluster models. In addition, we can find out the following three points were confirmed from these calculation results.

①As a general tendency, even in the same model, it was confirmed that the SCO can be generated easily by substituting an H_2O ligand (that has a high coordination ability) replaced with an axial position (Z-axis).

② In the case of [Fe^{3+} $(H_2O)_5Cl_1$] : Two-step SCO phenomenon can be confirmed by changing the ligand field splitting, that is changed from High spin → Intermediate spin → Low spin.

③As the ligand field splitting increases under the influence of the ligand with high coordination ability, the SCO energy should be larger. The threshold energy of the SCO phenomenon are summarized in Table 2.

Table 3. Ligand field splitting (⊿) when SCO is confirmed.

	[Fe^{3+} $(H_2O)_nCl_{6-n}$]	⊿(eV)			
n = 0	[Fe^{3+} Cl_6]	1.5			
n = 1	[Fe^{3+} $(H_2O)_1Cl_5$]	1.65	1.63		
n = 2	[Fe^{3+} $(H_2O)_2Cl_4$]	1.92	1.7	×	×
n = 3	[Fe^{3+} $(H_2O)_3Cl_3$]	1.98	1.85	×	×
n = 4	[Fe^{3+} $(H_2O)_4Cl_2$]	2.08	1.95	×	×
n = 5	[Fe^{3+} $(H_2O)_5Cl_1$]	2.08 / 2.14	×		
n = 6	[Fe^{3+} $(H_2O)_6$]	2.48			

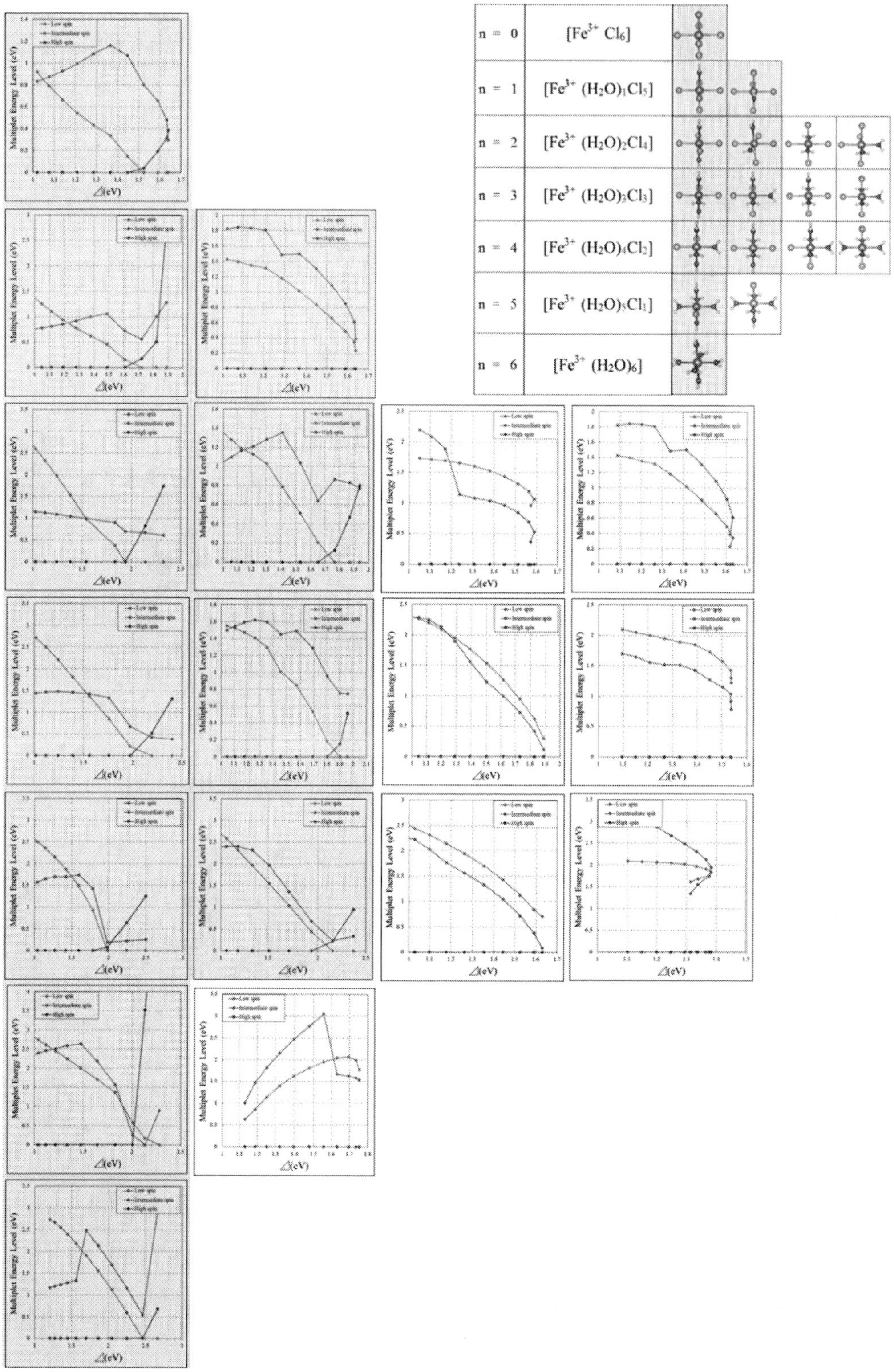

Fig 4. 18 calculated energy diagrams of [Fe^{3+} $(H_2O)_nCl_{6-n}$]

4. Conclusion

In this study, the electronic state calculations have been carried out an octahedral hexa-coordinated metal complexes ML_6, involving two types of different ligand molecules, H_2O and Cl. It was found that the spin state changes by the uniaxial Jahn-Teller distortion in an octahedral hexa-coordinate complex. The 11 kinds of the metal complexes are expected the spin state changing, so there is a chance to obtain the novel spin crossover properties. There is a correlation among the spin crossover, the coordinated position, and the ligand field splitting. In the future, we will expand the calculation targets and also study the relationship with other parameters. In the future, we will calculate much more kinds of hexa-coordinate complexes, and we want to find the boundary between high- and low spins and the mechanism of the spin state changing.

References

[1] J W. Addison *et al., Dalton Trans*, 1349 -1356 (1984).
[2] H. Jahn and E. Teller, *Proceedings of the Royal Society of London. Series A, Mathematical and Physical Sciences*, **161** 220–235 (1937).

IOP Conf. Series: Materials Science and Engineering **835** (2020) 012015 doi:10.1088/1757-899X/835/1/012015

Single-Crystal X-Ray Structural Analysis and Electronic Structures Calculation of Rare Sugar

Bui Quoc Huy,[1] Tomohiko Ishii,[1] Norito Taniguchi,[1] Natsumi Nagayama,[1] and Genta Sakane[2]

[1]Department of Advanced Materials Science, Faculty of Engineering, Kagawa University
Hayashi-cho, Takamatsu, Kagawa 761-0396, JAPAN.
[2]Department of Chemistry, Faculty of Science, Okayama University of Science 1-1 Ridaicho,
Kita-ku, Okayama 700-0005, JAPAN.

s18g580@stu.kagawa-u.ac.jp

Abstract. There are two types of mono-saccharide sugars in nature, which are natural monosaccharide and rare sugar. The sugar that exists in extreme few amount in nature is called as rare sugar e.g. D-psicose and D-allose. The reasons we choose the rare sugar as a research target are because the rare sugar has almost no calories and is very useful for diabetic and dieter patients. Specifically, D-psicose is a no-calorie rare sugar with the sweetness of about 70 % of a sucrose. It has been confirmed that taking the D-psicose with meals suppresses the rise of sugar into the blood. Further, it is very effective in improving and preventing the diabetes. In order to investigate a research on the rare sugars, the first step must be to obtain the correct molecular structure. However, few crystal structures of rare sugars have been reported up to now. Therefore, the purpose of this research is to analyze the single-crystal X-ray structure of unknown rare sugar 'L-glucose'. In addition , the DV-Xα molecular orbital method is used to compare the differences in a hydrogen bonding between α-D-glucose and α-L-glucose, and to make a novel supramolecular rare sugar.

1. Introduction

The abundance of rare sugars is small and very expensive, so research of rare sugar is difficult. Therefore, much structural analysis of rare sugar and mechanism of synthesis are not yet understood. However, in recent year, researching of Kagawa University has led to the discovery of the isomerization enzyme DTE (D-tagatose -3 epimerase), which isomerizes the natural monosaccharide D-Fructose into the rare sugar D-Psicose, leading to the large-scale synthesis of rare sugars. The discovery of DTE led to the development of a strategic monosaccharide synthesis pathway called 'Izumoring' that systematized the relationship between monosaccharides and their molecular structures and enzymes into a ring. In this research, we will focus on research on the 6-membered ring of monosaccharides. There is no difference in chemical formula between rare sugars and natural monosaccharides, both of which are represented by $C_6H_{12}O_6$ (e.g. D-glucose, L-glucose) and are simply structural isomers. Therefore, the purpose of this study is mainly to analyze the X-ray crystal structure of rare sugar whose crystal structure (absolute coordinates) has not yet been reported. In our laboratory, we succeeded in synthesizing a new rare sugar called supramolecular rare sugar (SRS), which is synthesized from several kinds of rare sugars. However, the mechanism by which different types of sugar molecules combine has not yet been elucidated. Therefore, it is another purpose to clarify the mechanism of supramolecular rare sugar by using a theoretical calculation method called DV-Xα for calculating and comparing the differences in the hydrogen bonds between α-D-Glucose and α-L-Glucose.

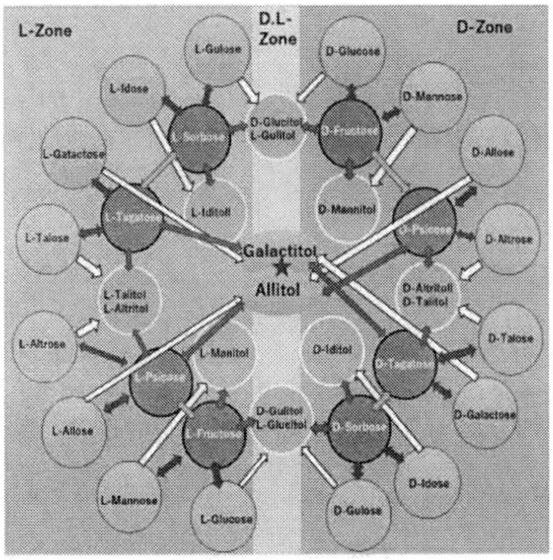

Figure 1. Izumoring.

2. Experimental

2.1 Material and methods

The single-crystal X-ray structure analysis can be divided into two stages. The first, X-ray diffraction experiments are performed to collect *hkl* diffraction data. The second, the obtained data is analyzed to determine the crystal structure. Each process is described below.

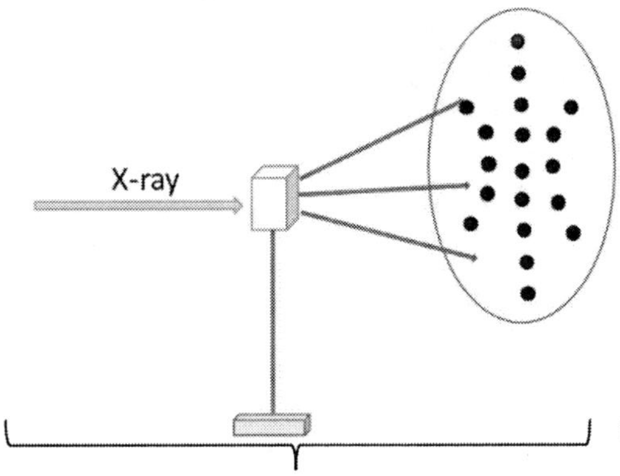

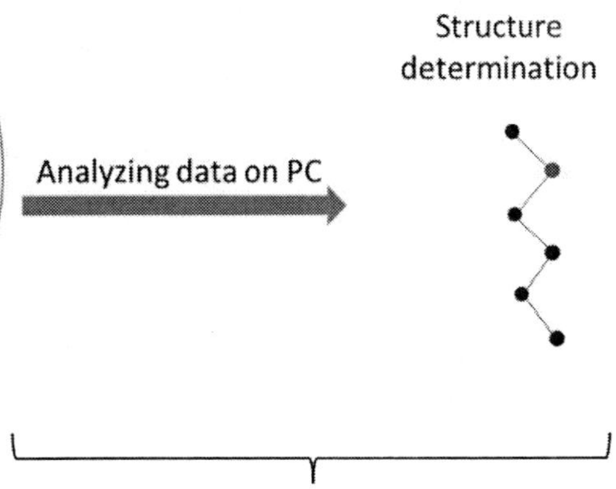

Figure 2. The first stage **Figure 3.** The second stage

In the X-ray diffraction experiment, a single crystal is cut with a design knife to a size suitable for measurement (the dimensions of the sample are 0.2 mm x 0.2 mm x 0.2 mm). Then, the sample was set in the experimental apparatus, and an X-ray diffraction experiment was performed.

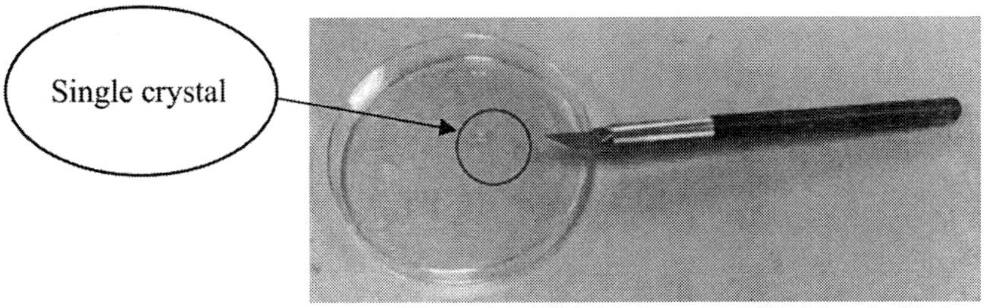

Figure 4. Single crystal and design knife

Then, when a suitable sample is obtained, it is fixed to the tip of the capillary with an adhesive.

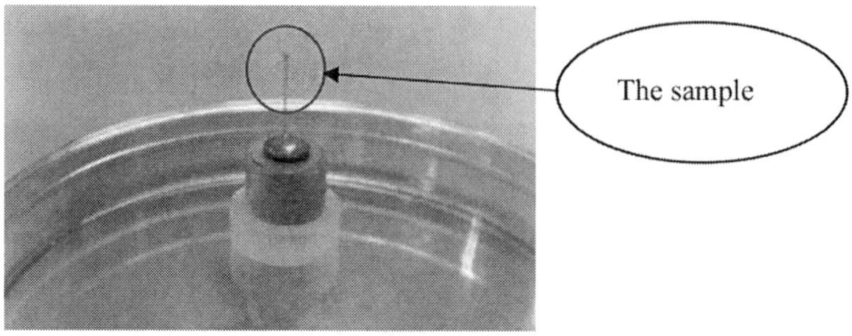

Figure 5. The sample is fixed.

The sample is placed in screw bottle to dry and allowed to harden for about a day. Next, an X-ray diffraction experiment is performed. The sample prepared earlier was mounted on a goniometer head of an X-ray diffractometer and centered so that the X-ray radiate the entire crystal. Then, basic information such as the sample name, crystal shape, color, size, and composition formula was input to the data file, and measurement conditions such as voltage, current, and 2θ range to be measured were set, and the measurement program was executed.

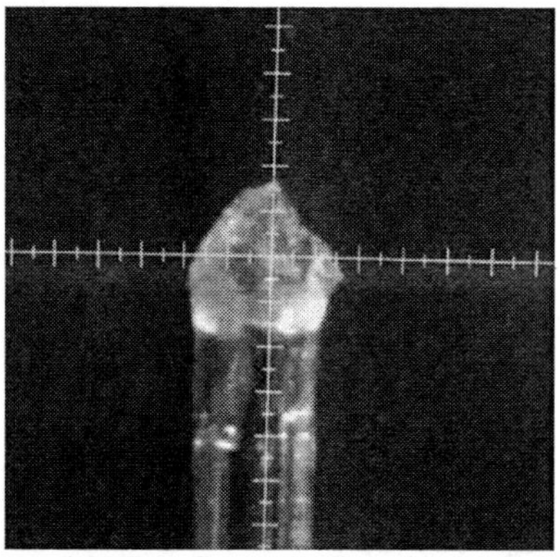

Figure 6.X-Ray is applied to the sample.

2.2. Analysis

The reflection data obtained by the X-ray diffraction experiment were subjected to structural analysis using a structural analysis program package "Crystal Structure 3.7" made by Rigaku. The following is a simple flow to the structure determination.

1) **Evaluate** data determines the space group and composition. **Unit Cell Transformation** replace the cell length and cell angle of the a, b, c-axes of the unit cell by the appropriate.
2) **Formula**. If the composition is inappropriate in evaluate data, change the chemical composition.
3) **Space group.** Changes are made when the space group is not appropriate in evaluate data or when unit cell transformation is performed.
4) **Solve.** The peak is obtained using the direct method (We use Sir 2014 for this graduation study).
5) **Default 99** uses the predicted structure to find out the non-hydrogen atom if it does not appear
6) **Least square.** Let you refine the parameters of atoms (coordinates, temperature factors, etc.)
7) **Refinement menu.** This sets the refinement method for all non-hydrogen atoms and the temperature factor of hydrogen. Shelxl 2014 was used in this study.
8) **Fourier.** The hydrogen atom is found by the difference Fourier synthesis.
9) **Least square.** Turn on secondary attenuation effect correction and change weights to sigma for precision. At this time, since maximum shift/error indicates the convergence state of the refinement, when this value becomes 0.01 or less, the structural analysis is finished.

Two types of sugar structures, α-D-Glucose and α-L-Glucose were used in this calculation. The electronic states are calculated and discussed using structures which have not been reported in the Cambridge Structure Database (CSD). We use DV-Xα molecular orbital method to compare the differences in a hydrogen bonding between α-D-glucose and α-L-glucose, and to make a novel supramolecular rare sugar.

3. Result of this study

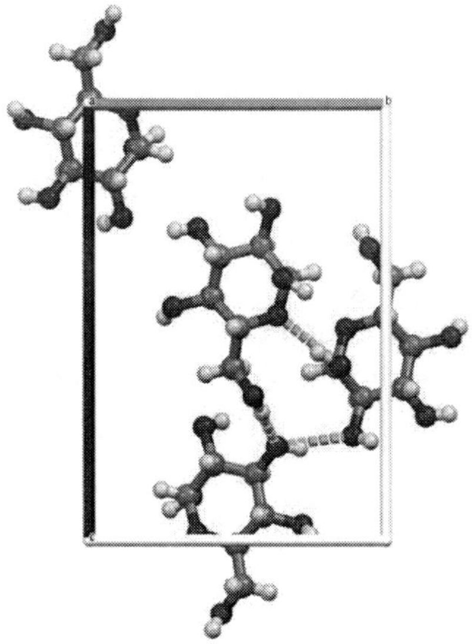

Figure 7. α-D-glucose

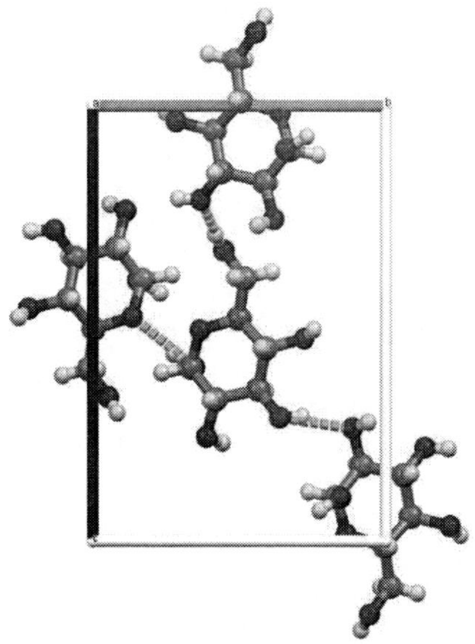

Figure 8. α-L-glucose

Table 1. BOP in a unit of α-D-Glucose.

A···H − *D*	*A*···H [Å]	H···*D* [Å]	A···D [Å]	A···H − D [°]	BOP
O2···H3A−O3	1.909	0.820	2.699	161.44	0.0391
O3···H6A−O6	1.811	0.888	2.699	176.72	0.0409
O5···H1A−O1	1.908	0.955	2.835	163.26	0.0405

Table 2. BOP in a unit of α-L-Glucose.

A···H − D	A···H [Å]	H···D [Å]	A···D [Å]	A···H − D [°]	BOP
O2···H3A−O3	1.891	0.827	2.698	174.95	0.0329
O3···H6A−O6	1.861	0.846	2.704	167.23	0.0298
O5···H1A−O1	1.917	0.946	2.836	163.07	0.0314

Total BOP (Bond Overlap Population) of α-D-Glucose was 0.1204 and BOP of α-L-Glucose was 0.0941. We can reveal the hydrogen bonding of α-D-Glucose in a unit stronger than α-L-Glucose. At first glance, the arrangement of hydrogen bonds in the packed molecule is different, but in three-dimensional space, the hydrogen bond pair is the same. Specifically, there are three hydrogen bonds in both, and the positions of hydrogen bonds are the same. (For example, both have O2···H3A−O3 hydrogen bonds).

Table 3. Experimental results of α-D-glucose and α-L-glucose.

Identifier	GLUCSA03 Former work	α-D-Glucose This work	α-L-Glucose New data
Formula	$C_6H_{12}O_6$	$C_6H_{12}O_6$	$C_6H_{12}O_6$
Crystal System	Orthorhombic	Orthorhombic	Orthorhombic
Space Group	$P2_12_12_1$	$P2_12_12_1$	$P2_12_12_1$
Cell Length [Å]	4.9464(7)	4.9526(3)	4.9536(4)
	10.3440(15)	10.3209(6)	10.3185(7)
	14.8472(22)	14.7792(9)	14.7798(11)
Cell Angles[°]	90.000	90.000	90.000
	90.000	90.000	90.000
	90.000	90.000	90.000
Cell Volume [Å^3]	759.665	755.443	755.451
Z	4	4	4
R-Factor [%]	3.7	2.85	3.33

4. Conclusion

In this study, we succeeded in single crystallization and structural analysis of a new monosaccharide of α-L-glucose. As for α-D-glucose, that structure has already been reported, more accurate results have been obtained. In our laboratory, we used the DV-Xα method to calculate hydrogen bonds in both parked selected molecules. As a result, both contain three hydrogen bonds. Although the arrangements of the hydrogen bonds appear to be different at first, the hydrogen bond pairs are actually the same. The Flack parameters for both are very small 2.85 % and 3.33 %, respectively. The R-value is the most important number. The R-value represents the crystal precision at the International Crystallographic Society. The lower the R-value, the higher the precision of the crystal. If the R-value falls below 5 %, the results can be submitted to an international paper. In this study, α-D-Glucose has the lowest R-value of 2.85 %, it the most accurate substance in the world. The R-value of α-L-Glucose is also very small (R-value is 3.33 %), and it is the first substance in the world. The calculated hydrogen bonds in each lattice show that α-D-Glucose has stronger hydrogen bonds than α-L-Glucose. Based on the results, we are thinking of making supramolecular rare sugars by adding a little bit more α-D-Glucose instead of the same amount.

Reference

[1] A. Mostad, *Acta Chem.Scand.* (1994), 48, **276**.
[2] T. Ishii, G. Sakane, A. Yoshihara, K. Fukada and T. Senoo, *Acta Crystallogr, Sect. E: Cryst. Commun.* (2015), 71, **o289**.

Data Mining Application with Fuzzy Logic Method for Monitoring Vehicle Position Based on Android

Munaf Ismail[1], Sri Arttini Dwi Prasetyowati[2], Ida Widihastuti[3]

[123]Department of Electrical Engineering, Universitas Islam Sultan Agung, Jl. Kaligawe Km 4 Semarang, Central Java 50112, Indonesia

munaf@unissula.ac.id

Abstract. Cases of vehicles theft in the Central Java region occurred 869 cases, the percentage was 57.86% against other crimes. To prevent vehicles theft, the police need public participation by increasing awareness of their vehicles. Vehicle theft can occur due to owner negligence and lack of safety locks on the vehicle. This study aims to create data mining application as monitoring vehicle position. The transmitter WiFi 2.4 GHz placed on a vehicle, then measured the RSSI signal by an android application based on fuzzy logic. Indication of theft is that the vehicle is positioned away from the owner, so the signal RSSI is less than -90 dBm. Fuzzy logic monitors this change in distance. When the membership function is lost, fuzzy logic instruct android to activated the alarm . Fuzzy logic was able to make decision give a alarm to user when RSSI signal less than -90 dBm. RSSI signal is less than -90 dBm when the distance is 20 meters if there is a wall barrier, without obstacle the distance is up to 80 meters. RSSI signal is influenced by the measurement environment.

Keywords: Data Mining Application, Fuzzy Logic, monitoring vehicle position

1. Introduction

The high number of crime of theft of motorized vehicles in general makes people anxious, leading to large material and non-material losses for the victims. According to the National Police Chief in 2018 the number of cases of theft of motorized vehicles was more than the cases of crimes of theft, fraud and violence [1]. Cases of theft of motorized vehicles in the Central Java region occurred 869 cases, the percentage was 57.86% against other crimes [2]. Theft can occur due to negligence of vehicle owners and lack of vehicle safety locks.

Motorbikes are the type of motorized vehicles most targeted by thieves because motorbikes are more easily carried away than cars so they are easier to steal. One of the factors causing the high rate of motorcycle theft is the neglect of motorized vehicle owners by parking in any place without supervision.

Additional safety systems for motorized vehicles are additional safeguards in addition to the starter buttons on motorized vehicles. For that we need a better security system. The need for a vehicle security system can make it difficult for criminals to steal vehicles to be able to run motorbikes and give warnings to motor vehicle owners.

2. Data Mining

Data processing needs are getting added value from the collected data, encouraging the application of data processing techniques from various fields of knowledge such as statistics and artificial intelligence. It turns out that the application of this technique presents a new challenge that ultimately leads to a new method called data mining. There are several definitions of mining data that are known from various sources, including:

1. Data mining is a search and analysis of large data techniques to find meaningful patterns and rules.
2. Data mining is the process of finding patterns and relationships in a data.
3. Data mining is an automatic or semi-automatic process for discovering new and potential knowledge from a set of data [3].

Based on some of these definitions, it can be concluded that data mining is closely related to the discovery of information or knowledge that is new, unpredictable in the database, both automatically and semi-automatically.

3. Receive Signal Strength Indicator (RSSI)

RSSI is a technology used to measure signal strength indicators received by a wireless device . In fact, there are many limitations obtained in direct mapping of an RSSI value measured by distance, because in the process, RSSI is affected by noise, multi-path fading, transmit power, interference, etc. that make fluctuations in the power received. A power received at an antenna (Pr) is separated by the distance d from a transmitter antenna with the known transmission power (Pt), then given the Friis equation as shown below.

$$P_r = P_t G_r G_t \left(\frac{\lambda}{4\pi d}\right)^2 \quad (1)$$

where Gt is the gain / gain of the transmitter antenna, Gr is the gain / gain of the receiving antenna and λ is the wavelength. States that channel models such as log normal shadowing provide RSSI values of the distance d from the transmitter given in the equation :

$$\text{RSSI}(d) = P_t(d_0) - P_L(d_0) - 10n_p \log_{10}\left(\frac{d}{d_0}\right) + X_\sigma \quad (2)$$

where Pt is transmission power, while PL is path loss to distance reference and is path loss this exponent is influenced by the environment around the transmission media. Random variations on RSSI are modeled as random Gaussian variables where $X_\sigma = N(O,\sigma^2)$ np and σ values can be set according to the propagation environment conditions [4].

4. Fuzzy Logic

The fuzzy logic processes the measurements of RSSI using fuzzy method, resulting in the measurement recommendations used based on the position vehicle and the highest degree of fuzzy membership. Fuzzy is one of the best ways to map an input space into an output space. The starting point of the modern concept of uncertainty is that it is introduced about a theory that has objects of a fuzzy set that has imprecise boundaries and membership in a fuzzy set, and not in the form of logical right or wrong, but expressed in degrees . The function of membership is a curve showing the mapping of data input points into their membership value (also often called as membership degree) which has intervals between 0 and 1.

4.1. Triangle Curve Representation

The representation of triangle curve is essentially a combination of two linear representations (increase and decrease) as shown in Fig. 3.

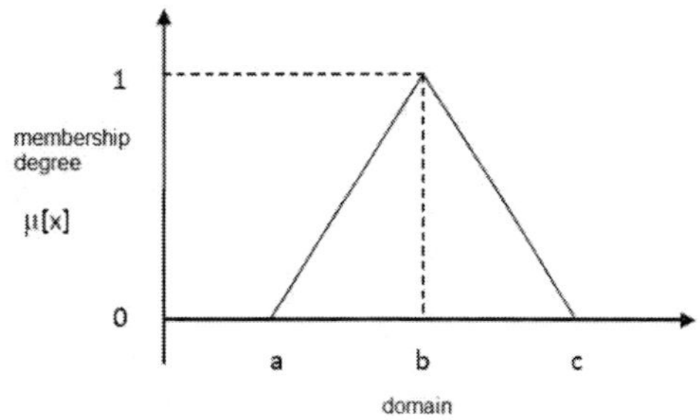

Figure 1. Triangle Curve

The function of membership is:

$$\mu[x] = \begin{cases} 0; & x \leq a \; or \; x \geq c \\ \dfrac{(x-a)}{(b-a)}; & a \leq x \leq b \\ \dfrac{c-x}{c-b}; & b \leq x \leq c \end{cases} \tag{3}$$

Where:
a = the lowest value of domain which has membership degree of 0
b = the value of domain which has membership degree of 1
c = the highest value of domain which has membership degree of 0

5. Research Method

Research prevents vehicle thieves from turning on vehicles and detecting vehicle positions. So that if the thief will take the vehicle, he will lead the vehicle away from the parking location. If there is an indication of theft as it is called, the android application will monitor the position of the vehicle with fuzzy logic based on the RSSI signal received. The block diagram image will be like this.

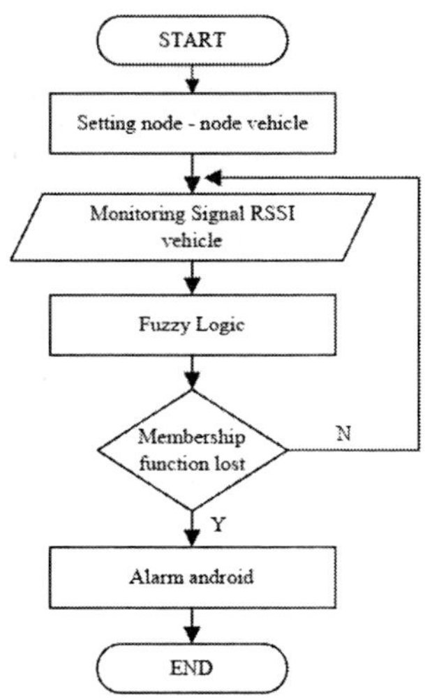

Figure 2. Block diagram system

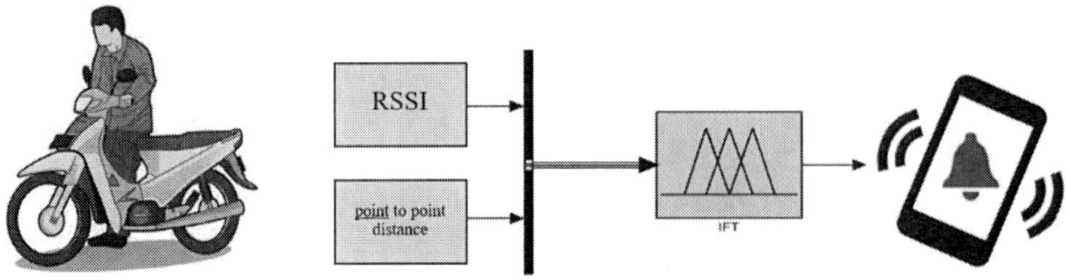

Figure 3. Data mining application system for monitoring vehicle position

The thief must guide the vehicle away from the parking location because the vehicle has an additional safety lock above. 2.4 GHz wifi transmitter placed on the vehicle, then the vehicle is monitored position by point to point android application with fuzzy logic by measuring RSSI.

5.1. Measurement location

a b

Figure 4. a. The measurement location vehicle position with the obstruction of the wall.
b. Locations of measurements without obstruction in the parking area.

The research was carried out in the Sultan Agung Islamic University campus. The measurement of the first RSSI as shown in Figure 3 is carried out in the wall space to monitor the vehicle. Figure 3.b. is an unobstructed RSSI measurement carried out in the parking lot. The time of measurement is done in the afternoon when there is not much traffic activity around the measurement area.

5.2. Measurement samples

The measurement sampling technique determines :

a. The initial coordinates of the access point placed on the vehicle as a 2.4 GHz WiFi signal transmitter, for setting point to point.
b. Determination of the maximum range of WiFi signals that can be received by the RSSI measuring application with a wall obstacle and without obstruction.
c. RSSI (Receive Signal Strength Indicator), analyzed by fuzzy logic to map into membership as monitoring vehicle position.

The RSSI measurement sampling test drive is useful for monitoring the position of a vehicle with the fuzzy logic method, the signal received will be determined in the category of very secure, secure, alert or lost. Determination of fuzzy membership if there is an indication of theft of the vehicle's position of the vehicle will change so that the RSSI signal decreases until it is lost. After membership is in the lost category, it means that the vehicle moves away until the lost RSSI signal then alarms the Android application activated. Previous research found that the Android application compared to the Windows application has a difference in the measurement size below 5% [5]. This study uses a wifi R206 mobile transmitter, then measured the RSSI with an Android application based on fuzzy method.

6. RESULTS AND DISCUSSION

6.1. Display of Android Applications

The wireless sensor network application for monitoring the position of the vehicle will result in some views being generated as shown below.

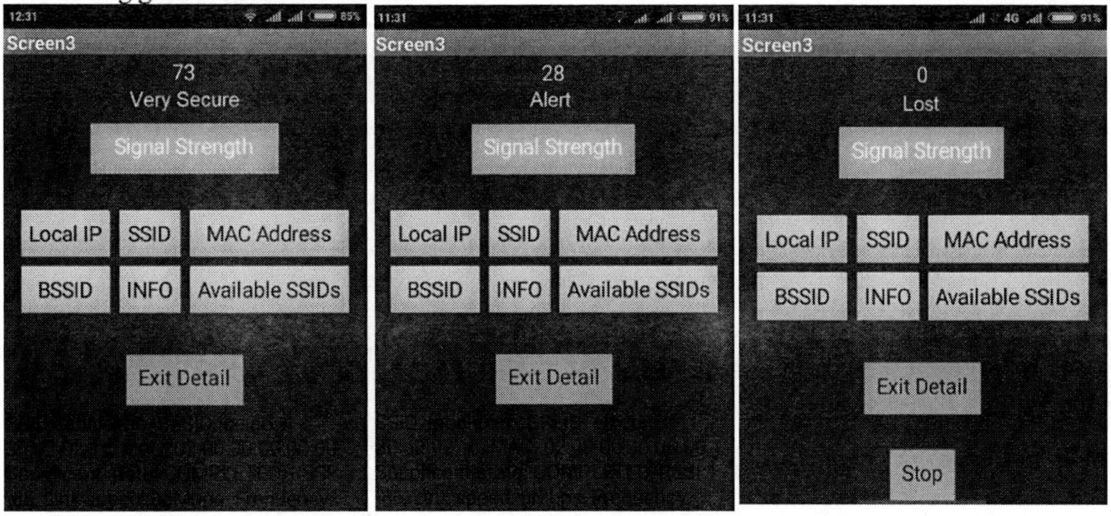

Figure 5. a. Display of membership Fuzzy Very Secure membership with RSSI -65 dBm
b. Fuzzy Alert membership display with RSSI -86 dBm
c. The membership display of Fuzzy Lost with RSSI is smaller than -90 dBm

Figure 5.a. is the measurement of the display of the adroid monitoring application of the vehicle in a position close to the owner's smartphone, measured by the RSSI of -65 dBm and membership fuzzy very secure. Figure 7.b. distance of 15 meters with RSSI of -86 dBm, membership fuzzy alert. Figure 3.c. is a distance of 20 meters so that the RSSI is smaller than -90 dBm and fuzzy lost ownership. Fuzzy lost membership will sound an alarm from the android application. The above display is the result of monitoring the position of the vehicle with the obstruction of the wall of the room. Measurements with locations without obstruction have a maximum distance of 80 meters with fuzzy lost membership and RSSI below -90 dBm.

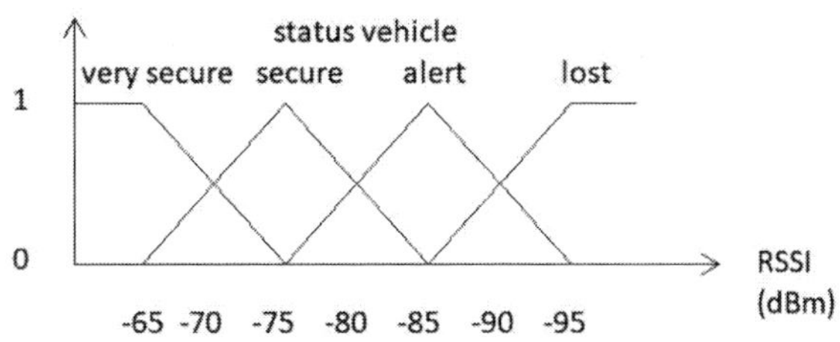

Figure 6. Membership function fuzzy monitoring vehicle position

6.2. Data Appearance

The first measurement was carried out in the room with the vehicle at a maximum distance of 20 meters from the user, the second measurement carried0 out in the parking area without obstruction of the wall obtained maximum measurements of RSSI reception as far as 80 meters.

Table 1. Membership Function of RSSI point to point with wall obstacles

RSSI (-dBm)	Distance (m)	Membership function fuzzy
-56	3	Very secure
-65	6	Very secure
-77	12	Secure
-86	15	Alert
-89	19	Alert
	≥ 20	Lost

From table 1 is the measurement of RSSI point to point with fuzzy logic. Measurements are made with wall obstacles. The results obtained are a maximum range of 20 meters and RSSI point to point is -90 dBm. When the distance ≥ 20 meters the membership function is in the "Lost" category so that it sounds an alarm on the android application to alert the vehicle owner.

Table 2. Membership Function of RSSI point to point with no obstacles

RSSI (-dBm)	Distance (m)	Membership function fuzzy
-60	15	Very secure
-66	30	Very secure
-73	45	Secure
-86	75	Alert
-89	78	Alert
	≥ 80	Lost

Table 2 shows the measurement of RSSI point to point without obstacles. When the distance ≥ 80 meters and RSSI -90 dBm, the membership function is in the "Lost" category, this condition turns on the android alarm so that the vehicle owner checks the position of the vehicle.

7. CONCLUSION

The conclusions obtained are as follows :

1. Data mining with fuzzy method is able to give a warning if there is an indication of vehicle theft.
2. Monitoring vehicle with obstacle walls has the farthest distance of 20 meter and RSSI -90 dBm.
3. Monitoring vehicle without obstacle walls has the farthest distance of 80 meter and RSSI -90 dBm.

8. ACKNOWLEDGMENTS

Acknowledgments to the Indonesian Ministry of Research, Technology and Higher Education for funding community partnership program grant scheme 2019.

References

[1] Medcom, "Jumlah Kejahatan Menurun di 2018," *medcom.id*, Jakarta, Dec-2018.
[2] Jawapos, "2018, Jumlah Kejahatan di Jateng Diklaim Turun 17,6 Persen," *jawapos.com*, Semarang, Dec-2018.
[3] F. Kurniasih *et al.*, "Analisa Dan Perancangan Data Mining Dengan Metode Market Basket Analysis Untuk Analisa Pola Belanja Konsumen pada Tendencies Store," *Stud. Inform. J. Sist. Inf.*, vol. 5, no. 1, pp. 1–10, 2012.
[4] P. K. Sahu, E. H. Wu, and J. Sahoo, "DuRT: Dual RSSI Trend Based Localization for Wireless Sensor Networks," *IEEE Sens. J.*, vol. 13, no. 8, pp. 3115–3123, 2013.
[5] M. Ismail, "Rancang Bangun Pengukur Rssi (Receive Signal Strength Indicator) Berbasis

Aplikasi Android Menggunakan App Inventor," *SNST.,* vol 1, no 1 (2018) pp. 25–30, 2018.

IOP Conf. Series: Materials Science and Engineering **835** (2020) 012017 doi:10.1088/1757-899X/835/1/012017

Coulomb blockade transport emerged in quasi one-dimensional PEDOT: PSS fiber

Megumi Akai-Kasaya, Naoya Ogawa, Soichiro Kakinoki

[1] Precision Science & Technology, Graduate School of Engineering, Osaka University, 2-1 Yamadaoka, Suita, Osaka 565-0871, Japan,

kasaya@prec.eng.osaka-u.ac.jp

Abstract. In organic materials, peculiar nonlinearity to current voltage appears, thought a general and comprehensive explanation of them is still controversial. Conductive segments in poorly conductive organic materials are expected to have a smaller electrical capacity, leading to a higher critical temperature for the blockade effect. Here we show an experimental evidence of Coulomb blockade taking place on quasi one-dimensional conductive polymer, PEDOT:PSS [poly (3, 4-ethylenedioxy-thiophene) doped with poly (styrene sulfonate) anions], fibers. The PEDOT:PSS wire grows through electro-polymerization, and bridges between electrodes immersed in EDOT monomer solution. Conducting measurement for the dried fibers shows clear nonlinear behaviour in the current-voltage characteristics as temperature decreases. The non-zero threshold voltage, which increased with decreasing temperature, appears in the current flows through a thinnest fiber. The effective percolative transport passes in thin fiber is able to consists of the connection of the Coulomb blockade islands. By considering both the charge blockade effect and the influence of structural disorder and dimensionality, it is hoped that a clear understanding of charge transport in organic materials can be achieved.

1. Introduction

Good organic conductors including the polymers and monomer materials often have a low-dimensional configuration, e.g., quasi-one-dimensional structures or two-dimensional (2D) layers. Nevertheless, a large number of fundamental questions remain to be answered regarding the charge transport mechanism, particularly in low-dimensional structures, One such question concerns the nonlinear behavior that is often observed in the current-voltage (I-V) characteristics of organic conductors. Even for materials that exhibit good linear I-V characteristics near room temperature (RT), nonlinearity can occur as the temperature is reduced. This effect has been interpreted using a variety of mechanisms, such as charge hopping, trapping, tunneling and emission, either within the organic material or at the interfaces. However, the observed nonlinearity cannot be fully explained using these conventional models or combinations of them.

Recently, it has been reported that the I-V characteristics obey a power-law relationship in low-dimensional organic materials such as polymer nanofibers [1], nanotubes [2,3] and films [4-6]. The observed power-law relationship for polymer materials was sometimes put forward as evidence for tunneling into a one-dimensional (1D) Luttinger liquid because of the quasi-1D structure of these materials [4]; however, power-law behavior was also observed for a three-dimensional (3D) organic films [5,6]. Meanwhile, in inorganic granular materials, the power-law dependence of the I-V characteristics has commonly been attributed to dissipative tunneling processes, such as that associated with a Coulomb blockade (CB) [7-11]. CB transport occurs in systems consisting of an array of small conductive islands connected by narrow junctions, provided the tunneling resistance between neighboring sites is significantly larger than the quantum resistance (>>h/e2), the capacitance associated with each island is sufficiently small, and the energy corresponding to an additional electron charge at each site is large compared to kBT. Here, h is Planck's constant, e is the charge of an electron, and kB is Boltzmann's constant.

We have reported that CB transport investigated in a 2D conjugated polymer monolayer exhibits nonlinear behavior in the current-voltage characteristics, and a non-zero threshold voltage appeared, which increased with decreasing temperature [12] as predicted by Middleton and Wingreen [13]. The CB effect has been confirmed during charge transport through a single molecule spanning adjacent electrodes, although it has rarely been suggested as the origin of nonlinear conduction in larger condensed organic conductor systems [5,14,15]. In this paper, we investigate charge transport through quasi one-dimensional conductive polymer, PEDOT:PSS [poly (3, 4-ethylenedioxy-thiophene) doped with poly (styrene sulfonate) anions], fibers. The polymer fibers with different diameter show different charge transport feature at lower temperature regime. The typical CB transport feature were observed for the thin polymer fiber with about110nm diameter, whereas the thicker fiber shows only conductance decrease as hopping conduction. We associate the percolative transport passes connecting the conductive segment with the origin of CB charge transport in the quasi one-dimensional conductive fibers.

2. Methods

2.1. Polymer fiber growth

Single and straight conducting polymer fiber grows between electrodes with sharp apex immersed in monomer solution when the gap distance is short and absolute bipolar square-wave AC voltage with a 50% duty cycle were applied [16]. We performe the wire grows of PEDOT:PSS in 0.135M EDOT and 0.02M PSS in acetonitrile and ultrapure water ratio of 1:1. Liquid EDOT and PSS solution were obtained from Sigma-Aldric. Electrodes were fabricated by conventional photo lithography technique. Cr (10 nm) and Au (190 nm) were deposited on a glass substrate by using EB1100 (Canon Anelva). A solution trough made of Polydimethylsiloxane (PDMS) was placed on the substrates with electrodes and filled with EDOT and PSS solution. Bipolar AC voltage with square shape are made by using waveform generator WF1973 (NF) connected to high speed bipolar amplifier HSA4101 (NF). The AC voltage, which frequency from 500kHz to 1MkHz and peak-to-peak voltage (V_{pp}) of 8V, is applied to one side of electrode whereas an opposite electrode is connected to GND. After the wire growth connecting between electrodes, the wires are dried in ambient condition. Wire images are observed by scanning electron microscopy (SEM) S-4700 (Hitachi) after the current measurement.

2.2. Current measurement

Measurements are carried out using a variable temperature probe (TPP-4, Lakeshore Co., Ltd.) and a semiconductor characterization system (Keithley 4200) in a vacuum of about 1×10^{-5} Pa under dark conditions. Temperature is controlled by the cryogenic system with liquid helium flow from room temperature (RT) to 4.2 K. The *I-V* characteristics are obtained from high to low temperatures keeping restrictive temperatures.

3. Results and discussion

3.1. Polymer fibers

It has been reported that PEDOT grows in wire shape through electro-polymerization by applying bipolar AC voltage between electrodes immersed in EDOT solution. The previously reported wire's diameter was ranging from sub to few micrometres, thus they were observed by optical microscopy usually. We tried to fabricate single PEDOT wire connecting between electrodes with thinner dimeter to confirm the low dimensional structural effect on the charge transport feature. Figure 1 shows SEM images for two types of fabricated PEDOT:PSS wire. An AC voltage was applied for 0.1s for (a) and for 10s for (b), respectively. The wire diameter and number of bridging wires depends on the both frequency and time span of the applied voltage.

IOP Conf. Series: Materials Science and Engineering **835** (2020) 012017 doi:10.1088/1757-899X/835/1/012017

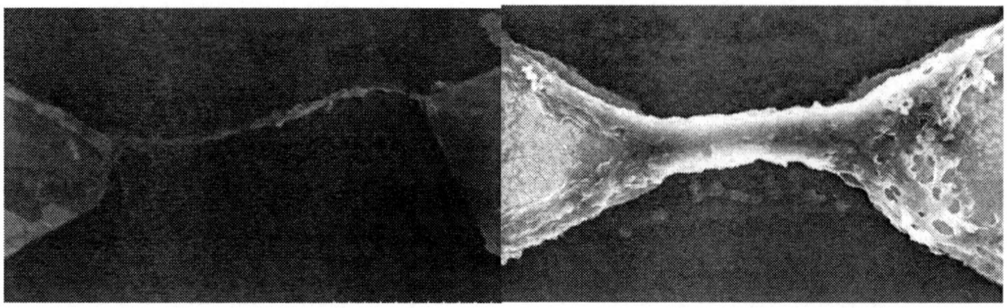

Figure 1. SEM images of grown PEDOT:PSS wire (a) thin wire with diameter of ca. 118 nm and (b) thick wire with diameter of ca. 800 nm.v

3.2. I-V measurement

The current flows through the PEDOT wire was measured at temperatures from RT to 4.2 K. Figure 2(a) and (b) shows the I-V characteristics observed for the thin and thick wires shown in Fig. 1(a) and (b). Both temperature depending I-V characteristics shows similar rapid current decreases with temperature decreases, Subtle nonlinearity, which is typical characteristic feature observed for organic materials frequently, are seen in the I-Vs observed in all temperature range. The nonlinearity can be emphasised in a double logarithmic plot as shown in Fig. 1(c) and (d). The solid black lines, which are overlapped on the data at low voltage region, have gradient 1 in the double logarithmic plot, whereas the data at higher region deviate from the black line to have higher gradient. This means that the I-Vs have power law dependence with higher exponent at higher voltage region, thought it has linear relationship at low voltage region.

There is a conductance difference of about 10 times at RT between these fibers, which is a reasonable value based on the cross-section area ratio. Although, the range of current decrease as temperature decrease is more than four order of magnitude for thin wire (Fig. 1(c)), whereas the conductance decrease only in two order for thick wire (Fig. 1(d)). The I-Vs at the lower temperature in Fig. 1(c) show the blockaded region with null current and subsequent power law current increase as voltage increase, which is a typical feature of CB charge transport. It is generally accepted that the thermal hopping transport in organic materials is origin of monotonical temperature dependence of conductance and subtle I-V nonlinearity as the features seen in Fig. 1(d). However, the power low with threshold observed at low temperatures in Fig. 1(c) never be explained by using the thermal hopping model.

Our previous study suggested that the percolative charge transport pass explained well the emergence of CB charge transport in low-dimensional organic materials. The charge blockade effect, i.e., the difficulty of charge injection from one conducting segment into another, becomes significant, when the temperature is low and the conducting segment has poor electrical capacity. If the hopping barrier is enough low and the charge transport pass only consists of the high capacitive conductive segments, charges will easily hop into subsequent segment. In the thick wire and three-dimensional PEDOT:PSS wire, such hopping transport passes survives even at lower temperatures. However, in quasi one-dimensional thin fiber, the electrical capacitive conductive segments might be basically poor and also the multiplicity of the pass would be low. Thus, the thermal activated hopping conduction vanishes and CB charge transport emerges at lower temperature in the thin fiber. By such considering both the charge blockade effect and the influence of structural disorder, it is hoped that a clear understanding of charge transport in organic materials can be achieved.Fitting by CB model

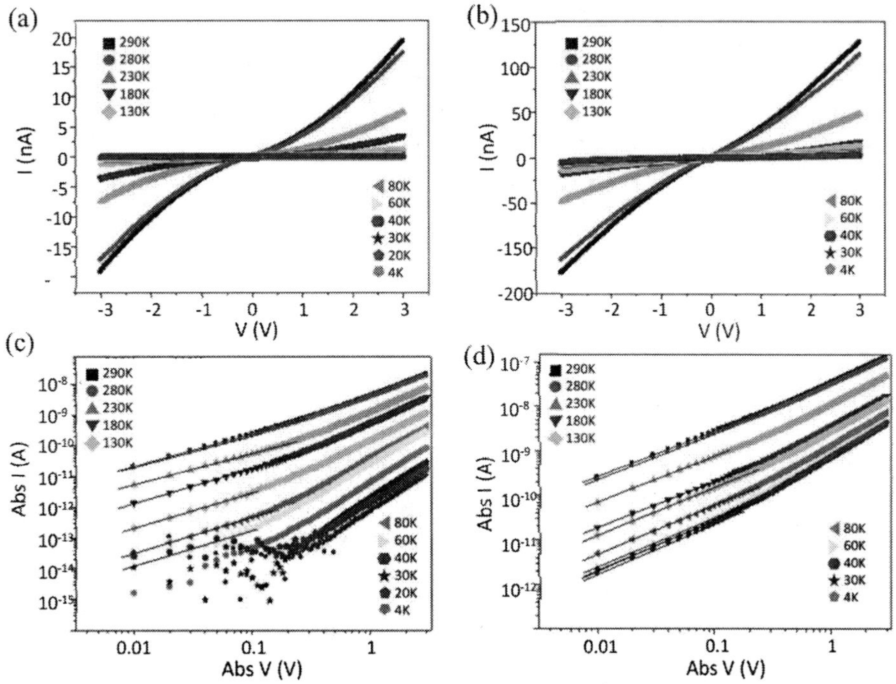

Figure 2. *I-V* characteristics observed for (a) thin wire and (b) thick wire from RT to 4.2 K. (c) and (d) *I-V* curves of (a) and (b) shown on log-log scale, respectively. The solid black lines indicate linear conductance increase at lower voltage region.

The nonlinear CB current with a clear threshold voltage that shifts to higher values with decreasing temperature can be well fitted using a charge transport model based on a 2D CB array,

$$I = \alpha(V - V_T)^{\xi} \qquad (1)$$

where α and exponent ξ are coefficients, and V_T is threshold voltage, where V_T has the temperature dependence

$$V_T(T) = V_T(0) \times (1 - \beta T) \qquad (2).$$

The CB parameters ξ and $V_T(0)$ can be estimated from the fitting results. In fact, curve fitting was carried out from low temperature to RT using the expression,

$$I = aV^b + \alpha(V - V_T)^{\xi} \qquad (3),$$

where a and exponent b are coefficients also.

We perform data fitting by using Eq. (3) for all *I-V*s observed in th PEDOT:PSS fiber shown in Fig. 2(c). Through the fitting, the following conditions are applied:

A) ξ should be independent of temperature;
B) *α* is better to be a constant especially at temperatures below *T**;
C) Above *T**, the coefficient *b* should be 1,

IOP Conf. Series: Materials Science and Engineering **835** (2020) 012017 doi:10.1088/1757-899X/835/1/012017

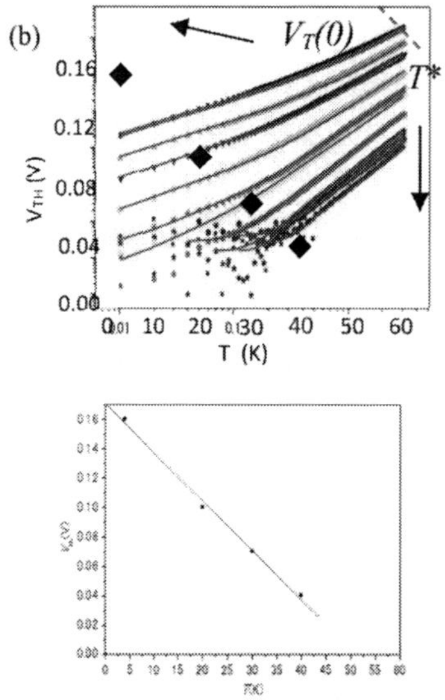

Figure 3. (a) Fitting curves for *I-Vs* and (b) V_T-*T* plot.

here the T^* is critical temperature for the blockade effect. Figure 3 shows fitting curves overlayed on the *I-Vs* and resulted V_T-*T* plot. The fitting parameters are denoted in

Table 1. This sample has ξ of 2.2, which is smaller than that of polythiophene two-dimensional CB network [12]. We empirically know that the smaller value of ξ is observed in better conductor, whereas the larger value is observed in material with lower conductance. The $V_T(0)$ of 0.16 and T^* 51K are derived from the linear regression line in the V_T-*T* plot as shown in Fig. 3(b). The relatively small values ξ and T^*suggest that the PEDOT:PSS is basically good conductor in an organic conductor.

Table 1. Fitting parameters

T (K)	*a*	*b*	α	V_T	ξ
290	2.3×10^{-9}	1	1.4×10^{-9}	-	2.2
280	2×10^{-9}	1	1.1×10^{-9}	-	2.2
230	5.5×10^{-10}	1	5.3×10^{-10}	-	2.2
180	1.5×10^{-10}	1	2.9×10^{-10}	-	2.2
130	2.2×10^{-11}	1	1.1×10^{-10}	-	2.2
80	2.8×10^{-12}	1	3.7×10^{-11}	-	2.2
60	8×10^{-13}	1	2.3×10^{-11}	-	2.2
40	4×10^{-14}	0.1	7.7×10^{-12}	0.04	2.2
30	2×10^{-14}	0.1	2.6×10^{-12}	0.07	2.2
20	5×10^{-14}	0.1	1.6×10^{-12}	0.1	2.2
4	3×10^{-14}	0.1	1.0×10^{-12}	0.16	2.2

In order to observe CB conduction in an inorganic system, it is necessary to prepare the quantum dots, which diameter almost less than 15 nm [11]. Meanwhile, conductive segments in poorly conductive organic materials are expected to have a smaller electrical capacity, leading to a higher T* for the blockade effect. Harnessing the nonlinear response in complex networking is now actively investigated in the exciting future challenge for physical implementation of informational processing [14]. The blockade effect emerges in the organic complex structure might be utilized in signal processing system interconnecting to the electrical circuit and biological systems, so on.

IOP Conf. Series: Materials Science and Engineering **835** (2020) 012017 doi:10.1088/1757-899X/835/1/012017

4. Conclusion

Charge transport through quasi one-dimensional conductive polymer, PEDOT:PSS fibers, were investigated. The typical CB transport feature were observed for the thin polymer fiber, whereas the thicker fiber showed only hopping conductance as we expected. It is demonstrated that the control of the structural dimensionality provides different charge transport regime. In 3D materials i.e, bulks and thicker films or thicker wires, CB transport should be difficult to be observed. Because due to its high variability of the conduction paths, the thermal activating hopping paths still keep its connection at low temperatures. Nevertheless, low-dimensional structure narrows down their conductive latitude, thus the hopping pass would be disconnected . Localization by disorder is a common cause of insulating behaviour of low-dimensional electron systems. The presence of disorder in organic materials will obscure the emergence of a distinguishable conduction threshold, although ironically, such disorder is a requirement for percolative transport. By considering both the charge blockade effect and the influence of structural disorder, it is hoped that a clear understanding of charge transport in organic materials and expanding to applications can be achieved.

References

[1] A. N. Aleshin, H. J. Lee, Y.W. Park, and K. Akagi, Phys. Rev. Lett. 93, 196601 (2004).
[2] A. Rahman and M. K. Sanyal, J. Phys. Condens. Matter 22, 175301 (2010).
[3] Z. Zhou, K. Xiao, R. Jin, D. Mandrus, J. Tao, D. Geohegan, and S. Pennycook, Appl. Phys. Lett. 90, 193115 (2007)..
[4] J. D. Yuen, R. Menon, N. E. Coates, E. B. Namdas, S. Cho, S.T. Hannahs, D. Moses, and A. J. Heeger, Nature Mater. 8, 572 (2009).
[5] A. J. Kronemeijer, E. H. Huisman, I. Katsouras, P. A. van Hal, T. C. T. Geuns, P. W. M. Blom, S. J. van der Molen, and D. M. de Leeuw, Phys. Rev. Lett. 105, 156604 (2010).
[6] J. H. Worne, J. E. Anthony, and D. Natelson, Appl. Phys. Lett. 96, 053308 (2010).
[7] R. Parthasarathy, X.-M. Lin and H. M. Jaeger, Phys. Rev. Lett. **87**, 186807 (2001).
[8] C. Reichhardt and C. J. Olson Reichhardt, Phys. Rev. B 68, 163305 (2003).
[9] H. Fan, K. Yang, D. M. Boye, T. Sigmon, K. J. Malloy, H. Xu, G. P. López, C. J. Brinker, Science 304, 567 (2004).
[10] D. Greshnykh, A Frömsdorf, H.Weller and C. Kilinke, Nano Lett. 9, 473 (2009).
[11] K. Xu, L. Qin, J. R. Heath, Nat. Nanotechnol. 4, 368 (2009).
[12] M. Akai-Kasaya, Y. Okuaki, S. Nagano, T. Mitani and Y. Kuwahara, Phys. Rev. Lett. 115, 196801 (2015).
[13] A. A. Middleton and N. S. Wingreen, Phys. Rev. Lett. 71, 3198 (1993).
[14] N. Aleshin, H. J. Lee, S. H. Jhang, H. S. Kim, K. Akagi and Y.W. Park, Phys. Rev. B 72, 153202 (2005).
[15] Y. Hirano,Y. Segawa, T. Kawai and T. Matsumoto, J. Phys. Chem. C 117, 140 (2013).
[16] P. S.Thapa, *et al.*, Appl. Phys. Lett. **94**, 33104, (2009).
[17] S. K. Bose, C. P. Lawrence, Z. Liu, K. S. Makarenko, R. M. J. van Damme, H. J. Broersma and W. G. van der Wiel, Nature Nanotechnol. **10**, 1048, (2015).

IOP Conf. Series: Materials Science and Engineering **835** (2020) 012018 doi:10.1088/1757-899X/835/1/012018

Prediction of emission energy of Cr^{3+} in oxides based on first-principles calculations and machine learning

Tsubasa Hori, Kazuyoshi Ogasawara

School of Science and Techonology, Kwansei Gakuin University, 2-1 Gakuen Sanda, Hyogo 669-1337 Japan

fye08548@kwansei.ac.jp

Abstract. The first-principles calculations of the energies of the emission level (^2E_g) of Cr^{3+} in 10 oxide crystals were performed using the first-principles configuration-interaction calculations. In order to improve the accuracy of the prediction, a machine learning model was created by using the results of the first-principles calculations and the experimental ^2E_g energies as the training data. The predicted values using this model showed good correlation with the experimental values, where the correlation coefficient is 0.92. The obtained predicted model indicated that the Cr 3d component of t_{2g} molecular orbital is the most important quantity for prediction of the ^2E_g energy.

1. Introduction

Transition-metal ions with d^3 electronic configuration such as Cr^{3+} and Mn^{4+} are utilized as emission centers in various luminescent materials such as ruby (Cr^{3+}-doped alumina) [1] or alexandrite (Cr^{3+}-doped chrysoberyl)[2]. Since it is difficult to investigate the multiplet energy levels of Cr^{3+} in a wide variety of crystals experimentally, the theoretical prediction of the emission energy is indispensable for the efficient development of novel Cr^{3+}-doped luminescent materials. In the high field environment, the emission of octahedrally coordinated Cr^{3+} occurs from ^2E_g state. Since both this state and the ground state belong to the same $(t_{2g})^3$ electronic configuration, the Stokes shift due to the structural relaxation in the excited state is negligibly small. Therefore the ^2E_g energy calculated in the ground state can be regarded as the emission energy. In this study, we predicted the multiplet energy of the emission level (^2E_g) of Cr^{3+} in oxides by first-principles calculations based on the discrete variational multi-electron (DVME) method [3] using relatively small clusters consisting of seven atoms. In order to improve the accuracy of the prediction, we also performed a machine learning modeling using the Waikato environment for knowledge analysis (WEKA) [4].

2. Computational Methods

2.1. *First-principles calculation*

We performed the first-principles calculations of the multiplet energies of Cr^{3+} in oxides using the discrete variational multi-electron (DVME) method [3], which is a configuration-interaction (CI) calculation program based on the discrete-variational Xα (DV-Xα) molecular orbital (MO) method. We considered 10 oxide crystals listed in **Table 1**. In these crystals the impurity Cr^{3+} ions occupy the Al sites coordinated with six O^{2-} ions. Therefore, we constructed 7-atom clusters centred at Al^{3+} ion based on the crystal structures of the host crystals [6-15]. Then the model clusters for Cr^{3+} in oxides were constructed by substituting Cr^{3+} for Al^{3+}. The space group, lattice constants, the symmetry of the model clusters are summarized in **Table 1**.

Table 1. Crystal structure data considered in this work and the symmetry of the model cluster.

Host crystal	Space group	Lattce constant / Å	Symmetry(7 atomic cluster)
Al_2O_3 [6]	$R\bar{3}cH$	a=b=4.7589, c=12.9919	C_3
Al_2SiO_5 [7]	Pnmn	a=7.719, b=7.587, c=5.533	C_2
$Y_3Al_5O_{12}$ [8]	$Ia\bar{3}d$	a=b=c=12.004	C_i
$YAl_3(BO_3)_4$ [9]	R32H	a=b=9.295, c=7.243	C_i
$YAlO_3$ [10]	Pbnm	a=5.172, b=5.327, c=7.361	C_i
$GdAl_3(BO_3)_4$ [11]	R32H	a=b=9.031, c=7.257	C_i
$MgAl_2O_4$ [12]	$Fd\bar{3}m$	a=b=c=8.081	C_i
$CsAl(MoO_4)_2$ [13]	$P\bar{3}m1$	a=b=5.551, c=8.037	D_{3d}
$KAl(MoO_4)_2$ [14]	$P\bar{3}m1$	a=b=5.545, c=7.070	D_{3d}
$NaMgAl(MoO_4)_5$ [15]	$P\bar{1}$	a=6.874, b=6.931, c=17.575	C_i

2.2. Machine learning

The machine learning was performed using the software developed in the University of Waikato known as WEKA (Waikato Environment for Knowledge Analysis) [31]. The predictive model of ^2E_g energy was created based on the linear regression algorithm. As the attributes, the theoretical ^2E_g energy obtained by the first-principles calculations, the crystal field splitting $10Dq$, the Cr 3d compositions of the t_{2g} or e_g MOs were considered.

3. Results

The correlation between the theoretical ^2E_g energy obtained by the first-principles calculation and the experimental ones is shown in **Figure 1**. The correlation coefficient is 0.52 and the theoretical values are significantly overestimated. The main cause of the poor correlation is the small size of the model clusters while the main cause of the overestimation is underestimation of the electron correlation due to the finite number of the Slater determinants in the CI calculation. The accuracy of the first-principles calculation can be improved by using larger clusters and introducing some corrections such as configuration-dependent correction (CDC) and correlation correction (CC) [3]. However, in this work, we tried to improve the accuracy of the prediction by creating a machine learning model.

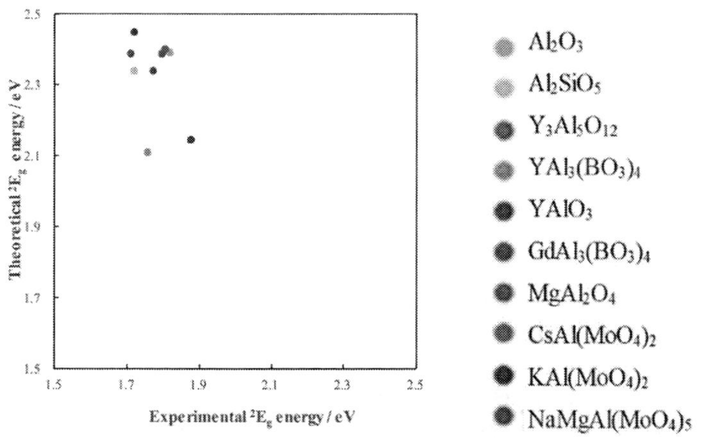

Figure 1. Correlation between the theoretical ^2E_g energies of Cr^{3+} in oxides obtained by the first-principles calculations and the experimental ones [16].

Using the experimental ^2E_g energies of Cr^{3+} in oxides [16] as the training data, we obtained the predictive model of the ^2E_g energies of Cr^{3+} in oxides (y) as,

$$y = 1.2983a - 8.7732b - 1.6185c + 8.5079, \tag{1}$$

where a is the theoretical ^2E_g energy obtained by the first-principles calculation, b and c are the Cr 3d orbital component of t_{2g} and e_g MOs, respectively. In this model, $10Dq$ is not included since consideration of $10Dq$ did not improve the accuracy of the prediction. The correlation between the predicted ^2E_g energy using Eq (1) and the experimental ones [16] is shown in **Figure 2**. In this case, the correlation coefficient is 0.92. Therefore, the correlation is significantly improved compared to the simple CI calculations using small clusters.

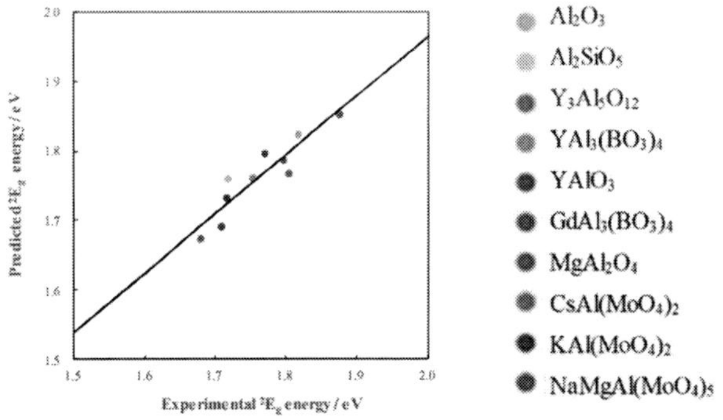

Figure 2. Correlation between the predicted ^2E_g energies of Cr^{3+} in oxides using the machine learning model and the experimental ones [16].

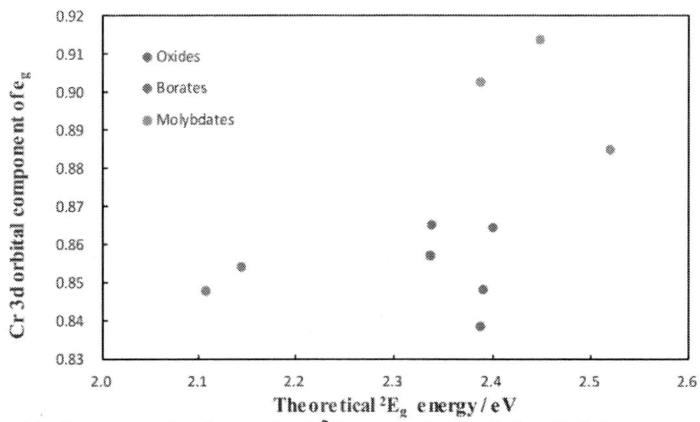

Figure 3. Relationship between the theoretical ^2E_g energies and the Cr 3d components of e_g orbitals obtained by the first-principles calculations.

4. Discussion

In order to investigate the physical meaning of Eq. (1), we have investigated the correlation between the theoretical ^2E_g energy and the Cr 3d orbital component of the MOs mainly consisting of Cr 3d orbitals. The relationship between the theoretical ^2E_g energy and the Cr 3d orbital component of e_g MO is shown in Figure 3. Although the correlation is not good, it should be noted that the data could be classified into three groups, namely, oxides, borates, and molybdates. They clearly appear in different regions in this map. This fact indicates that the Cr 3d orbital component of e_g MO is a good quantity to correct the computational errors depending on the group. The relationship between the theoretical ^2E_g energy and the Cr 3d orbital component of t_{2g} MO is also shown in Figure 4. There is a relatively high correlation between these quantities. This is due to the fact that the ^2E_g energy is mainly determined by the Coulomb repulsion among electrons within the $(t_{2g})^3$ configuration. In Eq. (1), the coefficient of the Cr 3d orbital component of t_{2g} MO (*b*) is much larger than that of the theoretical ^2E energy (*a*), indicating that it is the better quantity to predict the actual ^2E_g energy.

5. Conclusion

First-principles calculations of ^2E_g energy of Cr^{3+} in 10 oxides were calculated based on simple CI calculations using small clusters. The correlation coefficient between the calculated ^2E_g energies and the experimental ones was 0.52, indicating poor correlation. In order to improve the accuracy of the prediction, we created a machine learning model of the ^2E_g energy, using the results of the first-principles calculations and the experimental ^2E_g energies as the training data. The correlation coefficient between the predicted ^2E_g energies using this model and the experimental ones is 0.92, indicating significantly improved predictive capability. The obtained predictive model indicated that the Cr 3d component of t_{2g} MO is the most important quantity to predict the actual ^2E_g energy while the Cr 3d component of e_g MO is also effective for corrections of computational errors depending on the groups of the host crystals.

Acknowledgement
Supports from Kwansei Gakuin University are gratefully acknowledged.

References

[1] T. H. Maiman, *Nature*, **187**, 493 (1960).
[2] J. C. Walling, O. G. Peterson, H. P. Jenssen, R. C. Morris, and E. Wayne O'Dell, *IEEE J. Quantum Electron.*, **16**, 1302 (1980).
[3] K.Ogasawara, T.Iwata, Y.Koyama, T.Ishii, I.Tanaka, and H.Adachi, *Phys. Rev.* B **61**, 143 (2000).
[4] I.H. Witten, E. Frank, M.A. Hall, and C.J. Pal, *Data Mining: Practical Machine Learning Tools and Techniques* (Elsevier, Amsterdam, 2000).
[5] H.Adachi, M.Tsukada, C.Satoko, *J. Phys. Soc. Jpn.*, **45**, 875 (1978).
[6] G. F. Imbush and R. Kogelman, *Optical Spectroscopy of Electronic Centers in Solids*, in: W. A. Yen and P. M. Selzer (Eds.), *Laser spectroscopy of solids, Top. Appl. Phys. Vol. 49* (Springer, Berlin, 1981).
[7] J.Qin,C.Rudowicz,Y. M.Chang,andY. Y.Yeung,Phys. *Chem. Minerals*,**21**,532 (1994)
[8] D.-P. Ma and J.-R. Chen, *Commun. Theor. Phys.*, **44**, 165 (2005)
[9] M. G. Brik, A. Majchrowski, L. Jaroszewicz, A. Wojciechowski, and I. V. Kityk, *Philos. Mag.* B, **90**, 4569 (2010)
[10] M. Sugiyama, T. Yanagida, D. Totsuka, Y. Yokota, Y. Futami, Y. Fujimoto, and A. Yoshikawa, *J. Cryst. Growth*, **362**, 157 (2013)
[11] G. Wang, H. G. Gallagher, T. P. J. Han, and B. Henderson, *J. Cryst. Growth*, **183**, 272 (1996)
[12] A. R. Molla, C. R. Kesavulu, R. P. S. Chakradhar, A. Tarafder, S. K. Mohanty, J. L. Rao, B. Karmakar, and S. K. Biswas, *J. Alloys Compds.*, **583**, 498 (2014).
[13] G. J. Wang, L. Z. Zhang, Z. B. Lin, and G. F. Wang, *J. Alloys Compds.* **489**, 293 (2010).
[14] G. J. Wang, X. F. Long, L. Z. Zhang, G. F. Wang, S. Polosan, and T. Tsuboi, *J. Lumin.*, **128**, 1556 (2008).
[15] K. Hermanowicz, *J. Lumin.*, **109**, 9 (2004).
[16] M. G. Brik, S. J. Camardello, A. M. Srivastava, N. M. Avram and A. Suchockic , *ECS J. Solid State Sci.*, **5** (1) R3067-R3077 (2016)

IOP Conf. Series: Materials Science and Engineering **835** (2020) 012019 doi:10.1088/1757-899X/835/1/012019

Considerarion of Superatomic Hibrid Orbital in superatom M@Al_{12} (M = Be, Mg, Ca, Sr)

Teruyuki Toba ・ Masaya Yamasaki ・ Tomohiko Ishii ・ Genta Sakane

Department of Advanced Materials Science, Faculty of Engineering, Kagawa University

s18g573@stu.kagawa-u.ac.jp

Abstract. We introduce a superatom, which is regarded as a modern alchemy and has been studied around the world. The superatom is a nanocluster composed of multiple atoms, for example Al_{13}. Very interestingly, the "superatom" can mimic the chemical properties of "atom", for example, the superatom Al_{13} exhibits the character of halogen atom, such as Cl. If this superatom is synthesized to an appropriate size freely, there is the possibility to be placed as an "artificial atom" in a three-dimensional periodic table, such as the conventional two-dimensional periodic table. In recent years, the exhaustion of rare metals has become a big problem, so that in the next generation of materials and devices, the superatom play as an important role, and it is expected to be used for various materials or as a substitute for expensive metal catalysts such as Pt. From these points of view, the superatom have been studied all over the world up to now, but there is not much studies on their electronic structures. Therefore, in this study, we have analyzed the electronic states of the chalcogen-encapsulated type superatom M@Al_{12} (M = Be, Mg, Ca, Sr) by means of DV-Xα molecular orbital calculation. Our research revealed that the electronic configuration of the superatom, which changed only the central atom, depends largely on the kind of doped atom. And we can also confirm the "superatomic hybrid orbitals (SAHOs)" composed of superatomic orbitals (SAOs), and chalcogen characteristics.

1. Introduction

A nanocluster composed of fewer metal atoms and stably present, and having chemical properties like atoms is called a superatom. If these can be freely synthesis, it can be arranged as a three-dimensional periodic table, such as the conventional two-dimensional periodic table. A superatom that is structured to an appropriate size and synthesized to mimic the atomic chemistry can be considered an "artificial element". If we can perform such as the "modern alchemy" that produces the superatom having the same chemical character of an expensive metals from cheaper metals, we may be able to explore the new era of materials science.

The most famous example of the superatom is Al_{13} having an icosahedral structure (I_h), and it is well known that has the chemical properties similar to a halogen atom such as Cl atom. The main reason for having the same chemical property is explained by both "atomic orbital" and "electron configuration". In the case of an atom, it has atomic orbitals and electron configurations can be determined such as $1s^2$, $2s^2$, $2p^6$, … (written by small letter). In contrast, superatom have orbitals and the special configurations called "superatomic orbitals(SAOs)" according to the jellium model, such as $1S^2$, $1P^6$, $1D^{10}$, $2S^2$, $1F^{14}$, … (written by capital letter).

IOP Conf. Series: Materials Science and Engineering **835** (2020) 012019 doi:10.1088/1757-899X/835/1/012019

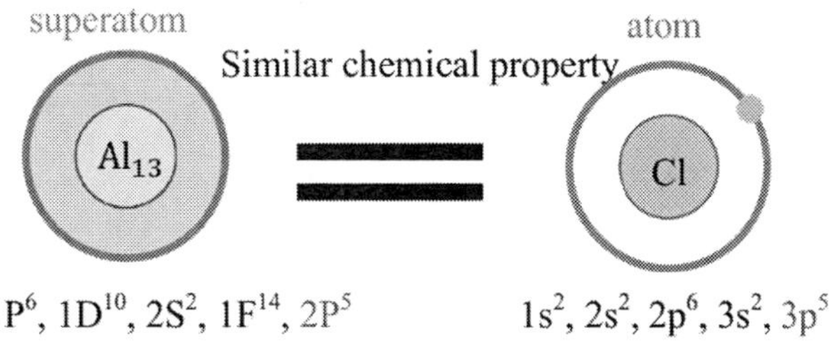

Fig.1. electronic configurations of Al_{13} and Cl.

For example, the III-group atoms like Al have 3 electrons as a valance electros. They are all trivalent atoms and their 13 atom cluster have, totally, 39 valence electrons. The electronic configurations of the superatom Al_{13} and Cl atom are exhibited in Fig. 1. In the figure, both the superatom Al_{13} and Cl atom have five electrons each in the outermost shells 2P and 3p orbitals. This indicates that obtaining a single electron results in a stable structure as a closed shell structure $2P^6$ or $3p^6$, respectively. Mainly from these points, the superatom is considered to have the chemical properties of the halogen atom Cl.

2. Computational details

We determined the coordinates of the calculation cluster models from the experimental results of a bond lengths and the molecular structures from past papers, and calculated the electronic state calculations by means of the DV-Xα molecular orbital method. Then, we conformed the spatial extent of the wave functions using the application "VESTA". The calculated cluster models M@Al_{12} (M = Be, Mg, Ca, Sr) used for the calculation is shown in Fig. 2. The molecular structural data of the most famous superatom Al_{13} has been employed. So this models have an icosahedral structure in which the distance between Al-Al is 2.80 Å, whereas the distance between M-Al is 2.66 Å, respectively. In this study, we replace only the central metal of this calculation model with M and compare the electronic states among them.

Fig. 2. Model of M@Al_{12} (M = Be, Mg, Ca, Sr).

3. Result and discussion

First, the result of the energy levels of Be@Al_{12} obtained by an electronic state calculation are shown in Figs. 3 and 4 (Fig. 4 is partially enlarged of Fig. 3). In these figures, the horizontal and vertical axes are the number of molecular orbitals and, the energy levels, respectively. From the result of Fig. 4, it is revealed that the typical energy level dispersion expected in the case of the superatom can be confirmed after around molecular orbital number #62. These molecular orbitals after #60 can be assigned the SAOs composed of valence electrons from Al and Be.

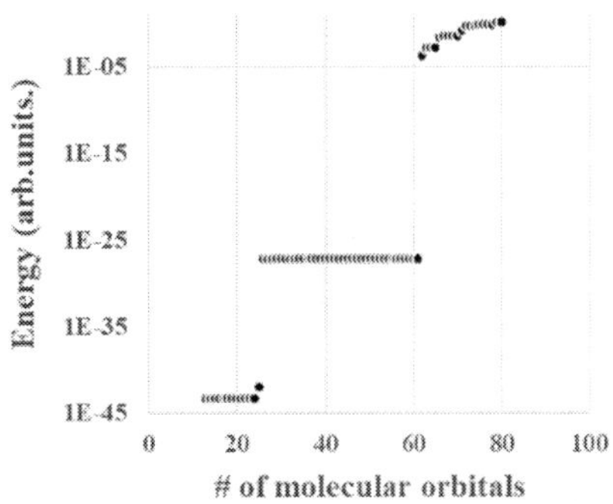

Fig. 3. Energy level of Be@Al_{12}.

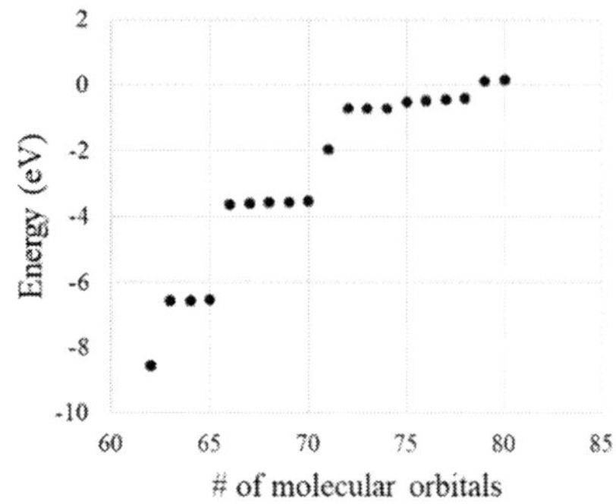

Fig. 4. Energy level of Be@Al_{12} (#62 ~ #80).

Then, we confirm the 3D spatial distribution of the wave functions of SAOs using VESTA. As a result of substituting with Be into Al_{13}, the SAOs and the electron configuration can be assigned as $1S^2$, $1P^6$, $1D^{10}$, $2S^2$, $1F^{14}$, $2P^4$ which completely follows the jellium model. Similarly, results of substituting the central metal of Mg, Ca, and Sr into Al_{13} are summarized in Fig. 5.

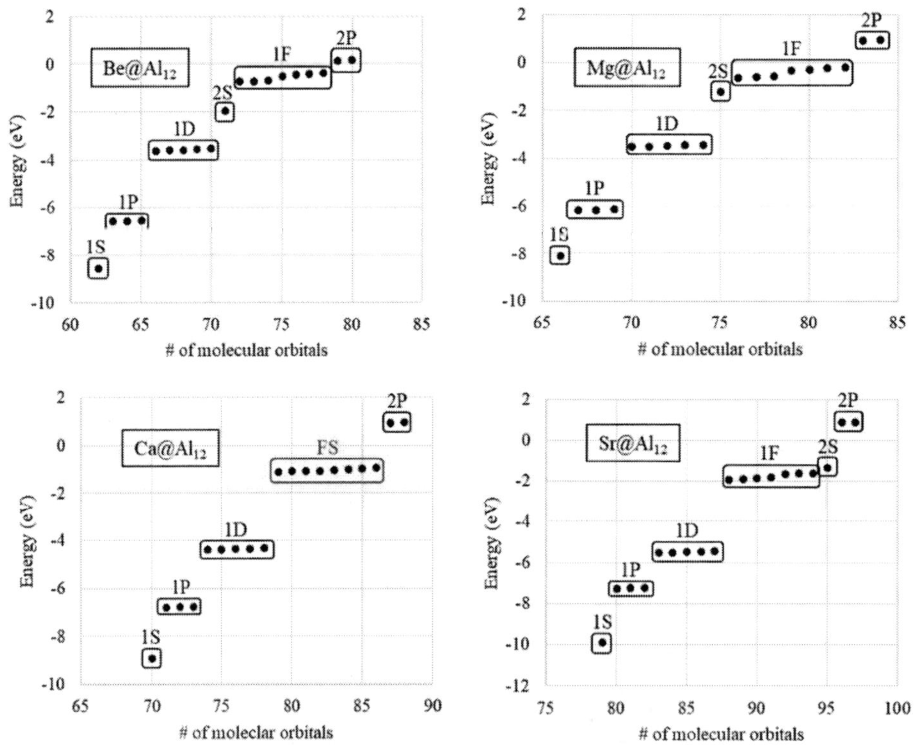

Fig. 5. Energy level of M@Al_{12} (M = Be, Mg, Ca, Sr)

In the case of Mg substituting, the electron configuration is assigned as $1S^2$, $1P^6$, $1D^{10}$, $2S^2$, $1F^{14}$, $2P^4$ follow the jellium model as well as the previous result of Be substituting. Next, in the case of Ca substitutin, we could confirm a significant phenomenon of the energy level structure, in which eight 1F orbitals appeared instead of 2S orbital. The images of the energy level and SAOs of Ca@Al_{12} are shown in Fig. 6.

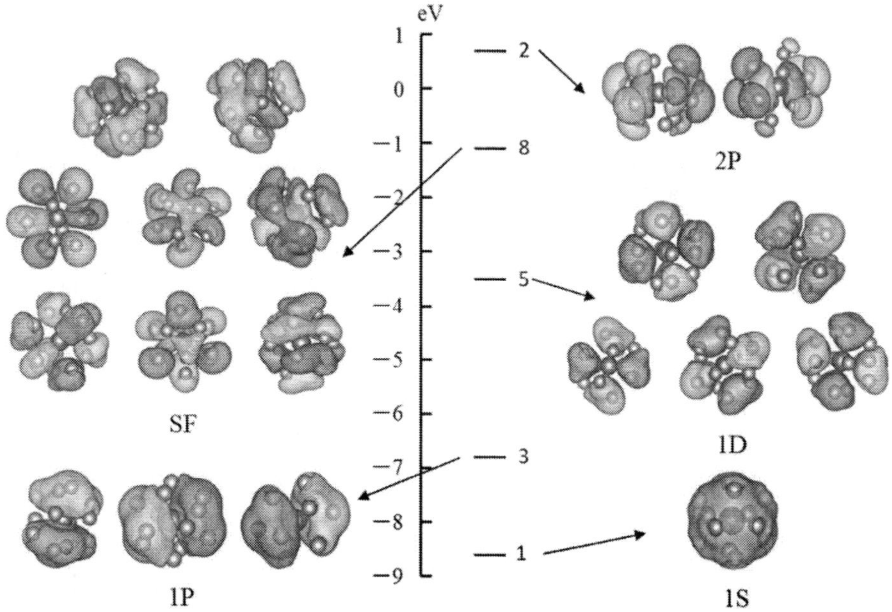

Fig. 6. superatomic orbitals and energy level of the Ca@Al$_{12}$. For each level of Ca@Al$_{12}$, the degeneracy has been marked. At the bottom of each picture, the electronic shell structure of the cluster is displayed.

For explaining the origin of these SAOs, Fig. 7 shows the Mulliken population analysis which show the component of each SAO. Generally, in the case of 2S, the 4s orbital of the Ca atom is expected to be appeared, whereas in this result the 4s orbitals has been dispersed into the molecular orbital number 83, 84, 85 and 86. From this result, the eight 1F orbitals can be realized as a result of the hybridized orbital between the 2S and 1F orbitals. Therefore, the SAOs consists of 1F and 2S of the Ca@Al$_{12}$ are named as SF superatomic hybrid orbitals (SAHOs), and the electron configuration is assigned as 1S^2, 1P^6, 1D^{10}, SF^{16}, 2P^4 in the result of Ca@Al$_{12}$. Finally, in the case of doping Sr, the order of 2S and 1F were switched. However, as well as in the case of doping Be and Mg, seven 1F and one 2S SAOs could be confirmed, and the electron configuration was assigned as 1S^2, 1P^6, 1D^{10}, 1F^{14}, 2S^2, 2P^4. Comparing the energy levels in Fig. 7, it can be revealed that the energy gap between 2S and 1F SAOs decreases as the atomic number of the central atom increase. When Be or Mg is doped, the 2S orbital has lower energy than that of 1F orbitals. But in case of Sr is substituted, the 2S orbital is higher than that of 1F orbitals. This is considered to be affected by the size of the s orbital between small s orbital such as Be and Mg atoms and large s orbital such as Sr atom. Therefore, when Ca with medium size is substituted, the energy of 1F and 2S orbitals are considered to form a

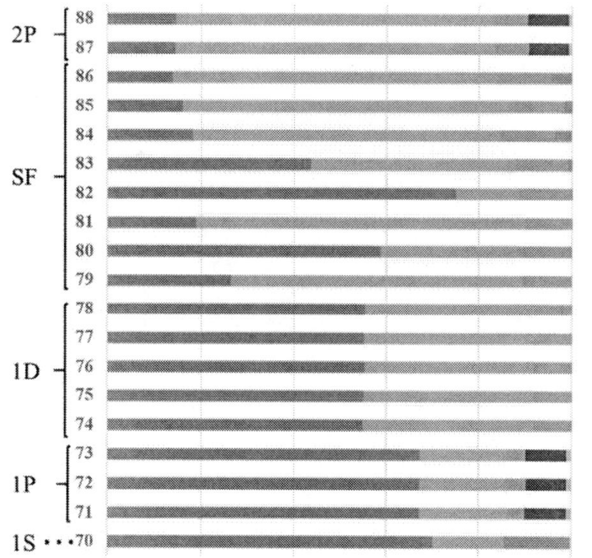

Fig. 7. Mulliken population analysis of Ca@Al$_{12}$.

hybridized orbitals (that is SF SAHO). In addition, all electron configurations of $M@Al_{12}$ calculated in this work are occupied up to the $2P^4$, and it is possible to achieve a closed shell structure by acquiring two more electrons. This indicates that these $M@Al_{12}$ superatoms have the possibility of ability to mimic the chalcogen atoms.

4. Conclusion

We clarified the electronic states such as superatomic orbitals and electronic configurations for $M@Al_{12}$ (M = Be, Mg, Ca, Sr) which changed only the central metal. We confirmed the SAOs according to the jellium model, especially in the cases of the Be and Mg substituted $M@Al_{12}$. However, when Ca was substituted, the significant SF hybridized orbitals consisting of 2S and 1F orbitals were confirmed. And it was also found that novel these superatoms could be a mimic of chalcogen atoms due to the relationship of electron configuration. In the future, it is necessary to discuss the SAHO in other various superatoms.

References

[1] Bergeron, D. E.; Roach, P. J.; Castleman, A. W., Jr.; Jones, N. O.; Khanna, S. N. (2005) "*Al Cluster Superatoms as Halogens in Polyhalides and as Alkaline Earths in Iodide Salts*" Science, 307 231−235.

[2] Sun, Wei-Ming; Wu, Di; Li, Xiang-Hui; Li, Ying; Chen, Jing-Hua; Li, Chun-Yan; Liu, Jia-Yuan; Li, Zhi-Ru (2016) "*Quasi-Chalcogen Characteristics of $Al_{12}Be$: A New Member of the Three Dimensional Periodic Table*" Journal of Physical Chemistry, 120 (4), 2464-2471.

[3] Jena, Puru (2013) "*Beyond the Periodic Table of Elements: The Role of Superatoms*" Journal of Physical Chemistry Letters, 4 (9), 1432-1442.

IOP Conf. Series: Materials Science and Engineering **835** (2020) 012020 doi:10.1088/1757-899X/835/1/012020

Mechanism of Optical Rotation of Amino Acids Using Electronic State Calculation

Shuichi Katsumoto,[1] Manami Mimura,[1] Tomohiko Ishii,[1] Genta Sakane,[2] and Kazuyoshi Ogasawara[3]

[1]Program of Advanced Materials Science, Department of Engineering and Design, Faculty of Engineering and Design, Kagawa University, 2217-20 Hayashi-cho, Takamatsu, Kagawa 761-0396, JAPAN.
[2]Department of Chemistry, Faculty of Science, Okayama University of Science, 1-1, Ridaicho, Kita-ku, Okayama 700-0005, JAPAN.
[3]School of Science and Technology, Kwansei Gakuin University, 2-1 Gakuen Sanda, Hyogo 669-1337, JAPAN.

s19g558@stu.kagawa-u.ac.jp

Abstract. Because of the chiral nature of the building blocks of living matter, an optical phenomena associated with the chirality constitute an important topic in physical chemistry. The specific optical rotation, which is a parameter for the characterization of the natural optical activity, depends strongly on the conformation of the molecule. In this study, we investigated the dependence of the optical rotation on a molecular conformation in the gas phase by calculating the electronic states of seven kind of chiral amino acids using the DV-Xα method. As a result, we can confirm the existence of an antibonding orbital on the side chain and the optical rotations are strongly related.

1. Introduction

Optical rotation is a physical property that causes linearly polarized light to rotate left or right when it passes through a substance, and many biological substances such as proteins and sugars exhibit the optical rotation. Furthermore, enantiomers such as L-alanine and D-alanine have the property that most of the physical and chemical properties are the same, except for the sign of the optical rotation.

On the other hand, with regard to biological properties, one enantiomer may be a drug and the other a poison such as thalidomide. Therefore, identification of the enantiomers is an important factor, and optical rotation is a key to determine the conformation of a chiral molecule. However, there are few theoretical studies of the optical rotation mechanism, and it is hard to say that everything is clear.

Therefore, in this study, it is assumed optical rotation is caused by spirality of charge density located on asymmetric carbon, and electronic state calculation was performed on amino acids that show optical rotation while having a simple molecular structure. After that, comparison with experimental values was performed.

2. Method

In this study, a software "Chem3D" was used to create a computational model of 15 kinds of amino acids with only one asymmetric carbon. Based on the calculation model, we used the DV-Xα molecular orbital method to calculate the BOP of the bond to the asymmetric carbon and the overlap of the bonding orbitals and the antibonding orbitals. Here, since the spin direction is reversed due to the presence of the antibonding orbital, the spirality of the space charge density is considered to be reversed, and this is used as an index of optical rotation. From the calculation results obtained, the areas of overlap of all antibonding orbitals around HOMO, LUMO, and less than HOMO were plotted and compared in order of experimental optical rotation.

3. Results and Discussion

Table 1 shows the experimental values of the optical rotation ($[\alpha]_D$) of amino acids calculated in this work and the calculated BOP values. Next, from the results obtained from the overlap population diagram, the area of the antibonding orbital around the HOMO is extracted and shown in Figure 1. Similarly, Figure 2 shows the results at around LUMO, and Figure 3 shows all overlaps results below HOMO. The horizontal axis of each graph plots the experimental value of an optical rotation, and the vertical axis plots the area of each corresponding antibonding orbital.

According to the result in Table 1, it can be seen the values of BOP at the asymmetric carbon of each amino acid take similar values. From this, it seems difficult to explain the factor of different optical rotation by the value of BOP.

Next, according to the result in Figure 1, around LUMO, antibonding orbitals exist for hydrogen and carboxyl groups, but there is almost no antibonding orbital for side chains and amino groups. It can be seen the contribution to the side chain of the antibonding orbital due to the change in the side chain of the amino acid is small, affecting others.

In Figure 2, the antibonding orbital of the amino group which didn't influence in Figure 1 is largely shown around LUMO. As a result of component analysis, it is found that the antibonding orbital appears around the HOMO of the amino group due to the 2s orbital of the amino group and the 2p

IOP Conf. Series: Materials Science and Engineering **835** (2020) 012020 doi:10.1088/1757-899X/835/1/012020

orbital of the asymmetric carbon. There is no side chain effect on this, and it is thought an antibonding orbital always exists between the amino group and the asymmetric carbon around the HOMO of the amino acid.

Finally, as shown in Figure 3, it is found the overlap of antibonding orbitals is divided into four regions shown by the highlighted ellipsoid, in the order of hydrogen, side chain, carboxyl group and amino group. Furthermore, as the dextrorotatory property increases, the overlap of the antibonding orbitals between the carboxyl group and the side chain increases. This is considered to be due to the fact the spirality of the charge in the right rotation is increased by the antibonding orbital. However, in this study, it couldn't confirm the switch of levorotatory to dextrorotatory or dextrorotatory to levorotatory from the change in the antibonding orbital.

Table 1. Experimental optical rotation values and BOP values of amino acids.

			BOP		
substance	−R	$[\alpha]_D$ in H_2O	N(-NH_3)	C(-COOH)	C(-R)
L-cysteine	CH_2SH	-16.5	0.693	0.759	0.733
L-leucine	$CH_2CH(CH_3)_2$	-11	0.686	0.748	0.744
L-methionine	$CH_2CH_2SCH_3$	-10	0.690	0.758	0.738
L-serine	CH_2OH	-7.5	0.701	0.756	0.743
L-asparagine	CH_2CONH_2	-5.6	0.700	0.759	0.741
L-alanine	CH_3	+1.8	0.676	0.754	0.753
L-norleucine	$CH_2CH_2CH_2CH_3$	+4.7	0.689	0.751	0.736
L-aspartic acid	CH_2COOH	+5.05	0.689	0.760	0.735
L-valine	$CH(CH_3)_2$	+5.6	0.692	0.758	0.738
L-glutamine	$CH_2CH_2CONH_2$	+6.3	0.684	0.763	0.733
L-norvaline	$CH_2CH_2CH_3$	+7.0	0.698	0.765	0.740
L-amino butyric acid	CH_2CH_3	+9.3	0.700	0.765	0.735
L-glutamic acid	CH_2CH_2COOH	12	0.698	0.750	0.735
L-ornithine	$CH_2CH_2CH_2NH_2$	12.1	0.698	0.750	0.739
L-arginine	$CH_2CH_2CH_2NHC(NH)NH_2$	12.5	0.679	0.751	0.735

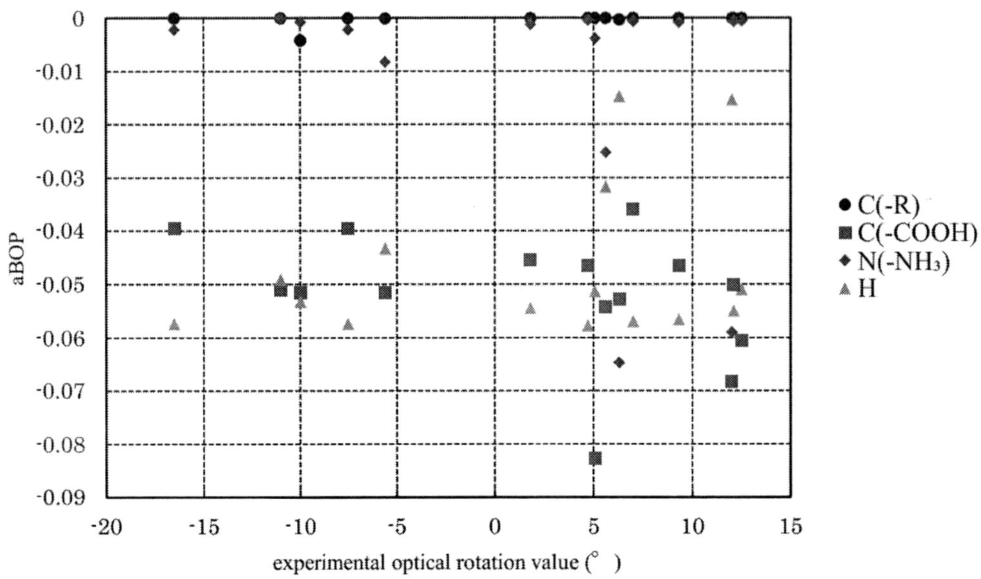

Figure 1. Relationship between the optical rotation value and the antibonding orbital around the HOMO.

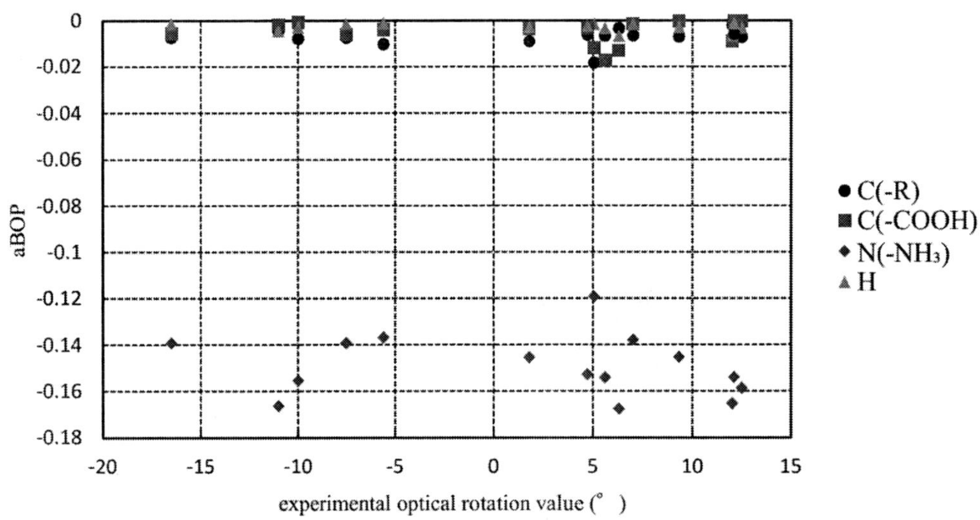

Figure 2. Relationship between the optical rotation value and the antibonding orbital around the LUMO.

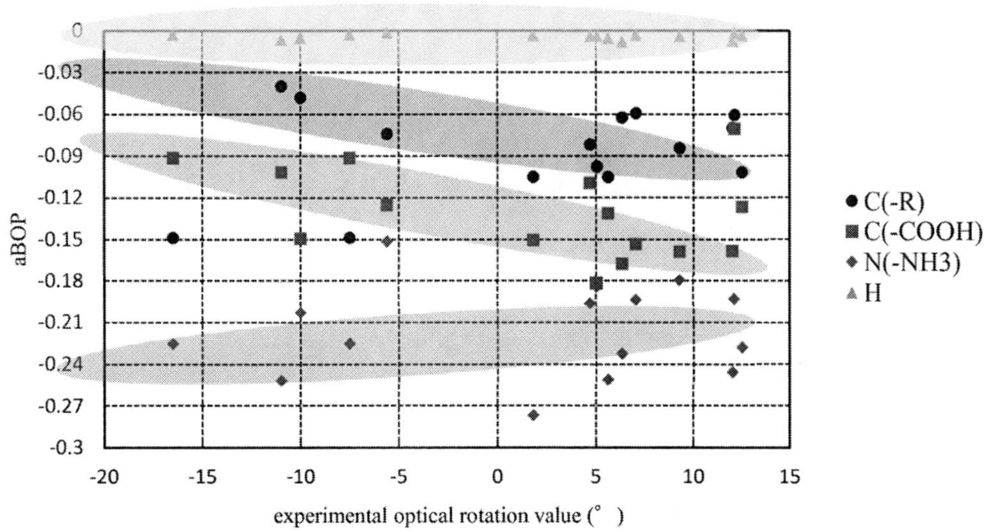

Figure 3. Relationship between the optical rotation value and the antibonding orbital below HOMO.

4. Conclusion

In this study, in order to clarify the mechanism of optical rotation, we calculated the electronic states of amino acids having only one asymmetric carbon, confirmed the BOP and each overlap population diagrams, and confirmed their relationship with optical rotation. From the calculation results, it was possible to confirm the regular change in the antibonding orbital by the change in the side chain of amino acids. In the future, we will consider the influence of the size of the bonding orbital, and calculate the case where the amino acid is zwitterion. In addition, we'll examine whether the effect on physical properties can predict according to the calculation by collecting data.

References

[1] Magdalena Pecul, "Conformational Effects on the Optical Rotation of Alanine and Proline", *Journal of Physical Chemistry A*, **108**, 4269-4276, 2004.

[2] Shokouh Haghdani, "Optical Rotation from Coupled Cluster and Density Functional Theory: The Role of Basis Set Convergence", *Journal of Chemical Theory and Computation*, **12, 2**, 535-548, 2016.

IOP Conf. Series: Materials Science and Engineering **835** (2020) 012021 doi:10.1088/1757-899X/835/1/012021

Generation of two-dimensional spectrochemical series in tetrahedral complexes

Takuhiro Yamamoto,[1] Takeyoshi Oguma,[1] Yuki Tsutumi,[1] Shino Tsuboi,[1] Tomohiko Ishii,[1]

[1] Department of Advanced Materials Science, Graduate School of Engineering, Kagawa University . JAPAN

E-mail: s18g586@stu.kagawa-u.ac.jp

Abstract. The physical properties of metal complexes that exist in many places are due to the ligand field splitting and electronic configuration of the d-orbital of the central transition metal, so changing the ligand field splitting of the d-orbital allows to control the physical properties. The arrangement of metals and ligands in order of the ligand field splitting is called a spectrochemical series. There is already a spectrochemical series only in the case of octahedral hexa-coordination complexes, but no equivalent index is shown for any other symmetry such as tetrahedral tetra-coordination complexes. In this study, we have investigated the creation of spectrochemical series in tetrahedral tetra-coordinated metal complexes. As a result, it is possible to accurately grasp the ligand field splitting of a tetrahedral tetra-coordinated metal complex, and to more effectively apply energy for expressing a spin crossover phenomenon in which the magnetism reversibly changes.

1. Introduction

A metal complex is a generic name for compounds in which a ligand present as an anion or neutral molecule is coordinated to a centrally located metal ion (which may be neutral or anionic), and is used for photosynthesis and production of plants or haemoglobin in the body is essential to our lives. The metal complex is located between the inorganic compound and the organic compound because it has the element diversity / functionality of the inorganic compound, the excellent molecular property of the organic compound, and the designability. And the physical properties of the metal complex originate from the ligand field splitting and the electron configuration of the d-orbital of the central transition metal. That is, its physical properties can be freely controlled by changing the ligand field splitting of the d-orbital. The electronic structure of the d-orbital varies depending on the coordination number and symmetry of the complex, but in the case of d^3 to d^6 tetrahedral tetra-coordinated complex, a plurality of spins such as a high spin state and a low spin state exists. The reversibility of the spin state interchanges the magnetism of the metal complex. This phenomenon is called spin crossover (SCO).

Spectrochemical series and Tanabe-Sugano diagrams (TSD) are indicators of ligand field splitting and electron configuration. The spectrochemical series is a sequence in which the ligand and the metal ion are arranged in the order of the magnitude of the energy difference of the d-d transition of the metal complex. Although it only holds in the case of the combination of a specific metal and ligand of an octahedral hexa-coordination complex, it is well shown that the ligand field splitting of a metal complex depends on the kind of ligand. The spectrochemical series of octahedral hexa-coordination complexes are shown below.

(Metal): $Mn^{2+} < Ni^{2+} < Co^{2+} < Fe^{2+} < Fe^{3+} < Co^{3+} < Rh^{3+} < Ru^{2+} < Ru^{3+} < Pd^{4+} < Ir^{3+} < Pt^{4+}$
(Ligand): $I^- < Br^- < S^{2-} < SCN^- < Cl^- < NO_3^- < F^- < OH^- < ox^{2-} < H_2O < NCS^- < CH_3CN < NH_3 <$ en $<$bpy $<$ phen $< NO_2^- < PPh_3 < CN^- < CO$

On the other hand, the Tanabe-Sugano diagram shows the relationship between ligand field splitting and multiplet energy of the d-orbitals, and is known as an indicator of the spin crossover. From these two indexes, it is considered that controlling of ligand field splitting and spin crossover of the metal complex, that is, controlling of magnetism is possible. However, in the tetra-coordinated complex, neither index equivalent to octahedral hexa-coordinated has been shown yet. In this study, we aimed to create the spectrochemical series and the Tanabe-Sugano diagram of tetrahedral complex, to understand the ligand field splitting correctly and to apply energy for expressing the spin crossover phenomenon more effectively. In this DV-Xα conference, we will report on the spectrochemical series of tetrahedral metal complexes.

2. Research procedure

In this research, first create a metal complex model to be calculated, then calculate the ligand field splitting of the 3d orbital of the central metal ion by the DV-Xα method, then make a two-dimensional spectrochemical series to predict ligand field splitting at a glance. However, the number of reported coordination data of the tetrahedral metal complexes is small, so the number of models that can be created is small simply by examining the metal-ligand bond distance from the Cambridge Structural Database (CSD) and applying the average to the metal complex model to be calculated. Therefore, for complexes that cannot be modelled from CSD, modelling is carried out using two methods: Method of distort a planar four-coordinated complex into a tetrahedral shape and Method of finding regularity of bond distance by comparing each complex. I will explain it in detail.

2.1 (Method to distort planar four-coordinated complex into tetrahedral form)
We searched for a planar four-coordination model in CSD, and created a regular tetrahedral complex model using distortion parameter τ_4 (Fig. 1) which can express the degree of distortion of four-coordination complex with the bond angle without changing the bond distance.

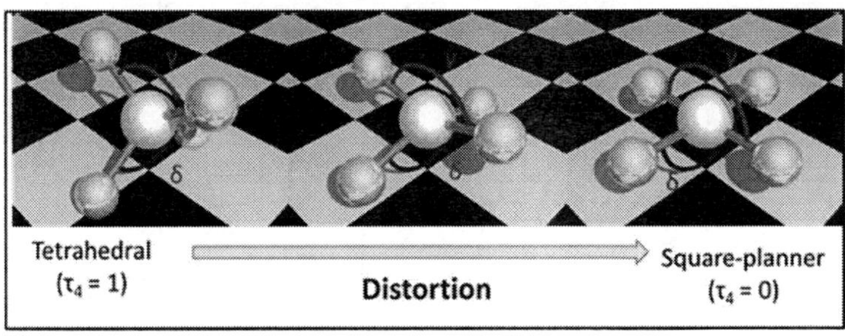

$$\tau_4 = \frac{360° - \gamma - \delta}{141°}$$

Figure.1 Distortion parameters τ_4.

2.2 (Model creation by comparison)
There are two ways to make a model by finding the regularity of bond distance by comparison. Comparison between tetrahedral complex and comparison with octahedral hexa-coordinates. First, we will show an example of model creation by comparison with a tetrahedron complex in Table 1. I will find the bond distance of Pd^{4+}-I comparing with the same d^6 metal complex. The product average of the bonding distance was 1.0795 when the ligand changes from bromine to iodine in each metal complex. In order to obtain the bonding distance of Pd^{4+}-I, 2.3550 was multiplied by 1.0795.

Table 1. Bonding distance in each complex

L \ M	Pd^{4+}	Co^{3+}	Rh^{3+}	Pt^{4+}	Ru^{2+}	Ir^{3+}
L	4 Coordination	4 Coordination	4 Coordination	4 Coordination	4 Coordination	4 Coordination
Br	2.3550	2.3184	2.4206	2.3761	2.4609	2.3969 ×1.0868
I	?	2.5123	2.6021	2.5777	2.6270	2.6050

Subsequently, we will show an example of model creation by comparison with an octahedral six-coordinate complex in Table 2 we will find the bond distance of Rh^{3+}- CO comparing with the same d^6 metal complex. The product average of the bonding distance was 0.9745 when the structure changes to tetrahedral complex from octahedral six-coordinate complex whose ligand is a carbonyl group. In order to obtain the binding distance of Rh^{3+}- CO, 1.88925 was multiplied by 0.9745 to obtain the binding distance.

Table 2. Bonding distance in each complex.

L \ M	Fe2+		Ru2+		Rh3+	
L	4 coordination	6 cordination	4 coordination	6 coordination	4 coordination	6 coordination
CO	1.777	1.825	1.816	1.862	?	1.88925

×0.9737 ×0.9753 ×0.9745

Average=0.9745

3. Result

A tetrahedral complex model created from a combination of 12 metals and 13 ligands is shown in Table 3.

Table 3. Combination of 12kinds of metal with 13 kind of ligand in CSD.

	I^-	Br^-	S^{2-}	SCN^-	Cl^-	NO_3^-	F^-	OH^-	H_2O	NCS^-	NO_2^-	CN^-	CO
Mn^{2+}	1	1	×	×	1	×	2	×	×	1	×	1	×
Ni^{2+}	1	1	×	×	1	×	×	×	3	1	3	3	1
Co^{2+}	1	1	×	×	1	×	×	×	3	1	×	3	×
Fe^{2+}	1	1	×	×	1	×	2	×	×	1	×	3	1
Fe^{3+}	1	1	×	×	1	×	4	×	×	1	×	×	×
Co^{3+}	4	4	×	×	1	×	2	×	×	×	×	4	×
Rh^{3+}	4	4	×	×	4	×	2	×	×	×	×	4	4
Ru^{2+}	4	4	×	×	4	×	4	×	×	×	×	4	1
Ru^{3+}	2	4	×	×	4	×	4	×	×	×	×	×	×
Pd^{4+}	2	4	×	×	4	×	2	×	×	×	×	×	×
Ir^{3+}	4	4	×	×	4	×	2	×	×	×	×	4	4
Pt^{4+}	4	4	×	×	4	×	4	×	×	×	3	4	×

1 : Model created as it is from CSD 2 : Model created by comparison between tetrahedral complex
3 : Model created by distorting a plane tetra-coordinate into a tetrahedron
4 : Model created by comparison with octahedral hexa-coordinates

The two-dimensional spectrochemical series created this work is shown in Figure 2. It was found that the color changed continuously from red (small) to black (large) as the numbers of the metal series and the ligand series increased. It was also found that the spectrochemical series of the conventional octahedral hexa-coordination complex tends to hold for tetrahedral complexes. By using this figure, we think it is possible to predict the magnitude of ligand field splitting to a certain extent even for combinations of metals and ligands that have not been calculated in this work.

IOP Conf. Series: Materials Science and Engineering **835** (2020) 012021 doi:10.1088/1757-899X/835/1/012021

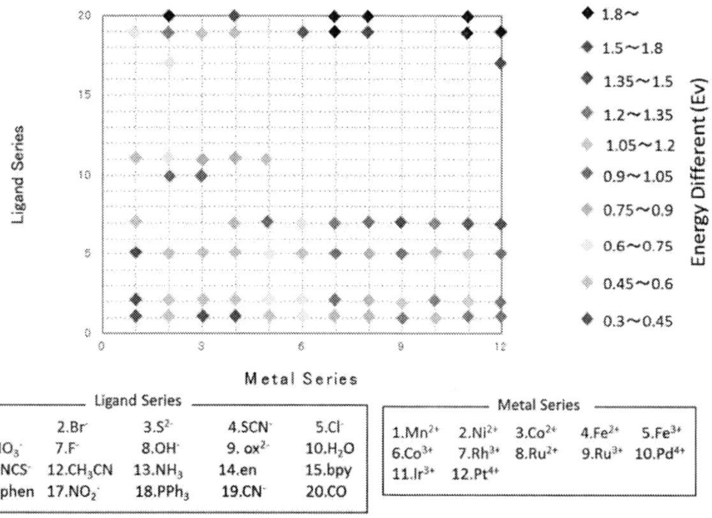

Figure.2 Two-dimensional spectrochemical series in four-coordinated metal complex.

4. Conclusion

In tetrahedral metal complexes, one electron calculation by DV-Xα molecular orbital method was performed to calculate ligand field splitting. As a result, for the combination of 12 kinds of metals and 13 kinds of ligands, the ligand field splitting of the tetrahedral structure complex was created as a two-dimensional spectrochemical series. At that time, it has been found that the spectrochemical series of tetrahedral structure complex also holds to some extent in the conventional octahedral hexa-coordinated complex. Therefore, we think that the system design by two-dimensional visualization of spectrochemical series has made it easier to discuss the design of functional materials. In the future, it is thought that new complexes will continue to be synthesized through research activities that are constantly advancing. Therefore, by registering new metal complexes in CSD, it is possible to add combinations of metals and ligands that could not be modeled in this study. As future research, we will work on the creation of the Tanabe-Sugano diagram for each metal complex created in this work. Thereby, the tendency of spin crossover in each metal complex can be shown.

Reference

[1] Shino Tsuboi (2007) "An electronic study of ligand field splitting and physical properties of metal complexes". Kagawa University master's thesis.

Effect of Molecular Distortion of Ligand Field Splitting in Five-Coordinated Metal Complex

Yoshiki Fujikawa,[1] Dan Tsuneda,[1] Yuki Tsutsumi,[1] Momoe Nakano,[1] Takeyoshi Oguma,[1] Tomohiko Ishii,[1] Genta Sakane,[2] And Kazuyoshi Ogasawara[3]

[1]Program of Advanced Materials Science, Department of Engineering and Design, Faculty of Engineering and Design, Kagawa University, 2217-20 Hayashi-cho, Takamatsu, Kagawa 761-0396, JAPAN.
[2]Department of Chemistry, Faculty of Science, Okayama University of Science, 1-1 Ridaicho, Kita-Ku, Okayama 700-0005, Japan
[3]School of Science and Technology, Kwansei Gakuin University, 2-1 Gakuen Sanda, Hyogo 669-1337, JAPAN.

s19g572@stu.kagawa-u.ac.jp

Abstract. Metal complex indicates the significant physical properties such as magnetism, electrical conductivity and an optical property due to the electronic interaction between the ligand molecule and the central metal ion. We have evaluated the magnetism of the metal complex by controlling the ligand field splitting, which is the energy splitting of the d-orbitals. The magnetic phenomenon of such compounds are realized as a result of the electronic state based on the transition metal coordination complexes, and which can be switched between high spin (HS) and low spin (LS), that is called a spin-crossover (SCO) phenomenon. Our group have been investigated the way how to control the spin state according to the concept of a molecular distortion. The five-coordinated metal complex ML_5 has been experimentally confirmed to be a molecular distortion called the Berry pseudorotation (BPR). In the BPR process, both the axial and equatorial ligands can move at the same rate of increasing the angle between the other axial or equatorial ligands in five-coordinated metal complex. The classical BPR mechanism changes its molecular symmetry between two independent trigonal bipyramidals (TBPs) of D_{3h} symmetry via a square pyramidal (SP) of C_{4v} symmetry. The distortion parameters in the five-coordinated metal complexes τ_5 are proposed in order to represent the distortion of the BPR process quantitatively. In this work, we predicted the possibility of the SCO phenomenon in intramolecular exchange mechanism through the BPR process in a five-coordinated metal complexes.

1. Introduction

In recent years, there are a wide variety of metal complexes, and in addition to the significant functional properties (such as magnetism, conductivity and optical property) often found in inorganic compounds, they have significant molecularness and design of organic compounds. Because of this, a number of studies are underway. As factors to determine the physical properties of metal complexes, it is necessary to consider ligand field splitting and electron configuration, which are the energy splits among five d-orbitals of the central metal ion. This ligand field splitting differs depending on the central metal and the ligand, and this changes the spin state, and it is possible to take two states of high spin or low spin. A phenomenon in which the spin states are interchanged by applying external field stimulation (for example, heat, magnetic field, pressure or light) if the energy difference between the two spin states is slightly is called a spin-crossover (SCO) phenomenon. Therefore, our group have been investigated the way how to control the spin state according to the concept of a molecular distortion. In five-coordinate complexes, an intramolecular transfer phenomenon called Berry pseudorotation (BPR) has been reported experimentally (Scheme 1). The classical BPR mechanism changes its molecular symmetry between two independent trigonal bipyramidals (TBPs) of D_{3h} symmetry via a square pyramidal (SP) of C_{4v} symmetry. The distortion parameters in the five-coordinated metal complex τ_5 are proposed in order to represent the distortion of the BPR process quantitatively. In TBP, $\tau_5 = 1$ and SP becomes $\tau_5 = 0$.

$$\tau_5 = \frac{\theta_{ax} - \theta_{eq}}{60^\circ} \quad (0 \le \tau_5 \le 1)$$

θ_{ax} : Bonding angle of ax to ax
θ_{eq} : Bonding angle of eq to eq
W. Addison *et al.*, *Dalton Trans.*, 1349 -1356 (1984).

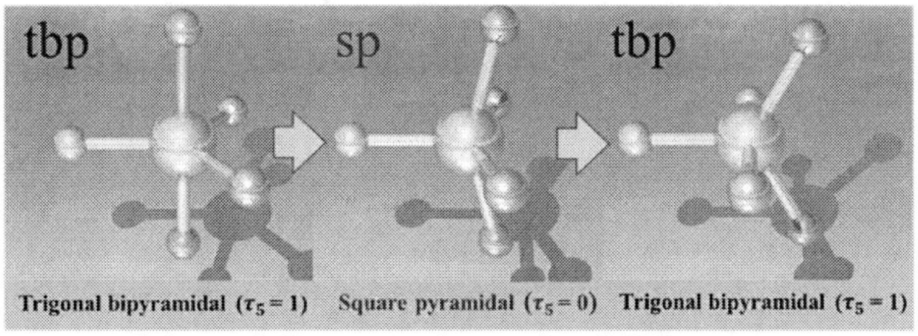

Scheme 1. Berry pseudoroutation (BPR) process.

However, In the five-coordinated complex where all the ligands are equivalent ML_5, the SCO couldn't confirmed in BPR process. Therefore, in this work, in the monosubstituted five-coordinate complex ML_4L' in which one of the ligands is substituted, the change of the spin state can be confirmed, and the spin state can be controlled from the bulkiness of the substituted ligand molecule.

2. Calculating Method

2.1. monosubstituted five-coordinate complex

Two patterns are possible only if the mono-substituted five-coordinate complex undergoes a structural change by Berry pseudorotation. This is distinguished by whether the central axis of the pseudo-rotation is a substituent. In the monosubstituted five-coordinates of this study, there are two patterns of pseudorotation, in which the central axis ligand is a cyano group (-CN) or a group of 15 pnictogen element substituents. In these two kinds of pseudorotations, the former is called a non-degenerate type Berry pseudorotation because the position of the ligand changes before and after the pseudo-rotation, while the latter is called a degenerate type Berry pseudorotation because the position of the ligand does not change before and after the pseudorotation (see Figure. 1).

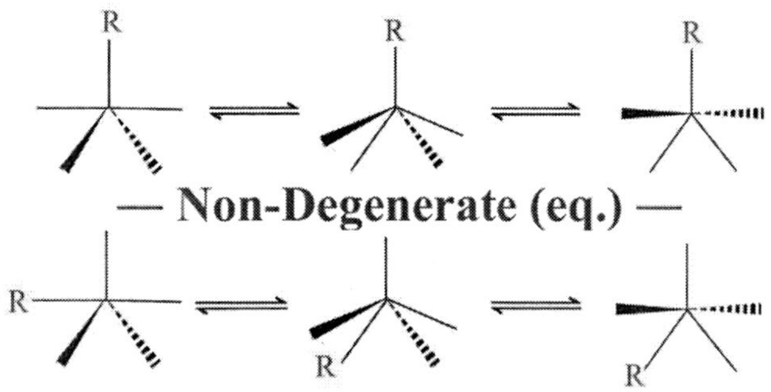

Figure 1. The Berry pseudorotation of a monosubstituted five-coordinate complexes (Upper: degenerate process, Lower: non-degenerate process)

IOP Conf. Series: Materials Science and Engineering **835** (2020) 012022 doi:10.1088/1757-899X/835/1/012022

2.2. Calculation model

The energy state and the steric barrier of the five-coordinate complex in Berry pseudorotation are different depending on the substituted ligand. In this work, a total of 15 cluster models were created, in which a pentacyano complex $[Fe^{2+}(CN^{-})_5]$ was used as the basic structure, and XH_3, XMe_3, and X $^tBu_3$ were coordinated to five kinds of group 15 pnictogen atoms (X denotes a pnictogen element, X = N ,P ,As ,Sb and Bi) respectively. In making the model, the carbonyl (-CO) complex, which was studied by Tsuneda *et al.* in our group, was used as a reference, and the atomic bond distances are shown in Table 1.

Table 1. Atomic distances of monosubstituted five-coordinate complexes.

$Fe(CN^{-})_4RX_3$ Complexes Fe – R distance (Å)			
	RH_3	RMe_3	R $^tBu_3$
R=N	2.120	2.200	2.275
R=P	2.219	2.250	2.363
R=As	2.320	2.296	2.455
R=Sb	2.420	2.490	2.547
R=Bi	2.520	2.600	2.627

3. Results and discussion

3.1 ML_5 five-coordinate complex with equivalent ligands

First, we show the results of the ligand field splitting and spin states of $[Fe^{2+}(CN^{-})_5]$ ML_5 with the equivalent ligand with no substituted. Then, the ligand field splitting Δ in the BPR process which becomes TBP again from TBP via SP is Δ = 2.329, 2.509 and 2.330 eV, respectively, and it is expected that SCO will not occur and keep the intermediate spin state (Figure. 2)

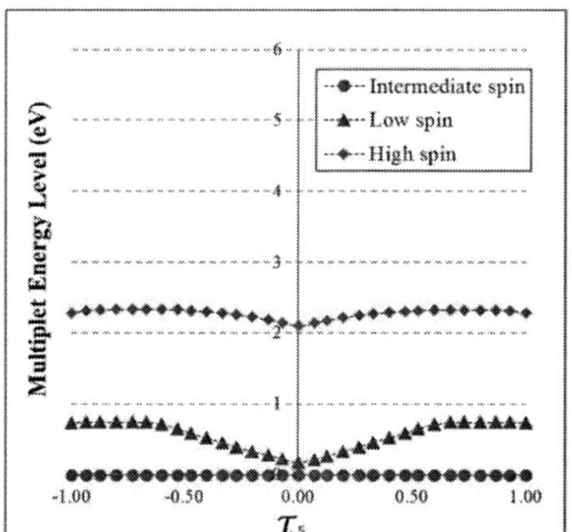

Figure 2. Multiplet energy diagram of $[Fe^{2+}(CN^{-})_5]$ ML_5.

Figure. 2 shows the multiplet energy diagram of $[Fe^{2+}(CN^{-})_5]$ ML_5 in the Belly pseudorotation process. In Figure. 2, the horizontal axis is the distortion parameter τ_5, and the vertical axis is the multiplet energy level. The spin states in the process of changing the distortion parameter from −1 to 1 were plotted by the DVME method. In this complex, by using a CN group having a very strong coordination ability as a ligand, the ligand field splitting became larger, and the spin state continued to keep the intermediate spin.

3.2. ML_4L' Monosubstituted five-coordinated complexes substituted with group 15 pnictogen.
The ligand field splitting and the spin state in the structural change of BPR in each substituent are shown in Tables 2 and 3 for the substituted five-coordinate complex ML_4L', respectively.

Table 2. Spin state and ligand field splitting in each substituent, and SCO τ_5 values (Degenerate type).

	TBP ($\tau_5 = 1$)		SP ($\tau_5 = 0$)		TBP ($\tau_5 = 1$)		
Substituent	**Spin**	**Δ(eV)**	**Spin**	**Δ(eV)**	**Spin**	**Δ(eV)**	**SCO τ_5**
NH_3	Intermediate	1.988	Intermediate	2.239	Intermediate	1.987	–
PH_3	Intermediate	2.113	Intermediate	2.689	Intermediate	2.107	–
AsH_3	Intermediate	1.987	Intermediate	2.614	Intermediate	1.984	–
SbH_3	Intermediate	2.159	Intermediate	2.652	Intermediate	2.163	–
BiH_3	Intermediate	2.112	Intermediate	2.624	Intermediate	2.115	–
NMe_3	Intermediate	1.927	Low	2.254	Intermediate	1.924	−0.47 and 0.47
PMe_3	Intermediate	2.344	Low	2.777	Intermediate	2.332	−0.33 and 0.47
$AsMe_3$	Intermediate	2.247	Low	2.722	Intermediate	2.246	−0.07 and 0.20
$SbMe_3$	Intermediate	2.367	Low	2.637	Intermediate	2.372	−0.07 and 0.13
$BiMe_3$	Intermediate	2.428	Low	2.570	Intermediate	2.392	−0.13 and 0.20
N $^tBu_3$	Intermediate	1.956	Intermediate	2.197	Intermediate	2.005	–
P $^tBu_3$	Intermediate	2.087	Low	2.603	Intermediate	2.075	−0.13 and 0.13
As $^tBu_3$	Intermediate	1.963	Intermediate	2.535	Intermediate	1.978	–
Sb $^tBu_3$	Intermediate	2.625	Intermediate	2.590	Intermediate	2.620	–
Bi $^tBu_3$	Intermediate	2.233	Intermediate	2.545	Intermediate	2.242	–

Table 3. Spin state and ligand field splitting in each substituent, and SCO τ 5 values (Non-degenerate type).

	TBP ($\tau_5 = 1$)		SP ($\tau_5 = 0$)		TBP ($\tau_5 = 1$)		
Substituent	**Spin**	**Δ(eV)**	**Spin**	**Δ(eV)**	**Spin**	**Δ(eV)**	**SCO τ_5**
NH_3	Intermediate	2.147	Low	2.320	Intermediate	2.262	0.13 and 0.33
PH_3	Intermediate	2.018	Intermediate	2.206	Intermediate	2.217	–
AsH_3	Intermediate	1.921	Intermediate	2.106	Intermediate	2.095	–
SbH_3	Intermediate	2.047	Intermediate	2.231	Intermediate	2.214	–
BiH_3	Intermediate	1.987	Intermediate	2.179	Intermediate	2.157	–
NMe_3	Intermediate	1.565	Intermediate	1.855	Intermediate	1.907	–
PMe_3	Intermediate	2.037	Intermediate	2.209	Intermediate	2.252	–
$AsMe_3$	Intermediate	1.977	Intermediate	2.152	Intermediate	2.170	–
$SbMe_3$	Intermediate	1.811	Intermediate	1.973	Intermediate	1.976	–
$BiMe_3$	Intermediate	1.597	Intermediate	1.766	Intermediate	1.711	–
N $^tBu_3$	Intermediate	1.353	Intermediate	1.685	Intermediate	2.311	–
P $^tBu_3$	Intermediate	1.867	Intermediate	2.090	Intermediate	2.055	–
As $^tBu_3$	Intermediate	1.769	Intermediate	2.015	Intermediate	1.995	–
Sb $^tBu_3$	Intermediate	1.777	Intermediate	1.979	Intermediate	1.954	–
Bi $^tBu_3$	Intermediate	1.689	Intermediate	1.889	Intermediate	1.854	–

In the axially substituted degenerate Berry pseudorotation, spin state switching (Intermediate spin → Low spin) is predicted for complexes substituted with Me_3 group. In addition, non-egenerate Berry pseudorotation substituted at equatorial position could not predict the SCO. It could be predicted that in either pattern it is easy to take the low spin state in the SP structure. This can be explained as controlling the spin state using chemical pressure caused by the steric bulk of the substituted ligand. Next, Figure 3. shows the multiplet energy diagram of each complex using the DVME method.

IOP Conf. Series: Materials Science and Engineering **835** (2020) 012022 doi:10.1088/1757-899X/835/1/012022

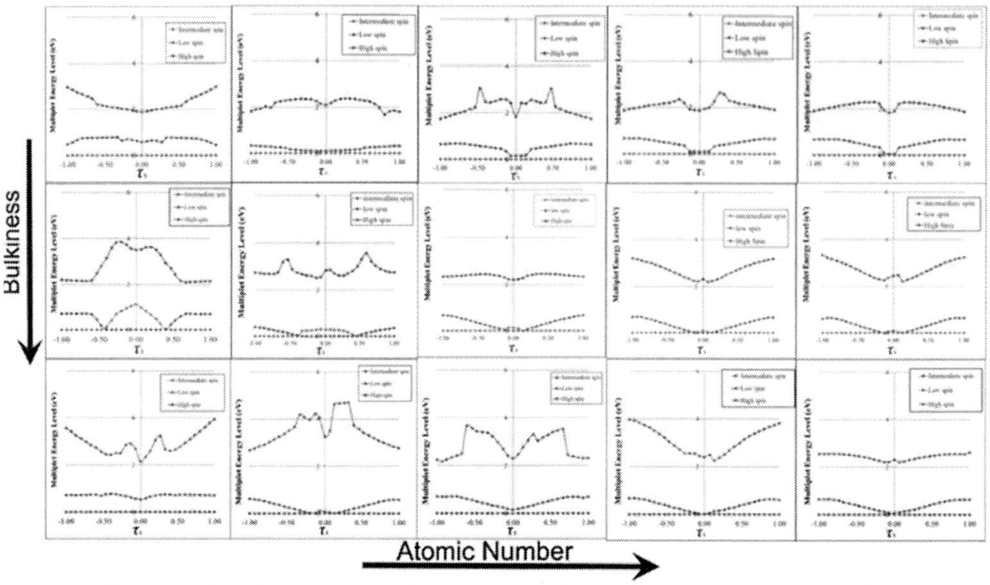

Figure 3. Multiplet energy diagram of all degenerate Berry pseudorotations.

The degenerate type Berry pseudorotations in the axial position were arranged in the order of XH_3, XMe_3, and X $^tBu_3$, and in the order of increasing atomic numbers of the group 15 punictogen atoms (X = N ,P ,As ,Sb and Bi) from the left.

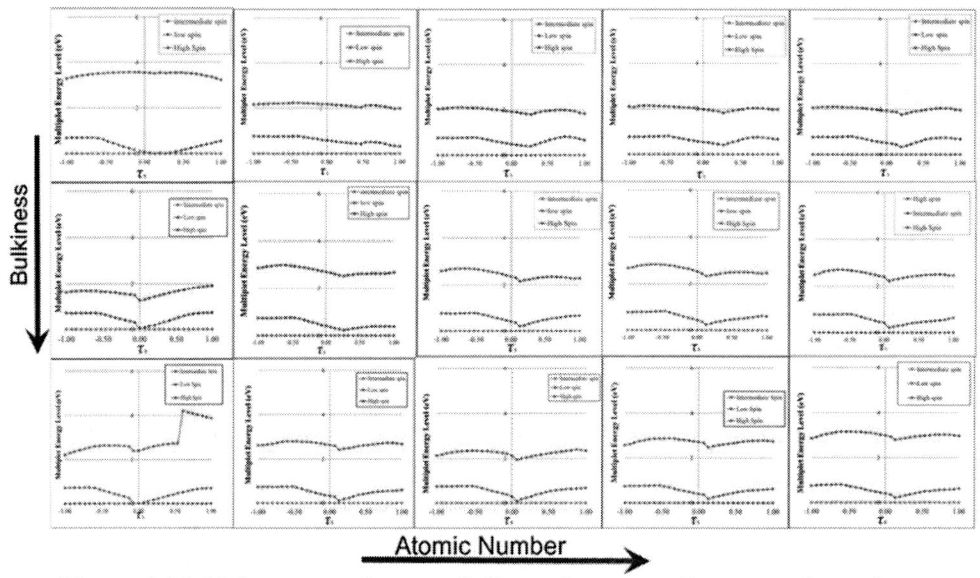

Figure 4. Multiplet energy diagram of all non-degenerate Berry pseudorotations.

Substituents containing XMe_3 may be expected to the SCO phenomenon, but even if other complexes have SP structures, it seems that SCO may be possible with a little stimulation.

4. Conclusion

The SCO phenomenon was predicted for the spin state in the Belly pseudorotation using the molecular distortion of the five-coordinate complex ML_5 with equivalent ligands and the five-coordinate complex monosubstituted ML_4L' with a group 15 pnictogen atom . From the results, it was predicted that SCO would not occur in a five-coordinate complex with an equivalent ligand ML_5, but it is possible to be easy to switch low spin at the SP structure in both degenerate and non-degenerate pathways by monosubstitution ML_4L'. This explains the control of the spin state using chemical pressure causes by the steric bulk of the ligand. We can control the spin switching from intermediate spin to low spin from the bulkiness of the ligand by molecular distortion and found the possibility of SCO.

References

[1] W. Addison et al., Dalton Trans, pp. 1349 -1356, 1984.
[2] Dan Tsuneda, graduation thesis, 2015.

Single Crystal X-ray Structural Analysis and Electronic Theory of Sugar Molecules

Norito Taniguchi[1], Tomohiko Ishii[1], Bui Quoc Huy[1], and Genta Sakane[2]

[1] Department of Advanced Materials Science, Graduate School of Engineering, Kagawa University, 2217-20 Hayashi-cho, Takamatsu, Kagawa 761-0301, Japan E-mail : s19g567@stu.kagawa-u.ac.jp, Tell : +81-87-864-2414

[2] Department of Chemistry, Faculty of Science, Okayama University of Science, 1-1 Ridaicho, Kita-ku, Okayama 700-0005, Japan

E-mail : s19g567@stu.kagawa-u.ac.jp

Abstract. Rare sugar is a monosaccharide that there are trace amounts in nature, natural sugars present in large amounts in nature world is referred to as a naturally monosaccharide. Recently, the isomerase called D-tagatose-3-epimerase (DTE) was discovered, and it has a possibility to synthesize quantity of the rare sugar from a natural sugar through the enzymatic reaction. And all of hexoses can be synthesized by four kinds of enzymatic reactions (oxidoreductase, aldose isomerase, aldose reductase and DTE reactions). In this study, single-crystal X-ray structure analysis of a sugar alcohol and its derivative have been performed in order to determine the absolute coordinates of these sugar molecules. We have also investigated the electronic state calculation by means of DV-Xα method using the obtained data of the absolute coordinates. We also consider that mechanism of the enzymatic reaction and intermolecular energy properly such as a hydrogen bond in order to synthesize the supramolecular rare sugars (SRSs) which can be controlled structures and the hydrogen bonds.

1. Introduction.

1.1. Rare sugars and supramolecular rare sugars (SRSs).

Sugar is a kind of carbohydrate and an indispensable substance for most living organisms living on the earth. Sugars are broadly classified into monosaccharides and polysaccharides. A monosaccharide is a sugar that cannot be further hydrolyzed, and a polysaccharide refers to a monosaccharide linked by a glycosidic bond. Among the polysaccharides, two monosaccharides with glycosidic bonds are called disaccharides, three monosaccharides with glycosidic bonds are called trisaccharides, and ten to twenty monosaccharides with glycosidic bonds are called oligosaccharides.

Monosaccharides are finely classified according to their carbon number and structure, have aldehyde groups or ketone groups. As shown in Fig. 1, aldoses are those with an aldehyde group attached to the end of a monosaccharide chain, and ketoses are those where the second carbon is a ketone group. Further, a product obtained by reducing these aldehyde groups and ketone groups is called a sugar alcohol. A typical example of aldose is D-Glucose, a typical example of ketose is D-Fructose, and a typical example of sugar alcohol is D-Sorbitol. Although rare sugars are present in low amounts, there are more than 50 types of derivatives. Since rare sugars are rare and very expensive, they have not been studied much, and many of their properties such as structure, optical activity, and

IOP Conf. Series: Materials Science and Engineering **835** (2020) 012023 doi:10.1088/1757-899X/835/1/012023

physiological activity remain unknown. However, in recent years, D-fructose, a natural monosaccharide, was discovered by the discovery of an enzyme that catalyzes the reversible epithelization of the third carbon of ketose, called D-tagatose-3-epimerase (DTE), by Prof. Izumori Kagawa University, that can be converted to D-psicose, a rare sugar. Furthermore, the discovery of the DTE devised Izumoring, which systematizes enzyme reactions of various monosaccharides in a ring shape, enabling the strategic synthesis of the monosaccharides. Therefore, the main focus of this study is to elucidate the physical properties of the rare sugars located on the map of the Izumoring strategy.

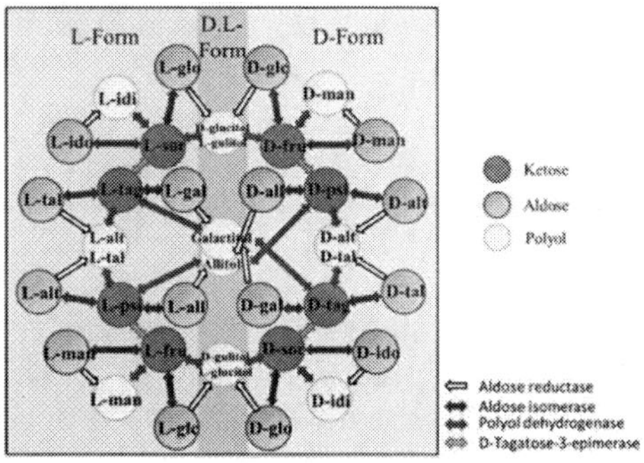

Figure. 1 Synthetic strategy of hexose by enzymatic reaction (Izumoring).

A supramolecule is a molecule in which multiple molecules are assembled in an orderly manner by bonds other than covalent bonds, that is, hydrogen bonds or hydrophobic interactions, electrostatic interactions (Coulomb interaction), van der Waals interaction, etc. The concept of the supramolecules was proposed by Jean-Marie Lehn *et al.* Initially, compounds that encapsulate molecules and ions through intermolecular interactions, such as crown ethers and cyclodextrins, were particularly studied.

Supramolecular rare sugar is an ordered arrangement of two or more different rare sugar molecules in a single unit cell, named by Professor Izumori.

In other words, supramolecular rare sugars are different from polysaccharides in which monosaccharides are glycoside-bonded molecules in which different molecules. They are regularly arranged with a relatively weak force due to hydrogen bonding. For this reason, it is easier to control the binding than polysaccharides, and it is considered that engineering applications such as an optical devices are easier.

Single crystal X-ray structural analysis of rare sugars.

In this study, crystal structure analysis is performed using an X-ray diffractometer. In 1895, X-rays were discovered by German physicist Wilhelm Conrad Röntgen, in 1912 Max Theodor Felix von Laue discovered the X-ray diffraction phenomenon. The discovery of Bragg's law by William Henry Bragg and William Lawrence Bragg established the theory of the structural analysis using X-ray diffraction. In 1916, Peter Joseph William Debye and Paul Scherrer developed the Debye-Scherrer method, which can perform structural analysis from powder crystals, and structural analysis using X-ray diffraction became active. X-rays have very strong energy and are capable of passing through materials, so they are also being used in the field of nondestructive inspection. From the 20th century, it began to be used in fields such as structural biology.

Many analytical instruments that use X-ray diffraction phenomena use Bragg's conditions. Many substances are made of extremely small crystals, and atoms and ions are regularly arranged inside the crystal to form a spatial lattice (unit cell). The spatial lattice has atoms arranged in parallel to each other, and when an X-ray with a wavelength λ is incident on the lattice plane with a surface interval d at an incident angle θ,

$$2dsin\theta = n\lambda \ (n = 1,2,3 \cdots).$$

The following relational expression holds. This is called Bragg's condition. When this conditional expression holds, X-rays interfere with each other and strengthen each other.

2. The purpose of this study.

Since research on rare sugars has recently started full-scale research, many physical properties such as the structure and properties of rare sugars are unknown. The purpose of this study is to find applicability of rare sugars in the field of materials. In order to achieve this goal, it is necessary to elucidate the chemical and physical properties of rare sugars in detail. Moreover, intermolecular interactions play an important role in designing organic materials, and the molecular structure often affects this interaction. Therefore, the molecular structure of rare sugars and the high-precision absolute coordinates of each atom are indispensable for future research on rare sugars. The purpose of this study is to investigate the mechanism of the hydrogen bonding network and optical activity of sugars by analyzing the crystal structure and electronic state of rare sugars, and to synthesis the supramolecular rare sugars.

3. Experimental Method.

In this study, a single crystal obtained by precipitating a 15 % concentration of Galactitol solution (mixed solution of 1.5 g of Galactitol and 10 mL of water) at 60 °C for 24 hours, and an Allitol solution of 35 % concentration (Allitol 1.05 g). A single crystal obtained by precipitating a mixed solution of 3.0 mL of water at 30 °C for 48 hours is subjected to crystal diffraction using an X-ray structural analyzer R-AXIS RAPID II (Rigaku Corporation). The diffraction data was subjected to structural analysis using structural analysis software (Crystal Structure). Furthermore, using the obtained crystal structural data, BOP (Bond Overlap Population), HOMO-LUMO gap, total energies of Galactitol and Allitol were calculated by using DV-Xα method. Besides we calculated four sugars (D-Fructose, D-Psicose, D-Galactose and D-Glucose) by using the obtained crystal structure data and derive these data to consider four kinds of enzymatic reactions which synthesize all of hexoses (oxidoreductase, aldose isomerase, aldose reductase and DTE reactions).

4. Results and Discussion.

4.1. Single crystal X-ray structural analysis of Galactitol and Allitol.

First, we have obtained more precise single crystal structure of Galactitol and Allitol. These polyols have no distinction between D and L forms and do not have an optical isomers. Since these molecules are located at the conversion points of D-form and L-form in the Izumoring strategy, they can be said to be important molecules among Izumoring molecules.

Figure. 2 Crystal structure of Galactitol.

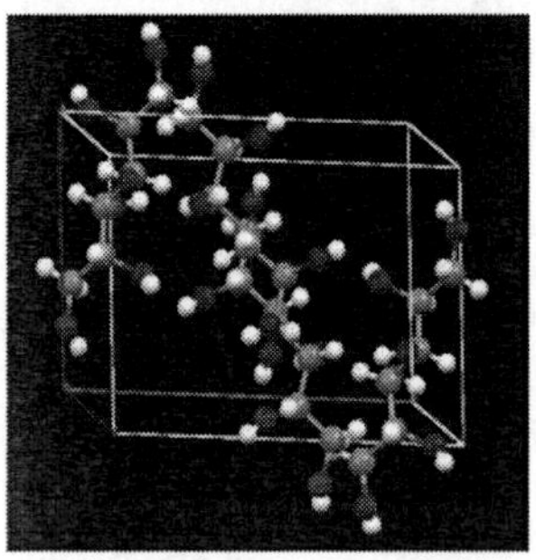

Figure. 3 Crystal structure of Allitol.

Table 1 Cell parameters for Galactitol and Allitol.

		Galactitol	Allitol
Formula		$C_6H_{14}O_6$	$C_6H_{14}O_6$
Crystal System		monoclinic	monoclinic
Space Group		$P2_1/c$	$P2_1$
Cell Length [Å]	*a*	8.4051(6)	4.6873(4)
	b	11.4534(8)	13.3361(10)
	c	9.0027(6)	6.5887(5)
Cell Angles [°]	*α*	90	90
	β	112.995(8)	100.027(7)
	γ	90	90
Cell Volume [Å^3]		797.892	405.56(5)
Z		4	2
R-Factor [%]		4.01	7.03

Figs. 2 and 3 show the crystal structures of Galactitol and Allitol obtained by the crystal structural analysis, and Table 1 shows their cell parameters. In Figs. 2 and 3, the red, green, and blue lines indicate the *a, b,* and *c*-axis of the unit cell, respectively, and each molecule is displayed in the Ball and Stick format. One molecule of Galactitol was confirmed to be present in the unit cell by four molecules by the symmetry operation of the space group $P2_1/c$. In Figs. 2 and 3, the twice (2_1 symmetric) helical axis extends in the direction parallel to the *b*-axis, and the Galactitol molecule has a projection plane perpendicular to the double helical axis, the *ab*-axis plane (*c*-axis direction) is exist parallel to the axial direction. The R values obtained from this analysis result, which are the standard for the accuracy of the structural analysis results, were 4.01 % and 7.03 %, and high-precision absolute coordinates were obtained.

4.2. Electonic state calculation of main hexoses.

Second, we calculated two polyols (Galactitol and Allitol), two ketoses (D-Fructose and D-Psicose) and two aldoses (D-Galactose and D-Glucose) by using the obtained crystal structure data (Table 2). Comparing Galactitol and Allitol, the BOP around C2 is generally larger for Galactitol, and even when comparing the symmetry and density of the space group, Galactitol is higher, and the crystallinity of the whole molecule is better than Allitol.

Comparing D-Glucose and D-Fructose, HOMO-LUMO gap is larger for D-Fructose (9.600 eV) and the BOP around C2 is roughly larger for D-Fructose. In nature, D-Glucose is predominantly present in natural monosaccharides. It is shown that the lower the value of the parameter (BOP and HOMO-LUMO gap) related to intramolecular stability, the more it can exist in nature. This is thought to be the fact that the higher the intramolecular stability, the lower the stability of the crystal, leading to a decrease in the abundance in nature.

Table 2. Structure data, BOP, HOMO-LUMO gap and Total Energy of main hexoses (Galactitol, Allitol, D-Fructose, D-Psicose, D-Galactose and D-Glucose).

	Space Group	Density (g/cm^3)	BOP C2-H	C2-C1	C2-OH	C2-C3	HOMO-LUMO (eV)	Total Energy (Hr)	Cavity (%)
Galactitol	$P2_1/c$	1.516	0.8121	0.7629	0.5728	0.7583	8.4399	-680.94	11.1
Allitol	$P2_1$	1.472	0.8080	0.7392	0.5849	0.7392	9.4880	-680.72	10.8
D-Fructose	$P2_12_12_1$	1.602	0.7945	0.7702	0.5812	0.7531	9.600	-679.49	10.8
D-Psicose	$P2_12_12_1$	1.606	0.8166	0.7459	0.5853	0.7587	9.447	-679.41	11.8
D-Galactose	$P2_12_12_1$	1.580	0.8228	0.7404	0.5821	0.7476	9.405	-679.69	12.3
D-Glucose	$P2_12_12_1$	1.575	0.7801	0.7316	0.5045	0.7536	8.864	-679.40	11.8

5. Conclusion

In the crystal structure analysis of Galactitol and Allitol, the R value was improved by succeeding in identifying the position of the hydrogen atom considering the direction of intermolecular hydrogen bonding, and the crystal structure that exists more stable than the crystal structure on the CSD Succeeded in finding out. In addition, more accurate electronic state calculations related to intermolecular interactions are possible. Galactitol and Allitol are the central molecules in Izumoring, and they have the common feature of sugar alcohols located at the conversion points of D-forms and L-forms. In addition, parameters related to the stability of molecules and crystals could be obtained by calculating the electronic structure using precise crystal structure data. The data obtained by these electronic state calculations will be the most useful means for elucidating the principles, phenomena and principles in nature.

References

[1] F.Longchambon, H.Gillier-Pandraud, R.Wiest, B.Rees, A.Mitschler, R.Feld, M.Lehmann, P.Becker, *Acta Crystallogr.Sect.B:Struct.Sci.*, **41**, 47, (1985).
[2] K.Fukada, T.Ishii, K.Tanaka, M.Yamaji, Y.Yamaoka, K.Kobashi, K.Izumori, *Bull.Chem.Soc.Jpn.*, **83**, 1193, (2010).

IOP Conf. Series: Materials Science and Engineering **835** (2020) 012024 doi:10.1088/1757-899X/835/1/012024

Cobalt Thiolates with Organic *N,N,S*-Tridentate Ligand: Conversion of Thiolate Complex to Sulfinate Complex via Sulfenate

Takanori Kotera[1], Atsushi Fujita[1], Masahiro Mikuriya[1*], Daisuke Yoshioka[1], Takashi Shibahara[2], Kichisuke Nishimoto[2], Makoto Handa[3]

[1]Department of Applied Chemistry for Environment and Research Center for Coordination Molecule-based Devices, School of Science and Technology, Kwansei Gakuin University, 2-1 Gakuen, Sanda 669-1337, Japan
[2]Department of Chemistry, Faculty of Science, Okayama University of Science, 1-1 Ridaicho, Kita-Ku, Okayama 700-0005, Japan
[3]Department of Chemistry, Graduate School of Natural Science and Technology, Shimane University, 1060 Nishikawatsu, Matsue 690-8504, Japan

junpei@kwansei.ac.jp

Abstract. Mononuclear cobalt(III) complex with 2-[(3-aminopropyl)amino]ethanethiol (Hapaet), $[Co(apaet)_2]ClO_4$ (**1**) and the sulfinate complex, $[Co(apaetO_2)_2]ClO_4$ (**2**) were synthesized and characterized by measurements of the elemental analysis, infrared and electronic spectra (UV-vis-NIR), and temperature dependence of magnetic susceptibilities (4.5—300 K), and DFT calculations. X-Ray crystal structure analysis of **1** and **2** revealed an octahedral CoS_2N_4 or $Co(SO_2)_2N_4$ chromophore with *meridional* chelation of two $apaet^-$ or $apaetO_2^-$. Conversion of **1** to the sulfinate complex **2** via the sulfenate complex was investigated by UV-vis spectra. Oxidation reaction of capsanthin with **1** was also examined.

1. Introduction

Organic thiols containing amino-nitrogen donors are useful to prepare metal thiolates [1-4]. Some *N,N,S*-tridentate thiolic ligands can be compared with the corresponding amino-alcoholic *N,N,O*-tridentate ligands which form dinuclear copper(II) species. Contrary to amino-alcoholic ligands, amino-thiolic ligands have a great affinity for many kinds of metal ions to form various metal thiolates from mononuclear to polynuclear species [5-20]. We focused on cobalt systems in the hope of obtaining new feature of metal thiolates, elucidating interesting crystal structures and spectral properties, although most metal thiolates are diamagnetic. An *N,N,S*-tridentate thiol, 2-[(3-aminopropyl)amino]ethanethiol (Hapaet), afforded mononuclear Co(III) and trinuclear $Co(II)_3$ species, $[Co(apaet)_2]ClO_4$ (**1**) and $[\{Co(apaet)_2\}_2Co](ClO_4)_2$ (**1'**), depending on the reaction condition [16]. Interestingly, the mononuclear Co(III) species underwent oxidation reaction with hydrogen peroxide to form another Co(III) species with the corresponding sulfinate ligand via the sulfenate ligand. Their conversion was monitored by UV-vis spectra and DFT calculations were performed. The oxidizing property of the Co(III) species was also tested for an orange-color dye, capsanthin.

2. Experimental

2.1. Synthesis

All the chemicals except for toluene and ethylenesulfide which were dehydrated and distilled, respectively, before using, were commercial products and were used as supplied. The thiolic ligand, 2-(3-aminopropylamino)ethanethiol (Hapaet), was synthesized according to a method reported in the literature [5]. Synthesis of $[Co(apaet)_2]ClO_4 \cdot CH_3OH$ (1). To a methanol solution (4 cm^3) of Hapaet (34 mg, 0.25 mmol) and two drops of triethylamine, a methanol solution (4 cm^3) of cobalt(II) perchlorate hexahydrate (47 mg, 0.13 mmol) was added, and the reaction solution was left at room temperature for several days. The deposited dark purple crystals were filtered off and desiccated in vacuo. Yield: 21 mg, 37% (based on the Hapaet ligand). Found C 28.79, H 6.58, N 12.37%. Calcd for $C_{11}H_{30}ClCoN_4O_5S_2$: C 28.92, H 6.62, N 12.26%. IR (KBr) □/cm^{-1}: □(NH) 3313(m), 3223(m), 3148(m); □(CH) 2949(m), 2800(m); □(ClO) 1100(s), 1076(s), 1040(s), 622(s). Diffuse reflectance spectrum $□_{max}$/nm: 270, 340, 480, 890, 1364, 1552, 1708, 1746 nm. Absorption spectrum in H_2O $□_{max}$/nm (□/$M^{-1}cm^{-1}$): 205 (20700), 276 (16500), 467 (590). Magnetic moment at room temperature μ_{eff}/μ_B: 0.68. Synthesis of $[Co(apaetO_2)_2]ClO_4 \cdot$dmac (2) (dmac = *N,N'*-dimethylacetamide). To a solution of Hapaet (34 mg, 0.25 mmol) and two drops of triethylamine in methanol (4 cm^3)/dmac (2 cm^3) was added a methanol solution (4 cm^3) of cobalt(II) perchlorate hexahydrate (93 mg, 0.25 mmol). The mixture was allowed to stand for several days at room temperature. After the resulting crystals removed by filtration, aqueous hydrogen peroxide (30%, 1 cm^3) was added to the filtrate. The mixture solution was allowed to stand for several days to give orange plates. These were collected by filtration (yield 16 mg, 22% based on the thiolic ligand). Found: C, 29.02; H, 6.43; N, 11.97%. Calcd for $C_{14}H_{35}N_5O_9S_2ClCo$: C, 29.19; H, 6.12; N, 12.16%. IR (KBr) □/cm^{-1}□(NH) 3302(m), 3140(m); □(CH) 2957(m), 2875(m); □(SO) 1200(s), 978(s); □(ClO) 1096(s), 1043(s), 621(s). Electronic absorption spectrum in H_2O $□_{max}$/nm (□ /$dm^3mol^{-1}cm^{-1}$): 232 (11300), 294 (16600), 310 (15900), 456 (370).

2.2. Measurements and Methods

Elemental analyses for C, H, and N were performed using a Thermo-Finnigan FLASH EA1112 series CHNO-S analyzer. Infrared spectra were measured with a JASCO MFT-2000 FT-IR Spectrophotometer in the 4000—600 cm^{-1} region. Diffused reflectance spectrum was measured with a Shimadzu UV-vis-NIR Recording Spectrophotometer Model UV-3100 in the 200—1500 nm region. Magnetic susceptibilities were measured with a Quantum Design MPMS-XL7 SQUID susceptometer over a temperature range of 4.5—300 K. Cyclic voltammetric measurements were carried out on a Hokuto Denko HA-501 potentiostat equipped with a Hokuto Denko HB-104 function generator. A three-electrode cell consisting of a glassy carbon electrode, a platinum-wire counter electrode, and an SCE electrode was used. Single-crystal diffraction data were measured on a Bruker Smart APEX CCD diffractometer equipped with a graphite crystal and incident beam monochromator using Mo Kα radiation (λ = 0.71073 Å). Crystal data and the data collection details are listed in Table 1. The structures of **1** and **2** were solved by an intrinsic phasing method, and refined by full-matrix least-squares methods. The hydrogen atoms were inserted at their calculated positions and fixed there. All of the calculations were carried out utilizing the SHELXT-2014/4 and SHELXTL-2014/7 [21]. Crystallographic data of **1** and **2** have been deposited with Cambridge Crystallographic Data Centre: Deposit numbers CCDC-1953681 and 1953680, respectively. Copies of the data can be obtained free of charge via http://www.ccdc.cam.ac.uk/conts/retrieving.html (or from the Cambridge Crystallographic Data Centre, 12, Union Road, Cambridge, CB2 1EZ, UK; Fax: +44 1223 336033; e-mail: deposit@ccdc.cam.ac.uk). DFT calculations were carried out by using the described method in the literature [22]. Single point energy calculation and time-dependent density functional theory (TD-DFT) calculation were conducted.

3. Results and discussion

When the reaction was done under the condition with the molar ratio of Hapaet:M = 1:1, trinuclear species $[\{Co(apaet)_2\}_2Co](ClO_4)_2$ (**1'**) was obtained [16]. Whereas, in the condition with the 2:1 ligand/metal ratio, mononuclear complex $[Co(apaet)_2]ClO_4 \cdot CH_3OH$ (**1**) was isolated. The crystal structure of **1** consists of a mononuclear cation, $[Co(apaet)_2]^+$, perchlorate ion, and methanol molecule in the asymmetric unit. A perspective view of $[Co(apaet)_2]^+$ is shown in Figure 1. The Co atom has a distorted octahedral geometry similar to those of the terminal octahedral cobalt atoms of the trinuclear complex **1'**. The Co atom is coordinated by two $apaet^-$ ligands in a *meridional* fashion with four amino-nitrogen atoms and two thiolate-sulfur atoms. The bond distances of Co-S [2.2612(19),

Table 1. Crystal and experimental data.

	$[Co(apaet)_2]ClO_4 \cdot CH_3OH$ (**1**)	$[Co(apaetO_2)_2]ClO_4 \cdot dmac$ (**2**)
Chemical formula	$C_{11}H_{30}ClCoN_4O_5S_2$-	$C_{14}H_{31}ClCoN_5O_9S_2$
Formula weight	456.89	571.94
T	293	293
Crystal system	monoclinic	monoclinic
Space group	$P2_1/c$	$C2/c$
a	14.039(16) Å	8.2072(12) Å
b	9.898(12) Å	27.374(4) Å
c	14.145(17) Å	10.9117(16) Å
□	94.657(16)°	98.898(2)°
V	1954(4) $Å^3$	2422.0(6) $Å^3$
Z	4	4
D_x	1.549 g/cm^3	1.569 g/cm^3
μ(Mo K□)	1.253 mm^{-1}	1.044 mm^{-1}
F(000)	960	1192
Crystal size	0.19×0.20×0.40 mm^3	0.10×0.32×0.40 mm^3
No. of reflections collected	11469	7575
No. of independent reflections	4496	2865
□ □range for data collection	2.51 to 28.46°	1.49 to 28.62°
Data/Restraints/Parameters	4496/0/219	2865/0/140
Goodness-of-fit on F^2	1.157	1.522
R indices [*I*>2□(*I*)]	*R*1 = 0.0533, *wR*2 = 0.1614	*R*1 = 0.0713, *wR*2 = 0.2262
R indices (all data)	*R*1 = 0.0739, *wR*2 = 0.1725	*R*1 = 0.0929, *wR*2 = 0.2368
$(□□□)_{max}$	0.001	0.000
$(□□)_{max}$	0.754 $eÅ^{-3}$	1.409 $eÅ^{-3}$
$(□□)_{min}$	−0.417 $eÅ^{-3}$	−0.572 $eÅ^{-3}$

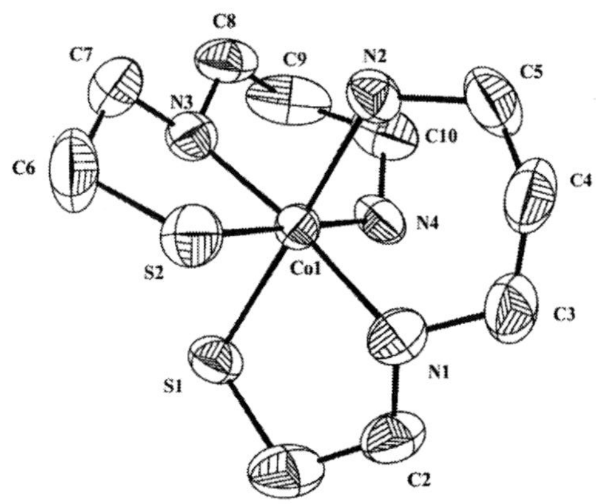

Figure 1. ORTEP drawing of the complex cation of **1**. Selected bond lengths (Å) and angles (°): Co1-S1 2.2612(19), Co1-S2 2.271(2), Co1-N1 2.010(4), Co1-N2 2.028(3), Co1-N3 2.002(3), Co1-N4 2.010(4); S1-Co1-S2 92.04(9), S1-Co1-N1 86.26(12), S1-Co1-N2 179.10(10), S1-Co1-N3 88.71(12), S1-Co1-N4 88.34(12), S2-Co1-N1 86.93(11), S2-Co1-N2 87.58(13), S2-Co1-N3 88.13(11), S2-Co1-N4 179.03(11), N1-Co1-N2 94.53(15), N1-Co1-N3 172.82(14), N1-Co1-N4 92.21(15), N2-Co1-N3 90.47(15), N2-Co1-N4 92.05(16), N3-Co1-N4 92.76(15).

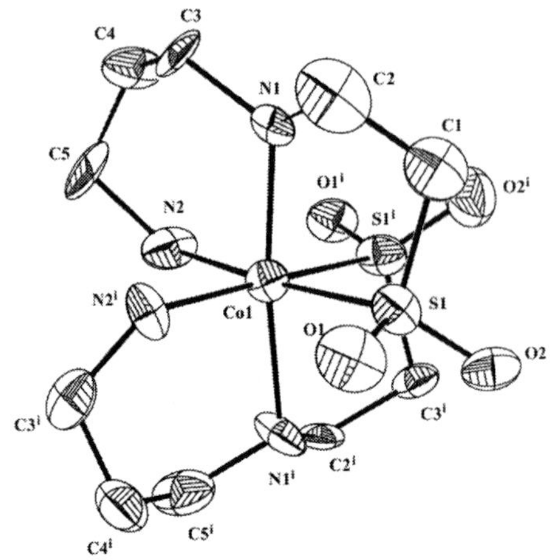

Figure 2. ORTEP drawing of the complex cation of **2**. Selected bond lengths (Å) and angles (°): Co1-S1 2.2164(13), Co1-N1 2.026(4), Co1-N2 2.014(4), S1-O1 1.469(4), S1-O2 1.461(4); S1-Co1-S1^{i} 91.36(7), S1-Co1-N1 85.20(12), S1-Co1-N1^{i} 90.74(12), S1-Co1-N2 176.85(12), S1-Co1-N2^{i} 88.82(14), N1-Co1-N1^{i} 174.2(2), N1-Co1-N2 91.66(16), N1-Co1-N2^{i} 92.41(16), N2-Co1-N2^{i} 91.2(3). Symmetry code; i $(1-x, y, 1/2-z)$.

2.271(2) Å] and Co-N [2.002(3)—2.031(4) Å] are significantly shorter than the corresponding distances for the terminal Co(II) atoms of **1'** [16]. These values are comparable to those of octahedral Co(III) complexes [4]. The bond distances of the Co-N located at *trans* positions of the sulfur donor are elongated, suggesting a *trans* influence of the thiolate-sulfur donor. The five-membered chelate rings, Co1-S1-C1-C2-N1 and Co1-S2-C6-C7-N3, take a *gauche* form which is common for five-membered ring, whereas the six-membered chelate rings, Co1-N1-C3-C4-C5-N2 and Co1-N3-C8-C9-C10-N4, take a *chair* form. The methanol molecule is located near the amino-nitrogen atom as indicated by the distance O5(MeOH)···N3 2.935 Å which suggests the occurrence of hydrogen bonding. The magnetic susceptibility measurement shows that the complex **1** is essentially diamagnetic throughout the 4.5—300 K temperature region, which is consistent with the low-spin cobalt(III) formulation. Although, the driving force of the oxidation of cobalt(II) is uncertain, the octahedral coordination environment around the Co atom may be important for the stable low-spin Co(III) electronic configuration. The complex **2** was obtained from a reaction solution of **1** by air oxidation or reaction with hydrogen peroxide. The crystal structure of **2** consists of mononuclear cations, $[Co(apaetO_2)_2]^+$, perchlorate ions, and dmac molecules. A perspective view of $[Co(apaetO_2)_2]^+$ is shown in Figure 2. The complex cation has a crystallographic C_2 axis. Now, each sulfur atom of thiolic ligands, apaet$^-$, was added by two oxygen atoms to form the sulfinato ligands, $apaetO_2^-$. The Co has a distorted octahedral geometry, coordinated by two sulfinate $apaetO_2^-$ ligands with four amino-nitrogen atoms and two sulfinate-sulfur atoms in a *meridional* fashion. The bond distances of Co-S [2.2164(13) Å], Co-N [2.026(4), 2.014(4) Å], and S-O [1.461(4), 1.469(4) Å] are comparable to those of octahedral Co(III) sulfinate complexes [23]. As the electron density of S atoms would be higher by addition of oxygen atoms, the bond distances of Co-S are slightly shorter than those of the thiolate complex **1**. The bond distances of Co-N located at *trans* positions of sulphur atoms are not elongated, in contrast with the case for the thiolate complex **1**, where the *trans* influence was observed. This reflects the weaker donor property of the sulfinate-sulfur compared with that of the thiolate-sulfur. The five-membered chelate rings, Co1-S1-C1-C2-N1, take an *envelope* form which is not common for five-memberd chelate ring [24], whereas the six-membered chelate rings, Co1-N1-C3-C4-C5-N2, take a *chair* form. This mononuclear Co(III) sulfinate complex may be a model for nitrile hydratases (NHase) which was shown to have noncorrinoid cobalt center with nitrogen and thiolate-sulfur donors. The Fe-containing NHase has the bound Cys-S centers, and at least, two of those are modified to Cys-sulfenic and -sulfinic groups [23]. The Co-containing NHase must have a comparable configuration to that of Fe.

IOP Conf. Series: Materials Science and Engineering **835** (2020) 012024 doi:10.1088/1757-899X/835/1/012024

3.1 Electronic spectra and reactions of complexes

The crystal structure of the mononuclear complex **1** is quite similar those of octahedral parts of the trinuclear complex **1'** except for the oxidation state. To clear the relationship between the structural configuration and oxidation states, we attempted to oxidize the aqueous solution of **1'**. The diffuse reflectance spectra of **1'** show absorption bands at 350, 675, 735, 783, and 1333 nm, where the absorption at 350 nm and the other absorptions in the visible and near-infrared region can be assigned to thiolate-to-cobalt(II) charge transfer transition and ligand field absorptions for high-spin tetrahedral cobalt(II) species $^4A_2 \rightarrow {}^4T_1(P)$ and $^4A_2 \rightarrow {}^4T_1(F)$, respectively [16]. The electronic absorption spectra in aqueous solution are essentially the same as the diffused reflectance spectra, which means that the trinuclear species in the solid is maintained in the aqueous solution. However, the absorption spectra of the aqueous solution of **1'** show a gradual change with time upon exposure to air at room temperature [16]. The spectra obtained after one day show a broad absorption at 470 nm and intense one at 274 nm similar to those of **1** (276 nm (□□□16500 $dm^3mol^{-1}cm^{-1}$), 467 nm (590)). This result shows that the trinuclear cobalt(II) species was oxidized to the mononuclear cobalt(III) species, where octahedral environment is stable for low-spin d^6 species rather than tetrahedral geometry. On the reaction of **1** with hydrogen peroxide, an interesting change occurred in the electronic absorption spectra as shown in Figure 3. At the beginning of the reaction, the absorption band at 276 nm (□□ 16500 $dm^3mol^{-1}cm^{-1}$) which can be assigned as thiolate-sulfur to cobalt charge transfer transition was decreasing and then increasing the intensity of the absorption band at 376 nm which is attributed to sulfenate-sulfur to cobalt charge transfer transition, molar absorption coefficient became up to 10500 $dm^3mol^{-1}cm^{-1}$. This reaction should be a first order and rate constants in decomposition of thiolate complex (8.□□10^{-5} s^{-1}, 25 °C) and formation of sulfenate complex (8.□□10^{-5} s^{-1}) are almost same. The intensity of this sulfenate-sulfur to cobalt charge transfer transition band was weakened with the successive time, the absorption at 294 nm, which may be identified as sulfinate-sulfur to cobalt charge transfer transition was arisen (□□□13300 $dm^3mol^{-1}cm^{-1}$) and is compatible with the absorption spectra of **2**. The rate constant for the formation of sulfinate complex is 6.1□10^{-6} s^{-1}, more than 10 times slower than that for the formation of sulfenate complex. In this manner, the sulfinate cobalt complex was obtained from oxidation of thiolate complex via sulfenate complex as an intermediate.

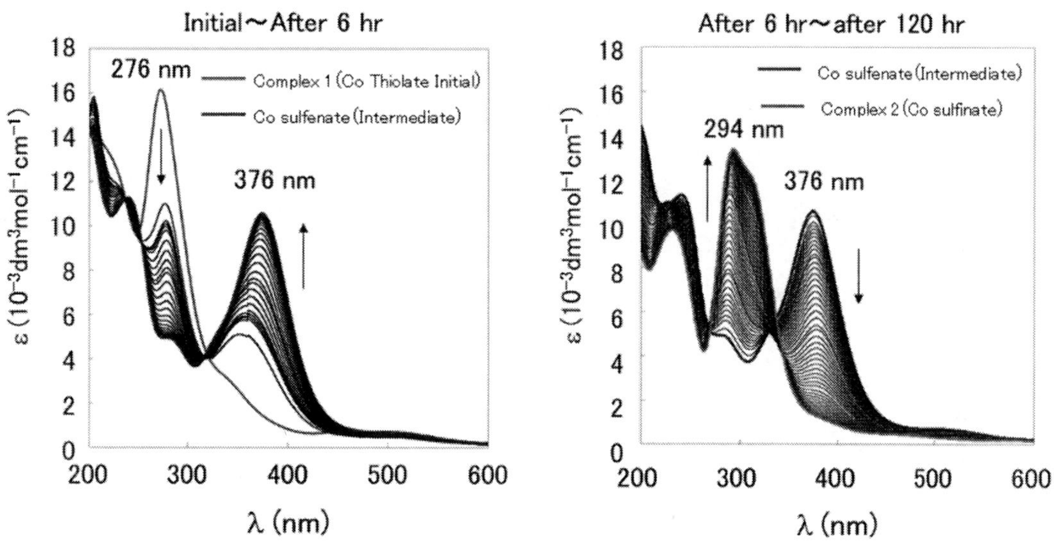

Figure 3. Spectral change for the reaction of **1** with hydrogen peroxide.

IOP Conf. Series: Materials Science and Engineering **835** (2020) 012024 doi:10.1088/1757-899X/835/1/012024

3.2 DFT calculations

To explain the mechanism of oxidation by hydrogen peroxide, we performed the DFT calculations for **1** and **2**. The MO energy diagrams are shown in Figure 4. In **1**, unoccupied Co(III) d-orbitals (dx^2-y^2, dz^2) are almost degenerated and are located at LUMO. While, the energy levels of occupied d-orbitals (d_{xy}, d_{zx}, d_{zy}) are separated and are divided into two parts by influences of coordinated S atoms. Since the HOMO of **1** exists at the lone pair of thiolate sulfur atoms, the sulfur atoms have considerable nucleophilic character. Whereas, the HOMO of **2** is not located at S atoms but also lone pairs of oxygen atoms, the sulfur atoms are no longer so called "activation sites". It is generally accepted that H_2O_2 oxidation of nucleophilic substrates occurs via nucleophilic attack of the substrate on the O-O bond of H_2O_2. The oxidation reaction scheme from thiolate complex to sulfinate complex seems to be as follows (Scheme 1). (1) H_2O_2 molecule reacts with nucleophilic attack to the coordinated thiolate sulfur atom and this is deprotonated to form terminal bonded S=O-OH species. (2) As the oxygen

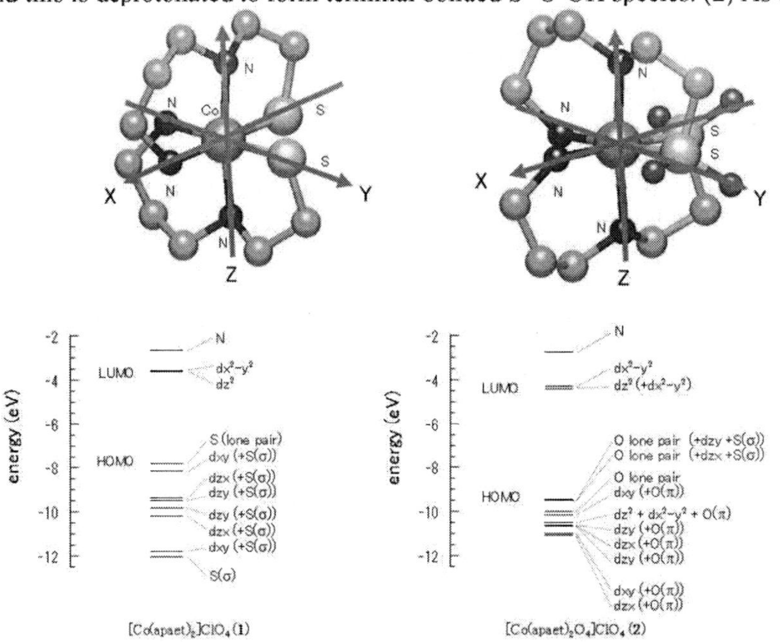

Figure 4. MO diagrams of **1** and **2**.

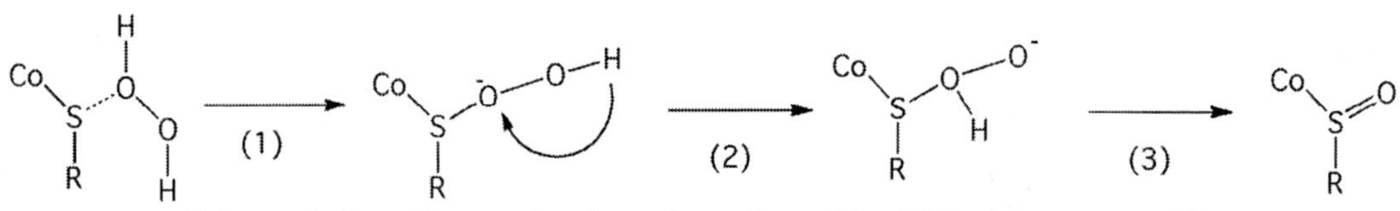

Scheme 1. Reaction mechanism of reaction of **1** with hydrogen peroxide.

atom connected to sulfur atom has a negative □ charge, terminal hydrogen atom moves to the oxygen atom connected to sulfur atom. (3) The O-O bonding of O_2^{2-} is weakened and terminal oxygen is disconnected. The electron of oxygen atom is localized toward sulfur atom, oxygen atom is deprotonated, and then sulfenate intermediate is formed. The leaving oxygen atom may oxidize other sulfur atom, generate dioxygen or H_2O molecule. (4) A similar nucleophilic attack of H_2O_2 molecule to sulfenate oxygen atom forms sulfinate complex. Because of the steric hindrance around sulfenate sulfur atom, the reaction from sulfenate to sulfinate complex may be slower than that from thiolate complex to sulfenate complex.

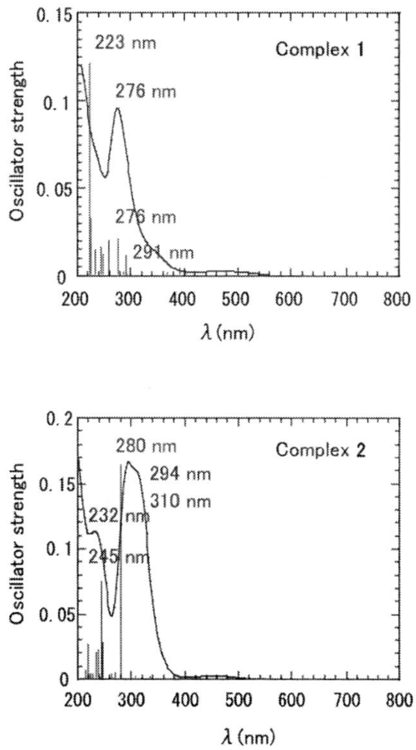

Figure 5. Diffused reflectance spectra of **1** and **2** with the TD-DFT calculated transition energies as absorption wavelengths in nm and oscillator strengths indicated by vertical lines for comparison.

To assign electronic absorption spectra, we perform time-dependent density functional theory (TD-DFT) calculations for **1** and **2**. The calculated and experimental spectra of the complexes **1** and **2** are shown in Figure 5. The calculated spectra exhibit good agreement with experimental spectra. For the electronic absorption spectra of **1** show two large absorption bands at 220(sh) and 276 nm, calculated peaks are located at 223 and 277 nm. The calculated peak at 223 nm may be corresponding to the experimental shoulder absorption at 220 nm is arisen mainly by the transition from HOMO-6 (consists of Co d_{xy} + s(□)) to LUMO+1 (Co dx^2-y^2) , which can be assigned S(□)□Co charge transfer. The calculated peaks around at 276nm are consists of many transitions. In particular, the peak at 291 nm can be attributed to S(□)□Co charge transfer. Because the calculated peaks around at 276 nm are divided some peaks in calculation, intensity of this peak would be lower than that of experimental peak at 276 nm. Whereas, the electronic absorption spectra of complex **2** show intense absorption bands at 232, 294 and 310 nm, calculated peaks are positioned at 245 and 280 nm. Although, the two intense calculated peaks are mainly consists of the same transition from HOMO-1 (lone pair of O + Co d_{zx} + S(□)) to LUMO (dz^2) and from HOMO (lone pair of O + Co d_{zy} + S(□)) to LUMO+1 (dx^2-y^2), the transition moments of them are pointed opposite or same directions. As the transition moments calculated at 245 nm are pointed opposite directions, the intensity of peak becomes weak, whereas those of calculated at 280 nm are pointed the same direction and strong intensity can be observed. Because of the electron configuration interaction, the two calculated peaks apart each other. The observed peak at 232, 294, and 310 nm can be assigned S(□)□Co charge transfer.

3.3 Reaction of 1 and H_2O_2 with carotenoids

As we assumed that an active O atom forms in the reaction of **1** and H_2O_2, we examined oxidation reaction of a dye like carotenoids by using **1**. The "orange color 100" is one of the carotenoid dyes used as food additives, which is a mixture of capsanthine, glycerin, and so on. When an aqueous solution of orange color 100 (ca. 140 □mol/L as capsanthine), was reacted with the complex **1** (24.3□□mol/L) and hydrogen peroxide (3.65 mmol/L) in alkaline aqueous condition (pH10.8), the electronic absorption spectra shows the decomposition of the carotenoid. The spectral change of orange color 100 with time upon the reaction is shown in Figure 6. The reaction constant of this reaction is 5.14□10^{-5} s^{-1} at 20°C, 25.0□10^{-5} s^{-1} at 30°C, and 59.5□10^{-5} s^{-1} at 40°C and activation energy for decomposition of orange color 100 was calculated to 216 kJ/mol. No spectral change was occurred in the reaction condition without **1** or with cobalt(II) chloride hexahydrate instead of **1** in the similar time range (□one week at 25°C). This result shows that the complex **1** reacts as an oxidant and may has a DTI (dye-transfer inhibition) effect during laundering.

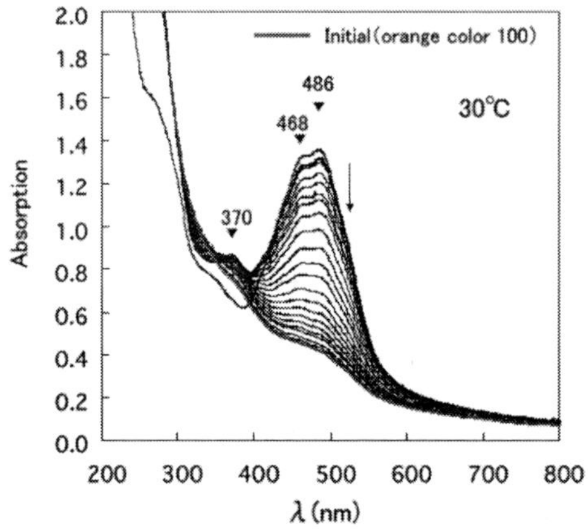

Figure 6. Spectral change for oxidation of orange color 100 by **1** and H_2O_2.

3.4 Electrochemistry

The cyclic voltammograms of **1** and **2** are shown in Figure 7. The first reduction processes have the characteristic of an electrochemically quasi-reversible couple and assigned for Co(III)/Co(II) redox couple. Half-wave potentials ($E_{1/2}$) of them are +1.19V (□*E* 210mV) vs SCE for **1** and +0.61V (□*E* 60mV) for **2**. This result show that Co(III) species of **1** is more stable than that of **2** and they go hand-in-hand with the LUMO of **1** (–3.6157 eV) has higher energy than that of **2** (–4.4651 eV) in DFT calculation.

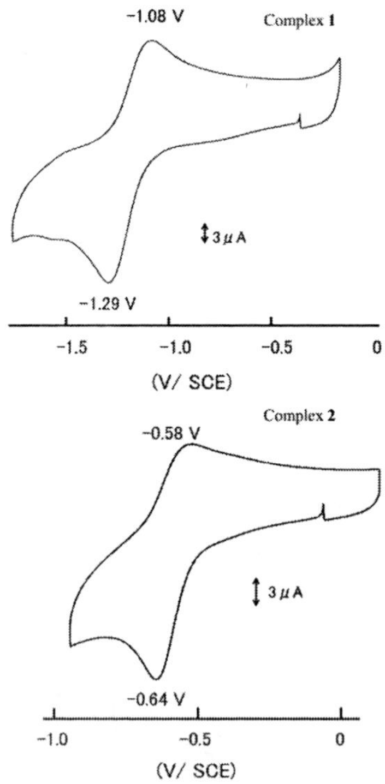

Figure 7. Cyclic voltammograms of **1** and **2** (1 mM) in MeOH [(*n*-$(C_4H_9)_4NClO_4$, scan rate 100 mV/s].

4. Conclusion

In this study, we have isolated and characterized two kinds of mononuclear Co(III) complexes with apaet$^-$ and apaetO_2^-. The mononuclear thiolate complex can react with hydrogen peroxide, resulting sulfinate mononuclear Co(III) complex via the sufenate Co(III) species, which may be relevant to a model for bacterial enzymes, nitrile hydratase (NHase).

Acknowledgement

The present work was partially supported by the MEXT-Supported Program for the Strategic Research Foundation at Private Universities and Grants-in-Aid for Scientific Research No. 17K05820 from the Ministry of Education, Culture, Sports, Science and Technology (MEXT).

References

[1] B. Krebs, G. Henkel, "Transition-metal thiolates: from molecular fragments of sulfidic solids to models for active centers in biomolecules," Angew. Chem. Int. Ed. Engl., vol. 30, 769-788, 1991.

[2] J. A. Garcia-Vazquez, J. Romero, A. Sousa, "Electrochemical synthesis of metallic complexes of bidentate thiolates containing nitrogen as an additional donor atom," Coord. Chem. Rev., vol. 212, 691-745, 1999.

[3] M. Mikuriya, S. Kida, S. Ueno, I. Murase, "Crystal and molecular structure of a triply-thiolate-bridged binuclear cobalt(III) complex, $[Co_2\{SCH(CH_2CH_2NH_2)_2\}_3](ClO_4)_3{\cdot}H_2O$," Bull. Chem. Soc. Jpn., vol. 58, pp. 1857-1858, 1985.

[4] M. Mikuriya, S. Kida, I. Murase, "Crystal and molecular structure of a thiolato-beridged binuclear nickel(II) complex, $[Ni_2\{SCH(CH_2CH_2NH_2)_2\}_2]Br_2$," Bull. Chem. Soc. Jpn., vol. 60, pp. 1180-1182, 1987.

[5] M. Handa, M, Mikuriya, Z.-J. Zhong, H. Okawa, S. Kida, "Synthesis, properties, and crystal structure of binuclear nickel(II) and palladium(II) complexes of tridentate thiolic ligands with NNS-donor set," Bull. Chem. Soc. Jpn., vol. 61, pp. 3883-3887, 1988.

[6] M. Mikuriya, F. Adachi, H. Iwasawa, M. Handa, M. Koikawa, H. Okawa, "A novel thiolate-bridged polynuclear manganese(II) complex with 2-[2-(2-pyridyl)ethylamino]ethanethiol," Inorg. Chim. Acta, vol. 179, pp. 3-4, 1991.

[7] M. Mikuriya, F. Adachi, H. Iwasawa, M. Handa, M. Koikawa, H. Okawa, "Synthesis and characterization of thiolate-bridged manganese(II) complexes with NNS-tridentate thiolic ligands," Bull. Chem. Soc. Jpn., vol. 67, 3263-3270, 1994,

[8] M. Mikuriya, T. Kotera, F. Adachi, M. Handa, M. Koikawa, H. Okawa, "Syntheses and structural characterization of dinuclear and trinuclear iron(II) complexes of tridentate thiolic ligands with an NNS donor set," Bull. Chem. Soc. Jpn., vol. 68, pp. 574-580, 1995.

[9] M. Mikuriya, H. Tsutsumi, R. Nukada, M. Handa, Y. Sayama, "Synthesis and characterization of thiolato-bridged trinuclear heteromatal complexes with 2-[(3-aminopropyl)amino]ethanethiol," Bull. Chem. Soc. Jpn., vol. 69, pp. 3489-3498, 1996.

[10] T. Kawahashi, H. Tsutsumi, M. Mikuriya, "A novel tetranuclear palladium(II) complex with 2-[(3-aminopropyl)amino]ethanethiol," Polyhedron, vol. 15, pp. 169-171, 1996.

[11] M. Mikuriya, X. Jian, S. Ikemi, T. Kawahashi, H. Tsutsumi, "Synthesis and structural characterization of thiolato-brideged zinc(II) complexes with NNS-tridentate ligands," Bull. Chem. Soc. Jpn., vol. 71, pp. 2161- 2168, 1998.

[12] T. Kawahashi, M. Mikuriya, R. Nukada, J.-W. Lim, "Synthesis and structural characterization of thiolato-bridged tetranuclear palladium(II) complexes with N,N,S-tridentate ligands," Bull. Chem. Soc. Jpn., vol. 74, pp. 323- 329, 2001.

[13] M. Mikuriya, X. Jian, S. Ikemi, T. Kawahashi , H. Tsutsumi, A. Nakasone, J.-W. Lim, "Synthesis and characterization of thiolato-brideged cadmium(II) complexes with NNS-chelating thiolic ligands," Inorg. Chim. Acta, vol. 312, pp.183-187, 2001.

[14] T. Kotera, M. Mikuriya, "Thilato-bridged hexanuclear Cu^ICu^{II} mixed-valence complex," Chem. Lett, vol. 31, pp. 654-655, 2002.

[15] T. Kotera, A. Fujita, M. Mikuriya, M. Handa, "Thiolato-bridged copper complexes with N,N,S-

tridentate ligands," Materials Science (Poland), vol. 21, pp. 171-179, 2003.

[16] T. Kotera, A. Fujita, M. Mikuriya, H. Tsutsumi, M. Handa, "Cobalt complexes with 2-[(3-aminopropyl)amino]ethanethiol," Inorg. Chem. Commun., vol. 6, pp. 322-324, 2003.

[17] T. Kotera, T. Sugimoto, M. Mikuriya, "A thiolato-brideged octanuclear copper(I,II) mixed-valence complex with N,N,S-tridentate ligand," Chem. J. Moldova, vol. 2, pp. 102-107, 2007.

[18] M. Mikuriya, T. Watanabe, A. Suyama, D. Yoshioka, "Rhenium complex with 2-[(2-pyridylmethyl)amino]ethanethiol," X-ray Struct. Anal. Online, vol. 31, pp.3-4, 2015.

[19] M. Mikuriya, K. Kusunoki, T. Kotera, D. Yoshioka, and K. Ogasawara, "Synthesis, crystal structure, and DFT calculation of a dioxido-bridged dinuclear oxidomolybdenum(V) complex with 2-(2-aminoethyl)aminoethanethiol," X-ray Struct. Anal. Online, vol. 33, pp.37-39, 2017.

[20] M. Mikuriya, K. Kusunoki, T. Kotera, D. Yoshioka, S. Takemura, and K. Ogasawara, "Synthesis, crystal structure, and relativistic DV-X□ calculation of a μ-oxido-μ-molybdato(VI)-bridged dinuclear oxidomolybdenum(V) complex with 2-(3-aminopropyl)aminoethanethiol," X-ray Struct. Anal. Online, vol. 34, pp.19-21, 2018.

[21] G. M. Sheldrick, "Crystal structure refinement with SHELXL," Acta Crystallogr., Sect. C, vol. 71, pp. 1-8, 2015.

[22] S. Wada, T. Kotera, M. Mikuriya, "Synthesis and structural characterizaytion of a dinuclear palladium(II) complex with *N,N',N'',N'''*-tetrakis(2-*p*-toluenesulfonamidoethyl)cyclam," Bull. Chem. Soc. Jpn., vol.81, pp.1454-1460, 2008.

[23] I. Kung, D. Schweitzer, J. Shearer, W. D. Taylor, H. L. Jackson, S. Lovell, J. A. Kovacs, "How do oxidized thiolate ligands affect the electronic and reactivity properties of a nitrile hydratase model compound?," J. Am. Chem. Soc., vol. 122, pp. 8299-8300, 2000.

[24] M. Mikuriya, H. Okawa, S. Kida, "Binuclear metal complexes. XLIV. Crystal structures and magnetic properties of alkoxo-bridged binuclear copper(II) complexes with 2-[2-(dialkylamino)ethylthio]ethanol," Bull. Chem. Soc. Jpn., vol. 55, pp. 1086-1091, 1982.

IOP Conf. Series: Materials Science and Engineering **835** (2020) 012025 doi:10.1088/1757-899X/835/1/012025

Contribution of edge-carbon atoms and non-benzenoid rings in graphitic carbons to π* peak profiles in C*K*-XANES spectra

Yasuji Muramatsu and Yuma Hirai

Graduate School of Engineering, University of Hyogo, 2167 Shosha, Himeji, Hyogo 671-2201 Japan

murama@eng.u-hyogo.ac.jp

Abstract In order to analyze the local structure of graphitic carbons by soft X-ray absorption spectroscopy, CK-XANES spectra of polycyclic aromatic compounds having 5-membered pentagonal or 7-membered heptagonal rings were theoretically analyzed by first-principles calculations. Pentagonal rings in the bent graphitic structure take different π* orbital states. Hence, pentagonal rings split the π* peak into two peaks in CK-XANES spectra. Heptagonal rings create electrostatic potential holes in graphitic structures, and the carbon atoms in the heptagonal ring can be regarded as conjugated edge carbon atoms. Hence, heptagonal rings also split the π* peak into two peaks. Edge-carbon atoms split the π* peak into two peaks in CK- XANES spectra. Consequently, carbon atoms in non-benzenoid rings and the edge-carbon atoms in graphitic structures make the π* peak profile wider when compared to graphite.

1. Introduction

Local structure analysis is important in improving the physical/chemical functions of graphitic carbons, which have been utilized for battery electrodes, catalysts, adsorbents, carbon fiber composites, carbon alloys, and nano-carbons. This is because physical/chemical properties of graphitic carbons depend on the electronic states of the graphitic structures. In identifying the local structure, X-ray absorption near-edge structure (XANES) spectroscopy is a powerful tool because XANES spectra can easily and finely provide element- and orbital-selective information based on the selection rule [1]. XANES spectra can also be well simulated by various theoretical calculation methods. Therefore, the combination of XANES measurements and theoretical calculations has been a powerful tool for finely determining the local structure of carbon materials at the atomic scale.

From local structure analysis of graphitic carbons using XANES measurements and the first-principle calculations, we have previously clarified that the π* peak profiles in CK-XANES spectra involve information on the edge carbon atoms in graphitic structures [2-5]. In this study, aiming to further extract the local structure information of non-benzenoid rings from the π* peak profiles, we have calculated the CK-XANES spectra of various graphitic structures having 5-membered pentagonal rings or 7-membered heptagonal rings and compared these results to those for structures with 6-membered hexagonal rings.

2. Methods

2.1. XANES and DOS calculations

CK-XANES spectra were theoretically simulated by the first-principles density-functional theory (DFT) calculation code, CASTEP [6]. An individual molecular model was placed in a sufficiently large super cell. After structure optimization by the Materials Studio code, calculations in the ground state were achieved in CASTEP. Then calculations in an excited state with a $C1s^{-1}$ core hole were performed for each carbon atom. The transition energy of the calculated XANES spectra were corrected by considering the energy differences at the valence and core levels in the ground and excited states [7].

The density of states (DOS), convoluted with a Lorentzian function having a 0.5 eV width, were calculated using the discrete variational (DV)-Xα method [8]. To assign the π* and σ* peaks in the calculated XANES spectra, unoccupied C2p- and C2s-DOS profiles were compared with the XANES profiles. Electrostatic potential surface maps were also obtained with the DV-Xα calculations.

2.2. Calculation Models

Three types of polycyclic aromatic hydrocarbon (PAH) models, as shown in Figure 1, were used for calculations: Coronenes, which consist of 6-membered rings; Corannulenes, whose center rings are 5-membered pentagonal rings; and [7]Circulenes, whose center rings are 7-membered heptagonal rings. To elucidate the electronic effects between the edge atoms and center atoms, cluster size was varied to circulenes, circumcirculenes, and circumcircumcirculenes. The smallest models of Coronene ($C_{24}H_{12}$), Corannulene ($C_{20}H_{10}$), and [7]Circulene ($C_{28}H_{14}$) are Circulenes. The larger models of Circumcoronene ($C_{54}H_{18}$), Circumcorannulene ($C_{45}H_{15}$), and Circum[7]circulene ($C_{63}H_{21}$) are Circumcirculenes. The largest models of Circumcircumcoronene ($C_{96}H_{24}$), Circumcircumcorannulene ($C_{80}H_{20}$), and Circumcircum[7]circulene ($C_{112}H_{28}$) are Circumcircumcirculenes. In the models, CK-XANES spectra of commercially available Coronene and Corannulene were experimentally measured using synchrotron radiation soft X-rays in BL-6.3.2 [9] at the Advanced Light Source (ALS).

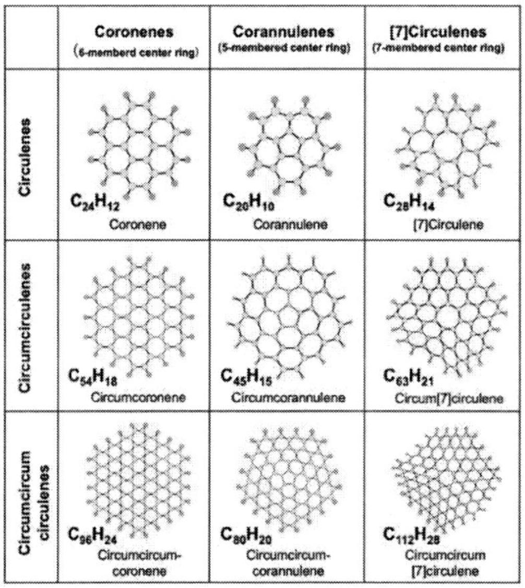

Figure 1. Models for calculations.

3. Results and discussion

3.1. Coronenes

Figure 2 shows the cluster model, calculated XANES spectra, C-DOS, and electrostatic potential surface map with the cluster model of Coronene (a), Circumcoronene (b), and Circumcircumcoronene (c). For Coronene, the center C atoms are labeled as $1C_{hex}$, the conjugated edge C atoms as $2C_{ed\text{-}C}$, and the H-terminated edge C atoms as $3C_{ed\text{-}H}$. The measured CK-XANES spectra can be well reproduced by the calculated XANES spectra (total). Therefore, the XANES spectra calculated by CASTEP are confirmed as being reliable for XANES analysis. In the C-DOS, three 2p*-DOS peaks, which are not hybridized with the 2s*-DOS, can be identified as π* peaks.

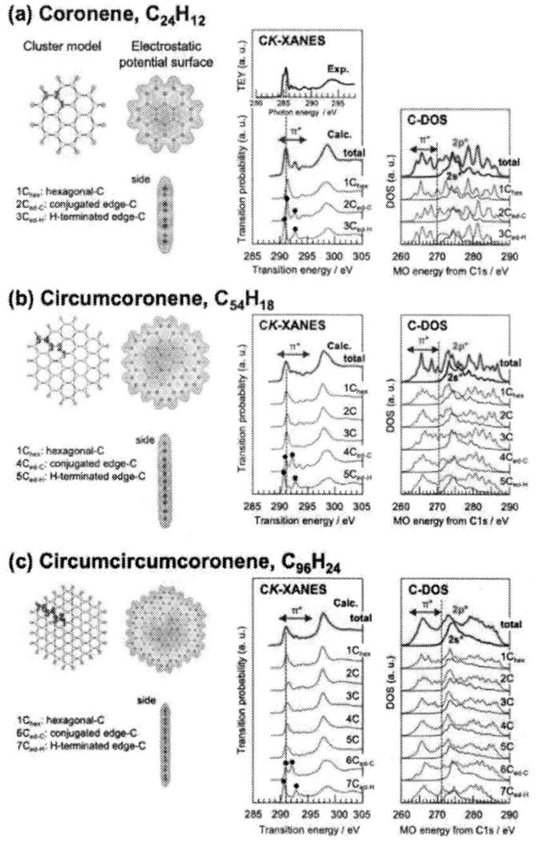

Figure 2. Cluster model, electrostatic potential surface, calculated CK-XANES spectra, and C- DOS of Coronenes. Measured CK-XANES spectrum is compared to the calculated XANES spectra in Coronene (a). In the calculated XANES spectra, split π* peaks are marked with solid circles.

Corresponding to the three π* peaks, the three threshold peaks in the calculated XANES spectra can be assigned to the π* peaks. In the calculated XANES spectra, the $1C_{hex}$ atoms in the center hexagonal ring exhibit a single π* peak, while the $2C_{ed\text{-}C}$ conjugated edge C atoms and $3C_{ed\text{-}H}$ H-terminated edge C atoms exhibit the split π* peaks. It can be clarified that such split π* peaks of the edge C atoms make the π* peak profile wider.

For the Circumcoronene and Circumcircumcoronene, similar π* peak splitting can be exhibited at the edge C atoms in the calculated XANES spectra. Such localized electronic states of edge C atoms can be clearly visualized in the electronic potential surface maps; the edge C atoms have a fairly positive potential while inner C atoms have a negative potential. Additionally, electronic states of the π* peak splitting of the edge C atoms cannot affect the inner C atoms, because the $1C_{hex}$~3C atoms in Circumcoronene and the $1C_{hex}$~5C atoms in Circumcircumcoronene exhibit the single π* peak profiles. These calculated results agree with those of previous reports [4, 5, 10].

3.2. Corannulenes

Figure 3 shows the results for Corannulene (a), Circumcorannulene (b), and Circumcircumcorannulene (c), which all have a 5-membered pentagonal ring at the cluster center.

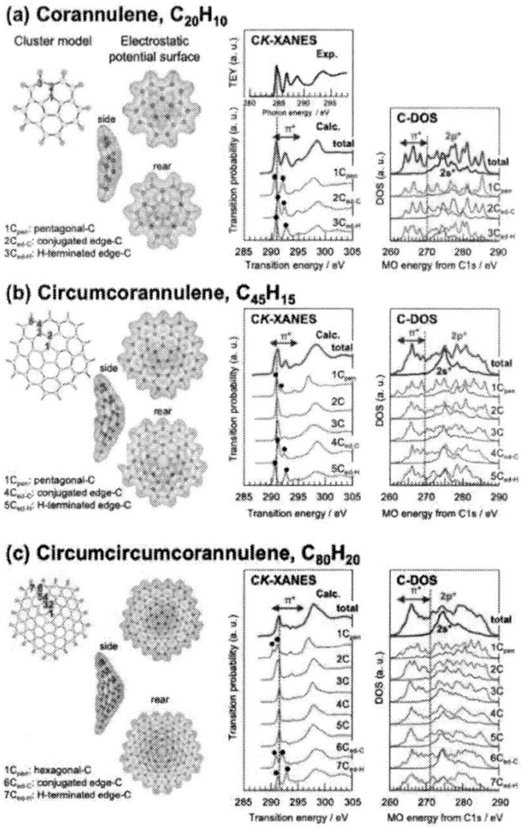

Figure 3. Cluster model, electrostatic potential surface, calculated CK-XANES spectra, and C- DOS of Corannulenes. Measured CK-XANES spectrum is compared to the calculatedXANES spectra in Corannulene (a). In the calculated XANES spectra, split π^* peaks are markedwith solid circles.

For Corannulene, the center C atoms of the pentagonal ring are labeled as $1C_{pen}$, the conjugated edge C atoms as $2C_{ed\text{-}C}$, and the H-terminated edge C atoms as $3C_{ed\text{-}H}$. The measured CK-XANES spectra of corannulene can be well reproduced by the calculated XANES spectra (total). Again, the XANES spectra calculated by CASTEP are confirmed as being reliable for XANES analysis. Corresponding to the three π^* peaks of C-DOS, the threshold three peaks in the calculated XANES spectra can be assigned to the π^* peaks. In the calculated XANES spectra, the $1C_{pen}$ atoms in the center pentagonal ring exhibit the split π^* peaks, which is different from the $1C_{hex}$ atoms in Coronene. This difference can be explained by considering the electronic states of π^* orbitals in Corannulene. Corannulene takes a bent plane structure due to the center pentagonal ring. The π^* orbitals in the convex side and in the concave side take different electronic states. The difference can be visualized in the electrostatic potential maps when viewed from the top and rear sides, which splits the π^* peak into two peaks. Hence, C atoms in a pentagonal ring make the π^* peak wider. The $2C_{ed\text{-}C}$, conjugated edge C atoms, and $3C_{ed\text{-}H}$, H-terminated edge C atoms, also exhibit the split π^* peaks.

In the calculated XANES spectra of Circumcorannulene and Circumcircumcorannulene, similar π^* peak splitting can be exhibited in the $1C_{pen}$ atoms. The conjugated edge C atoms, $4C_{ed\text{-}C}$, and H-terminated C atoms, $5C_{ed\text{-}H}$, in Circumcorannulene exhibit the split π^* peak profiles. The conjugated edge C atoms, $6C_{ed\text{-}C}$, and H- terminated C atoms, $7C_{ed\text{-}H}$, in Circumcircumcorannulene also exhibit the split π^* peak profiles. It can also be clarified that such split π^* peaks of the edge C atoms make the π^* peak profile wider. By contrast, inner 2C–3C atoms in Circumcorannulene and 2C–5C atoms in Circumcircumcorannulene exhibit the single π^* peak profiles.

It is therefore confirmed that C atoms in the pentagonal ring and the edge C atoms make the π^* peak profile wider.

3.3. [7]Circulenes

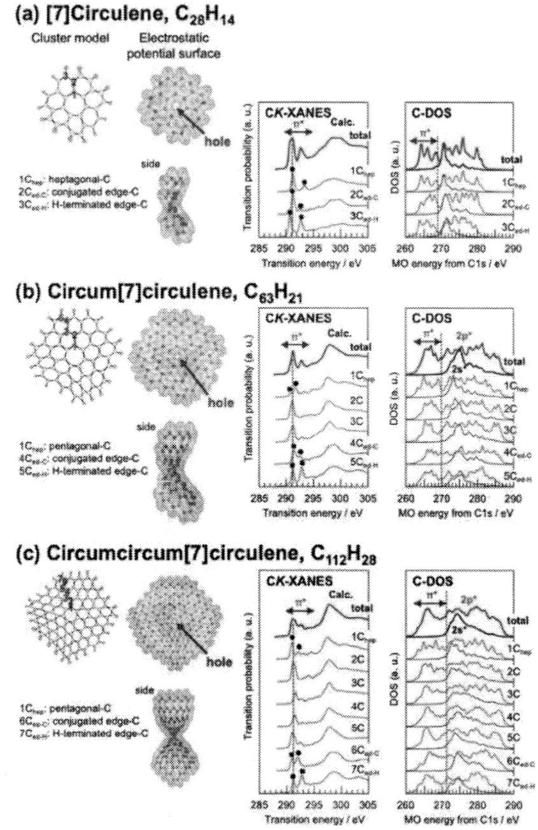

Figure 4. Cluster model, electrostatic potential surface, calculated CK-XANES spectra, and C- DOS of [7]Circulenes. In the calculated XANES spectra, split π^* peaks are marked with solid circles.

Figure 4 shows the results of [7]Circulene (a), Circum[7]circulene (b), and Circumcircum[7]circulene (c), which all have a 7-membered heptagonal ring at the cluster center.

In [7]Circulene, the center C atoms of the heptagonal ring are labeled as $1C_{hep}$, the conjugated edge C atoms as $2C_{ed\text{-}C}$, and the H-terminated edge C atoms as $3C_{ed\text{-}H}$. Corresponding to the three π^* peaks of C-DOS, the threshold three peaks in the calculated XANES spectra can be assigned to the π^* peaks. In the calculated XANES spectra, the $1C_{hep}$ atoms at the center heptagonal ring exhibit the split π^* peaks, which is different from the $1C_{hex}$ atoms in Coronene. This difference can be explained by considering the electronic states of π^* orbitals in the heptagonal ring. The electrostatic potential map of [7]Circulene shows an electrostatic potential hole at the heptagonal ring. This suggests that the $1C_{hep}$ atoms can be regarded as conjugated edge C atoms. Hence, C atoms in the heptagonal ring exhibit the split π^* peak profile. The $2C_{ed\text{-}C}$ of conjugated edge C atoms and $3C_{ed\text{-}H}$ of H-terminated edge C atoms also exhibit the split π^* peaks.

In the calculated XANES spectra of Circum[7]circulene and Circumcircum[7]circulene, similar π^* peak splitting can be exhibited in the $1C_{hep}$ atoms due to the electrostatic potential holes at the heptagonal rings. The conjugated edge C atoms, $4C_{ed\text{-}C}$, and H-terminated C atoms, $5C_{ed\text{-}H}$, in Circum[7]circulene exhibit the split π^* peak profiles. The conjugated edge C atoms, $6C_{ed\text{-}C}$, and H-terminated C atoms, $7C_{ed\text{-}H}$, in Circumcircum[7]circulene also exhibit the split π^* peak profiles. It can also be clarified that the split π^* peaks of the edge C atoms make the π^* peak profile wider. By contrast, inner 2C–3C atoms in Circum[7]circulene and 2C– 5C atoms in Circumcircum[7]circulene exhibit the single π^* peak profiles.

It is therefore confirmed that C atoms in the heptagonal ring and the edge C atoms make the π^* peak profile wider.

4. Conclusion

CK-XANES spectra of graphitic structures having non-benzenoid rings were theoretically calculated to analyze the local structure of graphitic carbons. The conjugated and H-terminated edge C atoms in graphitic structures exhibited split π^* peak profiles. C atoms in the 5-membered pentagonal rings also exhibited split π^* peak profiles, due to the different π^* orbital states of the convex and concave sides in the vent structures. C atoms in the 7-membered heptagonal rings also exhibited split π^* peak profiles, due to the electrostatic potential hole at the heptagonal ring. It is concluded that edge carbon atoms in the graphitic structures and in non-benzenoid rings make the π^* peak profile wider in CK-XANES spectra.

Acknowledgement

This work has been supported by a Grant-in-Aid for Scientific Research from the Ministry of Education, Culture, Sports, Science and Technology of Japan under contract No. 18H02007.

References

[1] J. Stöhr, "NEXAFS Spectroscopy", Springer, 1992.
[2] Y. Muramatsu, R. Harada, and E. M. Gullikson, "Soft x-ray absorption spectroscopy of high-abrasion-furnace carbon black", Proceedings of the X-ray Absorption Fine Structure-XAFS13, pp. 511-513, 2007.
[3] Y. Muramatsu, R. Harada, M. Motoyama, and E. M. Gullikson, "Soft x-ray absorption spectra in the C *K* region of carbon black and spectral analysis using the discrete variational Xα method", Tanso, 236, pp. 2-8, 2009.
[4] Y. Muramatsu, T. Ooe, T. Okada, and E. M. Gullikson, "Identification of edge- and inner-carbon atoms in hexagonal carbon layers by X-ray absorption near-edge structure (XANES) and the first principle calculations", Proceedings of the 9th International Symposium on Atomic Level Characterizations for New Materials and Devices '13, ALC'13, pp. 516-518, 2013.
[5] Y. Muramatsu, K. Murayama, and T. Okada, "Relationship between the π^* peak height and width in C *K*-XANES of graphitic carbons", Proceedings of the 10th International Symposium on Atomic Level Characterizations for New Materials and Devices '15, ALC'15, 27p-P-38, 2015.
[6] S. J. Clark, M. D. Segall, C. J. Pickard, P. J. Hasnip, M. J. Probert, K. Refson, and M. C. Z. Payne, "First principles methods using CASTEP", Krystallogr., 220, pp. 567−570, 2005.
[7] T. Mizoguchi, I. Tanaka, S.-P. Gao, and C. J. Pickard, "First-principles calculation of spectral features, chemical shift and absolute threshold of ELNES and XANES using a plane wave pseudopotential method", J. Phys. Condens. Matter, 21, 104204, 2009.
[8] H. Adachi, M. Tsukada, and C. Satoko, "Discrete Variational Xα Cluster Calculations. I. Application to Metal Clusters", J. Phys. Soc. Jpn. 45, pp. 875-883, 1978.
[9] J. H. Underwood, E. M. Gullikson, M. Koike, P. J. Batson, P. E. Denham, K. D. Franck, R. E. Tackaberry, and W. F. Steele, "Calibration and standards beamline 6.3.2 at the Advanced Light Source", Rev. Sci. Instrum., 67, 3372, 1996.
[10] K. Suenaga and M. Koshino, "Atom-by-atom spectroscopy at graphene edge", Nature, 468, pp. 1088-1090, 2010.

Smart Aquaponics Design Using Internet of Things Technology

R. Hafid Hardyanto[1], Prahenusa Wahyu Ciptadi[2]

[1,2] Informatics, Universitas PGRI Yogyakarta, Yogyakarta, Indonesia

[1] hafid@upy.ac.id, [2] nusa@upy.ac.id

Abstract. Smart Aquaponic is a system of planting plants and maintaining fish in one container with integrated Internet of Things technology. Aquaponic is the process by which plants utilize nutrients derived from fish feces which, if left in the pond, will become toxic to the fish. Plants function as a filter of vegetation which will break down these toxic substances into substances that are not harmful to fish. Plants also supply oxygen to the water used to maintain fish. With this cycle there will be a mutually beneficial cycle between plants and fish. This system is very profitable, because the land used will not be too large. The purpose of this study was to develop a smart aquaponic system using Internet of Things technology. In this system aquaponics are equipped with sensors that function as monitors and actuators. Actuators can be controlled by users using smart phones. Sensor monitoring can be observed by users using smart phones. The method used in this study uses the Microsoft Solution Framework (MSF) with the waterfall system development method and Object Oriented Development (OOD) method. Stages in this study includes problem identification, preliminary planning and design, piloting and implementation. Conclusion: This smart aquaponics design is ready to be implemented in small scale trials.

Keywords: Aquaponic, Internet of Things, Smart Aquaponic System

1. Introduction

Indonesia is known as an agrarian country where the majority of the population is earning a living by farming. In general, agriculture is carried out on people who live or live in villages because they have enough land to carry out crop cultivation activities. Different things are felt in urban areas. In the era of the industrial revolution 4.0 agricultural land in the urban areas increasingly narrowed. Urban communities who want to grow crops are constrained by the narrowness of their agricultural land. According to [4], Limited land for agriculture, yard and water is a problem that occurs in urban and dense residential environments. Narrow yards restrict the community's activities to cultivate vegetables and fruits with the aim of supporting the improvement of family nutrition and supporting food independence. Based on this, we need a farming method that can utilize narrow land. Suitable farming methods are farming using the aquaponic system. According to [1][2], Aquaponics is a sustainable method of raising both fish and vegetables, with this type of indoor farming, you grow substantially more food with less water, land and labor than traditional agriculture. Aquaponics, the combined culture of fish and plants in recirculating systems, Aquaponic systems offer several benefits. Dissolved waste nutrients are recovered by the plants, reducing discharge to the environment and extending wateruse.

In addition, the dense urban community activities can also limit the farming using aquaponic systems. Monitoring is needed that can be done anytime, anywhere without limited space and time. The solution that can be offered is to integrate IoT technology with aquaponic farming systems. Integration of IoT with aquaponic produces smart aquaponic. IoT can defined as an interconnection system among the Internet and other sensing and actuating devices which be able to operate information across various platforms via a unified framework such as cloud computing with seamless information representation, data analytics, and ubiquitous sensing [3]. It is expected that with smart aquaponic able to overcome the narrowness of existing agricultural land.

2. Literatur review

The research by [5] describe that integrating the agriculture field with Technology will make sound. Automatic soil features and condition fetching and decision taking can be done by using sensors and actuators, growing the seed and getting the yield is not the only thing, we can also provide security to farm land as well as to the product(obtained yield).

Presented [7] utilizing internet of things (IOT) model for agriculture, by victimization the projected approach, received updated info permits the farmers to address and even get pleasure from these changes. It's extremely difficult task that must give such data as a result of extremely localized nature of agriculture information specifically distinct conditions. The whole period of time and historical atmosphere info is anticipated to assist to attain economical management and utilization of resources

In the study conducted by [6], tell that the future of agriculture depends on strong partnerships between all stakeholders in the food producing industry. Developments related to agricultural research and smart technology must continue to advance in the fields of genetics, breeding, crop management, conservation and environmental protection as strategies to develop solutions to the increasing challenges of feeding the planet.

The research by [8] purpose to explore the possibility of applying IOT for agriculture to trace and track food quality and safety. Mobile application for food freshness investigation was successfully developed and the results showed that consumer mobile camera could be used to test the freshness of food. By applying the IOT technology this information could be shared with all the consumers and also the supervisors

3. Smart Aquaponics Infrastructure

Smart Aquaponics is one of the innovations that will be developed in UPY laboratory. The infrastructur is the main key of the smart aquaponics. The data of Smart Aquaponics can be accesed from mobile phone or other gadgets. Smart aquaponics consists of hardware and software. Hardware consists of water density sensor, light sensors, control for pumping motor, humidity sensor, power supplies (solar panels), wifi modules. The software consists of microcontroller programming software and user interface via the smartphone. Control system consists of atmega 328 microcontroller and ESP8266 module for wireless communication. Power supply consists of solar panels, batteries and control panels. The power supply for this system uses solar panels.

4. Proposed Work

4.1. Proposed Method

The communication in smart aquaponics consists on IoT unit that uses ethernet module with the capability of receiving and sending wireless connection. The component of IoT consists of hardware, software, cloud service. The design of hardware platform is built based on the needed. The hardware consists of microcontroller board that equipped by sensor module, ethernet module, water pump. The sensor module that being used water density sensor, light sensors, control for pumping motor, humidity sensor. Information being sent from the sensor, then being processed and saved in cloud that transmitted to android application. The application provides information data to the users about condition in smart aquaponics system.

4.2. System Design

The system design of smart aquaponics which described in figure 1 below. The sensor consists of a light sensor, humidity, and water density sensor. The water pump is used to circulate the water circulation in the system. The water pump can be controlled with a smartphone. Power supply uses solar panels. Water density sensor is placed in the pool to determine the condition of the water in the pond. Control system uses a microcontroller that is connected to the Ethernet module. Users can find out the status of light intensity, water density through a smartphone or other device.

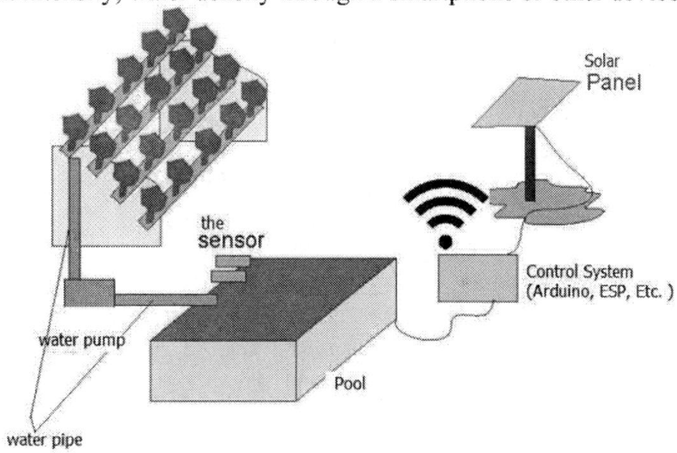

Figure 1. Design of smart aquaponics system.

4.3. Interface design

The interface design is made to facilitate users in observing information provided on the smart aquaponics system. This interface is designed according to the smartphone display. The interface consists of light information, water density, and can also control water pumps. The interface design is shown in Figure 2 below.

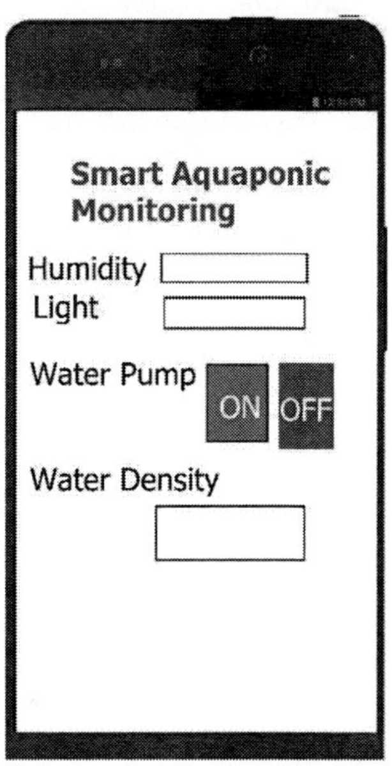

Figure 2. Interface design

5. Conclusion

IoT technology can be developed in agriculture field. The smart aquaponics system design consists of hardware and software systems. The hardware system consists of sensors, water pumps, control systems, and Ethernet modules. The software system consists of a mobile interface that functions as a user interface with the system. users can find out information from the smart aquaponics system via a smartphone. The results of this study are smart aquaponics system designs that can be tested and implemented on a small scale.

References

[1] Nelson and Pade, Inc.® , https://aquaponics.com/aquaponics-information/

[2] A Rizal, Y Dhahiyat, Zahidah, Y Andriani, A A Handakal and A Sahidin. 2017. *The economic and social benefits of an aquaponic system for the integrated production of fish and water plants*, ASEAN-FEN INTERNATIONAL FISHERIES SYMPOSIUM – 2017 IOP Conf. Series: Earth and Environmental Science 137 (2018) 012098 doi :10.1088/1755-1315/137/1/012098

[3] Gubbi, J., Buyya, R., Marusic, S., & Palaniswami, M. 2013. *Internet of Things (IoT): A vision, architectural elements, and future directions. Future Generation Computer Systems*, 29 (7), 1645–1660.Retrievedfrom http://www.sciencedirect.com/science/article/pii/S0167739X13000241

[4] T. Ferijal, Dewi Sri Jayanti, dan Diswandi Nurba. 2017. *Pemberdayaan Masyarakat Berbasis LEISA,* Prosiding Seminar Nasional, ISBN:978-602-1270-69-1, Banda Aceh, 10 September 2017, pp 41-57

[5] Laxmi S. Shabadi , Hemavati B. Biradar. 2018. *Design and Implementation of IOT based Smart Security and Monitoring for Connected Smart Farming* , International Journal of Computer Applications (0975 – 8887) Volume 179 – No.11, January 2018, pp 1-5

[6] Poonsri VATE-U-LAN, Donna QUIGLEY, and Panicos MASOURAS, *Internet of Things in Agriculture: a Case Study of Smart Dairy Farming in Ontario, Canada* , The 15th International Conference on Developing Real-Life Learning Experience: Smart Education for Sustainable Development

[7] Abhishek Singh, Sunil Kumar, Harish Nagar, 2018, *A Smart Agricultural Model by Integrating IoT* , International Journal for Research in Applied Science & Engineering Technology (IJRASET), ISSN: 2321-9653; IC Value: 45.98; SJ Impact Factor: 6.887 Volume 6 Issue IX, Sep 2018- Available at www.ijraset.com pp 729-732

[8] Gunawan Witjaksono, Almur Abdelkreem Saeed Rabih , Noorhana bt Yahya , Sagir Alva, 2018, *IOT for Agriculture: Food Quality and Safety,* ICEAMM 2017 IOP Publishing IOP Conf. Series: Materials Science and Engineering 343 (2018) 012023 doi:10.1088/1757-899X/343/1/012023

Stability of carbon-vacancy complexes in α-Fe

Masataka Mizuno[1,2], Waka Yamada[2], Kazuki Sugita[1,2], Hideki Araki [1,2]

[1] Center for Atomic and Molecular Technologies, Graduate School of Engineering, Osaka University, 2-1 Yamadaoka, Suita, Osaka 565-0871, Japan
[2] Division of Materials and Manufacturing Science, Graduate School of Engineering, Osaka University, 2-1 Yamadaoka, Suita, Osaka 565-0871, Japan

mizuno@mat.eng.osaka-u.ac.jp

Abstract. We investigated the stability and structure of C_nV and N_nV complexes in α-Fe using first-principles calculations. In the case of C_nV complexes, C_2V is most stable due to the energy gain by forming the C-C bonding. However, N_2V is less stable than N_1V. This is due to the repulsive interaction between N atoms in N_nV complex. The overlap population analysis reveals that the C 2s – C 2s antibinding interaction is reduced by the C 2s – C 2p bonding interaction. The interaction between the N 2s and N 2p orbitals is not strong enough to cancel the N 2s – N 2s antibonding interaction because the energy difference between the N 2s and N 2p orbitals is larger than that between the C 2s and C 2p orbitals.

1. Introduction

Carbon is the most important element in steel. In α-Fe, which crystallizes in BCC structure, carbon atoms occupy octahedral sites (O sites). The O site in BCC structure is distorted along the <100> directions. Therefore, a carbon atom strongly binds to a vacancy because of the energy gain associated with the release of distortion. Recently, theoretical studies suggest that C_2V complex, in which strong C-C bond is formed, is stable in α-Fe [1, 2]. However, the origin of C-C bond is still unclear. In this work, we have evaluated the stability and structure of C_nV complex in α-Fe using first-principles calculations. In addition, to clarify the origin of the C-C bond, we have analyzed the chemical bonding using first-principles molecular orbital calculations. For comparison N_nV complexes have been also calculated.

2. Computational Method

In order to obtain the energies and stable structures of C_nV and N_nV complexes, we employed the Vienna ab initio simulation package (VASP) [3, 4] with the generalized gradient approximation of Perdew-Burke-Ernzerhof (PBE) [5]. Potential based upon the all-electron projector augmented wave (PAW) method were used [6, 7]. We employed a supercell composed of 54 lattice sites with a BCC structure (3 × 3 × 3 unit cell). All calculations were spin-polarized. Structural relaxation of the atoms was continued until the forces on all atoms became less than 0.01 eV/Å. The stability of C_nV and N_nV complexes was evaluated by the formation energy $E^f_{X_n-V}$ with reference to the energy of an isolated C or N atom $E(X,0)$ and an isolated vacancy $E(0,\ V)$ in α-Fe matrix as follows:

$$E^f_{X_n-V} = [E(X_n, V) + nE(0,0)] - [nE(X,0) + E(0,\ V)]. \quad (1)$$

C and N atoms occupy O sites in α-Fe matrix. Around a vacancy in α-Fe matrix, there are six first nearest-neighbor (NN) and twelve second NN O sites. The formation energy of second NN C_1V, –0.046 eV, is 0.65 eV larger than that of first NN C_1V, –0.696 eV. Although the second NN O site is

IOP Conf. Series: Materials Science and Engineering **835** (2020) 012027 doi:10.1088/1757-899X/835/1/012027

less stable than the first NN O site, the second NN sites were also considered as initial sites of C and N atoms for searching stable structures of C_nV and N_nV complexes. We performed first-principles calculations for all the possible structures of C_nV and N_nV complexes up to $n = 6$. In order to analyze the chemical bonding, the discrete-variational (DV) Xα cluster method using a program code SCAT [8, 9] is employed with a cluster model composed of 59 lattice sites.

3. Results and discussion

The formation energy of C_nV and N_nV complexes are shown in Figure 1. The C_nV complex shows the minimum value at C_2V while the formation energy of N_nV complexes increases with increasing number of N atom. The formation energy of C_nV complex is still negative up to C_4V and steeply increases at C_5V and C_6V. The most stable structures of C_nV and N_nV complexes are shown in Figure 2. In C_2V, C_3V and C_4V complexes, C atoms occupy the first NN O sites and move inwards by structural relaxation to form C-C bonds. On the other hand, N atoms occupy the separated first NN O sites not to form the N-N bond in N_2V. In N_3V, two N atoms occupy the unstable second NN O sites to avoid the formation of N-N bonds. In N_4V, all of four N atoms occupy the unstable second NN O sites and the N-N bonds are not formed. These stable structures clearly indicate that the C-C bonds are stable in C_nV complex while strong repulsive forces occur between N atoms in N_nV complex.

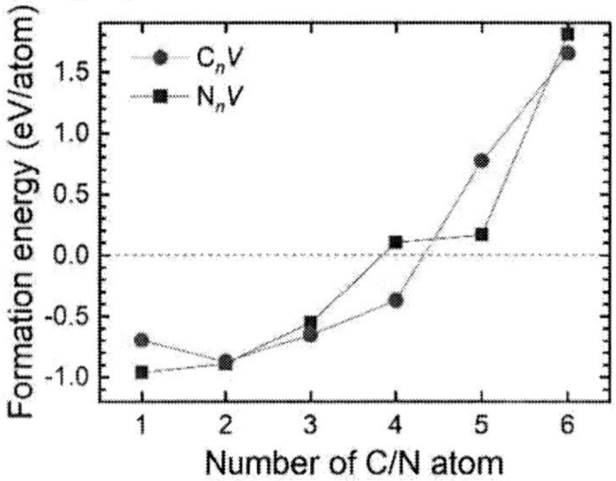

Figure 1. Formation energies of C_nV and N_nV complexes as a function of carbon or nitrogen atoms.

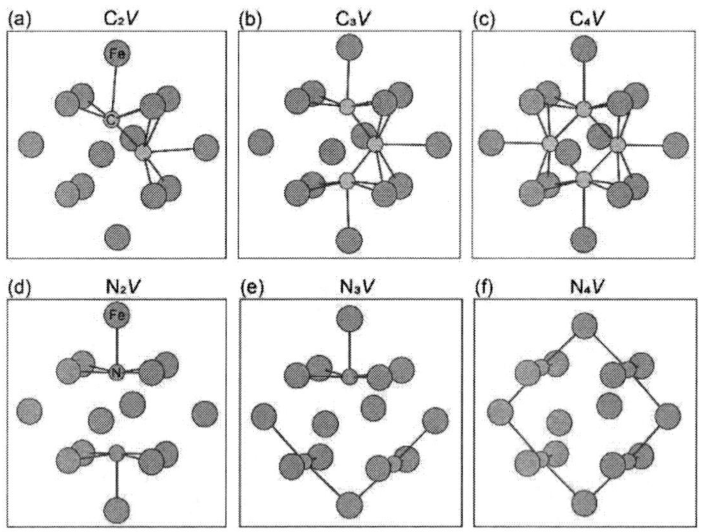

Figure 2. Stable structures of C_nV and N_nV complexes. (a) C_2V, (b) C_3V, (c) C_4V, (d) N_2V, (e) N_3V, and (f) N_4V.

\

Figure 3 shows the partial density of states (DOS) of C_2V and N_2V complexes. The most pronounced difference between C_2V and N_2V complexes is the positon of 2s orbitals. Therefore, contour plots of 2s orbitals are drawn in Figure 4. Regarding 2s component, there are two 2s orbitals in the valence band: one is a bonding and the other is an antibonding orbital. In N_2V, two N 2s orbitals appear at around -16 eV and the energy separation between the bonding and antibonding orbitals is very small. This is because two N atoms are well separated and the antibonding interaction between N 2s orbitals is not significant. On the other hand, the 2s antibonding orbital in C_2V is located just below the C 2p band because of the antibonding interaction between C 2s orbitals. As a result, C 2p component appears in the C 2s antibonding orbital.

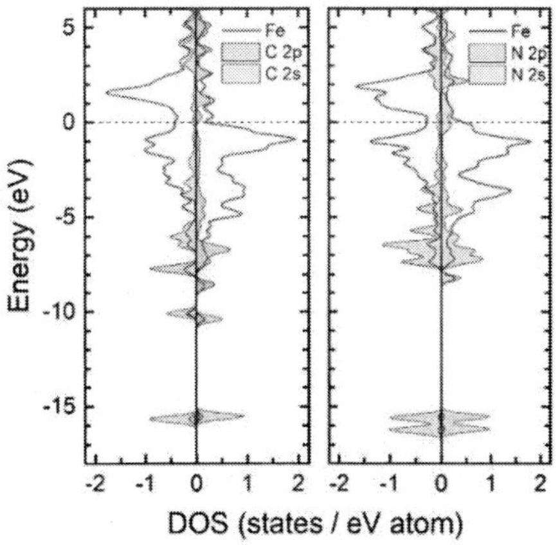

Figure 3. Partial DOS of C_2V and N_2V complexes.

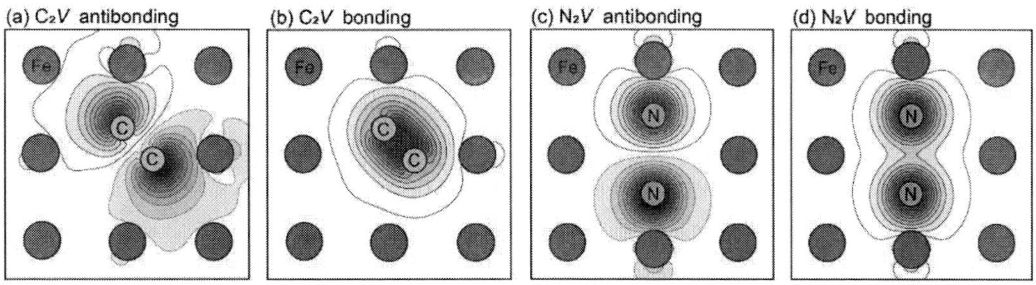

Figure 4. Contour plots of 2s-2s bonding and antibonding orbitals in C_2V and N_2V complexes.

(a) C 2s-C 2s antibonding in C_2V, (b) C 2s-C 2s bonding in C_2V, (c) N 2s-N 2s antibonding in N_2V, and (d) N 2s-N 2s bonding in N_2V.

In order to quantitatively evaluate the effect of 2p orbital, overlap populations at each molecular orbital for the C-C bond were calculated. Figure 5 shows the overlap population diagram for the C-C bond in C_2V. Regarding the C 2s antibonding orbital, a large negative component arises from the antibonding interaction between C 2s components and almost cancel the bonding component of the C 2s bonding orbital. The interaction between C 2s and C 2p components at the C 2s antibonding orbital is bonding and reduces the antibonding interaction between C 2s components. Therefore, the interaction between C 2s and C 2p orbitals play an important role to form the C-C bond.

In Figure 3, the DOS of C_2V and N_2V complexes with different structures were compared. In order to clarify the origin of the difference, the energy differences between 2s and 2p orbitals with the same bond length of C-C and N-N bonds are shown in Figure 6. The energy difference between the 2s bonding and 2s antibonding orbitals increases with decreasing bond length both in C_2V and N_2V complexes. This is because the antibonding interaction becomes stronger with decreasing bond length. The energy difference at the same bond length is almost comparable between C_2V and N_2V complexes. On the other hand, the energy difference between the 2s antibonding and the bottom of 2p band in N_2V is still more than 4 eV even if two N atoms become closer to each other. The energy difference between N 2s and N 2p orbitals is too large to interact with each other.

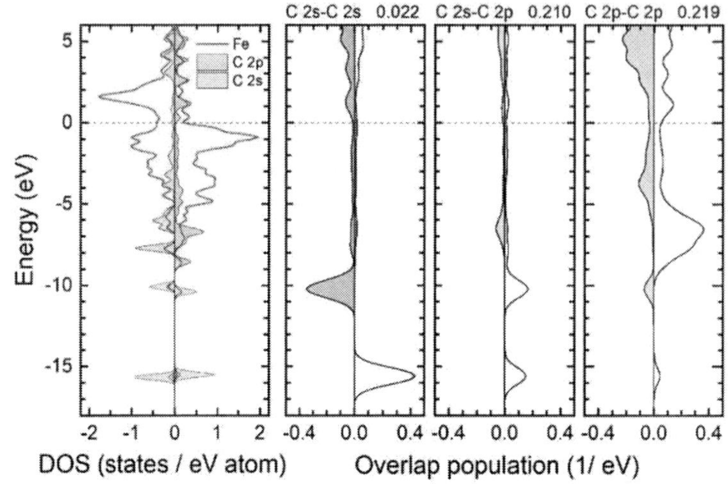

Figure 5. Overlap population diagram of C-C bond in C_2V complex.

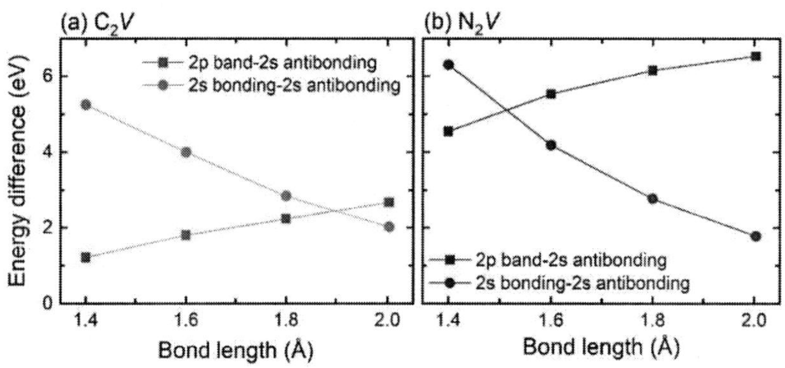

Figure 6. Energy differences between the 2s antibonding orbital and the bottom of 2p band, and the 2s bonding and antibonding orbitals in (a) C_2V and (b) N_2V complexes.

4. Conclusion

First-principles calculations were performed to investigate the stability and structure of C_nV and N_nV complexes in α-Fe. The antibonding interaction between C 2s orbitals in C_nV complex is reduced by the bonding interaction between C 2s and C 2p orbitals, which leads to the formation of C-C bonds in C_2V, C_3V and C_4V complexes. In N_nV complex, the interaction between N atoms is quite repulsive because of the antibonding interaction between N 2s orbitals. The energy difference between N 2s and N 2p orbitals is too large to reduce the antibonding interaction.

References

[1] Fu C -C, Meslin E, Barbu A, Willaime F and Oison V 2008 *Solid State Phenom.* **139** 157
[2] Barouh C, Schuler T, Fu C –C and Nastar M 2014 *Phys. Rev. B* **90** 054112
[3] Kresse G and Hafner J 1993 *Phys. Rev. B* **47** 558
[4] Kresse G and Furthmüller J 1996 *Phys. Rev. B* **54** 11169
[5] Perdew J P, Burke K and Ernzerhof M 1996 *Phys. Rev. Lett.* **77** 3865
[6] Blöchl P E 1994 *Phys. Rev. B* **50** 17953
[7] Kresse G and Joubert D 1999 *Phys. Rev. B* **59** 1758
[8] Adachi H, Tsukada M and Satoko C 1978 *J. Phys. Soc. Jpn.* **45** 875
[9] Ellis D E and Painter G S 1970 *Phys. Rev. B* **2** 2887

IOP Conf. Series: Materials Science and Engineering **835** (2020) 012028 doi:10.1088/1757-899X/835/1/012028

Recent applications of discrete variational multi-electron method

Kazuyoshi Ogasawara

School of Science and Technology, Kwansei Gakuin University, 2-1 Gakuen Sanda, Hyogo 669-1337 Japan

ogasawara@kwansei.ac.jp

Abstract. The discrete variational multi-electron (DVME) method is a first-principles many-electron calculation method based on the discrete-variational Xα (DV-Xα) molecular orbital method. There is also the relativistic version of the DVME method, which is based on the relativistic DV-Xα method. Since the explicit many-electron wave functions of the multiplet states of transition-metal (TM) or rare-earth (RE) ions in crystals can be obtained, the DVME method has been applied to the analyses of various luminescent materials such as solid-state lasers and phosphors. In this paper, as examples of the recent applications of the DVME method, the following researches will be briefly presented: (1) Visualization of the energy-structure relationship of Cr^{3+} in oxides, and (2) Creation of the predictive model of the multiplet energy using machine learning.

1. Introduction

The multiplet states of transition-metal (TM) or rare-earth (RE) ions in crystals are generally analyzed based on the semi-empirical approaches such as the ligand-field theory [1]. Such multiplet states cannot be analyzed by the ordinary first-principles electronic structure calculations based on the one-electron approximation. The discrete variational multi-electron (DVME) method [2] is a first-principles many-electron calculation method especially designed for non-empirical analyses of the multiplet states of TM and RE ions in crystals, which is a configuration interaction (CI) calculation method based on the discrete variational Xα (DV-Xα) molecular orbital (MO) method [3]. The DVME method has been applied to the nonempirical analyses of the optical spectra of TM-doped luminescent materials such as ruby [2], alexandrite [4], and Mn^{4+}-doped oxides [5, 6] / fluorides [7]. There is also the relativistic version of the DVME method [8] based on the relativistic DV-Xα method [9], in which CI calculations are performed using the fully relativistic four component molecular spinors. The relativistic DVME method has been utilized for analyses of core-excitation spectra such as TM $L_{2,3}$-edge x-ray absorption near-edge structures (XANES) [8,10]. Another remarkable applications of the relativistic DVME method are analyses of the f-f and f-d transition spectra of RE ions in crystals [11, 12]. The origins of the peaks in variety of spectra have been clarified based on configuration analyses of the explicit many-electron wave functions. Recently first-principles calculations of the ligand-to-metal charge transfer (LMCT) energies of TM ions in alumina [13] and RE ions in fluorite [14] were performed using the DVME method and the systematic variation of the LMCT energies were successfully reproduced. The DVME method was also utilized for nonempirical estimation of the crystal-field and Racah parameters of TM ions in crystals [15]. In this paper, as examples of the recent applications of the DVME method, the following researches will be briefly presented: (1) Visualization of the energy-structure relationship of Cr^{3+} in oxides, and (2) Creation of the predictive model of the multiplet energy by machine learning.

2. Computational method

The DVME method [2] is a first-principles CI calculation method based on the DV-Xα method [3]. The Slater determinants corresponding to the relevant electronic configurations are constructed using the MOs obtained by the DV-Xα cluster calculations and used as the basis functions to represent the many-electron states. By diagonalizing the many-electron Hamiltonian, the multiplet energies are obtained as the eigenvalues and the corresponding many-electron wave functions are obtained as the linear combinations of the Slater determinants. The physical properties such as optical absorption spectra can be calculated using the explicit many-electron wave functions.

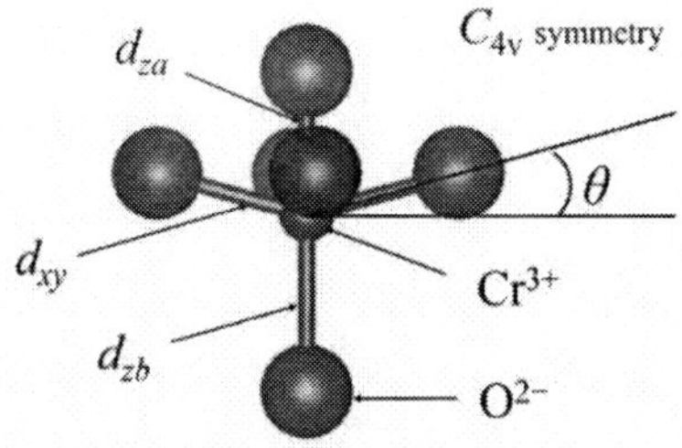

Figure 1. The CrO_6 model cluster with C_{4v} symmetry, which has four structural parameters d_{xy}, d_{za}, d_{zb}, and θ.

3. Results and discussion

3.1. Visualization of the energy-structure relationship of Cr^{3+} in oxides

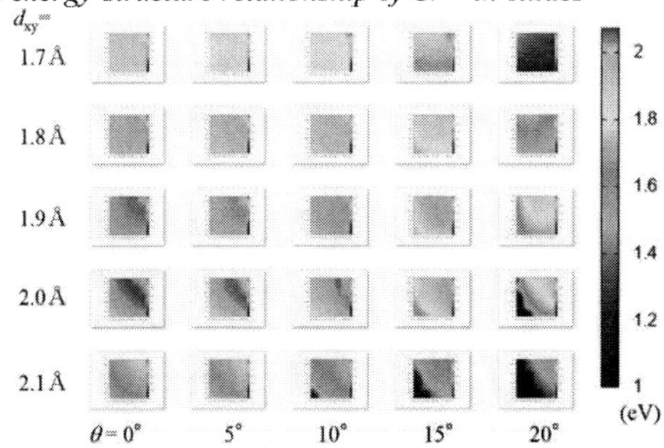

Figure 2. The matrix of the colored contour maps representing the structure dependence of the energy of the first excited multiplet state for CrO_6 cluster with C_{4v} symmetry. In each map, the variation of the energy depending on d_{za} and d_{zb} is represented, where the horizontal and vertical axes represent d_{za} (from 1.7 Å to 2.1 Å) and d_{zb} (from 1.7 Å to 2.1 Å), respectively, while d_{xy} and θ are fixed as the indicated values.

Cr^{3+}-doped crystals have been utilized as luminescent materials such as solid-state lasers and phosphors. However, for the theoretical design of novel Cr^{3+}-doped luminescent materials with controlled absorption and emission wavelengths, it is indispensable to clarify the relationship between the multiplet energies and the local structures. If we consider a CrO_6 cluster with O_h symmetry, the Cr-O bond length is the only structural parameter. Therefore the energy-structure relationship can be represented by the energy level diagrams where the horizontal axis represents the bond length while the vertial axis represents the multipelt energies [16]. In the case of CrO_6 clusters with D_{4h} and D_{3d} symmetries, there are two structural parameters and the energy-structure relationship can be represented by a colored contour map. However, for lower symmetries with three or more structural parameters, construction of similar map is not straightforward. In this work, in order to visualize the energy-structure relationship of Cr^{3+} in oxides with C_{4v} symmetry, first-principles calculations using CrO_6 clusters with various combination of the structural parameter values were performed. Then the energy-structure relationship depending on the four structural parameters was represented by the combination of the colored contour maps. As shown in **Figure 1**, the structure of CrO_6 cluster with C_{4v} symmetry can be specified with four structural parameters: three bond lengths d_{xy}, d_{za}, d_{zb}, and the elevation angle θ. For the systematic calculations, 625 model clusters were constructed by changing each bond length form 1.7 Å to 2.1 Å with 0.1 Å interval and θ from 0° to 20° with 5° interval. The energies of the first excited states for these 625 clusters were calculated by the DVME method. The number of sampling points were 100,000 for chromium and 10,000 for each oxygen. In order to obtain better prediction, the configuration-dependent correction (CDC) and the correlation correction (CC) were considered [2]. The colored contour maps of the energies of the first excited states depending on only d_{za} and d_{zb} were constructed for all possible combinations of the fixed values of d_{xy} and θ. These maps are arranged as a matrix in **Figure 2**. In this way, the energy-structure relationship depending on four structural parameters was clearly visualized. Since the emission usually occurs from the first excited state, this map would be useful for theoretical design of Cr^{3+}-doped luminescent materials with controlled emission energy. For example, the maximum value of the first excited energy was found for the structure with $d_{xy} = d_{za} = d_{zb} = 2.0$ (Å), and $\theta = 0$ (°), which indicates that the cluster with O_h symmetry has the highest emission energy.

3.2. Creation of the predictive model of the multiplet energy using machine learning

Recently machine learning technique has been extensively utilized for materials discovery and design [17]. In this work, in order to create an efficient predictive model of the theoretical energy of the first excited state, the systematic data calculated for the creation of the energy-structure maps were used as the training data. The machine learning was performed using the software known as the Waikato environment for knowledge analysis (WEKA) [18]. The predictive model was obtained using the multilayer perceptron algorithm including two hidden layers, each with 10 sigmoid nodes, considering the four structural parameters d_{xy}, d_{za}, d_{zb}, and θ as the attributes. The correlation between the predicted energy of the first excited state (eV) estimated by the machine learning model and the theoretical energy of the first excited state calculated by the DVME method (eV) is shown in **Figure 3**. The correlation coefficient was 0.9984. Even in the case of the 10-fold cross-validation [18], the correlation coefficient was relatively high (0.9748). Therefore the theoretical energy of the first excited state can be reasonably predicted by the predictive model obtained by the machine learning.

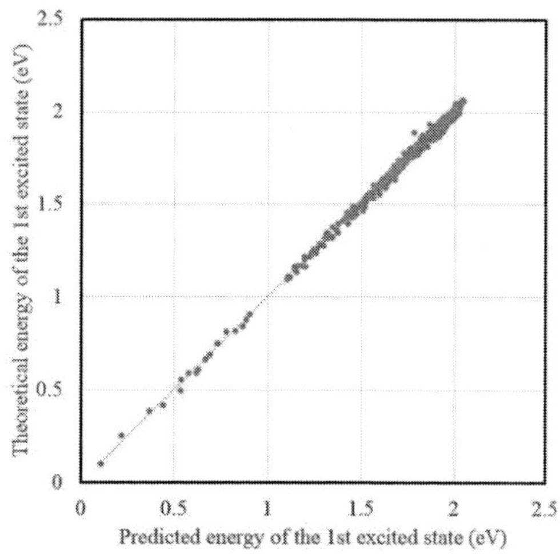

Figure 3. The correlation between the predicted energy of the first excited state (eV) estimated by the machine learning model and the theoretical energy of the first excited state calculated by the DVME method (eV).

4. Conclusion

By performing systematic first-principles calculations using the DVME method, colored contour maps representing the variation of the energy of the first excited state of Cr^{3+} in oxides with C_{4v} symmetry depending on the local structural parameters were constructed. The energy-structure relationship depending on four structural parameters were clearly visualized by the matrix of colored contour maps. Therefore, the DVME method is a powerful tool to provide guidelines for theoretical design of novel luminescent materials. In addition, by using the obtained systematic data as the training data for machine learning, an efficient predictive model of the theoretical energy of the first excited state was created. Therefore, the combination of the DVME method and machine learning is also a powerful tool for efficient prediction of the multiplet energies of TM and RE ions in crystals.

Acknowledgement
Supports from Kwansei Gakuin University are gratefully acknowledged.

References

[1] S. Sugano, Y. Tanabe, and H. Kamimura, *Multiplets of Transition-Metal Ions in Crystals* (Academic, New York, 1970).
[2] K. Ogasawara, T. Ishii, I. Tanaka and H. Adachi, Phys. Rev. B **61** ,143 (2000).
[3] H.Adachi, M.Tsukada, C.Satoko, *J. Phys. Soc. Jpn.*, **45**, 875 (1978).
[4] S. Watanabe, T. Sasaki, R. Taniguchi, T. Ishii, and K. Ogasawara, Phys. Rev. B **79**, 075109 (2009).
[5] M. Novita and K. Ogasawara, J. Phys. Soc. Jpn., **81**, 104709 (2012).
[6] M. Novita and K. Ogasawara, J. Phys. Soc. Jpn. **83**, 124707 (2014).
[7] M. Novita, T. Honma, B. Hong, A. Ohishi, K. Ogasawara, J. Lumin. **169**, 596 (2016).
[8] K. Ogasawara, T. Iwata, Y. Koyama, T. Ishii, I. Tanaka, and H. Adachi, Phys. Rev. B, **64**, 115413 (2001).
[9] A. Rosén, D.E. Ellis, H. Adachi, and F.W. Averill, J. Chem. Phys. **65**, 3629 (1976).
[10] S. Watanabe, T. Nagasaki, and K. Ogasawara, J. Appl. Phys. **110**, 123524 (2011).
[11] K. Ogasawara, S. Watanabe, H. Toyoshima, M.G. Brik, Handbook on Physics and Chemistry of Rare Earths, Vol. 37, Chapter 231 (Elsevier, 2007).
[12] S. Watanabe, K. Ogasawara, M. Yoshino, and T. Nagasaki, Phys. Rev. B **79**, 125128 (2010).
[13] S. Takemura and K. Ogasawara, Optical Materials: X **1**, 100005 (2019).
[14] S. Takemura and K. Ogasawara, J. Lumin. **214**, 116542 (2019).
[15] K. Ogasawara, K.C. Mishra, J. Collins, ECS J. Solid State Sci. Technol. **9**, 016011 (2020).
[16] K. Ogasawara, F. Alluqmani, and H. Nagoshi, ECS J. Solid State Sci. Technol. **5**, R3191 (2016).
[17] T. Lookman, F. Alexander, K. Rajan, *Information Science for Materials Discovery and Design* (Springer, Switzerland, 2016).
[18] I.H. Witten, E. Frank, M.A. Hall, and C.J. Pal, *Data Mining: Practical Machine Learning Tools and Techniques* (Elsevier, Amsterdam, 2000).

IOP Conf. Series: Materials Science and Engineering **835** (2020) 012029 doi:10.1088/1757-899X/835/1/012029

Alternative Solutions for Determining Village Funds Using Weight Product Method

Tri Hastono*, and Firdiyan Syah

Universitas PGRI Yogyakarta, Yogyakarta, Indonesia

*trihastono@upy.ac.id

Abstract. Based on the Minister of Home Affairs Regulation No. 113 of 2014 concerning Village Financial Management, the management of village funds must be carried out in a transparent, accountable, participatory, orderly, and budgetary discipline. Village fund management starts from January 1 to December 31 of the current year as outlined in the Village Budget (APBDesa). Village budget consists of village income, village expenditure, and Village Financing. The current village fund management mechanism is felt to be too convoluted and filled with obscurity. For example, the measurement for village fund recipients is still based on the results of deliberations. The results of the deliberations that are produced make it possible to have political elements or like-dislike factors. The result will be far from equal distribution and right on target. From this description, we need a tool that is used to help provide a fair decision regarding the recipient of village funds. The tool is a tool that is not influenced by things that cause no equalization and only requires a short time in its processing. So that the recipients of village funds are quickly identified by looking at several criteria. This study provides a solution for determining the recipients of village funds using the Weight Product method. Weight Product Method is one method often used to solve the problem of decision making with many attributes or criteria (MADM). This method is more efficient than some other methods in solving the problem of decision making with many attributes or criteria (Multi Attibute Decision Making). The criteria used in this study include: Poverty rates, area size, geographical difficulties, and infrastructure.

1. Introduction

Village fund management in the village becomes an important and fundamental aspect that must be understood and owned by stakeholders at the village government level (pemdes), especially village apparatus in realizing transparency and accountability of village finances. The basic principles of village financial management [1], starting from the planning stage to reporting and accountability of village finances and the duties and responsibilities of managing officials. Like PNPM Mandiri in Rural Areas this program aims to overcome the problems of public facilities and poverty alleviation [1].

Based on the Minister of Home Affairs Regulation No. 113 of 2014 concerning Village Financial Management, the management of village funds must be carried out in a transparent, accountable, participatory, orderly, and budgetary discipline. In the period of village fund management starting from January 1 to December 31 of the current year as outlined in the Village Budget (APBDesa). The village budget consists of village income, village expenditure, and village financing.

Planning in the management of village finances was prepared by the village secretary (sekdes). In the planning mechanism for managing village finances, the village secretariat first drew up a Village Regulation Plan (Raperdes) for the APBDesa. Then, the village head (village head) submits the APBDesa Raperdes to the Village Consultative Body (BPD) for discussion and mutual agreement. APBDesa Raperdes agreed upon, delivered to the regents or mayors through the camat. The regent or mayor then determines the results of the APBDesa Raperdes evaluation. The Camat can also evaluate the APBDesa Raperdes based on the delegation of authority from the regent or mayor.

The problem taken from this research is how to determine village funds to reach the right target, which is really in a location that really needs it. In many cases village funds that should have been received by the rightful recipient are even the unauthorized parties who enjoy them. Starting from the mechanism of village fund management which is still felt to be convoluted and full of obscurity. This started when determining the recipient of the village fund [2]. As far as researchers know, the recipients of village funds come from the results of deliberations. In the current political year, it is possible that the outcome of the deliberations is political. The like-dislike factor also has the potential to lead to mis-targeting to determine the recipients of village funds.

To overcome these problems requires a tool that is not affected by things that cause uneven distribution of village funds or misdirected. The tool is only independent and has the intelligence to help provide fair decisions regarding recipients of village funds by looking at several criteria. Based on references from several studies that have implemented this method, including research conducted applying the WP method to improve the effectiveness and efficiency of work for schools in determining student majors and providing reports on these majors [2].

This study provides a solution for determining the recipients of village funds using the Weight Product method. Weight Product Method is one of the methods often used to solve the problem of decision making with many attributes or criteria (MADM) [3]. This method is more efficient than some other methods in solving the problem of decision making with many attributes or criteria (Multi Attibute Decision Making). The criteria used in this study include: Poverty rates, area, geographical difficulty, and infrastructure. So that the results obtained obtained aids for determining the recipient of funds with the proposed method will be objective by looking at the specified criteria [4]. The results can provide decisions in determining the receipt of village funds.

2. Research Methods

2.1 Metode WP (Weight Product)

The WP method is one of the MADM (Multi Attribute Decision Making) methods [5]. MADM method is a method of decision making based on several attributes. The concept of the problem is evaluating alternative m A_i (i=1,2,...,m) in a set of attributes or criteria C_j (j=1,2,...,n), where each attribute is not interdependent with each other [6] [7]. This method makes a decision by determining the weight for each attribute [8] [9]. The WP method uses a normalization process, where the ranking of each attribute is raised first with the weight of the corresponding value attribute. This process is given in the following formula:.

$$S_i = \prod_{j=1}^{n} = {x_{ij}}^{w_j} \tag{1}$$

with i = 1, 2, ..., m, where :
S = stated alternative preferences,
x = states the value of the criterion,
w = states the criteria weights,
n = states the number of criteria.
w_j is a positive value for the profit attribute, and a negative value for the cost attribute.

IOP Conf. Series: Materials Science and Engineering **835** (2020) 012029 doi:10.1088/1757-899X/835/1/012029

2.2 Alternatif (Ai)

Alternative Ai with i = 1,2, ..., m is a number of different objects other than that it has the same opportunity to be chosen by the decision maker [10]. Data calculated by several villages that submitted proposals in the field of infrastructure activities in a number of 10 villages, namely: Geblagan, Gatak, Ngebel, Ngrame, Jetis, Jadan, Brajan, Gonjen, Pity, Twins, Relative preferences of each alternative are given as:

$$V_i = \frac{\prod_{j=1}^{n} X_{ij}^{\ w_j}}{\prod_{j=1}^{n} (X_j)^{w_j}} \tag{2}$$

dimana :

V : Alternative preferences, X : Criteria value,

w : Criteria weight.

The alternatives to be chosen are the top 5 that have the highest preference value..

2.3 Criteria (Cj)

To select a proposed priority village that has a proposed activity status "feasible" to receive BLM, then several decision-making criteria are needed [11].

C1: Compliance with the provisions on receiving village funds.
C2: Urge to be implemented.
C3: More useful for poor groups.
C4: Can be done by the community.
C5: The level of success and continuous development.
C6: Supported by existing resources.

The relationship data between alternatives and criteria is seen in TABLE I. These values are values given by TV for proposed activities proposed by alternative villages.

Tabel I. Alternative Value Table And Criteria

Alternatif (Ai)	**Kriteria (Cj)**					
	C1	**C2**	**C3**	**C4**	**C5**	**C6**
Geblakan	70	80	81	73	76	77
Gatak	77	45	80	80	66	87
Ngebel	79	90	82	81	80	79
Ngrame	78	80	90	84	80	80
Jetis	80	78	79	81	82	84
Brajan	80	85	82	81	79	83
Jadan	77	40	80	80	62	55
Gonjen	81	80	83	80	80	80
Kasihan	78	80	83	80	80	80
Kembaran	80	80	0,8	80	80	80

2.4 Bobot (w)

The weight is the value or level of relative importance of each criterion (Cj) given by the decision maker, in this case TV [12]. Weight values are given as:

$$W = \{w_1, w_2, w_3, \ldots, w_n\} \tag{3}$$

where is the value $\Sigma w_j = 1$

2.5 Algoritma

The steps of analyzing the problem with the WP method are illustrated in the flow chart in Figure 1 below.

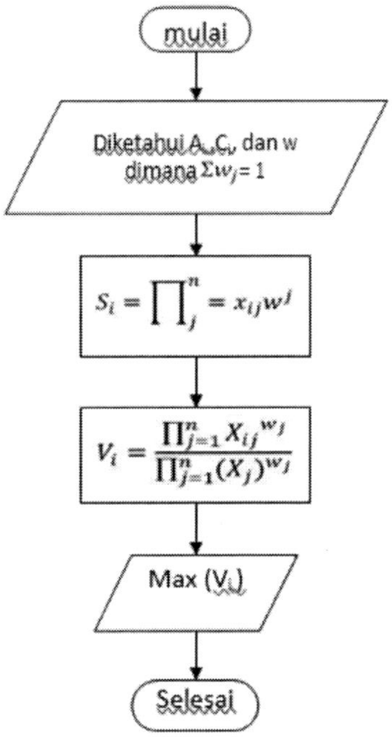

Figure 1. flowchart weight product method

The stages described by the flow chart in Figure 1 are explained as follows [13]:

1) Input data.
   Data entered, the data contained in Table 1. In addition, the weight determined by the TV is W = {w1, w2, w3, w4, w5, w6} = {0.18; 0.22; 0.23; 0.2; 0.08; 0.09} = 1. Then the values from the alternative village proposals based on their respective criteria are entered into the system.
2) Process S_i.
   This process is called the normalization process, the S vector value sought is the preference value for each alternative.
3) Process V_i.
   The process of finding vector V is used to make a ranking for each alternative.
4) V_i maksimum.
   Aiming to find the greatest value of the many alternatives or what is meant is, the largest alternative is the best alternative.

2.6 Entity Relationship Diagram

From the specified criteria and alternative data, the process carried out next is the database design of the proposed system. The picture below is a picture of the entity relationship diagram of the application.

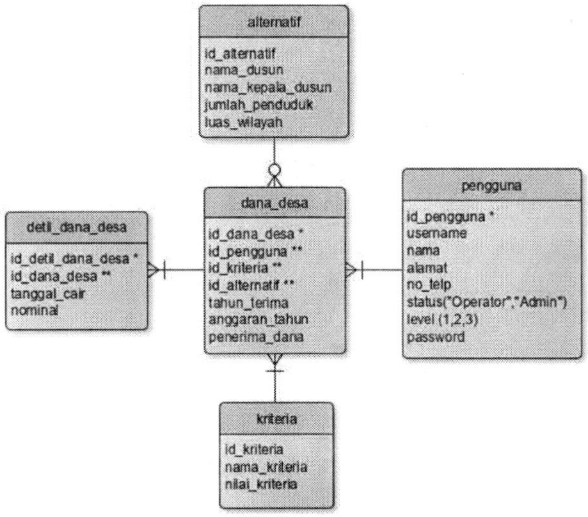

Figure 2 Entity Relationship Diagram of Application

2.7 Interface
From erd shown in the picture above, a prototype of the application was made.

3. Result and Discussion

3.1 The front page of the application
The front page of the application is the page that first appears when the application is run. This user validation will be performed on this front page. The components on this page have 2 components, namely: the content component and the button component. The content component on this front page is used to enter application user data which will later be matched with existing data in the database. The image below is the front page of the application.

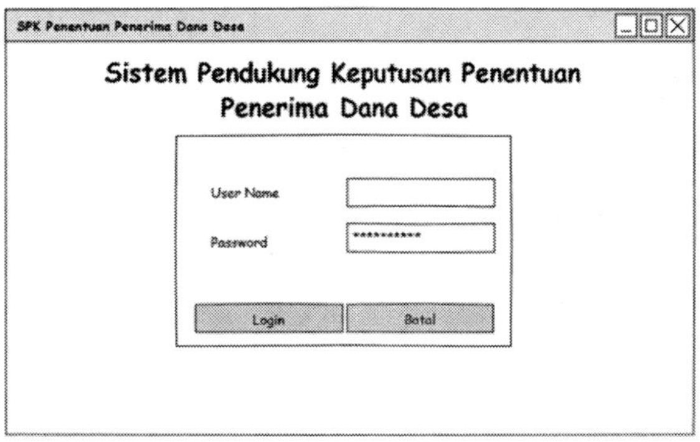

Figure 3 main page of application

3.2 Administrator page

Administrator page is the page used by administrators to update criteria data, alternatives, and officer data.

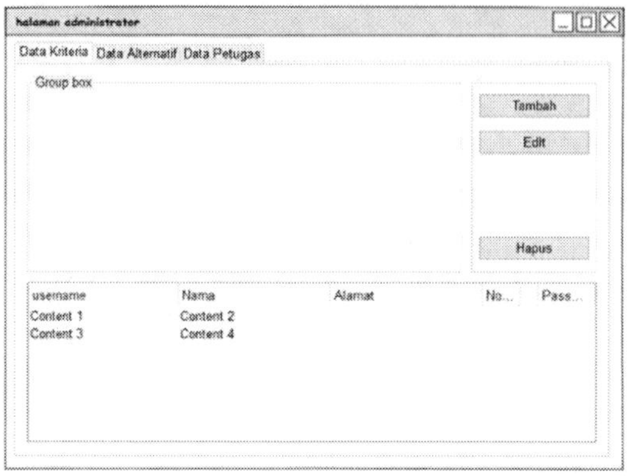

Figure 4 Administrator page

4. Conclusion

Village funds are funds used for village welfare. Determination of village funds should be done with a computing system. This is to avoid conflicts of interest and corrupt practices. A system is needed to enable the automation of determining the recipients of village funds. Many methods can be used to overcome these problems, one of which is the weight product. The system that will be built using the weight product method provides a solution for receiving village funds in the form of ranking.

References

[1] M. A. P. Muhammad Eko Atmojo, Helen Dian Fridayani, Aulia Nur Kasiwi, "Efektivitas dana desa untuk pengembangan potensi ekonomi berbasis partisipasi masyarakat di Desa Bangunjiwo Muhammad Eko Atmojo , Helen Dian Fridayani , Aulia Nur Kasiwi , Mardha Adhi Pratama Program Studi Ilmu Pemerintahan , FISIP Univeristas Muhammadiya," *Sos. Polit. Humaniora,* vol. 5, no. 1, pp. 126–140, 2017.

[2] J. D. Watts, L. Tacconi, S. Irawan, and A. H. Wijaya, "Village transfers for the environment: Lessons from community-based development programs and the village fund," *For. Policy Econ.*, no. December 2018, pp. 1–11, 2019.

[3] D. M. Khairina, D. Ivando, and S. Maharani, "47-1-139-1-10-20160929," *J. Infotel*, vol. 8, no. 1, pp. 1–8, 2016.

[4] J. Boonperm, J. Haughton, and S. R. Khandker, "Does the Village Fund matter in Thailand? Evaluating the impact on incomes and spending," *J. Asian Econ.*, vol. 25, no. 1, pp. 3–16, 2013.

[5] N. S. Fitriasari, S. A. Fitriani, and R. A. Sukamto, "Comparison of weighted product method and technique for order preference by similarity to ideal solution method: Complexity and accuracy," *Proceeding - 2017 3rd Int. Conf. Sci. Inf. Technol. Theory Appl. IT Educ. Ind. Soc. Big Data Era, ICSITech 2017*, vol. 2018-January, pp. 453–458, 2018.

[6] J. Kittur, "Optimal generation evaluation using SAW, WP, AHP and PROMETHEE multi - Criteria decision making techniques," *Proc. IEEE Int. Conf. Technol. Adv. Power Energy, TAP Energy 2015*, pp. 304–309, 2015.

[7] L. Menkhoff and O. Rungruxsirivorn, "Do village funds improve access to finance? Evidence from Thailand," *World Dev.*, vol. 39, no. 1, pp. 110–122, 2011.

[8] I. Meutia and Liliana, "Pengelolaan Keuangan Dana Desa," *J. Akunt. Multiparadigma*, vol. 8, no. 2, pp. 336–352, 2017.

[9] H. O. Maulana, I. M. Shofi, N. F. Rozy, and F. E. M. Agustin, "Application for determining mustahiq based on the priority using weight product method (case study: BAZIS DKI Jakarta)," *2017 5th Int. Conf. Cyber IT Serv. Manag. CITSM 2017*, 2017.

[10] D. M. Khairina, M. R. Asrian, H. R. Hatta, C. Science, and I. Technology, "Recruitment Using Weighted Product Method," *Int. Conf. Inf. Tech., Comput. Electr. Eng.*, pp. 297–301, 2016.

[11] A. Susanto, L. Latifah, Nuryasin, and A. Fitriyani, "Decision support systems design on sharia financing using Yager's fuzzy decision model," *2017 5th Int. Conf. Cyber IT Serv. Manag. CITSM 2017*, 2017.
[12] A. Ahmadi and D. T. Wiyanti, "Implementasi Weighted Product (WP) dalam Penentuan Penerima Bantuan Langsung Masyarakat PNPM Mandiri Perdesaan," *Semin. Nas. Apl. Teknol. Inf.*, pp. 19–22, 2014.
[13] F. V. Tangkumahat, V. V. J. Panelewen, and A. D. P. Mirah, "Dampak Program Dana Desa Terhadap Peningkatan Pembangunan Dan Ekonomi Di Kecamatan Pineleng Kabupaten Minahas," *Agri-Sosioekonomi*, vol. 13, no. 2A, p. 335, 2017.

IOP Conf. Series: Materials Science and Engineering **835** (2020) 012030 doi:10.1088/1757-899X/835/1/012030

Design Of Electricity Energy Sources In University Of PGRI Semarang Using Off-Grid Solar Panel Systems

Adhi Kusmantoro[1], Mega Novita[2], Th.Indriati W[3]

[1,2]Faculty of Engineering and Information, University of PGRI Semarang
[3]Faculty of Mathematics and Natural Sciences Information Technology, University of PGRI Semarang

[1]adhiteknik@gmail.com, [2]novita@upgris.ac.id, [3]Twindriati891@gmail.com

Abstract. The source of electricity in each university building in PGRI Semarang is obtained from each distribution transformer or power transformer. The power transformer used to distribute electricity in the central building (GP) has a capacity of 400 KVA, while the power transformer used in the main building (GU) has a capacity of 160 KVA. In the previous study, identification of the quality of electrical power in the PGRI University Semarang building, obtained a low power data factor and a large electrical network harmonics. The problem was solved by designing and installing bank capacitors and harmonic filters. To avoid dependence on PLN electricity and not using a generator when PLN electricity goes out, the source of electrical energy is designed by using solar energy sources through Photovoltaic. Solar Panel as an important component of solar power plants, converts sunlight into electricity. In designing off-grid solar panel systems it is only used at night or when PLN electricity goes out. In the design in the Central Building (GP) four solar panels and a battery unit are needed. Each solar panel unit consists of 14 solar panels and a battery unit consisting of an array of 110 batteries. The simulation results show that the simulation output power is almost the same as the design output power, which is 280,045 Watts. This shows the results of the design can be used to make electricity resources using solar panels. The inverter's output voltage is also in accordance with the voltage used in the load of the house, which is 220V AC.

1. Introduction

The high demand for electricity causes increased dependence on electricity, while the electricity source in Indonesia is increasingly decreasing, especially those that use fossil energy or non-renewable energy such as oil, coal and gas. In the study of identification of electrical quantities which include current, voltage, power factor, electric power and harmonic distortion (THD). Based on the results of the study there are several problems, namely the power factor that is still low, the imbalance of voltage (unbalanced voltage) and voltage increases (over voltage), unbalanced load sharing, and harmonics in the electricity network is quite large. The total percentage of harmonic distortion (% THD) of the central building is 67.9 and the main building is 76.3. In this study problem solving was emphasized in the issue of power factor and harmonics [1]. In DC network systems micro grid battery storage units are used to assist the main source of renewable energy (PV array) in providing stable voltage on the DC bus. Micro grid storage systems play an important role in providing services on demand. The lithium-ion batteries are used in this design, because these batteries have high energy density. In the micro-grid design is used three units of PV arrays and each unit consists of two PV arrays and two batteries. Battery units are used for storage of PV array sources and single phase gird AC. The PV array unit provides 22.8 V output voltage and the battery unit provides 22.78 V output voltage. Boost converter increases the voltage of the PV array and battery unit to a DC bus of 48 V. The PID controller is effectively used boost converter in raising or lowering the voltage. The advantage of using two battery storage systems is that if one unit is not sufficient to provide a voltage to the DC bus, then the battery unit on the AC grid side will also rapidly provide a voltage to the DC bus [2].

CLLC-type DC transformer (CLLC-DCT) is very popular in the hybrid AC/DC microgrid thanks to its high-power density advantage and good bidirectional power transfer capacity. In the hybrid AC/DC microgrid, the open-loop control is always utilized by the CLLC-DCT to cooperate with the bidirectional interlinking converter to realize the power and voltage conversion between the AC and DC bus. To solve this problem, a robust circuit parameters design scheme is proposed for the CLLC-DCT in this paper. With the proposed scheme, the designed CLLC-DCT exhibits good power transmission and voltage regulation ability in the hybrid AC/DC microgrid even when its actual inductors / capacitors values vary with the practical power and temperature [3][4]. This research presents a three-phase bidirectional DCAC converter suitable for operating as an interface between an energy storage system (ESS) based on a battery bank and an isolated microgrid with distributed generation. A control strategy is proposed where the DC-DC stage regulates the DC link voltage level while the DC-AC controls the three-phase output voltage and frequency, therefore acting as a grid-forming converter (GFC). A modified droop strategy is implemented to control the power generated inside the microgrid to avoid overcharging the ESS. Voltage unbalance reduction capabilities were implemented in order to deal with unbalanced loads. A 15kW prototype was successfully built and tested in different situations and the experimental results are shown [5].

This research investigates a specific type of converter based general interface, which can be used for AC microgrids integrating to the active distribution network. The interface, a Back-to-Back (BtB) converter, is used to connect the AC sides of the distribution network and the microgrid. The common DC link of the BtB converter is connected to one terminal of a bidirectional DC-DC converter whilethe other terminal of the DC-DC converter is connected with an energy storage system. By adopting proposed coordinated control within the interface, the distribution network and the microgrid can be decoupled in terms of the AC voltage and current. Furthermore, this interface can provide enhanced voltage stability and high power quality in the case of severe disturbances on the AC side. Hence, the microgrid can be integrated into the distribution network in a safe and flexible way. A testing system is built using RSCAD and the structure and control strategy proposed in this paper are verified through the Real Time Digital Simulator (RTDS) [6][7]. This research presents a closed loop digital controller of boost converter for renewable energy source like photovoltaic (PV) cell. PV cell efficiency varies according to weather conditions. The control loop is analyzed and the digital controller is designed using PI controller. The closed loop taking the sample values and continuously comparing reference signal with the output current and voltage to maintain a maximum constant output voltage. Even by changing the load value, output voltage and current kept constant. Also PWM switching frequency of MOSFET increases the performance of converter [8].

A novel multiple-input bidirectional DC-DC power converter to interface more than two dc sources of different voltage levels. Converter can be used to operate in both the buck and boost modes with bidirectional power control. It is also possible to independently control power flow when more than two sources are actively transferring power in either direction. This paper presents a power converter topology based on three switching legs of a standard 3-phase inverter module. The operation, analysis and design of the converter are presented with different modes of power transfer. Proposed converter is demonstrated for fuel cell vehicle application using real-time hardware-in-the-loop system. Results for a 5kW system are presented validating the theoretical analysis [9]. Due to the attractiveness of the Vehicle to Grid (V2G) technology, the present study focuses on designing a bidirectional DC charger of an EV. The DC-DC converter used in this study is the Dual Active Bridge converter. A design methodology as well as a control strategy is proposed [10][11]. In order to address the voltage control problem in a dc distribution system (dc-microgrid) with renewable sources and energy storage devices, in this research we propose a new Sliding Mode Controller for a dc-dc bidirectional power converter to control the dc bus voltage under instantaneous Constant Power Loads (CPLs). This type of loads introduce a destabilizing nonlinear effect on the converter through an inverse voltage term that can lead to significant oscillations in the dc bus voltage. A simplified scheme based on two cascaded converters was considered to validate the proposed voltage controller [12]. The energy generation from the renewable sources such as wind and solar are intermittent in nature. A battery port is mandatory for this purpose. The bidirectional power flow capability through a feedback back path to charge a battery from the load capacitor is also provided. Therefore, the battery is not only charged from the input source but also from the load capacitor. The number of outputs are independent from each other and the battery is charged from the second output. Steady state operation and theoretical analysis of the converter proposed is elaborated [13][14]. In chapter (1) discussed about previous research which is the background of this paper, chapter (2) discusses the PV system and the battery model, chapter (3) research methodology (4) results of research and discussion, and chapter (5) discuss the conclusions of the results of the testing and analysis.

2. Model Of a Photovoltaic System

To understand the PV system can be done by making a model in the form of an equivalent circuit, shown in fig 1. By looking at the equivalent circuit can be derived the PV current equation, in equation 1 [15].

$$i = i_{ph} - i_d - i_r \tag{1}$$

is the current (A) output of solar sell, I_{ph} is the current (A) which is affected by light intensity, I_d is current (A) of Shockley diodes, I_r is current (A) shunt, I_o is saturation current (A) of PV cells, q is electron charge (q = 1, 6 10^{-19} C), K is Boltzmann's constant (k = 1, 38 10^{-23} J / K), n is the ideal factor for solar sell, T is the temperature of solar sell, R_{sh} is the shunt resistance (Ohm) and R_s are internal series resistance (Ohm).

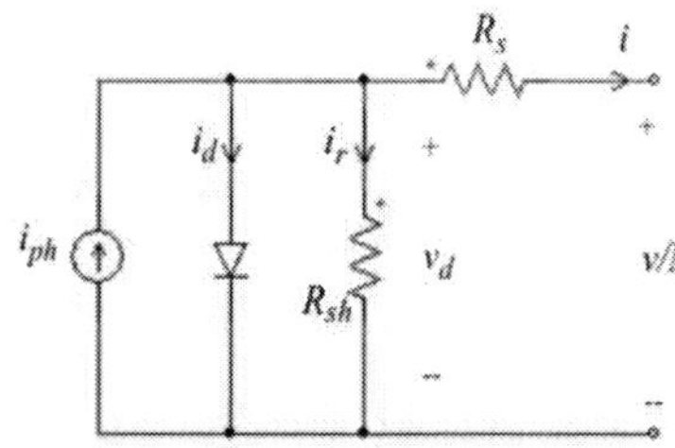

Figure 1. Solar sell equivalent circuit.

The I-V curve is basically affected by changes from two inputs which are solar insolation and array temperature. Practical arrays are composed of several connected photovoltaic cells and the observration of the characteristic the terminals of the photovoltaic array requires the inclusion of additional parameters [16].

$$I_{ph} = I_{sc0}.\frac{S}{S_0} + C_t.(T - T_{ref}) \tag{2}$$

$$I_d = I_0.\left\{e^{\frac{qvd}{AkT}} - 1\right\} \tag{3}$$

$$i_r = \frac{v_d}{R_{sh}} \tag{4}$$

$$v_d = \frac{v}{N_s} + i.R_s$$

The light generated current of the photovoltaic cell depends linearly on the solar irradiation and is also influenced by the temperature according to

$$I_{ph} = \left(I_{PV,n} + K_I \Delta T\right)\frac{G}{G_n} \tag{5}$$

where *Ipv,n* is the light-generated current at the nominal condition (usually 25 C and 10000W/m2), $\Delta T = T - Tn$ (being *T* and *Tn* the actual and nominal temperatures), G [W/m2] is the irradiation on the device surface, and *Gn* is the nominal irradiation. The diode saturation current *I0* and its dependence on the temperature may be expressed by equation :

$$I_0 = I_{s0} \cdot \left\{ \frac{T}{T_{ref}} \right\}^3 . e^{\frac{qEg}{Ak}\left(\frac{1}{Tref} - \frac{1}{T}\right)} \quad (6)$$

$$T = T_a + (k_s * S)$$

where Eg is the band gap energy of the semiconductor (*Eg* ˜1.12 eV for the polycrystalline Si at 25 C, and *I0,n*is the nominal saturation current.

3. Research methodology

In the research design of electrical energy sources using solar panels through the following stages:

a. Calculate the load power requirements per hour
b. Calculating power requirements per day
c. Calculating the total energy needed
d. Calculating the number of solar panels
e. Calculating the number of batteries
f. Creating a solar panel simulation model
g. Analyze the need for electricity with solar panels

4. Results and Discussion

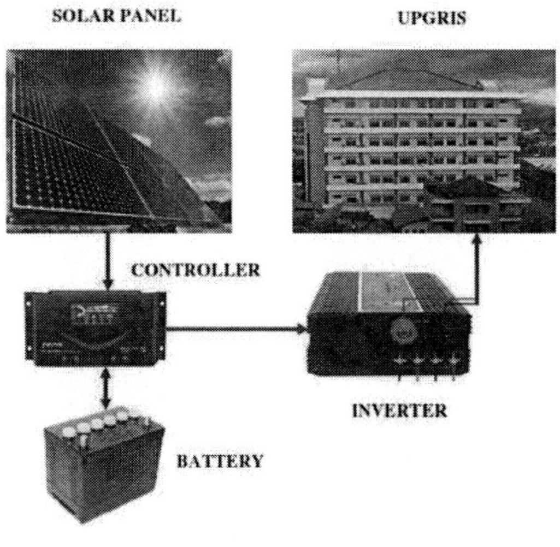

Figure 2. Design of UPGRIS solar panels.

In the use of solar energy sources for electrical power needs at the University of PGRI Semarang (UPGRIS) based on Figure 2, namely solar panels, charge controllers, batteries, and inverters. The work system used is an off-grid solar panel system, this system only works at night or if the PLN power source goes out.

4.1 Calculation of load power

In this discussion according to the power requirements of the Central Building (GP) of the University of PGRI Semarang in the amount of 26,120 Watts or hourly energy consumption of 26.12 KWh. This electricity load will turn on at 18.00 s / d 04.00, meaning that this electricity load will consume electricity for 10 hours. So the total energy consumed per day is 26.12 KW x 10 = 261.2 KW / day. So the total overall energy needed is 261.2 KWh.

4.2 Calculation of the Number of Solar Panels

From the calculation of the above energy consumption of 261.2 Kwh or equal to 261,200 Wh, we can choose the size of the solar panel that we need. In this design we use 500 Wp (Watt peak) solar panels, that is, this panel will produce a maximum of 500 Watts at 12 VDC per hour, so if these solar panels are installed a day (07.00-17.00 = 10 hours) assuming no cloudiness or intensity constant sunlight, this solar panel can produce 500 Wp x 10 hours of electricity = 5000 Wh or 5 KWh. The power output of a 500 Wp solar panel is produced at a temperature of 25 degrees Celsius, a pressure of 1.5 atm, with a light intensity of 1000 W / m2. In fact, on average 500 Wp solar panels will only produce an average electrical energy of around 600 Wh - 800 Wh in one day. This is because the intensity of sunlight is not the same throughout the day. In the design using 500 Wp solar panels, the following calculations are used:

261,200 / 500 x 10) = 56 solar panels. So the electricity produced is 56 units x 500 Wp = 28,000 Watts per one hour of heating at the peak of heating. In a day, more or less can produce electricity of 28,000 Wp x 10 hours Heating = 280,000 Wh.

4.3 Calculation of Battery Amount

For electrical energy storage systems 12 Volt batteries are used with a capacity of 200 Ah each. For battery needs, the calculation is carried out as follows:
Electric current per day is 261,200 / (12 x 10) = 2,177 Ampere
Number of batteries = 261,200 Watts / day: (12 x 200 Ah) = 110 batteries
Battery needs are considered to be able to serve every day without sunlight, so if we use a battery of 200 Ah 12 V, we need 110 batteries (200 x 12 x 110 = 264,000 watts).

4.4 Simulation Model

To see the results of the design of electrical resources using solar panels used a simulation model with a power simulator. In this simulation, it can be seen the accuracy of the number of solar panels and batteries on the electrical power produced.

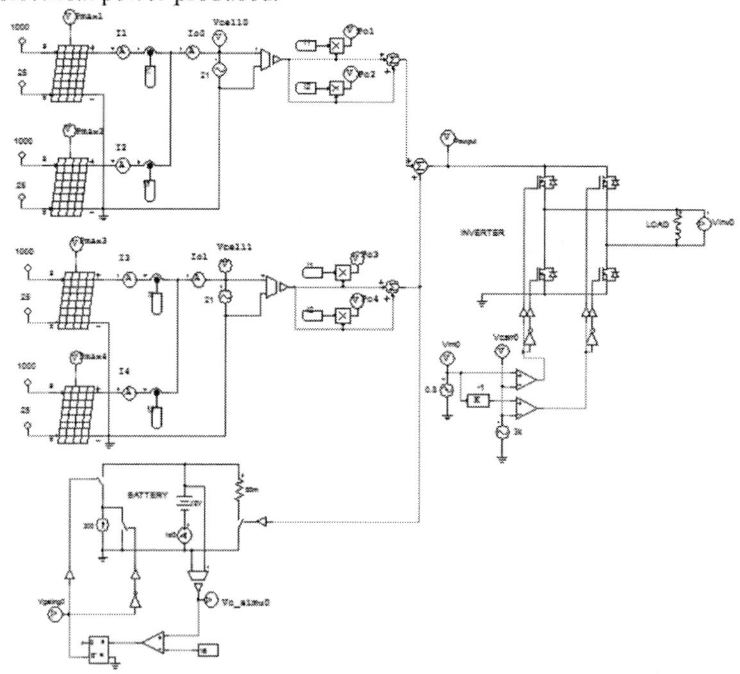

Figure 3 Design of solar panel simulations at the University of PGRI Semarang.

In the design of figure 3 it consists of four solar panel units and a battery unit. Each solar panel unit consists of 14 solar panels and a battery unit consisting of an arrangement of 110 12V DC batteries.

IOP Conf. Series: Materials Science and Engineering **835** (2020) 012030 doi:10.1088/1757-899X/835/1/012030

To change the DC output voltage from a solar panel or from a battery to an AC voltage, a single phase inverter is used.

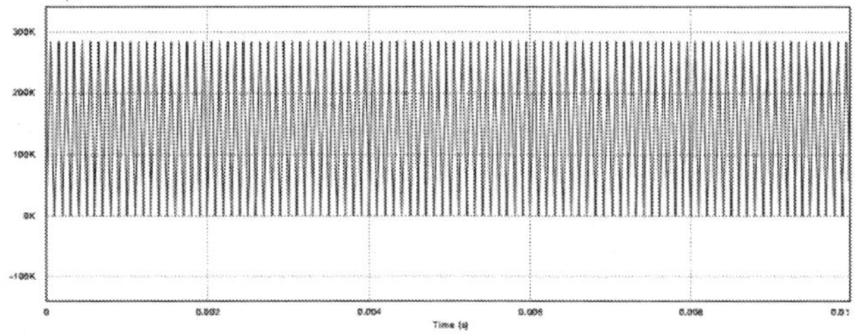

Figure 4. Output power at load.

Figure 4. shows the output power at the load, which is 280,045 Watts. The size of the simulation output power is almost the same as the design output power of 280,000 Watts, so the calculation of solar panel capacity and battery capacity is correct.

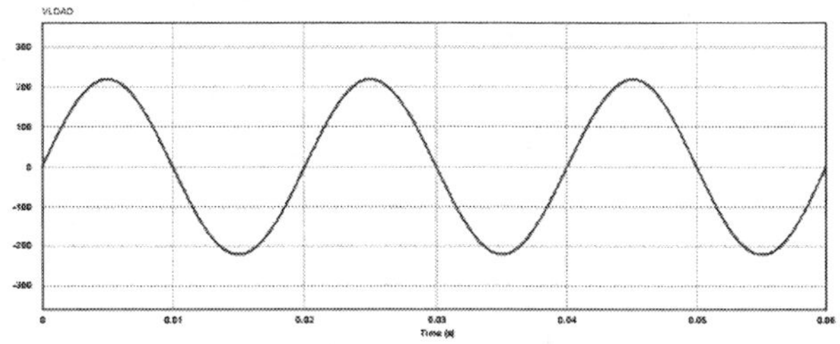

Figure 5. Voltage at load.

Figure 5. shows the shape of the voltage at the inverter's load or output voltage. The sinusoidal output voltage is in accordance with the voltage used in the housing load. The load voltage amplitude is 220V AC.

5. Results and Discussion

In designing an off-grid solar panel system it is only used at night or when PLN electricity goes out. In the design in the Central Building (GP) four solar panels and a battery unit are needed. Each solar panel unit consists of 14 solar panels and a battery unit consisting of an array of 110 batteries. The simulation results show that the simulation output power is almost the same as the design output power, which is 280,045 Watts. This shows the results of the design can be used to make electricity resources using solar panels. The inverter's output voltage is also in accordance with the voltage used in the residential load, which is 220V AC.

6. Conclusion

In designing off-grid solar panel systems it is only used at night or when PLN electricity goes out. In the design in the Central Building (GP) four solar panels and a battery unit are needed. Each solar panel unit consists of 14 solar panels and a battery unit consisting of an array of 110 batteries. The simulation results show that the simulation output power is almost the same as the design output power, which is 280,045 Watts. This shows the results of the design can be used to make electricity resources using solar panels. The inverter's output voltage is also in accordance with the voltage used in the load of the house, which is 220V AC.

References

[1] Adhi Kusmantoro, 2015, Sumber, Identifikasi Kualitas Daya Listrik Gedung Universitas PGRI Semarang, "Prosiding SNST ke-6 Tahun 2015 Fakultas Teknik Universitas Wahid Hasyim Semarang 1," pp. 1–7.

[2] Adhi Kusmantoro, 2018, "Voltage Stability in DC Micro Grid By Controlling Two Battery Units With Hybrid Network Systems," pp. 163–168.

[3] A. C. D. C. Microgrid, J. Huang, X. Zhang, Z. Shuai, and S. Member, 2018, "Robust Circuit

Parameters Design for the CLLC - Type DC Transformer in the Hybrid," vol. 0046, no. c.
[4] S. Singh, D. Fulwani, and V. Kumar, 2015, "Robust sliding-mode control of dc / dc boost converter feeding a constant power load," pp. 1–8.
[5] F. S. F. Silva, L. A. D. S. Ribeiro, J. G. De Matos, and S. Luis, 2014, "Bidirectional DC-AC Converter for Isolated Microgrids with Voltage Unbalance Reduction Capabilities," pp. 4985–4991.
[6] N. Deng, X. Zhang, S. Member, P. Wang, X. Gu, and M. Wu, 2013, "A Converter-Based General Interface for AC Microgrid Integrating to the Grid," pp. 1–5.
[7] K. Kurohane *et al.*, 2010, "A High Quality Power Supply System with DC Smart Grid."
[8] S. W. Mohod, 2013, "Closed Loop Digital Controller of DC-DC Converter for Renewable Energy Source (PV Cell)," pp. 112–116.
[9] A. Hintz, U. R. Prasanna, M. Ieee, K. Rajashekara, and F. IEEE, 2014, "Novel Modular Multiple-Input Bidirectional DC-DC Power Converter (MIPC)," no. d, pp. 2343–2350.
[10] I. Skouros and A. Karlis, 2018, "A bidirectional Dual Active Bridge Converter for V2G applications based on DC microgrid."
[11] J. A. Mane and P. A. M. Jain, 2015, "Design , Modelling and Control of Bidirectional DC-DC Converter (for EV)."
[12] V. Stramosk and D. J. Pagano, 2013, "Nonlinear Control Of A Bidirectional DC-DC Converter Operating With Boost-Type Constant-Power Loads," pp. 305–310.
[13] M. Yasir, A. Khan, L. Saeed, J. Saleem, M. Arif, and A. Majid, 2018, "A High Gain Multi-Port Bidirectional Non-Isolated DC-DC Converter for Renewable Integration."
[14] A. Frances, S. Member, A. Anvari-moghaddam, S. Member, J. M. Guerrero, and J. Uceda, 2018, "Dynamic Assessment of COTS Converters-based DC Integrated Power Systems in Electric Ships," vol. 3203, no. c, pp. 1–12.
[15] M. S. Hamad, A. M. Fahmy, and M. Abdel-Geliel, 2013, "Power quality improvement of a single-phase grid-connected PV system with fuzzy MPPT controller," *IECON Proc. (Industrial Electron. Conf.*, pp. 1839–1844.
[16] A. Hajizadeh, A. H. Shahirinia, N. Namjoo, and D. C. Yu, 2015, "Self-tuning indirect adaptive control of non- inverting buck – boost converter," pp. 2299–2306.

A Novel and Fast Memory Perturbation Method to Increase Exploration in Particle Swarm Optimization Algorithm

Muchamad Taufiq Anwar[1], Edy Winarno[1], Iswatun Chasanah[2], Saptono Nugrohadi[3]

[1]Universitas Stikubank, Semarang, Indonesia
[2]Satya Wacana Christian University, Salatiga, Indonesia
[3]Universitas PGRI Semarang, Semarang, Indonesia

muchamadtaufiqanwar@edu.unisbank.ac.id, edywin@edu.unisbank.ac.id, uswatunnana@gmail.com, sapton2000@gmail.com

Abstract Particle Swarm Optimization (PSO), one of the versatile nature-inspired optimization algorithm, continue to suffer from premature convergence despite the numerous amount of research trying to improve this algorithm. Many research had tried to address this issue but often use a complex algorithm which tax on computational time and complexity. This research introduced a novel perturbation method to mitigate premature convergence / to increase exploration while keeping the computational cost at a minimum. The particles' memories (i.e the position of personal and global best) are modified by a random multiplier which in turn will 'perturb' the particles' velocity. The implementation of this novel perturbation method in early iterations had resulted in 100% success rate in finding global optima in multimodal benchmark tests including the Rastrigin problem – whereas the original PSO failed in all benchmark tests – without adding a significant amount of computational complexity and time.

Keywords: PSO, premature convergence, memory perturbation, exploration

1. Introduction

Particle Swarm Optimization (PSO) is a versatile optimization algorithm used in several areas. The original PSO was introduced in 1995 by Eberhart and Kennedy, and had gained popularity in research to further improve the algorithm, customized it to particular problems, and combined it with other algorithms.

Some of the PSO enhancement strategies include the use of multi-sub-populations[1], particle restart[2], combining PSO with other search techniques[1], using different neighborhood topologies[3], and by having a better initial population [4]. Despite the rigorous research in PSO, one of the main problems which still remains for PSO is premature convergence, especially in problem with many local optima.

Many research had addressed this issue by incorporating perturbation to increase diversification/exploration. This kind of modification often incorporate complex computation and thus increase its computational and time cost. This research proposed a novel and simple perturbation method to mitigate premature convergence / to increase diversification/exploration while keeping the computation overhead at its minimum.

2. Literature Review

2.1. The Original PSO

PSO algorithm is inspired by how birds or fish flock for food. Individual / candidate solutions in PSO are called 'particles'. Each particle will move randomly in the search space to find the best solution. In each iteration, particles will move closer to the best solution by learning from the experience of the particle itself (cognitive learning, c1, applied to personal best, x*) and from other particle experiences (social learning, c2, applied to global best, xg). The c1 and c2 are positive constants which usually set to = 2. The formula of particle movement or "velocity update" vi at t+1 is shown in Equation 1.

$$v_i(t+1) = v_i(t) + c_1r_1[x_i^*(t) - x_i(t)] + c_2r_2[x^g(t) - x_i(t)]$$

The search capability of PSO in finding optimum solution lies in the tuning of particle's exploration and intensification in their movements[2], [5]. Exploration is when the particles explore a wider range of possibility, while intensification is when the particles move 'slowly' to find a better solution near a good solution (i.e the particles' memory of global best / personal best).

2.2. PSO problem: premature convergence

Research had shown that PSO tends to converge quickly (prematurely)[6], [7]. This characteristic of PSO will consequently cause a failure in finding an optimal solution, especially in a multimodal problem – a problem with multiple "good solutions" and one optimum solution. In a multimodal problem, PSO will tend to prematurely converge to / "trapped" in the "good solution" / local optima. This problem with the original PSO is called premature convergence[3], [4], [7]–[11]. Analysis showed that the premature convergence and the ability to find best solution / global optimum of PSO are both probabilistic[7], [12]. There had been numerous research which attempted to alleviate this issue such as[8], [9]. Some of the common strategies include the use of inertia weight and perturbation (explained later). These research often use a complex algorithm. The persistent premature convergence problem might mean that PSO needs exploration in general. It had been noted that the performance of the PSO mostly relies upon inertia weight and optimal parameter setting [10], [13], [14]. These parameters will affect the exploration and intensification ability of particles in PSO, thus research in solving premature convergence will focus on these parameter tuning.

2.3. Solving The Premature Convergence Problem: Exploration and Intensification Tuning

2.3.1. The Inertia Weight Strategy

The first attempt to tune the exploration and intensification is to incorporate inertia weight (IW)[15], which is not present in the original PSO. The inertia weight (α or w) refers to the amount of the contribution of the previous particle velocity to its current velocity. Inertia weight is intended to balance the exploitation and exploration when the particles are searching for optimal solution in a search space[13]. [15] used a randomized IW and in 1998, Shi and Eberhart used a linearly decreasing inertia weight (LDIW) (α) into the algorithm[16].

$$v_i(t+1) = \alpha v_i(t) + c_1r_1[x_i^*(t) - x_i(t)] + c_2r_2[x^g(t) - x_i(t)]$$

Where the value of α gradually decreases from α max to α min throughout the iterations:

$$\alpha(t) = \alpha_{max} - (\alpha_{max} - \alpha_{min})\frac{t}{T}$$

When w = 0, particle speed will only be affected by the personal best and global best position. This means that the particles will immediately change their position to the best position once the best position is known. A small value of w will increase intensification (local search). In contrast, when w is high, the current velocity of the particle will be affected by the previous velocity (inertia). Such particles maintain the previous speed even though a better position is known. A high value of w will increase diversification (global search). To successfully find the optimum values (maximum or minimum), high diversification is needed at the beginning of the iteration, while intensification is needed later in the iteration[17]. This can be achieved simply by using a linearly decreasing inertia weight. However, research had pointed out that LDIW strategy will cause PSO to prematurely converge to local optimum in multimodal problems[18], [19]. Other research also used an adaptive selection of inertia weight in which the value of inertia weight will be set according to certain criteria within the process of finding an optimum solution, such as in [20]. However, research also showed that LDIW is still a competitive strategy compared to other IW strategies if it's parameters are set properly[18].

Table 1. Research on inertia weight strategies in PSO

Reference	Inertia-weight strategy	Note
Y. Shi and R. Eberhart, 1998 [16]	Linearly decreasing	LDIW runs from 1.4 to 0
Bansal et al., 2011[19]	Linearly decreasing	Best LDIW runs from 0.9 to 0.4
Arasomwan, 2013[18]	Linearly decreasing	LDIW is still competitive against other variants
Farooq et al., 2017[4]	Linearly decreasing	LDIW in [19] applied twice
Nobile et al., 2018[14]	Self-tuning	Fuzzy logic
Agrawal, 2018[13]	Adaptive	Cumulative binomial probability
A. Agrawal and S. Tripathi, 2019 [15]	Adaptive	Binomial probability distribution

2.3.2. The Perturbation Strategy

Another attempt to alleviate premature convergence is by the use of 'perturbation'. A perturbation is a way to 'randomize' the particles' movement so that they can explore a more diverse candidate solution in the attempt to find the global optimum. The list of research in PSO perturbation strategy and the position of this research are shown in Table 2.

Table 2. Research on perturbation strategies in PSO

Reference	Object of perturbation	Perturbation method	Additional strategy
Zhihao Yuan et al., 2005[21]	Global best	Random number	Inertia weight jump threshold
S. Das, A. Konar, and U. K. Chakraborty, 2005[22]	Particles' position vectors	Differentially perturbed Velocity	-
A. M. Zavala, A. H. Aguirre, and E. V. Diharce, 2007[23]	Personal best	C-perturbation (Differential Evolution) and M-perturbation	-
A. H. Aguirre, A. M. Zavala, E. V. Diharce, and S. B. Rionda, 2007[24]	Personal best	Same as [23]	Constraint handling
E. Yang, A. Erdogan, T. Arslan, and N. Barton, 2007[25]	Global best and personal best	Random number	-
Maeda, 2007 [26]	Inertia weight	Simultaneous perturbation based on particles' gradient	-
Xinchao, 2010 [11]	Global best	Possibility theory	-
R. Kundu, S. Das, R. Mukherjee, and S. Debchoudhury, 2014[10]	Global best	Difference mean	Aging guideline, acceleration coefficients
L. Mengxia, L. Ruiquan, and D. Yong, 2016[27]	Swarm membership	Elitism	Anderson chaotic mapping
This research	Global best and personal best	Random number	Linear decreasing inertia weight

3. Method

In the standard PSO method, the "perturbation" is only applied to the difference between personal best and global best with the current particle position. The disadvantage of this method is that it limits the exploration capability of particles which can actually be greater. (Yang et. Al., 2007) proposed a new way to maintain diversification by applying direct 'perturbation' to the (current) personal best and global best in each iteration[25]. It uses two random number, r3 and r4, whinge range from -2 to 2, r [-2,2]. The formula for particles' velocity in [25] is shown in Equation 2.

$$v_i(t+1) = \alpha v_i(t) + \frac{[(1 + r_3)x_{pi}(t) - x_i(t)]}{3} + \frac{[(1 + r_4)x_g(t) - x_i(t)]}{3}$$

Inspired by[25], we propose a perturbation k*r which are applied to personal best and global best as shown in Equation 3. This modification will cause the particle to have a probabilistic movement within the enclosed area shown in Figure 2. With r is a random number between 0 to 1, r [0,1]. When k is set to 2, the (k*r) will produce perturbation that ranges from 0 to 2 and is applied as a multiplier for the personal best and global best. The perturbation is applied to both personal and global best as an attempt to further increase the exploration capability of the particles.

$$v_i(t+1) = \alpha v_i(t) + c_1 r_1 [k_1 r_3 x_i^*(t) - x_i(t)] + c_2 r_2 [k_2 r_4 x^g(t) - x_i(t)]$$

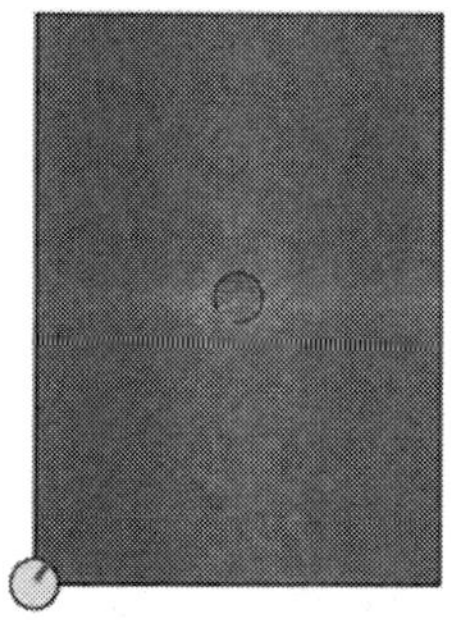

Figure 1. Comparison of particle movement in the original PSO (fuchsia arrow) vs the area of possible movement caused by perturbation (blue rectangle) where the yellow point is the previous best and the red point is the new best

We also experiment with a combination of IW proposed by Y. Shi and R.C. Eberhart [16] and use several range values of linear-decreasing iw. Here, we experiment with a w range that is somewhat wider (1-0) to see if the performance will get better. Experiments were done using Java programming language run in Netbeans. For all experiments, the parameters are set as follows: population= 50; max iteration = 100; x_{max} = 100; x_{min} = -100, v_{max} = 100, v_{min}= -100, C1 = 2, C2 = 2, k_1 = 2, k_2 = 2, and r[0,1]. IW is set with α_{max} = 1 and α_{min} = 0. Each experiment setting/type is repeated 1000 times. Then the success rate, min value and SD of min value for each test cases are calculated and compared.

4. Result

Based on 50 tests of each of the five function tests, the results of the combination of the Linearly-Decreasing Inertia-Weight (LDIW) and Memory Perturbation (MP) methods produced a 100% success rate in finding the minimum global for all test functions. This perturbation is only needed several times at the beginning of the iteration. Based on testing, the initial 10 iterations have given satisfactory results. The use of perturbation that is too long will actually weaken the power of LDIW's intensification. With 'nip' is the amount of perturbation we want at the beginning of the iteration (number of initial perturbation).

Table 3. Performance comparison between the original PSO, LDIW PSO, and LDIW-MP PSO

	Sphere	Schwefel 2.22	Rosenbrock	Rastrigin	Ackley
Standard PSO					
Mean	2,148973289	0,001496959	146,7146708	25,67040415	2,148973289
Std Dev	2,941967592	0,003330213	239,284953	17,058624	2,941967592
Success rate	0%	0%	0%	0%	0%
LDIW PSO					
Mean	0	0	0	0,119299392	0
Std Dev	0	0	0	0,323064281	0
Success rate	100%	100%	100%	88%	100%
LDIW MP-PSO					
Mean	0	0	0	0	0
Std Dev	0	0	0	0	0
Success rate	100%	100%	100%	100%	100%

The use of this LDIW-MP PSO method yielded satisfactory results with a 100% success rate in finding the minimum global in Sphere function, Schwefel Problem 2.22, Rosenbrock, and Ackley. But for Rastrigrin's function, the success rate is still around 88%. In the 12% of the experiments carried out, particles still trapped in local minima near the minimum global. This is understandable because the Rastrigin function does have many "traps" (local minima). This failure in the Rastrigin function is caused by the characteristics of the (standard) PSO where which the particles are “trapped” in a solution which appears to be the global minimum but is actually a local optimum.

To increase success in the Rastrigin function, an additional method is needed to maintain diversification so that particles can further explore the search space before determining the area for intensive searches. By maintaining diversification for some time, it is expected that particles can explore a more diverse search space and hopefully can visit an area near the minimum global – the intensification will proceed the digging into the minimum value.

The result showed that the best use of this perturbation method is to be applied only several times (10 times) in the initial iterations. Prolonged use of this perturbation will only result in particles moving randomly and failed to find global optimum, which is not what we want.

5. Conclusion

Despite a large amount of research, PSO still suffers from premature convergence where the particles are trapped in a sub-optimal solution in a multimodal problem. This research aims to explore a new strategy to increase the particles exploration capacity of particles in PSO in order to increase its capability to find the global optimum. Two perturbation factors are applied to global best and personal best to increase the exploration of search space in the hope of that the particle will stop at a point near the global optimum and continue to dig into the minimum value. The experimental result showed that this applying this perturbation method in early iteration, and combined it with LDIW can achieve 100% success rate in finding global optimum in five benchmark problems, i.e the Sphere function, Schwefel Problem 2.22, Rosenbrock, Rastrigin, and Ackley. Compared to other research, this research proposes a simple perturbation method which did not add a significant amount of computation complexity and time. The limitation of this study is that this method was only benchmarked against 2D problems. Further researches are open to experimenting on problems with a higher dimension.

References

[1] D. Wang, D. Tan, and L. Liu, “Particle swarm optimization algorithm: an overview,” *Soft Comput.*, vol. 22, no. 2, pp. 387–408, Jan. 2018.

[2] D. Tamayo-Vera, S. Chen, A. Bolufé-Röhler, J. Montgomery, and T. Hendtlass, “Improved Exploration and Exploitation in Particle Swarm Optimization,” 2018, pp. 421–433.

[3] K.-L. Du and M. N. S. Swamy, "Particle Swarm Optimization," in *Search and Optimization by Metaheuristics*, Cham: Springer International Publishing, 2016, pp. 153–173.
[4] M. U. Farooq, A. Ahmad, and A. Hameed, "Opposition-based initialization and a modified pattern for Inertia Weight (IW) in PSO," in *2017 IEEE International Conference on INnovations in Intelligent SysTems and Applications (INISTA)*, 2017, pp. 96–101.
[5] M. Oliveira, D. Pinheiro, M. Macedo, C. Bastos-Filho, and R. Menezes, "Better exploration-exploitation pace, better swarm: Examining the social interactions," in *2017 IEEE Latin American Conference on Computational Intelligence (LA-CCI)*, 2017, pp. 1–6.
[6] Y. Shi and R. C. Eberhart, "Empirical study of particle swarm optimization," in *Proceedings of the 1999 Congress on Evolutionary Computation-CEC99 (Cat. No. 99TH8406)*, pp. 1945–1950.
[7] G. Xu, Z. H. Wu, and M. Z. Jiang, "Premature convergence of standard particle swarm optimisation algorithm based on Markov chain analysis," *Int. J. Wirel. Mob. Comput.*, vol. 9, no. 4, p. 377, 2015.
[8] S. U. Khan, O. U. Rehman, N. Khan, A. Khan, S. A. A. Shah, and S. Yang, "Improving the Diversity of PSO for an Engineering Inverse Problem using Adaptive Inertia Weight," *Teh. Vjesn. - Tech. Gaz.*, vol. 25, no. 6, pp. 1631–1637, Dec. 2018.
[9] F. A. P. Paiva, J. A. F. Costa, and C. R. M. Silva, "A serendipity-based pso approach to delay premature convergence using scout particle," *Int. J. Innov. Comput. Inf. Control*, vol. 12, no. 4, pp. 1349–4198, 2016.
[10] R. Kundu, S. Das, R. Mukherjee, and S. Debchoudhury, "An improved particle swarm optimizer with difference mean based perturbation," *Neurocomputing*, vol. 129, pp. 315–333, Apr. 2014.
[11] Z. Xinchao, "A perturbed particle swarm algorithm for numerical optimization," *Appl. Soft Comput.*, vol. 10, no. 1, pp. 119–124, Jan. 2010.
[12] G. Xu and G. Yu, "Reprint of: On convergence analysis of particle swarm optimization algorithm," *J. Comput. Appl. Math.*, vol. 340, pp. 709–717, Oct. 2018.
[13] A. Agrawal and S. Tripathi, "Particle swarm optimization with adaptive inertia weight based on cumulative binomial probability," *Evol. Intell.*, pp. 1–9, Nov. 2018.
[14] M. S. Nobile, P. Cazzaniga, D. Besozzi, R. Colombo, G. Mauri, and G. Pasi, "Fuzzy Self-Tuning PSO: A settings-free algorithm for global optimization," *Swarm Evol. Comput.*, vol. 39, pp. 70–85, Apr. 2018.
[15] A. Agrawal and S. Tripathi, "Particle Swarm Optimization with Probabilistic Inertia Weight," Springer, Singapore, 2019, pp. 239–248.
[16] Y. Shi and R. Eberhart, "A modified particle swarm optimizer," in *1998 IEEE International Conference on Evolutionary Computation Proceedings. IEEE World Congress on Computational Intelligence (Cat. No.98TH8360)*, pp. 69–73.
[17] J. C. Bansal, P. K. Singh, M. Saraswat, A. Verma, S. S. Jadon, and A. Abraham, "Inertia Weight strategies in Particle Swarm Optimization," in *2011 Third World Congress on Nature and Biologically Inspired Computing*, 2011, pp. 633–640.
[18] M. A. Arasomwan and A. O. Adewumi, "On the performance of linear decreasing inertia weight particle swarm optimization for global optimization.," *ScientificWorldJournal.*, vol. 2013, p. 860289, Oct. 2013.
[19] J. C. Bansal, P. K. Singh, M. Saraswat, A. Verma, S. S. Jadon, and A. Abraham, "Inertia Weight strategies in Particle Swarm Optimization," in *2011 Third World Congress on Nature and Biologically Inspired Computing*, 2011, pp. 633–640.
[20] H. D. Purnomo and H.-M. Wee, "Particle swarm optimisation with adaptive selection of inertia weight strategy," *Int. J. Comput. Sci. Eng.*, vol. 13, no. 1, pp. 38–47, 2016.
[21] Zhihao Yuan *et al.*, "A Perturbation Particle Swarm Optimization for the Synthesis of the Radiation Pattern of Antenna Array," in *2005 Asia-Pacific Microwave Conference Proceedings*, vol. 3, pp. 1–4.
[22] S. Das, A. Konar, and U. K. Chakraborty, "Improving particle swarm optimization with differentially perturbed velocity," in *Proceedings of the 2005 conference on Genetic and*

evolutionary computation - GECCO '05, 2005, p. 177.

[23] A. M. Zavala, A. H. Aguirre, and E. V. Diharce, "Robust PSO-based constrained optimization by perturbing the particle's memory," in *Swarm intelligence, focus on ant and particle swarm optimization*, IntechOpen, 2007.

[24] A. H. Aguirre, A. M. Zavala, E. V. Diharce, and S. B. Rionda, "COPSO: Constrained Optimization via PSO algorithm," *Cent. Res. Math. (CIMAT). Tech. Rep. No. I-07-04/22-02-2007*, vol. 77, 2007.

[25] E. Yang, A. Erdogan, T. Arslan, and N. Barton, "An Improved Particle Swarm Optimization Algorithm for Power-Efficient Wireless Sensor Networks," in *2007 ECSIS Symposium on Bio-inspired, Learning, and Intelligent Systems for Security (BLISS 2007)*, 2007, pp. 76–82.

[26] Y. Maeda and N. Matsushita, "Simultaneous Perturbation Particle Swarm Optimization Using FPGA," in *2007 International Joint Conference on Neural Networks*, 2007, pp. 2695–2700.

[27] L. Mengxia, L. Ruiquan, and D. Yong, "The Particle Swarm Optimization Algorithm with Adaptive Chaos Perturbation.," *Int. J. Comput. Commun. Control*, vol. 11, no. 6, 2016.

[28] A. P. Windarto *et al.*, "Analysis of the K-Means Algorithm on Clean Water Customers Based on the Province," *J. Phys. Conf. Ser.*, vol. 1255, no. 1, 2019, doi: 10.1088/1742-6596/1255/1/012001.

[29] Budiharjo, T. Soemartono, A. P. Windarto, and T. Herawan, "Predicting School Participation in Indonesia using Back-Propagation Algorithm Model," *Int. J. Control Autom.*, vol. 11, no. 11, pp. 57–68, 2018.

[30] Budiharjo, T. Soemartono, A. P. Windarto, and T. Herawan, "Predicting tuition fee payment problem using backpropagation neural network model," *Int. J. Adv. Sci. Technol.*, vol. 120, pp. 85–96, 2018, doi: 10.14257/ijast.2018.120.07.

[31] Sumijan, A. P. Windarto, A. Muhammad, and Budiharjo, "Implementation of Neural Networks in Predicting the Understanding Level of Students Subject," *Int. J. Softw. Eng. Its Appl.*, vol. 10, no. 10, pp. 189–204, 2016.

IOP Conf. Series: Materials Science and Engineering **835** (2020) 012032 doi:10.1088/1757-899X/835/1/012032

3d Ions in Solids and Microscopic Crystal Field Effects: Theoretical Analysis and Relations with Experimental Spectroscopic Data

Mikhail G. Brik[1,2,3]

[1] College of Sciences, Chongqing University of Posts and Telecommunications, Chongqing 400065, People's Republic of China
[2] Institute of Physics, University of Tartu, W. Ostwald Str. 1, Tartu 50411, Estonia
[3] Institute of Physics, Jan Długosz University, Armii Krajowej 13/15, PL-42200 Częstochowa, Poland

mikhail.brik@ut.ee

Abstract. The transition metal ions with unfilled 3d electron shell are important activator ions for solid-state lasing and solid-state lighting. Since their spectroscopic properties depend significantly on the nearest environment, studies of the microscopic crystal field effects (influence of variation of the impurity center geometry) acquire additional importance. In the present paper, methods of calculating the crystal field strength $10Dq$ and its dependence on the interionic distances R are described. Relation between the $10Dq(R)$ functions and experimentally observed quantities such the Stokes shift is highlighted. The results of performed $10Dq(R)$ calculations for several crystals doped with various 3d ions are summarized and discussed.

1. Introduction

It is a well-known fact that the highly degenerated energy levels of free ions with unfilled d- and f-electron shells are split into a number of sublevels, if such ions are placed into crystalline solids. The number of the formed energy levels and their properties are determined by the crystal field created by the crystal lattice ions, whose geometrical arrangement and electrical charges are the most important factors affecting the optical properties of impurity ions [1].

The ions with the unfilled 3d electron shell are excellent probes for the crystal field effects, since their open external 3d shell makes them very sensitive even to small variations of the interionic distances and/or angles between the chemical bonds. In an ideal octahedral crystal field the five-fold degenerated 3d orbitals are split into two sets: three orbitals with the t_{2g} symmetry and two orbitals with the e_g symmetry. The energy interval between them is called the crystal field strength and is denoted by $10Dq$. The value of $10Dq$ increases with shortening interionic separations R and decreases otherwise. The distance dependence of $10Dq$ on R can be written in general as

$$10Dq = A/R^n , \qquad (1)$$

where A and n are some constants. The value of n if determined from the point charge model (PCM) of crystal field should be 5 [2]. However, numerical estimations of $10Dq$ for real systems with such n value do not yield good agreement with the experimental data because of oversimplified PCM assumptions, which completely neglects the quantum nature of the impurity ions and nearest neighbors, overlap of their wave functions etc. Various quantum chemical calculations allow for theoretical estimations of the n value; it appears to vary in a wide range from 3.5 to 7.3 for various systems [3-7]. It is also possible to estimate the n value from the experimental measurements of the absorption peaks shifts with pressure [8, 9].

Knowledge of n gives a deeper insight into the microscopic structure of the impurity ions and relations between the geometry of the impurity center and its energy level scheme. It also allows to estimate the experimentally observed quantities, such as Stokes shift (difference between the maxima of the absorption and emission bands related to the same electronic transition), thus linking together the microscopic crystal field effects with macroscopic characteristics of impurity centers.

In Section 2 the basic theoretical foundations needed for the microscopic crystal field effects studies are described briefly, and a short summary of the calculated results for a number of crystals containing the ions with unfilled 3d shell is given in Section 3. The paper is concluded with a short summary.

2. Calculations of 10*Dq* values: theoretical background

If the values of A and n in Eq. (1) are known, then it is possible to evaluate the constants of the electron-vibrational interaction with the a_{1g} and e_g normal modes of the octahedral complex as follows:

$$V_{a_{1g}} = -\frac{nA}{\sqrt{6}R^{n+1}}, V_{e_g} = \frac{V_{a_{1g}}}{\sqrt{2}}. \quad (2)$$

These estimated constants can be used for calculations of the energetical Stokes' shift $E_S(i)$, where i denoted the a_{1g} and e_g normal modes:

$$E_S(i) = 2S_i\hbar\omega_i = \frac{V_i^2}{M\omega_i^2}, \quad (3)$$

with M standing for the mass of a single ligand and S_i being the non-dimensional Huang-Rhys factor for ith normal mode with the frequency ω_i. The total Stokes shift, which is a result of interaction with both a_{1g} and e_g normal modes can be taken as a simple sum of the Stokes shifts coming from each mode separately [10, 11]. Moreover, if both pressure and distance dependences of $10Dq$ are known, it is possible to evaluate the bulk modulus B of a considered crystal from the following equation:

$$\left(\frac{\partial 10Dq}{\partial R}\right)_{R=R_0} = -\left(\frac{\partial 10Dq}{\partial P}\right)_{P=P_0}\frac{3B}{R_0}. \quad (4)$$

It is possible to use the crystal field theory or various ab initio methods for calculations of the $10Dq$ values. In the case of the ions with the $d^1(d^9)$ electronic configurations the $10Dq$ value is simply the separation between the t_{2g} and e_g states. If the ions with the $d^2(d^8)$ electronic configurations are considered, $10Dq$ equals to the energy interval between the $^3A_{2g}$ and $^3T_{2g}$ states originating from the ground ^3F term. If the ions with the $d^3(d^7)$ electronic configurations are studied, then $10Dq$ is the separation between the $^4A_{2g}$ and $^4T_{2g}$ states coming from the ground ^4F term [2].

The method of finding the distance dependence of $10Dq$ in Eq. (1) can be formulated as follows:

i) The energy levels of an impurity ion in a given crystal are calculated for different interionic distance with a small step (it is sufficient to take an equilibrium "impurity ion – ligand distance" R_0 and consider the interval from $0.9R_0$ to $1.1R_0$ with a step of $0.01R_0$).
ii) The extracted $10Dq$ values are plotted as a function of distance and then are fitted to the power law given by Eq. (1).
iii) Eqs. (2)-(4) can be used to estimate the Stokes shift and the bulk modulus, which then can be compared with the experimental data (if available) to check the validity of the performed calculations.

3. Results and discussion

The results of calculations of the $10Dq$ dependence on distance for a number of solids doped with the 3d transition metal ions are presented in this section. Different calculating techniques have been employed, e.g. discrete variational multielectron method (DVME) [12], exchange charge model (ECM) of crystal field [13], and plane-wave based CASTEP software [14]. The calculations were performed for varying interionic distances and the obtained $10Dq$ numerical values were approximated to the power laws, like in Eq. (1). The calculated $10Dq(R)$ distances are summarized below in Table 1. In all equations the distance R should be taken in Å, then the $10Dq$ value will be in cm^{-1}.

Table 1. Distance dependence of the $10Dq$ parameter for various solids and impurity ions.

Crystal and impurity ion	3d ion electron configuration	Calculated $10Dq(R)$ function, cm^{-1}	Method	Stokes shift, cm^{-1} (calc./exp.)	Reference
$Al_2O_3:T^{3+}$	d^1	$861347/R^{4.671}$	CASTEP	-	[7]
Rb_2CrF_6	d^2	$706048/R^{5.863}$	CASTEP	-	[15]
$Cs_2NaYF_6:Cr^{3+}$	d^3	$294650/R^{4.463}$	DVME	3652/3721	[5]
$Cs_2NaYCl_6:Cr^{3+}$	d^3	$557820/R^{4.374}$	DVME	1663/1612	[5]
$Cs_2NaYBr_6:Cr^{3+}$	d^3	$664190/R^{4.353}$	DVME	2256/2218	[5]
$SrTiO_3:Cr^{3+}$	d^3	$430120/R^{4.905}$	ECM	1706/-	[6]
$SrTiO_3:Mn^{4+}$	d^3	$880120/R^{5.799}$	ECM	3020/-	[6]
$SrTiO_3:Fe^{5+}$	d^3	$1612800/R^{6.550}$	ECM	4737/-	[6]
$KAl(MoO_4)_2:Cr^{3+}$	d^3	$477780/R^{5.623}$	ECM	3600/2800	[16]
$MgO:Cr^{3+}$	d^3	$1890500/R^{6.401}$	ECM	5200/4500	[17]
$ZnS:V^{2+}$	d^3	$193830/R^{4.421}$	ECM	667/-	[17]
$KZnF_3:Ni^{2+}$	d^8	$541029/R^{6.016}$	CASTEP	1793/1395	[18]

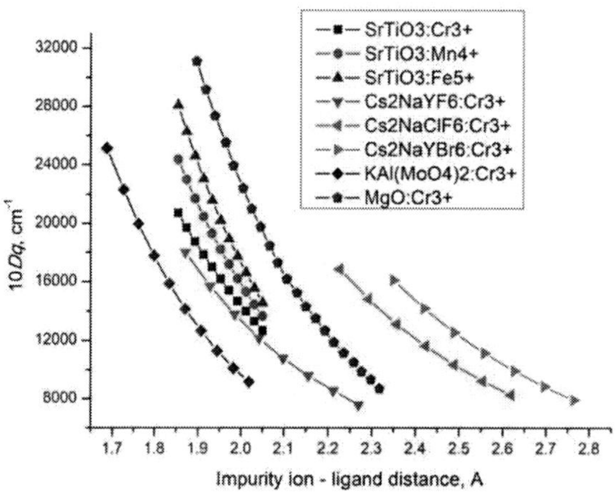

Figure 1. Calculated dependences of $10Dq$ on distance for a number of crystals from Table 1.

Figure 1 illustrates the calculated $10Dq(R)$ dependences for some of systems from Table 1. The decreasing trend of the crystal field strength with increasing interionic separation is clearly seen. The calculated $10Dq$ values are shown by symbols, they were approximated by the power law functions, whose equations can be found in Table 1 as well. The vast majority of the considered systems are the Cr^{3+}-bearing materials, which is explained by their importance for many applications (such as solid-state lasers, phosphors for the solid-state lighting etc) and availability of the experimental data on the absorption and emission spectra of these crystals.

The data in Table 1 show considerably wide range of the obtained n values, from about 4.3 to about 6.55. The deviation of n from the point charge value of 5 is explained by the influence of covalent effects, overlap of the wave functions of the impurity ions and ligands, formation of the molecular orbitals (rather than pure atomic states) etc. The distance dependence of $10Dq(R)$ appears to be essentially host- and impurity ion- dependent. It can be also seen from Table 1, that the value of n increases with increased charge of an impurity ion. This fact can be attributed to more pronounced variations of the electron density around highly charged ions, since they stronger attract the electron density of the s- and p-states of ligands.

The calculated Stokes shifts in Table 1 agree well with the experimental data. Additionally, using the $10Dq(R)$ dependence for $Cs_2NaYCl_6{:}Cr^{3+}$, its compressibility was estimated to be 9.88×10^{-4} $kbar^{-1}$, that is very close to the experimental value of 9.70×10^{-4} $kbar^{-1}$ [19], which serves as an additional argument confirming validity of the performed analysis of the microscopic field effects.

4. Conclusions

Several examples of calculations of the crystal field strength $10Dq$ for the transition metal ions with the unfilled 3d electron shell are given in the present paper. Importance of knowledge of the $10Dq(R)$ dependence is emphasized by the possibility of extracting the experimentally observed value of the Stokes shift and host's crystal compressibility. If the experimental and theoretical Stokes shifts are in good agreement, this circumstance can give an opportunity of predicting the Stokes shift for other systems, which may be of high importance for assessing the application perspectives of optical materials.

Acknowledgement

This research was supported by the Programme for the Foreign Experts (Grant No. W2017011) and Wenfeng High-end Talents Project (Grant No. W2016-01) offered by Chongqing University of Posts and Telecommunications (CQUPT), Estonian Research Council grant PUT PRG111, European Regional Development Fund (TK141) and NCN project 2018/31/B/ST4/00924.

References

[1] B. Henderson, R.H. Bartram, Crystal-Field Engineering of Solid-State Laser Materials, Cambridge University Press, (2000).

[2] S. Sugano, Y. Tanabe, H. Kamimura, Multiplets of Transition-Metal Ions in Crystals, Acad. Press, 1970.

[3] M. Moreno, M.T. Barriuso, J.A. Aramburu, "The Huang-Rhys factor $S(a_{1g})$ for transition metal impurities: a microscopic insight," J. Phys.: Condens. Matter, vol. 4, pp. 9481-9488, 1992.

[4] K. Wissing, J.A. Aramburu, M.T. Barriuso, M. Moreno, "Optical properties due to Cr^{4+} in oxides: density functional study," Solid State Commun., vol. 108, pp. 1001-1005, 1998.

[5] M.G. Brik, K. Ogasawara, "Microscopic analysis of the crystal field strength and lowest charge-transfer energies in the elpasolite crystals Cs_2NaYX_6 (X=F, Cl, Br)," Phys. Rev. B, vol. 74, p. 045105, 2006.

[6] M.G. Brik, N.M. Avram, "Microscopic analysis of the crystal field strength and electron-vibrational interaction in cubic $SrTiO_3$ doped with Cr^{3+}, Mn^{4+} and Fe^{5+} ions," J. Phys.: Condens. Matter, vol. 21, p. 155502, 2009.

[7] M.G. Brik, "Ab-initio studies of the electronic and optical properties of $Al_2O_3{:}Ti^{3+}$ laser crystals," Physica B, vol. 532, pp. 178-183, 2018.

[8] M. Grinberg, A. Suchocki, "Pressure-indiced changes in the energetic structure of the $3d^3$ ions in solid matrices," J. Lumin., vol. 125, pp. 97-103, 2007.

[9] M. Grinberg, "Principles of Energetic Structure and Excitation-Energy Transfer Based on High-Pressure Measurements," in: Phosphors, Up Conversion Nano Particles, Quantum Dots and Their Applications, R.-S. Liu (Editor), vol. 1, pp.67-151, 2017.

[10] A.M. Woods, R.S. Sinkovits, J.C. Charpie, W.L. Huang, R.H. Bartram, A.R. Rossi, "Computer modeling of the optical properties of substitutional chromium impurities in halide elpasolites,"

J. Phys. Chem. Solids, vol. 54, pp. 543-552.

[11] M.T. Barriuso, J.A. Aramburu, M. Moreno, "Coupling with the Jahn-Teller mode for triplet states of MF_6 ($M=Mn^{2+}$, Cr^{3+}) complexes: Dependence on the M-F distance and influence on the Stokes shift," Phys. Status Solidi B, vol. 196, pp. 193-208, 1996.

[12] K. Ogasawara, T. Iwata, Y. Koyama, T. Ishii, I. Tanaka, and H. Adachi, "Relativistic cluster calculation of ligand-field multiplet effects on cation $L_{2,3}$ X-ray-absorption edges of $SrTiO_3$, NiO, and CaF_2," Phys. Rev. B, vol. 64, p. 115413, 2001.

[13] B.Z. Malkin, in Spectroscopy of solids containing rare-earth ions, A.A. Kaplyanskii, B.M. Macfarlane, Editors, p. 13, North-Holland, Amsterdam, 1987.

[14] S.J. Clark, M.D. Segall, C.J. Pickard, P.J. Hasnip, M.J. Probert, K. Refson, M.C. Payne, "First principles methods using CASTEP," Z. Krist., vol. 220, pp. 567-570, 2005.

[15] M.G. Brik, N.M. Avram, C.-G. Ma, "First-principles calculations of structural, electronic, optical, elastic properties and microscopic crystal field effects in Rb_2CrF_6," Comput. Mater. Sci., vol. 50, pp. 2482-2487, 2011.

[16] M.G. Brik, C.N. Avram, "Exchange charge model and analysis of the microscopic crystal field effects in $KAl(MoO_4)_2:Cr^{3+}$," J. Lumin., vol. 131, pp. 2642-2645, 2011.

[17] N.M. Avram, M.G. Brik, I.V. Kityk, "Studies of variation of interionic distances and crystal field effects in $ZnS:V^{2+}$ and $MgO:Cr^{3+}$," Opt. Mater., vol. 32, pp. 1668-1670, 2010.

[18] M.G. Brik, G.A. Kumar, D.K. Sardar, "Ab initio, crystal field and experimental spectroscopic studies of pure and Ni^{2+}-doped $KZnF_3$ crystals," Mater. Chem. Phys., vol. 126, pp. 90-102, 2012.

[19] O.S. Wenger, R. Valiente, H.U. Güdel, "Influence of hydrostatic pressure on the Jahn-Teller effect in the $^4T_{2g}$ excited state of $CrCl_6^{3-}$ doped $Cs_2NaScCl_6$," J. Chem. Phys., vol. 115, pp. 3819-3826, 2001.

IOP Conf. Series: Materials Science and Engineering **835** (2020) 012033 doi:10.1088/1757-899X/835/1/012033

EKA v1: Emergency Call Auto-register, an Emergency Warning System based on Internet of Things for Intensive Care Patient at Hospital

Prisma Megantoro*, Hendra Ari Winarno

School of Advance Technology and Multidiscipline, [Universitas Airlangga]

*megantoro.prisma@gmail.com

Abstract. The emergency calling device for nurse is generally located in every ward of a hospital. It is an important facility for every inpatient. However, generally these devices are still in the form of conventional electrical components that uses long wiring. This is because this system must reach all inpatient rooms. In addition, the caller feature is also very limited, usually only in the form of an emergency button. The caller system is also not yet integrated with the hospital database, so it becomes an obstacle for nurses and doctors to understand the patient's condition in an emergency. This article discusses the design of a nurse calling device that is connected wirelessly via a wifi network on the ward. There are also several types of calling features so that the nurse or doctor immediately knows the type of assistance a patient needs. In addition, the client and server system is integrated with the hospital database, which shows a detailed patient history. This device uses the NodeMCU ESP 8266 board as a client and Raspberry Pi 3 as a server. The features placed on this device work well as a design and are successfully implemented for one ward per server.

1. Introduction

Industrial Revolution 4.0 provoked the development of many new technologies that simplify human work and can be used remotely without a physical cable connection. One of the technologies created in this era is the Internet of Things (IoT)[1][2][3]. IoT itself has grown so rapidly along with the development of telecommunications technology. With a fast wireless connection, it makes all electronic devices can be connected wirelessly and remotely. Utilization of IoT is very broad, almost in all areas of life can take advantage of this technology, especially in the field of medical services. Medical services in hospitals must also follow technological developments, especially with the use of this IoT [4]. With this, will improve hospital medical services for patients [5][6]. If this happens, the level of public health will also increase as more sophisticated technology is used. The relationship between patients and doctors and nurses is one of the most important services [7]. Handling quickly and precisely becomes a major factor in this regard. Nurse call feature is one of them, where the patient is needed when an emergency requires effective and efficient treatment. Then a nurse call device is needed that is mobile, concise, and has many features. On the other hand, this device must also be able to provide sufficient detailed information about the patient's history. This is done so that doctors and nurses can handle emergency conditions quickly and precisely.

2. Literature Review

This device design method includes designing the client side with arduino microcontroller [8]–[14] and server side. For the client side it is used specifically for 1 patient in 1 room on the ward. This client microcontroller [15] device is used as an emergency caller. And the server side is used as an interface or indicator for emergency calls by patients.

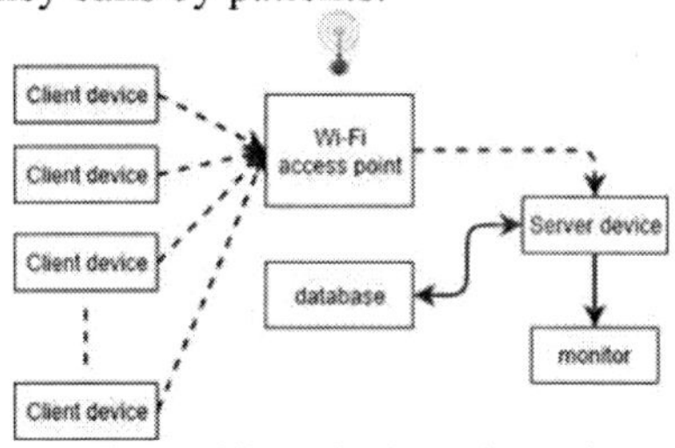

Fig. 1. Client device schematic

If the patient is in need of emergency treatment, then just press one of the feature buttons available on the client device. After that, the server will receive data on the type of assistance and the patient's room number. The data will appear in the server device interface layer along with detailed data about the identity and history of the patient's stay. The data was obtained from the hospital database. It is expected that with this, the doctor or nurse can prepare the equipment or medication needed before entering the patient's room.

2.1 Client device

This device is placed in each patient's room. The device is used if the patient needs the help of a doctor or patient. Caller features included are 3; emergency, infusion, and usual help. This device uses a NodeMCU 8266 microcontroller [16] board and the user interface is a push button. The power supply uses a Li-Po type battery, so the device can be carried by patients anywhere, including when going to the bathroom. As a battery power source, simply connect it to the adapter if the battery has run out.

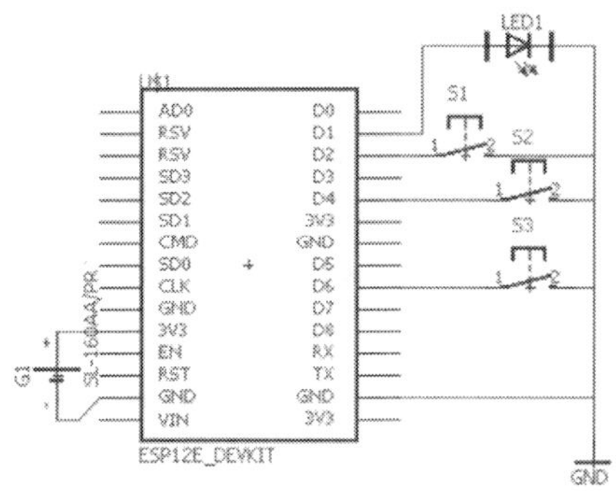

Fig. 2. Client device schematic

The NodeMCU board is a microcontroller [17] board specifically used for control that requires an internet connection wirelessly or with Wi-Fi media. This device can be connected to both local and public networks. In this article, both the NodeMCU microcontroller [18] board and the Raspberry Pi [19] must always be connected to a hospital or ward Wi-Fi network.

This program initialization is done by connecting to the access point device on the nearest Wi-Fi network. After that, initialize the push button and indicator features. In general, this device program is made to detect logic changes when push buttons for each of the caller features that have been mentioned. If one pugh button is pressed, the logic will change from HIGH to LOW. This change in push button logic will trigger a program to execute a sub program to send device ID data and help type codes to the server device. The data sent will be processed and displayed by the server in the nurse's or doctor's office. Thus, the client device packaging is made so that its use is easy, practical, and informative for patients.

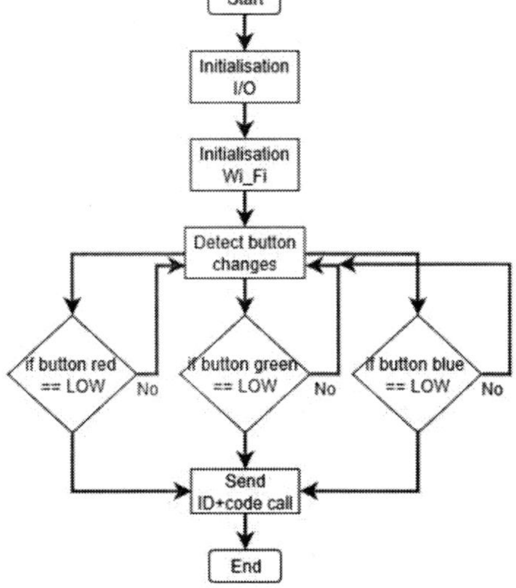

Fig. 3. Flow chart of client device

2.2 Server device

This device is placed in the nurse's or doctor's room in a ward. This device is used to process and display patient data while simultaneously displaying call signals from patients in the room and certain types of assistance. This device uses a Raspberry Pi 3 B type mini computer with a user interface using the Python programming language. This device is not mobile, so it requires a wired power supply. For power supply use an AC / DC adapter 5 V and 3 A. Use of this type of adapter is adjusted to the specifications of the electrical requirements of the Raspberry Pi and the monitor. This Raspbbery Pi type mini computer board is very easy to find on the market with various types. For this study using type 3 B. Raspberry Pi board in this type is equipped with a Wi-Fi adapter on board and an HDMI connection for the monitor.

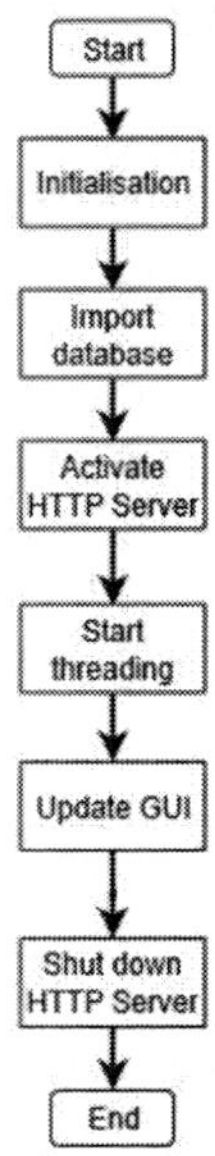

Fig. 4. Flow chart of server device

This program initialization is done by connecting the device to the access point locally. After that, making a graphical user interface (GUI) application at the monitor layer. This GUI application is used to monitor assistance calls from client devices held by patients in the treatment room. If there is data coming in from a client device, the monitor will show a rapid emergency accompanied by an alarm aimed at the doctor or nurse who is on standby in his room. In the GUI view, the application will show the room number and type of assistance needed by patients in a particular room. This message is accompanied by a "OK" button. The message will continue to appear and the alarm will continue to sound if the button has not been pressed. This is done as a warning to doctors or nurses on guard. After pressing the "OK" button, the application will search for patient data in the hospital database, then show it on the monitor.

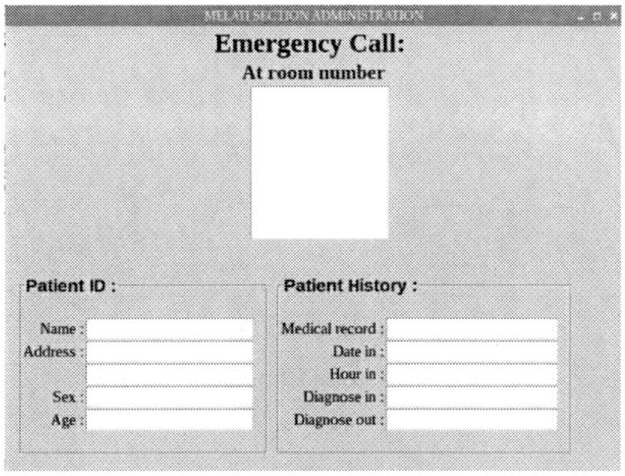

Fig. 5. GUI display

Figure 6 shows the patient identification data and patient history in detail. With this, it is expected that emergency treatment can be more swift and on target, without requiring doctors or nurses to go back and forth in the patient's room due to unclear information or medical equipment left behind.

2.3 System device test

The first test is done separately between the client device and server device. This test is conducted to determine the success of the design of both sides of the device's function. Testing on the client side is done to determine whether the data sent is correct and in accordance with changes in the logic of each push button. Electrical testing is also carried out to determine the reliability of the device and the usage period from full to empty battery conditions.

For testing the server device is also not much different. This is done to find out whether the operation of the program is in accordance with the features created, namely to process client data, access the database, and display in the application.

The second test is done by integrating both sides of the device via Wi-Fi network. This is done to find out whether the whole system is running well according to design. It also considers ergonomics and ease of use and information transfer, both by the patient and the doctor or nurse. All tables and figures will be processed as images. You need to embed the images in the paper itself. Please don't send the images as separate files.

3. Methodology

The method used in this study consisted of several stages, that were hardware design, software design, tool testing, and data retrieval.

A. Hardware Design

The hardware design on the TA module used several circuit modules including the minimum system of AT Mega 328P microcontroller [25]–[33], while the software used was Arduino programming software as a data processor in the device. The sensor used in this device was 808H5V5 sensors.

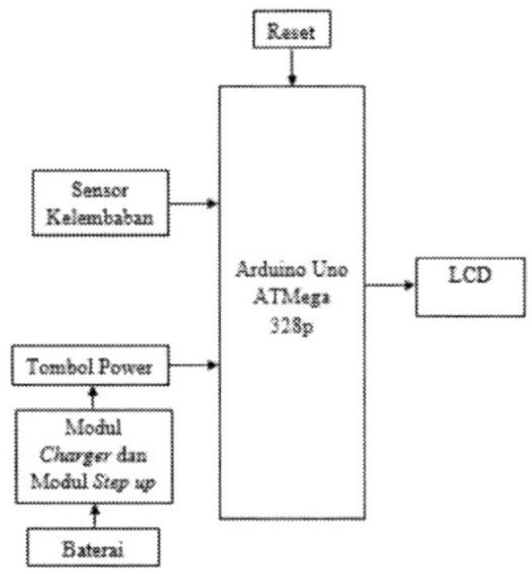

Figure 2.1 Block Diagram

When the power button is pressed, the power supply voltage will supply the overall voltage of the circuit so that the Humidity sensor will be active. After the power button is pressed, the sensor will start reading the area around it. The data from the reading of the humidity sensor will go to the ADC port on the Arduino IC [34]–[40]. Then the data received will be obtained by Arduino and will get the results (output) that will be displayed on one LCD simultaneously, if you want to make measurements again then press the reset button and press the start button again.

B. Software Design

The software on the device processing signals [41]–[52] obtained from the sensor uses the Arduino microcontroller as a data processor. The transmitter system flowchart can be seen in Figure 2.2 as follows:

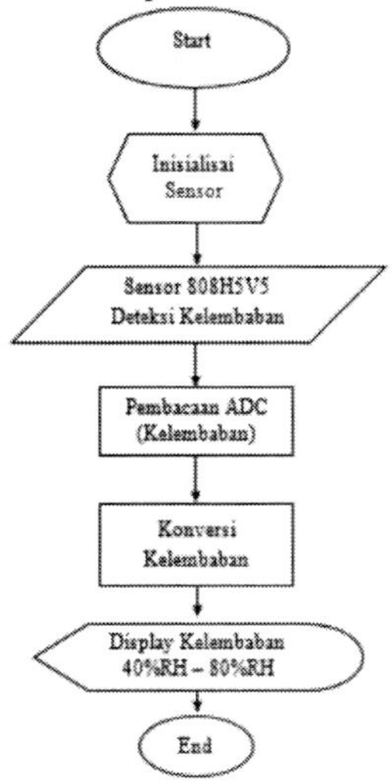

Figure 2.2 Humidity Measurement Flowchart

When the start button is pressed, the sensor will initialize. After that, it will start reading the ADC on the humidity sensor. Then the data from the sensor will be converted into digital data. After that, the program will convert the digital data to % RH units and be displayed to the LCD display. Then the End is to end the program.

4. Result and Discussion

3.1 Client device test

a. Sending data test

The data sent to the server is a string containing 2 types of variables, namely the device ID variable, which represents the caller's patient room number, and also the calling code. Call code 1 for emergencies. Call code 2 for conditions for additional infusion requests. Call code 3 for ordinary help. Both variable values are parsed into one variable and sent to the server device via the Local Wi-Fi network.This test uses a number of 3 client devices.

Table 1 shows the output data according to the design format. This format can only be sent from a programmed microcontroller [19]–[21], cannot be sent from other devices, such as a browser on Windows. Only data that matches the format can be processed by the server program.

Table 1. *Output data from client device*

Device ID	Call code	Pressed button	Output data
1	1	Red	1,1
	2	Green	1,2
	3	Blue	1,3
2	1	Red	2,1
	2	Green	2,2
	3	Blue	2,3
3	1	Red	3,1
	2	Green	3,2
	3	Blue	3,3

b. Power supply test

This client device uses a 18650 Li-Po battery which is arranged in 2 parallel as its power supply. Test results show that this device can be used for a maximum of 15 hours..

3.2 Server device test

a. Data receiving and processing

As the previous explanation, that the data sent by the client device is a string consisting of 2 variable values. This server program was first created for splitting data. With this the room number and call code data will be known. In this test, splitting data was successfully carried out with the indicator value of the two variables shown in Python Shell.

Table 2. *Data result from client*

Input data	Room number	Call code
1,1	1	1
1,2	1	2
1,3	1	3
2,1	2	1
2,2	2	2
2,3	2	3
3,1	3	1
3,2	3	2
3,3	3	3

This test is carried out with access to the database to find the data of patients who make calls and display it in the GUI application.

```
============ RESTART: /home/pi/Desktop/serverRaspiGresik_EN
['Prisma Megantoro', 'Puput Chandra', 'Arashena Humayra']
['Cerbonan', 'Matesih', 'Karangpandan']
['Karanganyar', 'Boyolali', 'Sragen']
['Male', 'Female', 'Female']
['28', '25', '12']
['20192456', '20193542', '20197889']
['14/04/19', '16/02/19', '18/01/19']
['20:13:00', '14:54:00', '04:15:00']
['headache', 'headache', 'stomacache']
['flu', 'PMS', 'maag']
Starting server, use <Ctrl-C> to stop
GET /1,2 HTTP/1.1
```

Fig. 7. Import from database

3.3 System test

This test is carried out in real time with a Wi-Fi network topology via the router access point. This is done whether the overall features of the nurse call system are made to function as a design.

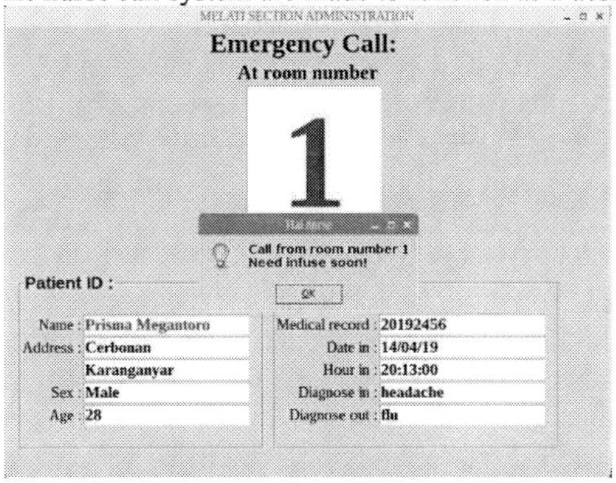

Fig. 8. Patient's daya display on GUI

Table 3. *Overall system test result*

Device ID	Pressed button	Patient's ID correct?	Patient's history correct?	Patient's request correct?
1	Red	Yes	Yes	Yes
	Green	Yes	Yes	Yes
	Blue	Yes	Yes	Yes
2	Red	Yes	Yes	Yes
	Green	Yes	Yes	Yes
	Blue	Yes	Yes	Yes
3	Red	Yes	Yes	Yes
	Green	Yes	Yes	Yes
	Blue	Yes	Yes	Yes

If a button from a client device is pressed, warnings and data that appear on the server application are correct. From table 3 it can be seen that the data processing and appearance of the data are in accordance with the design. Thus this will greatly help the doctor or nurse to improve the performance of medical services to patients for 24 hours.

5. Conclusion

From this study it can be concluded that the device made meets all the testing criteria, both from the client and server side, both from the program and hardware. This set of system can be implemented well in a hospital ward with adequate Wi-Fi network. The number of uses of client devices is unlimited, it only needs to be adjusted on the server side.

References

[1] B. Xu, J. Zheng, and Q. Wang, "Analysis and Design of Real-Time Micro-Environment Parameter Monitoring System Based on Internet of Things," Proc. - 2016 IEEE Int. Conf. Internet Things; IEEE Green Comput. Commun. IEEE Cyber, Phys. Soc. Comput. IEEE Smart Data, iThings-GreenCom-CPSCom-Smart Data 2016, pp. 368–371, 2017.

[2] A. Latif, K. Shankar, and P. T. Nguyen, "Legged Fire Fighter Robot Movement Using PID," J. Robot. Control, vol. 1, no. 1, pp. 15–19, 2020.

[3] S. Ziegler, S. Nikoletsea, S. Krco, J. Rolim, and J. Fernandes, "Internet of Things and crowd sourcing - A paradigm change for the research on the Internet of Things," IEEE World Forum

Internet Things, WF-IoT 2015 - Proc., pp. 395–399, 2015.

[4] C. Lee and A. Fumagalli, "Internet of Things Security - Multilayered Method For End to End Data Communications Over Cellular Networks," 2019 IEEE 5th World Forum Internet Things, pp. 24–28, 2019.

[5] K. Kunal, A. Z. Arfianto, J. E. Poetro, F. Waseel, and R. A. Atmoko, "Accelerometer Implementation as Feedback on 5 Degree of Freedom Arm Robot," J. Robot. Control, vol. 1, no. 1, pp. 31–34, 2020.

[6] Y. Shaikh, V. K. Parvati, and S. R. Biradar, "Survey of smart healthcare systems using internet of things (IoT) : (Invited paper)," Proc. 2018 Int. Conf. Commun. Comput. Internet Things, IC3IoT 2018, vol. 6, pp. 508–513, 2019.

[7] T. P. Tunggal, A. W. Apriandi, J. E. Poetro, E. T.Helmy, and F. Waseel, "Prototype of Hand Dryer with Ultraviolet Light Using ATMega8," J. Robot. Control, vol. 1, no. 1, pp. 7–10, 2020.

[8] T. Harjanto, D. S. E. W. Sumunar, and K. A. K. Putri, "E-Learning Implementation on Clinical Rotation Nursing Education: A Case Report of Universitas Gadjah Mada," Proc. - 2018 4th Int. Conf. Sci. Technol. ICST 2018, vol. 1, pp. 1–5, 2018.

[9] T. Tanioka et al., "Computerized electronic nursing staffs' daily records system in the 'A' psychiatric hospital: Present situation and future prospects," Proc. 6th Int. Conf. Nat. Lang. Process. Knowl. Eng. NLP-KE 2010, no. 978, pp. 0–3, 2010.

[10] N. H. Wijaya, Z. Oktavihandani, K. Kunal, E. T.Helmy, and P. T. Nguyen, "The Design of Tympani Thermometer Using Passive Infrared Sensor," J. Robot. Control, vol. 1, no. 1, pp. 27–30, 2020.

[11] X. Yang, Y. Li, L. Wang, and D. Li, "Development and application of nursing operation virtual simulation training system," Proc. - 2018 Int. Conf. Eng. Simul. Intell. Control. ESAIC 2018, pp. 185–187, 2018.

[12] D. Hardiyanto, I. Iswanto, D. A. Sartika, and M. Rojali, "Pedestrian Crossing Safety System at Traffic Lights based on Decision Tree Algorithm," Int. J. Adv. Comput. Sci. Appl., vol. 10, no. 8, pp. 375–379, 2019.

[13] Z. Dzulfikri, N. Nuryanti, and Y. Erdani, "Design and Implementation of Artificial Neural Networks to Predict Wind Directions on Controlling Yaw of Wind Turbine Prototype," J. Robot. Control, vol. 1, no. 1, pp. 20–26, 2020.

[14] K. Purwanto, I. -, T. Khristanto, and M. Yusvin, "Microcontroller-based RFID, GSM and GPS for Motorcycle Security System," Int. J. Adv. Comput. Sci. Appl., vol. 10, no. 3, pp. 447–451, 2019.

[15] Iswanto, S. Suripto, F. Mujahid, K. T. Putra, N. P. Apriyanto, and Y. Apriani, "Energy Harvesting on Footsteps Using Piezoelectric based on Circuit LCT3588 and Boost up Converter," Int. J. Electr. Comput. Eng., vol. 8, no. 6, pp. 4104–4110, 2018.

[16] Iswanto, J. Syaftriadi, A. Nur, N. Chamim, R. O. Wiyagi, and R. Syahputra, "LED and Servo Motor Control Via Bluetooth Based on Android Applications," Int. J. Recent Technol. Eng., vol. 8, no. 2, pp. 6227–6231, Jul. 2019.

[17] P. Megantoro, A. Widjanarko, R. Rahim, K. Kunal, and A. Z. Arfianto, "The Design of Digital Liquid Density Meter Based on Arduino," J. Robot. Control, vol. 1, no. 1, pp. 1–6, 2020.

[18] A. N. N. Chamim, M. E. Fawzi, I. Iswanto, R. O. Wiyagi, and R. Syahputra, "Control of Wheeled Robots with Bluetooth-Based Smartphones," Int. J. Recent Technol. Eng., vol. 8, no. 2, pp. 6244–6247, Jul. 2019.

[19] I. Iswanto, W. S. Agustiningsih, F. Mujaahid, R. Rohmansyah, and A. Budiman, "Accumulator Charging Control with Piezoelectric Based on Fuzzy Algorithm Scheduling," TELKOMNIKA (Telecommunication Comput. Electron. Control., vol. 16, no. 2, p. 635, Apr. 2018.

[20] Iswanto, K. Purwanto, W. Hastuti, A. Prabowo, and M. Y. Mustar, "Smart smoking area based on fuzzy decision tree algorithm," Int. J. Adv. Comput. Sci. Appl., vol. 10, no. 6, pp. 500–504, 2019.

[21] N. H. Wijaya, A. G. Alvian, A. Z. Arfianto, J. E. Poetro, and F. Waseel, "Data Storage Based

Heart and Body Temperature Measurement Device," J. Robot. Control, vol. 1, no. 1, pp. 11–14, 2020.,

IOP Conf. Series: Materials Science and Engineering **835** (2020) 012034 doi:10.1088/1757-899X/835/1/012034

Potential of Ketapang Seed Oil (*Terminaliacatappa Linn*) as Basic Material Mono-diglyceride Biodegradable Surfactant

Ery Fatarina Purwaningtyas, Khoir Fadilah

Chemical Engineering Department University of 17 Agustus 1945 Semarang.Indonesia.

ery_fatarina@yahoo.co.id

Abstract. Surfactants are widely used in pharmaceuticals, perfumes, cosmetics, food and beverages. One type of surfactant produced from the synthesis of palm oil is mono-diglyceride which can function as an emulsifier. Ketapang (Terminaliacatappa Linn) is a beach tree with a wide spread area, whose seeds have not been utilized optimally. The content of Ketapang seed oil has the potential to be converted into mono-diglyceride surfactant.This study uses experimental laboratory methods, while parameter optimization is done by the two level factorial design method. The raw materials used were Ketapang seeds from around the campus of Semarang 17 Agustus 1945 University, n-butanol solvents, glycerol, MgO catalysts. During the glycerolysis process, 8 run experiments were carried out with 3 variables which changed in temperature (60 & 90OC); MgO catalyst (2 & 4%) and solvent / 10 g oil volume 20 and 40 ml, while the variable is the weight of 25 gram ketapang seed oil; stirring speed of 400 rpm; reaction time of 4 hours; ratio of glycerol 2.5 ml / 10 g oil; and 24-hour deposition time.From the results of the study, the influential variable is temperature. The optimum results were obtained at conditions of 90OC, 4% MgO catalyst, solvent volume of 20 ml / 10g ketapang seed oil and yield 97.9% %. The resulting surfactant has the characteristics of acid number 57.2 mg KOH / gr, saponification number 218 mg KOH / gr. Surfactant has HLB value 14.75, meaning that the surfactant functions as an O / W type emulsifier or as detergent agent.

1. Introduction

Ketapang (Terminaliacatappa Linn) is a beach tree with a fairly wide distribution area. Originating from the tropics in India, then spread to Southeast Asia, North Australia, and Polynesia in the Pacific Ocean. Besides growing wildly on the beach, this tree is often planted as a shade tree in the lowlands. Therefore, ketapang trees are also planted as ornamental trees in cities. Ketapang tree is also one type of shade tree. Ketapang tree is one of the trees that are often found in the city of Semarang, one of which is in the UNTAG Semarang campus area as a parking area shade plant, or a road shade.

Until today, the utilization of ketapang trees has not been explored to the fullest, especially the fruit. The shape of the fruit of this ketapang tree is like an almond fruit, its fruit is roughly 4 - 5.5 cm. Ketapang seeds when unripe are green and turn brownish red when ripe. Utilization of seeds so far is only limited to direct consumption or processed into snacks, it is still very rarely extracted oil content [1]. Ketapang seeds contain 50% - 60% oil. Ketapang seed oil has not been used optimally, in some countries Ketapang seed oil is used as a substitute for Almond oil, the oil is tasteless and odorless so it is not tasty for consumption, its use is new as a remedy for skin diseases such as scabies. [1]

According to research by Ravensca I., et al. [5]. The making of surfactant based on Ketapang seed oil with Triethanolamine obtained a yield of 20%. The surface tension test results are 34,947 dyne / cm. In the chemical properties test for esteramine, the acid number is 5.61 mgKOH / gr, the saponification number is 98.175 mgKOH / gr and the HLB value is 5.25.

Based on research that has been done, it is possible that ketapang seed oil is converted into surfactants. This study aims to examine whether it is possible for Ketapang oil to be converted into "biodegradable" surfactant type Mono-diglyceride; know the most influential variables; know the optimum conditions of the glycerolysis process and determine the characteristics of the physico-chemical properties, the functional properties of the emulsifier based on HLB values.

Table 1. Composition of Ketapang seed fatty acids [6]

Fatty acid	Composition (%)
Palmitic Acid	35,26
Palmitoleic Acid	0,38
Stearic Acid	4,55
Oleic Acid	38,72
Linoleic Acid	20,57
Arachidic Acid	0,51

Glycerolysis is the reaction between glycerol and oil or fat to produce mono and di-acyl glycerol (MAG and DAG). The purpose of the catalyst in addition to getting a relatively short reaction time is also to direct the OH group toward the formation of MAG and DAG. Reactions with base catalysts (NaOH) usually run faster. Characteristics of the glycerolysis reaction from oils / fats will (predominantly) produce monoglycerides. Monoglycerides with a 2-acyl structure or β-monoglycerides will generally be obtained between 5-8% (at low temperatures) and above 30% (at very high temperatures) [10].

The main characteristic of surfactants is their surface activity. Surfactants can increase the ability to reduce surface tension and interface of a liquid, increase the ability to form oil emulsions in water, change the speed of aggregation of dispersed particles by inhibiting and reducing flocculation and coalescence of dispersed particles, so that the stability of dispersed particles increases.

Hydrophilic Lipophilic Balance (HLB) is a number that shows the comparison between hydrophilic groups and lipophils in surfactants. Griffin set up a HLB surfactant size scale that can be used to compose the optimum HLB efficiency region for each surfactant function. The higher the HLB value of a surfactant, its polarity properties will increase. Different Hydrophilic - Lipophilic Balance (HLB) values of surfactants, HLB values range from 0-20, each value has a different function [12].

Table 2. Function of surfactants based on HLB values [12]

Karakteristik dalam air	HLB	Rasio		Fungsi	
		Hidrophilic	Lipophilic		
Tidak terdispersi	0	0	100		
	1			Anti foaming agent	
Sedikit terdispersi	2-3	10	90		
	4-5	20	80		W/O emulsification
Milky dispersion	6	30	70		
	7			Wetting agent	
	8-9	40	60		O/W emulsification
Stable milky dispersion	10-11	50	60		
Transparant dispersion	12	60	40		
	13			Cleaning agent	
Larutan koloid	14-15	70	30		
	16-17	80	20	Solubilizing agent	
	18	90	10		
	19				
	20	100	0		

2. Research Method

Materials used: Ketapang seeds, MgO catalyst, aquadest. n-butanol, n-hexane, Glycerol. Equipment used: a series of extraction and glycerolysis devices, ovens, thermometers, electric scales, glass ware.

2.1. Fixed Variable

Glycerol Ratio = 2.5 ml / 10 gr of oil
Oil Weight = 25 gr
Solvent type = n-butanol
Separation Time = 24 hours

Changed Variable	High Level (+)	Low Level (-)
Temperature	90°C	50°C
Solvent Volume	40 ml/10 gr oil	20 ml/10 gr oil
Catalyst	4 %	2 %

2.2. Procedure

Extraction Process:

Ketapang seeds are dried, mashed, then weighed 30 grams. Then wrapped in filter paper and insert into the extractor tube. Extraction carried out for 1.5-2 hours with 250 ml n-hexane as solvent. Extraction stopped if the oil has been completely extracted and then the recovery solvent process. To weighed the result oil, roasted the extract to evaporate the remaining solvent.

Glycerolysis Process:

Ketapang seed oil is put with n-butanol solvent into a three-neck flask, in a certain amount according to the variable. Then it is heated to the specified temperature. A certain amount of glycerol is heated in a separate place until the operating temperature. Glycerol and MgO catalyst were put into a mixture of ketapang seed oil and n-butanol solvent in a three neck flask and stirring was carried out. Heating and stirring stopped after the reaction time was reached. Separate the catalyst by filtering. Then The product left in the separating funnel for 24 hours to create two perfect layers, the remaining glycerol is separated from the product and its weight and volume are measured.

3. Results and Discussion

The variable test results show that the temperature variable is the variable whose farthest point of effect shows that the temperature variable is the most influential variable on the glycerolysis process of ketapang oil. Providing the same results as research conducted by Purwaningtyas, et al. [11].

The optimization stages of the glycerolysis process gave the following results: at a glycerolysis temperature of 90°C, a solvent volume of 20 ml / 10 g of oil, a 4% MgO catalyst and a stirring time of ± 4 hours with a stirring speed of 400 rpm gave the largest yield of 97.7%. Figure 1 shows the relationship between the glycerolysis temperature and the % yield obtained, the results show a tendency to increase glycerolysis results at increasingly rising temperatures. But decreased after a temperature of 90°C. The cause is after reaching the optimum point the results of glycerolysis decrease with increasing temperature. An increase in temperature will cause the solvent n-butanol to be in the vapor phase so that the solubility of the ketapang seed oil in glycerol decreases, the temperature can increase the homogeneity of the reaction mixture.

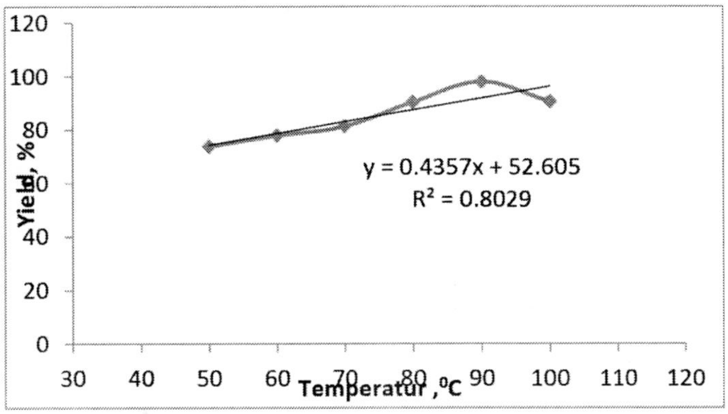

Figure 1. Graphic optimization of the relationship of temperature to yield.

The results of the HLB surfactant test values produced gave a value of 14.75. HLB values indicate that mono diglyceride surfactants can be used as O/W emulsifiers, namely at HLB values 8-18, these surfactants can dissolve in water and are used to make oil-in-water emulsions (O/ W) . Examples of Polyoxyethylenesorbitanmonooleate emulsifiers (Polysorbate 80 USP) or commonly referred to by the trade name tween 80, in addition to being used as an emulsifier are also used as cleaning agents (detergents) where HLB detergent 13-15.[12]

4. Conclusion

The most influential variable on the glycerolysis process of ketapang seed oil is temperature. The optimum conditions for the glycerolysis process are achieved in:

Solvent type : n-butanol
Content : 70%
Volume : 20 ml / 10 gr oil
Sample weight : 25 grams
Temperature : 90 ° C
Time : 4 hours
Stirring Speed : 400 rpm
Yield : 97.9%
MG content : 31.83%

Characteristics of surfactants produced:
Acid number : 57.2 Mg KOH / gr
Saponification Number : 218 Mg KOH / gr
HLB value : 14.75 (O/W emulsifiers being used as an cleaning agents).

References

[1] Handayani,M.P. and S.Wahyuono,. 2008. Analisis Biji Ketapang (Terminalia catappa L.) Sebagai Suatu Alternatif Sumber Minyak Nabati. Majalah Obat Tradisional, July-September.13/45. Page.101-107.

[2] Suwarso W.P., Gani I.Z., and Kusyanto.,1999 , Sintesis Biodesel Dari Minyak Biji Ketapang (Terminalia catappa L.) Yang Berasal Dari Tumbuhan Di Kampus UI Depok, Laboratorium Kimia Organik – Departemen Fakultas Matematika dan Ilmu Pengetahuan Alam - Universitas Indonesia, Depok.

[3] Balogun, A. M. and Fetuga, B. L., 1985, Fatty Acid Composition of Seed Oils of Some Membes of the Meliaceae and Combretaceae Families. JAOCS, Vol. 62, no. 3, Hal. 539 – 531. In Handayani,M.P. dan S.Wahyuono,. 2008. Analisis Biji Ketapang (*Terminalia catappa L.*) Sebagai Suatu Alternatif Sumber Minyak Nabati. Majalah Obat Tradisional, July-September.13/45. Page 101-107.

[4] Nuryanto E, 1997, Surfaktan yang Ramah Lingkungan dari Minyak Kelapa Sawit, Warta Pusat Penelitian Kelapa Sawit, volume 5 (1) , page 37-45.

[5] Ravensca I., Saleh C., and Daniel.,2017 , Pembuatan Surfaktan Berbahan Dasar Minyak Biji Ketapang (Terminalia catappa) Dengan Trietanolamina, Jurnal Atomik 02 (2) hal 183-189, Program Studi Kimia FMIPA Universitas Mulawarman, Samarinda.

[6] http://isjd.pdii.lipi.go.id/admin/jurnal/12084250.pdf downloaded on May 23rd, 2019.

[7] Oktaviany H., 2006.Analisis Mutu Minyak Biji Ketapang (Terminalia catappa Linn).FMIPA, Universitas Sriwijaya, Inderalaya.

[8] Negi, D.S ; Sobotka, F ; Kimmel, T; Wozny, G and Schomacker, R, 2007, Glycerolysis of Fatty Acid Methyl Esters : 1. Investigations in a Batch Reactor, Journal of American Oil Chemist's Society, Volume 84, Page 83 – 90. (www.springerlink.com)

[9] Hassanuddin A, 2001, Kajian Teknologi Pengolahan Minyak Sawit Mentah untuk Produksi Emulsifier Mono-diasilgliserol dan Konsentrat Karotenoid, Makalah Falsafah Sains, page -17, Universitas Tadulako, Palu.

[10] Anggoro ,D.D., Budi, F.S., Noviana, S.M and Hapsari, Y.S., 2008, Proses Gliserolisis Minyak Kelapa Sawit menjadi Mono-diacylgliserol dengan Pelarut n-butanol dan Katalis MgO, Prosiding Seminar Nasonal Rekayasa Kimia dan Proses, Teknik Kimia UNDIP, Semarang.

[11] Purwaningtyas E.F., Kasmiyatun M., Mulyaningsih S.MF and Indah,WN. ,2015 , Optimization of Glycerolysis Temperature Process for the Synthesis of Monoglyceride-Diglyceride Surfactants Derived from oil of Silkworm Pupae, Prosiding Seminar Nasional "Kejuangan", UPN Yogyakarta, 18 Maret 2015.

[12] Holmberg ,K., Jonsson, B., Kronberg ,B and Lindman, B., 2004, Surfactans and Polymers in Aqueous Solotion, 2nd edition, John Wiley & Sons Inc. USA.

Bioplastic from Jackfruit Seeds and Rice

Retno A.S. Lestari[1,2], Mega Kasmiyatun[1,2], Kevin Dermawan[1], Alfia N. Aini[1], Nur Riyati[1], Finka R. Putri[1]

[1]Chemical Engineering Department, University of 17 Agustus 1945 Semarang, Central Java, Indonesia
[2]Study Center of Environmental and Alternative Energy, University of 17 Agustus 1945 Semarang, Central Java, Indonesia

retnotengaran@gmail.com

Abstract. This study tried to explore characterization of biodegradable plastic from jackfruit seeds and rice waste. Jackfruit seeds and rice waste contain starch, so it can be used for plastic material with addition of several additives such as polyvinyl alcohol (PVA), glycerol, chitosan and sorbitol. Starch is one of polysaccharides that be used as raw material for biofilms. Polyvinyl alcohol is the most important plastic in making water-soluble films. It was ability to form films, emulsifiers, and its adhesive properties. Polyvinyl alcohol has high tensile strength, good flexibility, and good oxygen barrier. Sorbitol and glycerol are also an additive in the manufacture of biodegradable plastics from starch, which is a plasticizer that increase elasticity. Chitosan have effects to bioplastic, such as biodegradable, hydrophilicity, and anti bacterial. This research aims to determine the effect of addition of PVA, glycerol, chitosan and sorbitol to optimum characteristics of bioplastic. The parameters to be examined bioplastic include thickness, moisture content, tensile strength, and % elongation. Increasing of glycerol added to the plastic material effect to decrease of tensile strength of bioplastic and elongation, lower water resistance and more easily degraded. Increasing of chitosan added to the plastic material effect to higher tensile strength, lower elongation, higher resistance to water and lower degradation. Increasing of PVA added to the plastic material effect to increasing of the tensile strength of the bioplastic and decreasing of elongation of bioplastic, but the increasing of sorbitol effect to decreasing of tensile strength and increasing of elongation of bioplastic.

1. Introduction

Plastic is packaging that relatively strong, lightweight, and has a low price. This packaging has a role in holding and protecting food [1]. The packaging used by manufacturers to package their products is polymers of petroleum. Most of these products are used for packaging of disposable products. Waste of this packaging is largely discharged into the environment, thus affecting water quality and the environment [2]. The use of plastic materials has caused negative impacts on the environment, landscape, and human health [3,4]. For this reason, it is necessary to find a solution to overcome this problem, by using packaging materials that can decompose quickly.

Substitution of plastic materials with biodegradable plastic produced from natural polymers as the main material is an alternative to reduce the impact of environmental damage [5]. Reducing environmental impacts by using biodegradable polymers in the food industry is an alternative to removing packaging from petroleum [6,7]. Reducing environmental impacts by using biodegradable polymers in the food industry is an alternative solution [8]. This material can be used to reduce and resolve environmental pollution. Natural polymers using natural ingredients such as starch, chitosan, cellulose, proteins, and enzymes obtained from nature [9]. The degradation time of packaging which produced from starch material was shorter than other polymeric materials such as poly (lactic acid) (PLA) or poly (butylene adipate-co-terephthalate) (PBAT) [10]. The faster time to degradation of packaging at the landfill effect to reduce of the volume of waste to be compacted. A reduction in degradation time, a significant reduction in waste volume will certainly be very beneficial for the environment.

However, biodegradable polymers have lower physical properties, strength, and process capability compared to polymers from petroleum [11]. To overcome the disadvantages of applying natural polymer-based materials that can be degraded, various methods such as a mixture of natural polymers and polymers derived from petroleum, the addition of functional plasticizers, and the use of chemical agents have been tried to increase tensile strength, extension properties, waterproof properties and the application [12].

2. Methods

2.1. Materials

Waste of rice and jackfruit seeds as a bioplastic base material were obtained from restaurants and traditional markets. The plasticiser such as acetic acid, chytosan, glycerol, poly vinyl alcohol (PVA), sorbitol were obtained from Chemical store in Semarang, Central Java, Indonesia.

2.2. The Equipment of Research

The equipment of research include: beaker glass, blenders, porcelain cups, stirring rods, electric heater, thermometers, filter cloths, ovens, incubators, molds, basins, porcelain spoons, plastic spoons bioplastic molding tools from glass sheets.

Rice waste from restaurants were cleaned from other leftovers and then mashed using a blender until the texture like powder. The powder of rice waste and plasticizer materials and solvent were mixing in beaker glass by stirrer and heated at 50-60 ° C for 30 minute. Furthermore, the mixing of materials bioplastic was mold on sheet glass molding. While jackfruit seeds obtained from the traditional market at Semarang, Central Java, Indonesia were removed from the skin of the arrows and then washed. Furthermore, the jackfruit seeds which have been cut thinly crushed using a blender until smooth and the texture like powder

3. Results and discussion

3.1. Tensile Strength Test

Based on the results of the tensile strength test of plastic made from rice waste with the addition of chitosan and glycerol obtained the results shown in Figure 2. Figure 2 shows that the greater of the glycerol added to the plastic material, the plastic yield has a lower tensile strength, this is because glycerol functions as a *plasticizer* that can reduce intermolecular strength, so that the more amount of glycerol added causes the plastic tensile strength descend.

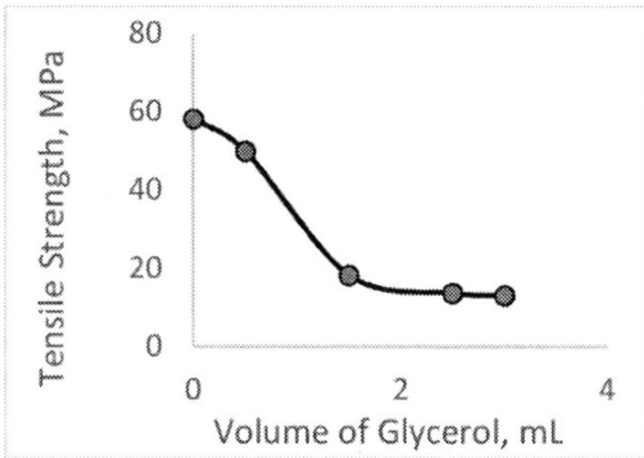

Figure 2. Effect of Glycerol to tensile strength of bioplastic from Rice Waste

While the effect of chitosan addition on tensile strength of bioplastic can be seen in Figure 3, which appears when chitosan is added 30%, tensile strength of bioplastic is still low or close to zero, when chitosan increases to 50% and 55%, tensile strength of bioplastic reaches up to 31.8 MPa. But when the chytosan increases to 60 percent, the tensile strength of bioplastics reduced to 3.33 MPa. The increased tensile strength of bioplastics when the percentage of chitosan increased due to the increase of hydrogen bonds in the plastic when the percentage of chitosan ascend. Increased hydrogen bonding in the plastic causes the plastic film be stronger and harder to break (Katili, 2013).

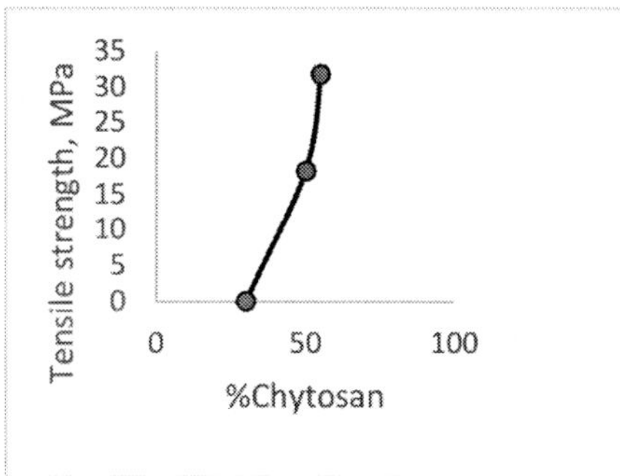

Figure 3. Effect of Chytosan to tensile strength of bioplastic from Rice Waste

For bioplastic made from jackfruit seeds with sorbitol plasticizer and the addition of poly vinyl alcohol (PVA), the effect of adding these ingredients to jackfruit seeds get the results of bioplastic with a strong tensile strength as follows (Figure 4). In Figure 4 shows the lowest *tensile strength* is obtained from the addition of PVA as much as 0 gr and 0 ml of sorbitol that is tensile strength of 0 MPa, while the highest *tensile strength* is obtained from the addition of PVA as much as 3 gr and 2 ml of sorbitol which is 2,2 MPa. The effect of the addition of PVA on the *tensile strength* value shows the greater the PVA, so the higher tensile strength possessed by bioplastics too. This is because PVA can form film forming from bioplastics, and strengthen hydrogen bonds formed among bioplastic polymer chains. While the effect of the addition of sorbitol volume on Tensile Strength value shows that the more volume of sorbitol, the tensile strength value is lower. This is caused by the addition of the volume of sorbitol which will reduce the tensile strength between polymers during water evaporation which results in decreased resistance to mechanical treatment of plastics. According to Lieberman and Gilbert (1973), *plasticizers* can change the physical characteristic of plastics by reducing cohesion and mechanical resistance of polymer chains.

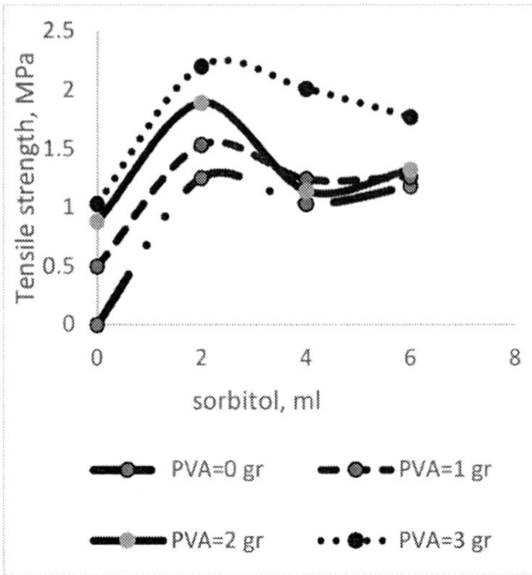

Figure 4. Effect of Sorbitol to tensile strength of bioplastic from Jackfruit seeds

3.2. Elongation Test

Based on the elongation test results on plastics made from rice waste with the addition of chytosan and glycerol, the results obtained as in Figure 5 and Figure 6. Based on Figure 5, it appears the greater of glycerol added to the plastic material, get the yield of plastics which has higher elongation, it's caused by glycerol functions as a plasticizer which increasing the flexibility of the plastic. Besides the addition of glycerol can impoverish hydrogen bonds, so the distance among the biopolymer molecules becomes tenuous. The state of strain among biopolymer molecules increase the flexibility of plastic samples.

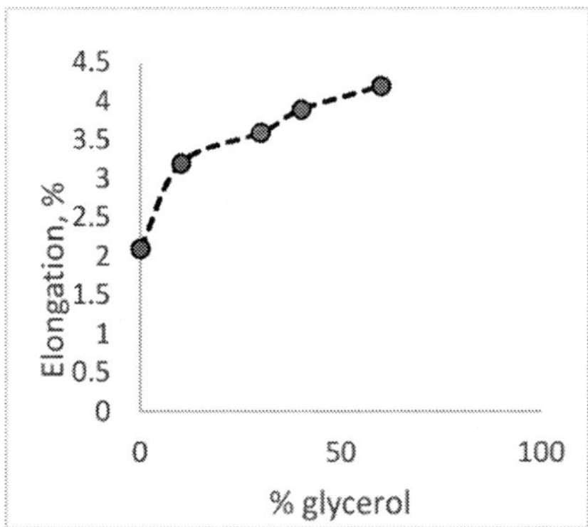

Figure 5. Effect of Glycerol to Elongation of bioplastic from Rice Waste

As shown in Figure 6, it appears the greater the% of chitosan added to plastic manufacturing materials, take the bioplastics which have a lower elongation percentage. This is because the addition of chitosan as an additive will close the distance among the stretching biopolymer molecules. The decrease in the distance among degradable plastic biopolymer molecules caused by the saturation point being exceeded so that the excess plasticizing molecules are in a separate phase outside of the polymer phase and will decrease intermolecular forces between chains (Laila, 2014).

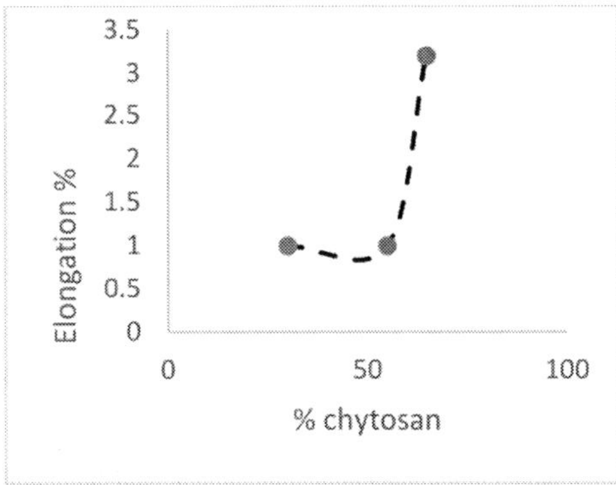

Figure 6. Effect of Chytosan to Elongation of bioplastic from Rice Waste

In Figure 7, it shows the lowest % of elongation value is obtained from the addition of 0 gr PVA and 0 ml of sorbitol with the value of% elongation is 0%, while for the highest% of elongation value obtained from the addition of 1 gr PVA and 6 ml sorbitol with% elongation of 6.8%.

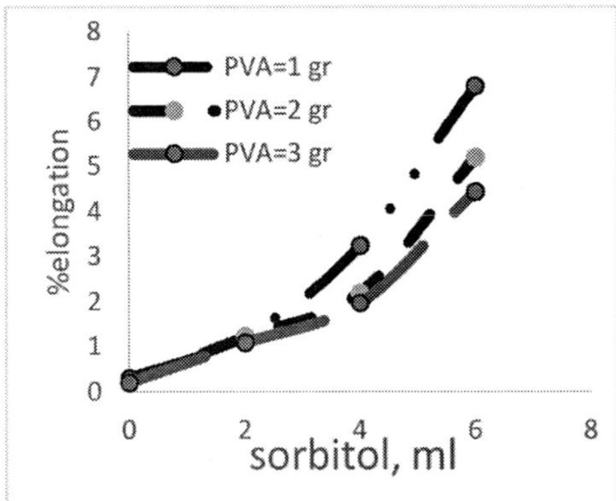

Figure 7. Effect of Sorbitol to Elongation of bioplastic from Jackfruit seeds

The effect of adding PVA to the% elongation value appears the more PVA added, the lower of the% elongation produced. This is because PVA can reduce the plasticity or flexibility of the plastic due to the strengthening of the hydrogen bonds in the formed polymer. While the effect of increasing the volume of sorbitol on the% elongation value shows the more volume of sorbitol, the% elongation is the higher. This is caused by the addition of the volume of sorbitol causing a decrease in intermolecular forces, as a result the level of mobility among molecular chains increase and results in OH groups will form intermolecular bonds with polymer chains reduce.

The interaction between PVA and Sorbitol was seen in the value of tensile strength and the % elongation. In outline, increasing the amount of PVA that is more, can increase the value of tensile strength but reduce the value of the % elongation, conversely increasing the amount of sorbitol can increase the value of the % elongation and reduce the value of tensile strength of bioplastics.

3.3. Bioplastic Resistance to Water -Test

Based on Figure 8, it appears that the higher levels of Based on Figure 8, it appears that the higher levels of glycerol added to the bioplastic mixture get the results of bioplastics which have lower water absorption. This is due to the increase in the amount of glycerol in bioplastics so that more hydroxyl groups of glycerol in bioplastics in order to increase the absorption of water by bioplastics. Glycerol as a plasticizer will increase the flexibility of the plastic. With the increase in glycerol, will create free volume in bioplastics, so that it will increase the gap in bioplastics to be occupied by water molecules.

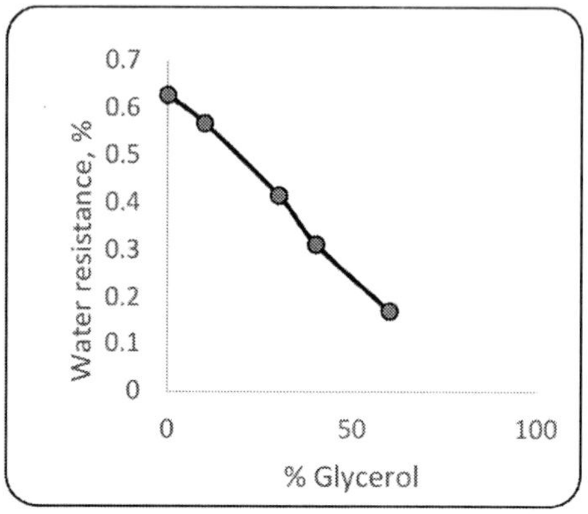

Figure 8. Effect of Glycerol to Water resistance of bioplastic from Rice Waste

The effect of the addition of chitosan to the resistance of bioplastics to water can be seen from Figure 9, which illustrates that the greater chitosan added to bioplastics causes the resistance of bioplastics to water also increases.

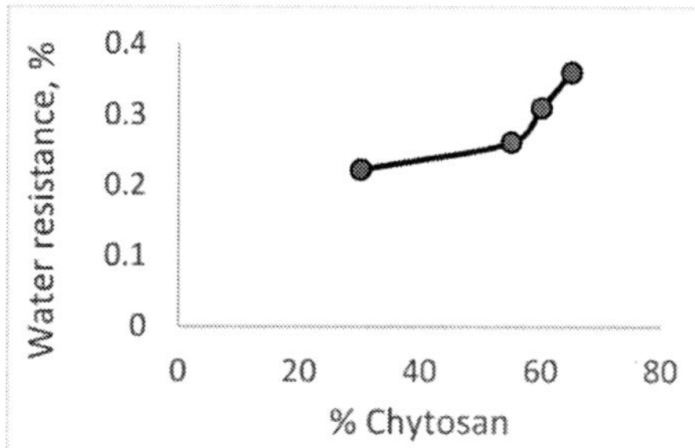

Figure 9. . Effect of Chytosan to Water resiatance of bioplastic from Rice Waste

For bioplastic made from jackfruit seeds, the effect of the addition of sorbitol and PVA to the bioplastic water content can be seen in Figure 10. In Figure 10, it can be seen that the lowest water content in bioplastic is obtained from the addition of PVA of 0 gr and sorbitol 0 ml that is water content of 0.3%, while the highest water content is obtained from the addition of PVA of 3 grams and 6 ml of sorbitol, which water content is 2.39%. From this appears the more amount of PVA is added, the more water is absorbed by bioplastic particles because PVA can bind water and make a film forming on bioplastics and inhibit the evaporation of water from bioplastics. Besides increasing the amount of sorbitol as a plasticizer essence which is increasing also increases the volume of water contained in bioplastics because sorbitol has a function as a humectant that can bind water and also has hydrophilic characteristic so that during the drying process, water is hard to evaporate. Bioplastics with low water level will get quality better because the less water level is obtained, the bioplastics will be more difficult to be damaged by microbes. This bioplastic water level shows the resistance of bioplastic in absorbing water, thus the more water level also shows the low resistance of bioplastic in absorbing water. Thus it can be said that the more the amount of PVA, so, the resistance of bioplastic in absorbing water will be lower because PVA can bind water and make a film forming on bioplastics and inhibit the evaporation of water from bioplastics. Likewise, the increase of the amount of sorbitol added to the bioplastic mixture from jackfruit seeds also decrease the resistance of bioplastics in absorbing water. Because sorbitol has a function as a humectant that can bind water and also has hydrophilic characteristic so that during the drying process, water is hard to evaporate.

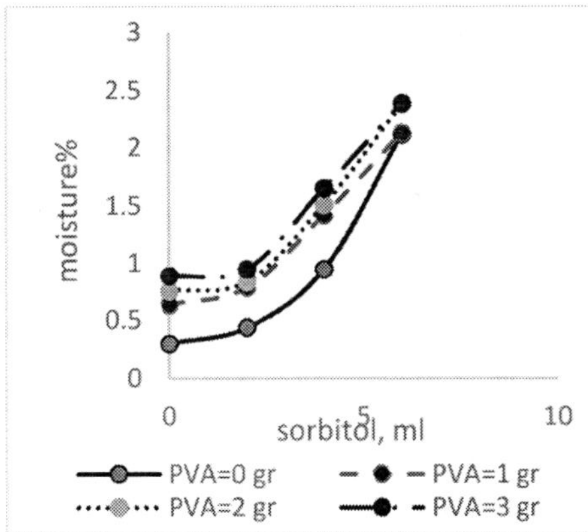

Figure 10. Effect of Sorbitol to moisture of bioplastic from Jackfruit seeds

3.4. Bioplastic FT-IR Characteristics Test

The characteristics of FTIR on bioplastic from rice waste based on FTIR images (Figure 11) generated several wave peaks in each vulnerable region. In region I (range from 4000 - 2,500) there are peaks with wave numbers 3448.72 cm^{-1} and 2924.09 cm^{-1}. The peak corresponds to the absorption caused by C-H bonds (alkane compounds type), O-H (phenol compound types (hydrogen bond alcohol)) and N-H (amine compound types (amines)). In region II (range from 2,500 - 2,000) there is a peak with a wave number of 2368.59cm^{-1}. The peak corresponds to the absorption caused by the triple bonds C=C (type of alkyne compound) bond. In region III (range from 2,000 - 1,500) there is a peak with wave number 1635,64cm^{-1}. The peak is in accordance with the absorption caused by the C = O bond (type of aldehyde compound, ketone, carboxylic acid, ester) and C=C (type of alkene compound). In region IV (streches of space of 1,500 - 400), there are many peaks with wave numbers from 1095.57cm^{-1} to 354.90cm^{-1}. The peak corresponds to absorption caused by C-O bonds (type of alcohol compound, carboxylic acid, ester). Biodegradable plastic from the rice waste and chitosan with glycerol as plasticizer, and acetic acid as a catalyst, has a functional cluster that is a combination of specific functional groups that have constituent components such as CH, OH, NH, C≡C, C = O, and C=C, and also have amide and ester functional groups in biodegradable plastic film samples, so that plastic from the rice waste can be degraded and can be said to be green-plastic.

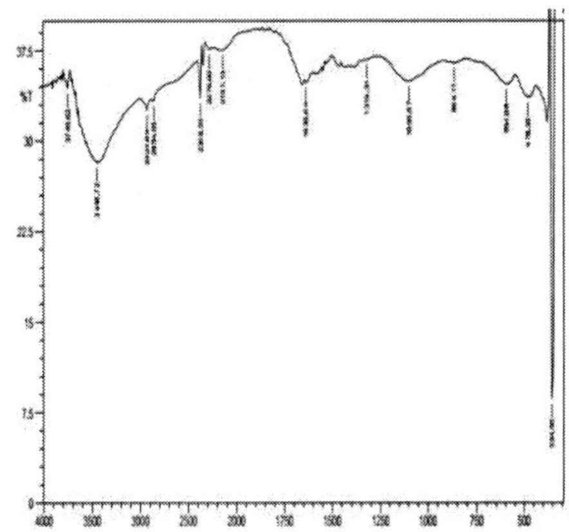

Figure 11. FTIR of bioplastic from Rice

3.5. Bioplastic Degradation Test

Bioplastic degradation test is done by *Soil burial test* which is planting biodegradable plastics in the soil for 14 days, the results obtained as in Figure 12. Figure 12 shows the higher glycerol levels added to the bioplastic material, the bioplastic is obtained with the greater percentage of degradation. This is due to the higher levels of glycerol in bioplastics causing high absorption of water by bioplastics, because glycerol is hydrophilic in order to accelerate the rate of water absorption which making easier for bacteria to decompose plastic samples further. and/or tables must be included in the results of the research. Article must be a research results. Therefore, figures and/or tables must be included in the results of the research. Article must be a research results. Therefore, figures and/or tables must be included in the results of the research. Article must be a research results. Therefore, figures and/or tables must be included in the results of the research.

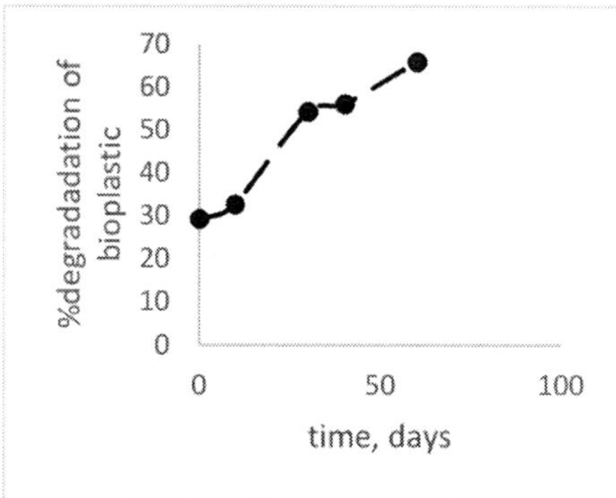

Figure 12. %Biodegradation of Bioplastic from Rice Waste

4. Conclusion

1. Biomass derived from rice and jackfruit seeds waste can be made into bioplastics.
2. The greater glycerol added to the plastic material, giving bioplastic results which have lower tensile strength, higher elongation, lower water resistance and degraded more easily.
3. The greater chitosan added to the plastic material, giving bioplastic results which have higher tensile strength, lower elongation, higher water resistance and lower degradation.
4. The addition of the amount of PVA affects in tensile value raising but reduces the produced % elongation of the bioplastics and the resistance of bioplastic in absorbing water.
5. The addition of the amount of sorbitol increases % elongation, but decreases the tensile strength value of the bioplastics produced, and decreases the resistance of bioplastic in absorbing water.
6. The result of bioplastic products meet SNI standards, with a tensile strength of 49.87 MPa obtained in the addition of 10% glycerol of the weight of rice waste and 50% of the weight of rice waste. This bioplastic has 32.18% degradation percentage in 14 days.

Acknowledgement

Please acknowledge anyone who contributed towards the study by making substantial contributions to conception, design, acquisition of data, or analysis and interpretation of data, or who was involved in drafting the manuscript or revising it critically for important intellectual content, but who does not meet the criteria for authorship. Please also include source(s) of funding. Please also acknowledge anyone who contributed materials essential for the study.

References

[1] L. Elgaaïed-Gambier, " Who buys overpackaged grocery products and why? Understanding consumers' reactions to overpackaging in the food sector". Journal of Business, 2014.

[2] S. Panseria, P.A. Martino, P. Cagnardi, G. Celano, D.Tedesco, M. Castrica, " Feasibility of biodegradable based packaging used for red meat storage during shelf-life: A pilot study". *Food Chemistry, 249*, 22–29, 2018.

[3] C.M. Rochman, E. Hoh, T. Kurobe, S.J. Teh, Ingested plastic transfers hazardous chemicals to fish and induces hepatic stress, Sci. Rep. 3 (2013) 1-7.

[4] E. Huerta, J. Mendoza Vega, V.K. Quej, A. De los, J. Chi, L. Sanchez del Cid, C. Quijano, G. Escalona-Segura, H. Gertsen, T. Salanki, M. Van der Ploeg, A.A. Koelmans, V. Geissen, Bioaccumulation of microplastics in the terrestrial food chain: an example from home gardens in SE Mexico.19th EGU Gen. Assem. EGU2017, in: Proc. From Conf. Held 23-28 April. 2017 Vienna, Austria, vol. 19, 2017, p. 15847.

[5] F. Gu, J. Guo, W. Zhang, P.A. Summers, P. Hall, "From waste plastics to industrial raw materials: a life cycle assessment of mechanical plastic recycling practice based on a real-world case study", Sci. Total Environ. 601-602 (2017) 1192-1207.

[6] Bonilla, J., Talón, E., Atarés, L., Vargas, M., & Chiralt, A. "Effect of the incorporation of antioxidants on physicochemical and antioxidant properties of wheat starch–chitosan films. Journal of Food Engineering", 118(3), 271–278, 2013.

[7] Seligra, P. G., Nuevo, F., Lamanna, M., & Famá, L. Covalent grafting of carbon nanotubes to PLA in order to improve compatibility. Composites Part B: Engineering, 46, 61–68, 2013.

[8] Shah, A. A., Hasan, F., Hameed, A., & Ahmed, S. "Biological degradation of plastics: A comprehensive review". *Biotechnology Advance, 26*, 246–265, 2018.

[9] Tian, H., Tang, Z., Zhuang, X., Chen, X., & Jing, X. "Biodegradable synthetic polymers: Preparation, functionalization and biomedical application". *Progress in Polymer Science, 37*, 237–280, 2012.

[10] Weng, Y. X., Jin, Y. J., Meng, Q. Y., Wang, L., Zhang, M., & Wang, Y. Z. (2013). "Biodegradation behavior of (PBAT) (PLA), and their blend under soil conditions.", Potter, N. N., & Hotchkiss, J. H. (2012). Food science. Springer Science & Business Media, Polymer Test Journal, 32(5), 918–926.

[11] Park, N. H., Kim, D. H., Park, B., Jeong, E., & Lee, J. W. (2013). The industry trend analysis and perspectives of biodegradable polymers. *Biomaterials Research, 17*, 114–120.

[12] Xie, F., Pollet, E., Halley, P. J., & Avérous, L. "Starch-based nano-biocomposites". *Progress in Polymer Science, 38*, 1590–1628, 2013.

Detecting Hoaxes in Indonesian News Using TF/TDM and K Nearest Neighbor

Eri Zuliarso[1], Muchamad Taufiq Anwar[1], Kristophorus Hadiono[1], Iswatun Chasanah[2]

[1]Universitas Stikubank, Semarang, Indonesia
[2]Satya Wacana Christian University, Salatiga, Indonesia

eri299@edu.unisbank.ac.id,muchamadtaufiqanwar@edu.unisbank.ac.id,
kristophorus.hadiono@edu.unisbank.ac.id, uswatunnana@gmail.com

Abstract The presence of the internet and the rapid growth of social media had given rise to the blossoming of hoax creation and distribution through it. A hoax can cause anxiety and reactivity to its readers and could harm a certain party. Thereby, it is important to detect and report hoaxes to stop its spreading as soon as possible. This research aims to utilize the K Nearest Neighbor (KNN) classification algorithm to detect whether a piece of news is a hoax or not. Experiments were done by using 74 hoaxes compiled from Indonesian hoax-debunking community websites and were being compared against 74 real news from various reputable news websites in Indonesia. The result showed that the model could give detection/classification accuracy up to 83.6% and that the model is prone to false positives detections. The characteristics of the resulted model and further research directions are then discussed.

Keywords: hoax detection, fake news detection, Indonesian hoaxes, hoaxes in Bahasa, K Nearest Neighbor, KNN, Term Frequency, Term Document Matrix

I. Introduction

Hoaxes (noun) can be simply defined as fake news, or as defined in the WordNet Dictionary, "Something intended to deceive; deliberate trickery intended to gain an advantage". Hoaxes had been present throughout the history of mankind to the nowadays internet era. The purpose of hoaxes often includes persuading or manipulating people to do or prevent pre-established actions, usually using threats or deceptions[1]. Aside from its 'traditional' kind, recent research had also identified other kind of hoaxes such as the humorous fakes or satire/parody [2].

The presence of the internet and the rapid growth of social media gave rise to the blossoming of hoax creation and distribution through it [3][4][5][6], as stated in [3] that "the boundaries between news production and information creation and sharing are gradually blurring in the current online news environments and social media" and [4] that social media are also increasingly used as vectors for the diffusion of misinformation and hoaxes. The spread of hoaxes has the potential to give extremely negative impacts on individuals and society[5]. The amount of disseminated information and the rapidity of its diffusion make it practically impossible to assess reliability in a timely manner [4]. Thereby, it is important to detect and report hoaxes to stop its spreading as soon as possible.

The K Nearest Neighbor KNN is one of the versatile and well-known clustering/classification algorithm which often used in data mining. The KNN has lower calculation time compared to other algorithms such as the Random Forest, which is useful for fast automatic detections. The KNN assigns class membership to data based on the majority vote of its neighbors, with the object being assigned to the class most common among its k nearest neighbors (k is a positive integer, typically small).

This research aims to utilize the KNN classification algorithm to detect whether a piece of news is a hoax or not. The model could be implemented within a social media application or even a browser or as an add-on to automatically warn the users if the news appeared in their screen is identified as a hoax.

2. Related Works

Several attempts to build automatic hoax detection had been flourishing in the last decades. The object of detection varied from e-mails (spam or scam) [7][8][9][10][11], social media [5][6][12], news [2][9][13], phishing websites [9], and specific topics like Wikipedia articles [14]. The methods for detections also varied, from the well-known and established method such as the 'traditional' features [13], logistic regression [4], neural networks [11], Levenshtein Distance Method [7], Support Vector Machine (SVM) [13][15][16], Decision Tree / C4.5 [13][15], Linear Regression (LR) [15], KNN [8][9][15], Poisson probabilistic theory [8], Bayesian Classifier [9], Naïve Bayes [8][9][13]; while some also introduced novel methods like the adaptation of boolean crowdsourcing algorithms [4] or combinations of several methods [8][9]. However, the research of hoax detection in the Indonesian language (Bahasa) is still very limited. One notable research is the work by [13] which tested three algorithms, i.e the SVM, C4.5, and Naïve Bayes to detect hoax news in the Indonesian language. The position of this research is to explore the utilization of KNN method in hoax detection for news articles written in Bahasa.

3. Methods

Hoaxes were retrieved from an Indonesian hoax-debunking community website. All articles labeled as [HOAX] and containing narration published between July 31, 2015 - November 22, 2017, was copied into individual text files (or documents). Photo and video hoaxes were excluded. The hoaxes were then being compared against 74 real news from various reputable news websites in Indonesia within a similar range of publication dates. The articles for real news were randomly selected, but we also purposely added articles with similar themes/topics as found in the hoaxes. This is intended for a wildcard. Most of the collected hoaxes were in Indonesian language (Bahasa), but a small number of them were written in English. Thus, we used a similar approach for the number of English articles in the real news articles.

The next step is the creation of corpus. This includes the removal of capitalizations, punctuation marks, and stopwords. Stemming, which turns the words into their basic form, is not applied in this research since stemming could remove the lexical characteristics of the hoaxes vs non-hoaxes, and thus decrease the model accuracy[13]. Since we used two kinds of languages, we also incorporated two kinds of stopwords collections for those languages. The corpus was then being transformed into Term Document Matrix (TDM) which uses unigram (the sentences were split into individual words). The TDM consists of the word count (frequency) found in a document (article, text file) for each word listed in the corpus. The TDM is a matrix with the size of m x n, where m is the number of columns which correspond to the number of words found in the corpus, while n is the number of rows resembling the number of documents (one row for each document). Not all terms found in the corpus is listed into the TDM. We used a certain limit to exclude words (terms) which have a low occurrence (high sparseness). We called this the Sparse Terms Limit (STL). The TDM was then being added an additional column to store the label for each document, whether it is 'hoax' or 'not hoax'.

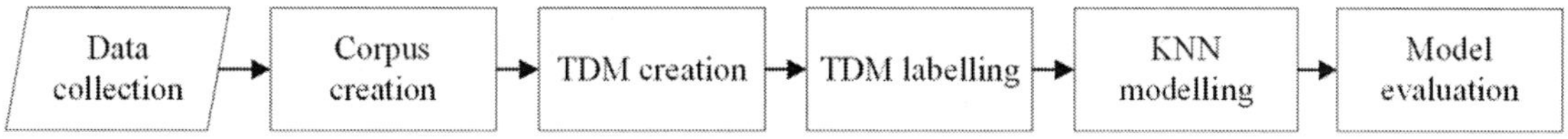

Figure 1. Research methodology

The final form of the TDM was then being randomly grouped into training data sets and the testing datasets using a certain proportion (we called this the Training/Testing Proportion, TTP). The training datasets were then being fed into the KNN algorithm with k=1 (Nearest Neighbor Rule). The model generated by the KNN was then being tested using the testing datasets. The accuracy of its prediction was then calculated based on its confusion matrix.

We did nine experiments with varied STL and TTP, and each experiment was repeated 1000 times to find the average accuracy for each corresponding setting of STL and TTP. The STL values used are: 0.8, 0.85, and 0.9; whereas the TTP values are: 0.7, 0.8, and 0.9. TTP value of 0.7 means that 70% of the data were used as the training sets, while the other 30% were used as the testing sets. Since the hoax articles generally have a smaller amount of word counts, in this research, the STL value below 0.8 would not give any terms, so we set the experiments with the STL value started from 0.8. STL value of 1 means that every word found in the corpus will be included in the TDM. Experiments were done using the RStudio and utilized the 'class' package.

4. Result and Discussion

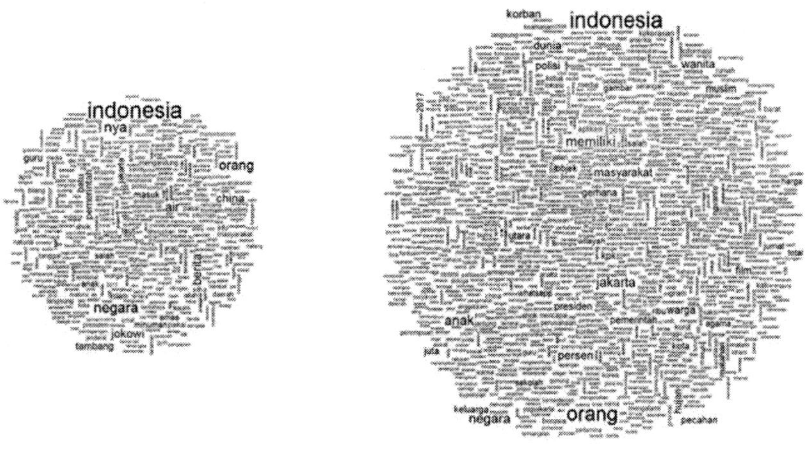

Figure 2. The comparison between word clouds for the hoaxes (left); and not hoaxes (right)

The hoax articles we had collected generally have a smaller amount of words compared to the real news articles, as shown in Figure 2, which also means that the hoaxes have a smaller amount of used vocabulary (and thus fewer words in the corpus). This agreed with the finding of [10]. We identified some most recurring themes in the hoaxes articles as shown in Table 1 (some article has overlapping themes). We also found that some of the articles themed religion, politics, and racism/ethnicity have a sentimental tone towards other groups. This finding agrees with the situation in Indonesia, which has recurring issues with religions, and the recent political 'gossips' regarding the religion and ethnicity of the politicians. The sentimental religion-themed hoax might not directly confront certain group, but by incorporating certain religious symbolism, to be combined with the presented bad image. The other religion-themed hoaxes did not have sentimental tone but they tried to 'convince' the reader to strengthen their belief towards their religion by presenting false claims. The hoax articles often use a less standardized form of written language, 'common' vocabularies, provocative words, capitalizations, and a lot of exclamation marks.

Table 1. The most recurring themes in the Indonesian hoax articles

Theme	Count	Percentage
Religion*	27	36%
Politics*	19	26%
Health	10	14%
Racism / Ethnicity*	9	12%
Technology	8	11%
Other	22	30%

The experimental result showed that the model could achieve up to 83% accuracy in determining whether an article is a hoax or not. The accuracy of all experiment settings is shown in Table 2. The table showed that the higher value of TTP had led to slightly better accuracy. This, of course, is not surprising, since the more data is fed into the model, the better it will learn and recognize the articles. On the other hand, the higher value of the STL led to worse prediction accuracy. This can be explained since the higher value of STL will make the TDM include more sparse words, by which that the sparse words from both 'hoaxes' and 'not hoaxes' collections would appear together in the TDM and consequently 'confused' the model and led to false detections. The accuracy of the model is also affected by the value of K, where the smaller value of K gives a better detection accuracy as shown in Table 3.

Table 2. The experiment results showing the evaluation of model accuracy (averaged from 1000 experiments) in predicting whether a piece of news is a hoax or not, for varied STL and TTP

	Training/Testing Proportion		
Sparse Terms Limit	0.7	0.8	0.9
0.80	81.4%	82.4%	83.6%
0.85	79.1%	79.7%	80.5%
0.90	67.4%	68.5%	69.4%

When we looked at the confusion matrices, we found that the models often fall for false-positive detections (the article were actually not hoaxes, but falsely identified/predicted as hoaxes). An example of the confusion matrices is shown in Table 3. The cause of false-positive detections in this research has not been investigated.

Table 3. The effect of K on model accuracy

Value of K	Model Accuracy (%)
1	83.6
2	77.9
3	75.9
4	76.4
5	74.1

Table 4. Example of Confusion Matrices for STL = 0.80 and TTP = 0.7; with the accuracy of 93% (left); and for STL = 0.90 and TTP = 0.7; with the accuracy of 64% (right) showing that the model is prone to false positive detections

	Actual			Actual	
Predictions	Not Hoax	Hoax	Predictions	Not Hoax	Hoax
Not Hoax	13	0	Not Hoax	7	2
Hoax	3	28	Hoax	14	21

5. Conclusion and further research directions

From the 74 hoaxes compiled from Indonesian hoax-debunking community websites, it was found that the most recurring themes of Indonesian hoaxes include religion, politics, health, racism/ethnicity, and technology, by which some of the religious, political, and ethnical hoaxes has some sentimental tone in them. Those hoaxes were then being compared against 74 real news from various reputable news websites in Indonesia. The KNN modeling result shows that the model is able to give detection/classification accuracy up to 83.6%. The accuracy of the detection model is affected by the value of K, Sparse Terms Limit, and the number of training data. Result also showed that the model is prone to false positives detections.

The cause of false-positive detections, which affect the model accuracy, has not yet investigated. This might be affected by the number of words included in the TDM (which is affected by article length in the article collections), the value of K, and the number of data used in the experiments. Thus, further research could experiment with varied ratios of words in TDM for both words in the hoax and non-hoax collections, experiment with different values of K, and a varied number of data. The TDM could also be fed into other classification algorithms to compare their performance. The TDM could also be fed into other machine learning algorithms such as the apriori association rules to give additional insight into the characterization of hoaxes vs real news.

References

[1] Hernandez, J. C., Hernandez, C. J., Sierra, J. M., & Ribagorda, A. (2002). A first step towards automatic hoax detection. In Security Technology, 2002. Proceedings. 36th Annual 2002 International Carnahan Conference on (pp. 102-114). IEEE.

[2] Rubin, V. L., Chen, Y., & Conroy, N. J. (2015). Deception detection for news: three types of fakes. Proceedings of the Association for Information Science and Technology, 52(1), 1-4.

[3] Chen, Y., Conroy, N. J., & Rubin, V. L. (2015). News in an online world: The need for an "automatic crap detector". Proceedings of the Association for Information Science and Technology, 52(1), 1-4.

[4] Tacchini, E., Ballarin, G., Della Vedova, M. L., Moret, S., & de Alfaro, L. (2017). Some Like it Hoax: Automated Fake News Detection in Social Networks. arXiv preprint arXiv:1704.07506.

[5] Shu, K., Sliva, A., Wang, S., Tang, J., & Liu, H. (2017). Fake News Detection on Social Media: A Data Mining Perspective. ACM SIGKDD Explorations Newsletter, 19(1), 22-36.

[6] GUNASEKARAN, K., GANESAN, G., RAMANUJAN, S. S., & SRINIVASAN, B. Fake News Detection in Social Media.

[7] Chen, Y. Y., Yong, S. P., & Ishak, A. (2014). Email Hoax Detection System Using Levenshtein Distance Method. JCP, 9(2), 441-446.

[8] Saberi, A., Vahidi, M., & Bidgoli, B. M. (2007, November). Learn to detect phishing scams using learning and ensemble? methods. In Proceedings of the 2007 IEEE/WIC/ACM International Conferences on Web Intelligence and Intelligent Agent Technology-Workshops (pp. 311-314). IEEE Computer Society.

[9] Mokhtari, M., Saraee, M., & Haghshenas, A. (2008). A novel method in scam detection and prevention using data mining approaches. Proceedings of IDMC2008, 1-11.

[10] Horvitz, E., Heckerman, D. E., Dumais, S. T., Sahami, M., & Platt, J. C. (2000). U.S. Patent No. 6,161,130. Washington, DC: U.S. Patent and Trademark Office.

[11] Vuković, M., Pripužić, K., & Belani, H. (2009). An Intelligent Automatic Hoax Detection System. Knowledge-Based and Intelligent Information and Engineering Systems, 318-325.

[12] Wang, A. H. (2010, July). Don't follow me: Spam detection in twitter. In Security and Cryptography (SECRYPT), Proceedings of the 2010 International Conference on (pp. 1-10). IEEE.

[13] Rasywir, E., & Purwarianti, A. (2016). Eksperimen pada Sistem Klasifikasi Berita Hoax Berbahasa Indonesia Berbasis Pembelajaran Mesin. Jurnal Cybermatika, 3(2).

[14] Kumar, S., West, R., & Leskovec, J. (2016, April). Disinformation on the web: Impact, characteristics, and detection of Wikipedia hoaxes. In Proceedings of the 25th International

Conference on World Wide Web (pp. 591-602). International World Wide Web Conferences Steering Committee.

[15] Dakwala, A., & Lavingia, K. A Machine learning approach to improve the efficiency of Fake websites detection Techniques. International journal of Computer Science and Communication (IJCSC) Vol, 7, 236-243.

[16] Ahmed, H., Traore, I., & Saad, S. (2017, October). Detection of Online Fake News Using N-Gram Analysis and Machine Learning Techniques. In International Conference on Intelligent, Secure, and Dependable Systems in Distributed and Cloud Environments (pp. 127-138). Springer, Cham.

[17] A. P. Windarto *et al.*, "Analysis of the K-Means Algorithm on Clean Water Customers Based on the Province," *J. Phys. Conf. Ser.*, vol. 1255, no. 1, 2019, doi: 10.1088/1742-6596/1255/1/012001.

[18] S. Sundari, Karmila, M. N. Fadli, D. Hartama, A. P. Windarto, and A. Wanto, "Decision Support System on Selection of Lecturer Research Grant Proposals using Preferences Selection Index," *J. Phys. Conf. Ser.*, vol. 1255, no. 1, pp. 1–8, 2019, doi: 10.1088/1742-6596/1255/1/012006.

[19] S. R. Ningsih, R. Wulansari, D. Hartama, A. P. Windarto, and A. Wanto, "Analysis of PROMETHEE II Method on Selection of Lecturer Community Service Grant Proposals," *J. Phys. Conf. Ser.*, vol. 1255, no. 1, pp. 1–7, 2019, doi: 10.1088/1742-6596/1255/1/012004.

[20] Sudirman, A. P. Windarto, and A. Wanto, "Data mining tools | rapidminer: K-means method on clustering of rice crops by province as efforts to stabilize food crops in Indonesia," *IOP Conf. Ser. Mater. Sci. Eng.*, vol. 420, no. 1, 2018, doi: 10.1088/1757-899X/420/1/012089.

[21] T. Imandasari, M. G. Sadewo, A. P. Windarto, A. Wanto, H. O. Lingga Wijaya, and R. Kurniawan, "Analysis of the Selection Factor of Online Transportation in the VIKOR Method in Pematangsiantar City," *J. Phys. Conf. Ser.*, vol. 1255, no. 012008, pp. 1–7, 2019, doi: 10.1088/1742-6596/1255/1/012008.

IOP Conf. Series: Materials Science and Engineering **835** (2020) 012037 doi:10.1088/1757-899X/835/1/012037

Decision Support System Museum Ambassadors Using Topsis Method

Setyoningsih Wibowo[1], Maichifa Ayuning Tyas[2], Noora Qotrun Nada[3], Mega Novita[4]

[1,2,3,4]Faculty of Engineering and Informatics, Universitas PGRI Semarang, Jl. Sidodadi-Timur No.24 Semarang, Central Java 50232, Indonesia

setyoningsihwibowo@upgris.ac.id

Abstract. Generally, museum ambassador is based on various criteria from several assessment stages. The evaluation criteria for the audition stage are: interview, communication, appearance. The assessment criteria for the quarantine stage are: creativity, ability, liveliness, loyalty, responsibility, and written test. The final stage assessment criteria are: ideas, communication, and appearance. In the selection process, the committee used manual method so it was prone to errors. The decision support system used TOPSIS method. SDLC (System Development Life Cycle) was used as a design method. The research showed a score of 100 which means that the system is accepted by users. Meanwhile the user testing to four respondents obtained a percentage of 89.24% which means that the system is worth using. The system was developed in the web. So that is need more research to be developed in mobile.

1. Introduction

A museum ambassador is a person who represents or becomes an icon, and his/her task is to educate the general public to be familiar with the museum. The presence of museum ambassador aims to help improve museum marketing, invite people to be willing to visit museum, make people aware of the importance of museum as a source of knowledge, a source of inspiration, a recreation place, and as a place of national character education [1]. The word "Museum" is derived from the word *Muze*, which in the Ancient Greek language is defined as a group of nine goddesses, a symbol of arts [2]. To date, the audition of museum ambassadors in Ranggawarsita Museum in Semarang City still uses paper-based conventional or manual selection processes. However in practice, there are many weaknesses of such paper-based system, for examples: paper-based data are damaged or lost, data processing is difficult or takes a long time, the tabulation process is error-prone, and the process of report writing takes a relatively long time. Therefore, the authors wanted to minimize the use of paper-based manual system by introducing a computer-based decision support system in the selection of the ambassadors of Central Java Museum.

2. Methods

2.1. Decision Support System

The concept of Decision Support System (DSS) was first introduced in the early 1970s by Michael S. Scott Morton using the term Management Decision System [3]. Decision Support System (DSS) is an information-producing system [4] which aims to solve various problems [5] with the help of computer technology in decision making process [6].

2.2. TOPSIS Method

Topsis method, which stands for Technique for Order Preference by Similarity to Ideal Solution, is one of decision making methods which can be applied in this problem. This method searches for the best alternative with the shortest distance from the positive ideal solution and the farthest distance from the negative ideal solution, with certain criteria and weight of each of these criteria determined by a company. Based on the test results obtained in this study, the decision support system being developed could help or support the museum in selecting the best museum ambassadors [1].

Topsis is one of the multi-criteria decision-making methods which was first introduced by Yoon and Hwang (1981) [3]. Topsis has the principle that the selected alternatives must have the shortest distance from the positive ideal solution and the farthest distance from the negative ideal solution from a geometrical point by using the Euclidean distance between two points to determine the relative proximity of an alternative [7].

In general, Topsis procedures are as follows [8]:

a) Making a normalized decision matrix
b) Making a weighted normalized decision matrix
c) Determining a matrix of positive ideal solution and negative ideal solution
d) Determining the distance between the values of each alternative with a matrix of positive and negative ideal solutions
e) Determining the preferred value for each alternative

3. Results and discussion

Topsis algorithms were used in the three stages of selection, i.e. the audition stage, the quarantine stage, and the final stage. Topsis calculation resulted in the ranking of each alternative (candidate) and the winner or the best alternative was determined based on such ranking. The number of shortlisted candidates of each of the selection stages was determined as needed. The assessment criteria for each of the selection stages were different and shown in Table 1 – Table 3.

Table 1 Assessment Criteria for Audition Stages

Criteria	**Descriptions**
Questions and Answers	Questions about the program that will be held, being creative in answering
Communication	Speaking fluently and focusing on the materials
Appearance	Tall, beautiful, expressive speech

Table 2 Assessment Criteria for Quarantine Stage

Criteria	Descriptions
Creativity	Being creative in describing the socialization programs of the museum
Ability	Ability to offer solutions to museum development and how to promote the museum
Liveliness	Being active in groups during quarantine/training
Loyalty	Being loyal, i.e. being serious in being an ambassador and loyal to the museum
Responsibility	Being responsible for any tasks given by the judges
Written test	Each judge was required to make questions, then the questions were put together to be given to each candidate

Table 3 Assessment Criteria for Final Stage

Criteria	Descriptions
Ideas	Ideas in promoting the museum to the general public without spending money
Communication	Fluently communicating and focusing on the materials
Appearance	Tall, beautiful, expressive speech

The scores that the judges could give to the candidates ranged from 2, 3, 4, and 5 or using a description, i.e. poor, fair, good, and very good.

Table 4. Score of Criteria

Score	Descriptions
2	Poor
3	Fair
4	Good
5	Very Good

The steps in Topsis calculation are as follows:

a) Making decision matrix (X)

Decision matrix contains data of alternatives and criteria listed in the form of table or matrix. The decision matrix of the final stage of Central Java museum ambassador selection can be seen in Table 5.

Table 4 Decision Matrix

Alternatives	Criteria		
	K_1	K_2	K_3
Abdurrouf Khodamul Umam	5	5	5
Anita Sholikha	5	4	3
Aulia Azzumardila Adzra	5	4	5
Michael Rivaldy Mewengkang	5	5	4
Ika Audiyah Andriyani	5	4	4
Jacika Pifi Nugraheni	4	3	3
Ayu Sabrina	4	2	2
Mega Purnomo	3	3	2
Rio Awang Maulana	3	3	2
Riki Supriyadi	4	4	3

b) Making weighted criteria matrix (W)

Weighted criteria matrix contains the data of criteria weights listed in the form of table or matrix. The weighted criteria matrix of the final stage of Central Java museum ambassador selection can be seen in Table 6.

Table 5 Weighted Criteria Matrix

Code	Criteria	Type	Weight
K_1	Ideas	Benefit	0.5
K_2	Communication	Benefit	0.3
K_3	Appearance	Benefit	0.2

c) Making normalized decision matrix (R)

Normalized decision matrix contains the data of the result of normalized calculation listed in the form of table or matrix.

Table 6 Normalized Decision Matrix

Alternatives	Criteria		
	K_1	K_2	K_3
Abdurrouf Khodamul Umam	0.362	0.415	0.455
Anita Sholikha	0.362	0.332	0.273
Aulia Azzumardila Adzra	0.362	0.332	0.455
Michael Rivaldy Mewengkang	0.362	0.415	0.364
Ika Audiyah Andriyani	0.362	0.332	0.364

Alternatives	Criteria		
	K_1	K_2	K_3
Jacika Pifi Nugraheni	0.289	0.249	0.273
Ayu Sabrina	0.289	0.166	0.182
Mega Purnomo	0.217	0.249	0.182
Rio Awang Maulana	0.217	0.249	0.182
Riki Supriyadi	0.289	0.332	0.273

d) Making weighted normalized decision matrix (Y)
Normalized decision matrix contains the calculation result of normalized matrix and weighted matrix listed in the form of table or matrix.

Table 7 Weighted Normalized Decision Matrix

Alternatives	Criteria		
	K_1	K_2	K_3
Abdurrouf Khodamul Umam	0.1809	0.1246	0.0909
Anita Sholikha	0.1809	0.0997	0.0545
Aulia Azzumardila Adzra	0.1809	0.0997	0.0909
Michael Rivaldy Mewengkang	0.1809	0.1246	0.0727
Ika Audiyah Andriyani	0.1809	0.0997	0.0727
Jacika Pifi Nugraheni	0.1447	0.0747	0.0545
Ayu Sabrina	0.1447	0.0498	0.0364
Mega Purnomo	0.1085	0.0747	0.0364
Rio Awang Maulana	0.1085	0.0747	0.0364
Riki Supriyadi	0.1447	0.0997	0.0545

e) Determining the matrix of positive ideal solution and negative ideal solution (A+ and A-)
Matrix of positive ideal solution and negative ideal solution (A^+) and (A^-) is the calculation result of positive ideal solution and negative ideal solution listed in the form of table or matrix. Benefit attribute is an attribute given the highest score to obtain the shortest distance from the positive ideal solution and the farthest distance from the negative ideal solution. On the other hand, cost attribute is an attribute given the lowest score to obtain the farthest distance from the positive ideal solution and the shortest distance from the negative ideal solution.

Table 8 Weighted Normalized Decision Matrix

	K_1	K_2	K_3
A^+	0.1809	0.1246	0.0909
A^-	0.1085	0.0498	0.0364

f) Determining the distance between the scores of each alternative with the matrix of positive ideal solution and negative ideal solution (D+ and D-)
Matrix of distance from positive ideal solution and distance from negative ideal solution ($D^+$and D^-) is the calculation result of distance from positive ideal solution and distance from negative ideal solution listed in the form of table or matrix.

Table 9 Matrix of Distance from Positive Ideal Solution and Negative Ideal Solution

Alternatives	D^+	D^-
Abdurrouf Khodamul Umam	0.0000	0.1175
Anita Sholikha	0.0441	0.0897
Aulia Azzumardila Adzra	0.0249	0.1034
Michael Rivaldy Mewengkang	0.0182	0.1102
Ika Audiyah Andriyani	0.0308	0.0951

Alternatives	D^+	D^-
Jacika Pifi Nugraheni	0.0715	0.0475
Ayu Sabrina	0.0993	0.0362
Mega Purnomo	0.1034	0.0249
Rio Awang Maulana	0.1034	0.0249
Riki Supriyadi	0.0570	0.0642

g) Determining preferred value (V) of each alternative

Preferred value (V) of each alternative. The highest preferred value (V_i) shows that the *i-th* alternative deserves to be selected as the best solution.

Table 10 Preferred Value

Alternatives	Vector (v)
Abdurrouf Khodamul Umam	1.0000
Anita Sholikha	0.6705
Aulia Azzumardila Adzra	0.8059
Michael Rivaldy Mewengkang	0.8584
Ika Audiyah Andriyani	0.7551
Jacika Pifi Nugraheni	0.3993
Ayu Sabrina	0.2669
Mega Purnomo	0.1941
Rio Awang Maulana	0.1941
Riki Supriyadi	0.5296

Table 12 Conclusions of Winners of Museum Ambassador

	Alternatives
Winner	Abdurrouf Khodamul U
1st Runner-up	Michael Rivaldy M
2nd Runner-up	Aulia Azzumardila Adzra
3rd Runner-up	Ika Audiyah Andriyani
4th Runner-up	Anita Sholikha
5th Runner-up	Riki Supriyadi

4. Conclusion

a. Topsis method which is a decision support system which is able to solve various multi-criteria decision making problems can also be used to solve the problems in selecting a Museum Ambassador.

b. An application has been successfully created using the SDLC (System Development Life Cycle) method and a system using a decision support system with Topsis method.

c. Manual Topsis calculation results are the same as computer-based calculation results.

d. User acceptance testing (UAT) uses questionnaires filled out or evaluated by museum staffs and by experts (lecturers). The results of this test show that the software has a percentage of 89.24% thus receiving a predicate of "The system is feasible to use

References

[1]. K. Andryan, S. Effendi, E. Santoso, and N. Hidayat, "Implementasi Metode Topsis Untuk PenentuanFinalis Duta Wisata Joko Roro Kabupaten Malang (Studi Kasus : Paguyuban Joko Roro), "Jurnal Pengembangan Teknologi Informasi dan Ilmu Komputer" vol. 2, no. 2, pp. 469-478, 2017, ISSN 2548-964X.

[2]. M. Sholeh and C. Iswahyudi, "E-Museum : Informasi Museum di Yogyakarta Berbasis Location," Prosiding Seminar Nasional Aplikasi Sains & Teknologi (Snast)no. November, 2014.

[3]. D. L. Kurniasih, "Sistem Pendukung Keputusan Pemilihan Laptop Dengan Metode Topsis Sumber: Pelita Informatika: Informasi dan Informatika jurnal.stmik-budidarma vol. 3, no. 2, 2013.

[4]. S. Eniyati, "Perancangan Sistem Pendukung Pengambilan Keputusan untuk Penerimaan Beasiswa dengan Metode *SAW (Simple Additive Weighting)*," Jurnal Teknologi Informasi dinamik vol. 16, no. 2, pp. 171–177, 2011.

[5]. H. Nurdiyanto and H. Meilia, "Sistem Pendukung Keputusan Penentuan Prioritas Pengembangan Industri Kecil dan Menengah di Lampung Tengah Menggunakan Analitical Hierarchy Process (*AHP*)," Jurnal Seminar Nasional Teknologi Informasi dan Multimedia pp. 6–7, 2016, ISSN: 2302-3805.

[6]. D. Isi, Y. Legok, R. F. Ningrum, and D. Hartanti, "Sistem Pendukung Keputusan Promosi Jabatan Struktural Dosen Menggunakan AHP (Analytical Hierarchy Process)". *Jurnal Teknik*, vol. *6, no.* 2, 2017.

[7]. Windarto, A. P, "Implementasi metode topsis dan saw dalam memberikan reward pelanggan," Kumpulan jurnal Ilmu Komputer, vol. 04, no. 01, pp. 88–101, 2017.

[8]. Sanada, A. B. (2015), "Perancangan Sistem Pendukung Keputusan Pemilihan Pelanggan Terbaik Dengan Metode TOPSIS (Studi Kasus: PD. Istana Duta) ". Jurnal Sistem dan Teknologi Informasi (Justin), vol 3, no 2, pp. 33-38, 2015.

[9]. Budiharjo, T. Soemartono, A. P. Windarto, and T. Herawan, "Predicting tuition fee payment problem using backpropagation neural network model," Int. J. Adv. Sci. Technol., vol. 120, pp. 85-96, 2018, doi: 10.14257/ijast.2018.120.07.

[10].Budiharjo, T. Soemartono, A. P. Windarto, and T. Herawan, "Predicting School Participation in Indonesia using Back-Propagation Algorithm Model," Int. J. Control Autom., vol. 11, no. 11, pp. 57-68, 2018.

[11].A. P. Windarto et al., "Analysis of the K-Means Algorithm on Clean Water Customers Based on the Province," J. Phys. Conf. Ser., vol. 1255, no. 1, 2019, doi: 10.1088/1742-6596/1255/1/012001.

IOP Conf. Series: Materials Science and Engineering **835** (2020) 012038 doi:10.1088/1757-899X/835/1/012038

Kinetics of Oxidation Decomposition on *Jatropha Curcas* Biodiesel

Supriyono[1*], Sumardiyono[1], Enny Purwanti[2], Retno Ambarwati S.L.[3]

[1]Chemical Engineering Department, Engineering Faculty, Setia Budi University, - Surakarta – Indonesia

[2]Agricultural Technology Faculty, University of 17 Agustus 1945 Semarang, Semarang - Indonesia

[3]Chemical Engineering Department, University of 17 Agustus 1945 Semarang, Semarang - Indonesia

suprisuwito@gmail.com

Abstract. High free fatty acid jatropha curcas oil was use as a feedstock of biodiesel through two step esterification, first by reaction of Jatropha curcas oil with methanol catalyzed by sulfuric acid, and second step is transesterification process with methanol catalyzed by sodium hydroxide. After purification step, 120 ppm pyrogallol was introduced into Jatropha curcas biodiesel. Accelerated controlled oxidation process was performed using rancimat test equipment with the dry airflow to the reaction tube is 10 L/min. Assuming that oxygen only reacts with double bond of the compound that compose biodiesel, measurement the quantity changes of double bond on biodiesel by the changes of the value of iodine number will brought to the rate of reaction. While reaction rate constant determine by Arrhenius law was calculated from the data of time to reach induction period in a certain temperature. It was found that total rate of oxidation process was determined by chemical reaction rate rather than mass transfer rate of oxygen to the body of biodiesel, both on biodiesel with or without pyrogallol. While in a presence of 120 ppm pyrogallol on jatropha curcas biodiesel, Arrhenius constant was determined by the value of frequency factor (A) = 28.47 x 106/ sec and activation energy (E) = 97,31 kJ/mol. This mean that on the ambient temperature 30 oC, Jatropha curcas biodiesel in a presence of 120 ppm pyrogallol will decrease their iodine value 3.4% after 2.3 years.

1. Introduction

One of the most promising sources for petrodiesel substitution is biodiesel, which is almost similar on physical and chemical properties to petrodiesel. For this reason, biodiesel could apply as fuel on the conventional internal combustion engine with a minor modification. There are some feedstocks for biodiesel production, as vegetable oil, waste from the slaughter house and fish processing, and also from microalgae [1-7] The final biodiesel product characteristic will depend on the feedstock source. The difference between biodiesel and petrodiesel is on the stability due to oxidation process during storage. In petrodiesel there is no significant influence on the oxidation process to their stability and quality [8] while oxidation process on biodiesel will affect significantly to at least 6 parameters on biodiesel standard (kinematics viscosity, cetane number, total acid number, flash point, cooper strip corrosion, iodine number) [9]. Oxidation on biodiesel will occur on double bond in fatty acid chain and highly correlated to unsaturated fatty acid that composed the feedstock of biodiesel. Feedstock contain high unsaturated fatty acid will produce biodiesel with high unsaturated fatty acid methyl ester. In a presence of oxygen, the unsaturated fatty acid compound on biodiesel will oxidize to free fatty acid which will perform simultaneous and complex reaction on a secondary oxidation process. The final products of the oxidation process are aldehyde, carboxylic acid, alcohols and hydrocarbon compound [10].

To determine the rate of oxidation process on biodiesel, measurement of changes on the component that compose biodiesel could conduct on the different temperature of the oxidation process. Total double bond on the fatty acid component in biodiesel could be determine by iodine number method [11] in case that double bond was oxidized it will show by decreasing the total iodine number of biodiesel. Determination of oxidation process in fatty acid by iodine number method is simple and accurate compare to determination of increasing acid number or any other substance as a result of oxidation on biodiesel because the product of the oxidation process is unstable and tend to decompose to any other substance [10].

There are some methods to reduce and prevent oxidation process on biodiesel for example removing substance that trigger and boosting oxidation process (water, metal and other contaminant substance), hydrogenation of biodiesel that converted unsaturated compound into saturated compound even this could increase viscosity of biodiesel, and the other method is introducing antioxidant into biodiesel [12-14].

The term Induction period (IP) on rancimat test analysis show the time when the oil starts to oxidize and produce volatile organic compound that dilute in the water and change significantly the conductivity of the water. The time when the conductivity changes namely Induction Period (IP). In this point a small amount of the double bond in biodiesel compound start react with oxygen and shown by decreasing of biodiesel iodine number. Initial oxidation reaction will form free radical that trigger chain reaction as shown in the equation (1) to (5) [15,16].

1.1. Biodiesel Oxidation

There are several sources for biodiesel decomposition, biological, physical and chemical source. Microorganism activity is main source for biodiesel decomposition by biological source, while heat, metals and light are physical source. Water, oxygen and free fatty acid are chemical source for biodiesel decomposition [17,18]. These sources could work separately or simultaneously on biodiesel and resulted on decompose of biodiesel. As one of the chemical source on biodiesel decomposition, oxygen could react with double bond on unsaturated fatty acid that composed biodiesel. Decomposition due to oxidation reaction on biodiesel could bring a serious problem on the engine because biodiesel could change viscosity, acid value, water content, cetane number, and also its caloric value.

Initiation:

$$RH + O_2 \rightarrow R^* + H_2O \quad (1)$$

$$RH + O_2 + RH \rightarrow R^* + H_2O + R^*$$

Propagation:

$$R^* + O_2 \rightarrow ROO^* \quad (2)$$

$$ROO^* + RH \rightarrow ROOH + R^* \quad (3)$$

Peroxide decomposition

$$ROOH \rightarrow RO^* + {}^*OH$$

$$2\ ROOH \rightarrow RO^* + ROO^* + H_2O$$

Termination:

$$ROO^* + ROO^* \rightarrow \text{inactive product} \quad (4)$$

$$ROO^* + IH \rightarrow ROOH + I \quad (5)$$

where

RH = organic matter (Alkyl lipid)
ROO^* = Peroxide radical
ROOH = Hydroperoxide; I = Stable radical or inactive radical
IH = free radical inhibitor

In the oxidation process, oxygen will react with the components that have double bond on the molecular chain, thus mean that the oxidation reaction to biodiesel will reduce the number of double bond in biodiesel compound. While double bond compound on biodiesel was found on the unsaturated chain on biodiesel component. The changes on the percentage of unsaturated chain could be analyzed by evaluation on Iodine number.

Antioxidant
Most of the oil and lipid came with their original antioxidant, it was namely natural antioxidants. Tocopherols, ascorbic acid, ascorbic acid ester, and carotenoids are example of natural antioxidant that exists in lipid [19] Natural antioxidant was decrease significantly in biodiesel due to the process steps, from oil extraction, pretreatment to remove gums and phospolipid, also esterifcation process to reduce high free fatty acid in case that biodiesel production use transesterification route with alkali catalyst, purification of biodiesel to fulfill the standard requirement. For this reason, introducing new antioxidant should be done to improve biodiesel quality in a term of oxidation stability.

Synthetic antioxidants are a manmade and mostly are phenols based [10]. Phenolic derivatives usually contain more than one hydroxyl or methoxyl group. Synthetic phenolic antioxidants are ***p*** substituted, while natural phenolic antioxidants are ***o*** substitutes.

There are two classification of antioxidant, primary and secondary antioxidant. Primary antioxidant refers to chain breaking antioxidants, because the chemical nature of these molecules they can act as free radical acceptors and delay or inhibit the initiation step or interrupt the propagation step of auto oxidation. Mechanism of primary antioxidant is shown in the equation (6) to (8) [20]

$$R\bullet + AH \rightarrow RH + A\bullet \quad (6)$$
$$RO\bullet + AH \rightarrow ROH + A\bullet \quad (7)$$
$$ROO\bullet + AH \rightarrow ROOH + A\bullet \quad (8)$$

Where: R•, RO•, ROO• = free radical

Radical from the antioxidant will react with radical from fatty acid to form stable complex compound as shown in equation (9) to (11).

$$R\bullet + A\bullet \rightarrow RA \quad (9)$$
$$RO\bullet + A\bullet \rightarrow ROA \quad (10)$$
$$ROO\bullet + A\bullet \rightarrow ROOA \quad (11)$$

1.2. Measurement method of biodiesel stability.

There are several methods to measure stability of biodiesel. Some of them base on the assumption that biodiesel stability is depend on oxidation on the biodiesel due to the presence of oxygen. These include the weight gain and headspace oxygen uptake method for oxygen absorption, chromatographic analysis for changes in reactants, iodometric titration, ferric ion complexes, FTIR method for peroxide value, spectroscopy for conjugated dienes and trienes. While for expressing antioxidant activity, there are some methods, ie. Induction Period (IP), Oxygen Stability Instrument (OSI) and also PetroOXY test [10]. European Standard for biodiesel (EN 14214), and Indonesian Standard (SNI-04-7182) were applied Induction Period to measure biodiesel stability. The requirement of Induction Period in EN 14214 standards was increase from 6 hours in EN 14214-2003 to 8 hours in EN 14214-2012.

1.3. Mathematical Model

According to one-layer mechanism of reaction, Oxygen (B) will diffuse to the interface of gas-liquid system, and then chemical reaction step will occur in a liquid body of biodiesel (A) as seen in Fig. 2 [21]

IOP Conf. Series: Materials Science and Engineering **835** (2020) 012038 doi:10.1088/1757-899X/835/1/012038

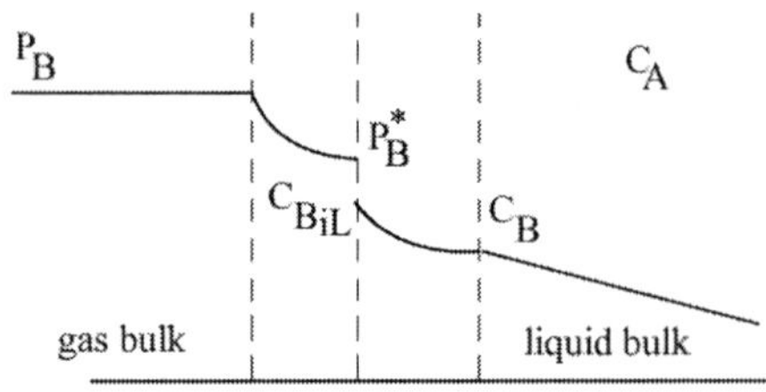

Figure 1. Mechanism of oxidation reaction in biodiesel according to quasy steady state reaction.

Unsaturated fatty acid component on biodiesel were dominated by oleic acid ($C_{18:1}$), linoleic acid ($C_{18:2}$) and linolenic acid ($C_{18:3}$). Oxidation reaction on each component could write as shown in equation (12) to (15).

$C_{18:1}$ + O_2 →P, written as $A_1 + B \rightarrow P$ (12)

$C_{18:2}$ + 2 O_2 →P, written as $A_1 + 2B \rightarrow P$ (13)

$C_{18:3}$ + 3 O_2 →P, written as $A_1 + 3B \rightarrow P$ (14)

Equation (12) to (14) then could be written as

$$A + 6B \rightarrow P \quad (15)$$

In a presence of antioxidant (M) on biodiesel, antioxidant will react as shown in equation (16)

M+ O_2 → written as $A_4 + B \rightarrow P$ (16)

Equation (5) and (16) then could written as

$$A + 7B \rightarrow P \quad (17)$$

Solving for both mass transfer and chemical reaction for the equations (15) and (17) under quasy steady state condition will give equation (18) and (19) as follows

$$r_A = \frac{k.\frac{P_B}{H}\alpha C_A}{6k.C_A - \alpha} \quad (18)$$

$$r_A = \frac{k.\frac{P_B}{H}\alpha C_A}{7k.C_A \quad \alpha} \quad (19)$$

Where

$$k^* = k.\frac{P_B}{H}\alpha \quad (20)$$

$$\alpha = \frac{1}{k_g a H} + \frac{1}{k_L a} \quad (21)$$

2. Methods

2.1. Materials

A source of the biodiesel feedstock is high free fatty acid (FFA) jatropha curcas oil which come with 18% of FFA was purchase from PT Pura Green Energy, Kudus Indonesia. Methanol was use as alcohol source for esterification and transesterification. Sodium hydroxide was use as transesterification catalyst, while sulfuric acid was use as catalyst in esterification process. All chemicals use in biodiesel production and analysis are reagent grade.

2.2. Methods

Biodiesel source for the experiment was prepare refer to Supriyono et al. [22] High free fatty acid jatropha curcas oil was treated by esterifcation process to decrease free fatty acid. The process was performed under operation condition 65 oC, 120 minute, methanol to oil volume ratio is 40% and sulfuric acid to oil volume ratio is 2%. Oil phase which contain jatropha curcas oil and fatty acid methyl ester was further process on transesterification step which conduct on 60 oC, 60 minute, methanol to oil volume ratio is 40% and sodium hydroxide is 2% of oil weight. Biodiesel phase was purifying on the following step. The rest of methanol on biodiesel product was remove by vacuum distillation on 65 oC, 200 mmHg, sodium hydroxide on biodiesel phase was neutralizing by HCl 0,5%, rinse by distilled water until pH of the waste water close to pH of distilled water. The rest of the water on biodiesel adsorb by Na2SO4. The final product then analyzed according to EN 14214-2012 standard method.

Controlled oxidation process was performed using rancimat test equipment as shown in Fig. 1, which is operated in 10 L/min air flow on the various temperature and time. About 3 grams of biodiesel sample was use in each test. Previous experiments work by Supriyono et al. [22] show that pyrogallol was the best antioxidant for jatropha curcas biodiesel compare to TBHQ, propyl gallate and baynox plus on the standard induction period by rancimat test. For the purpose to anticipate longer induction period requirement, the work also present equation that correlate between concentrations of pyrogallol on Jatropha curcas biodiesel to the time of induction period of Jatropha curcas biodiesel. Base on the result on previous work, pyrogallol utilized as antioxidant in jatropha curcas biodiesel and further analysis the rate of oxidation stability was performed. After certain time of controlled oxidation process, biodiesel sample from reaction tube of rancimat equipment was analyzed the iodine number. Base on the changes of iodine number of biodiesel, the rate of oxidation reaction on 110 oC could be calculated. To determine Arrhenius constant of the oxidation reaction, similar experiment was performed in a difference temperature of rancimat equipment on the range of 90 °C to 140 °C.

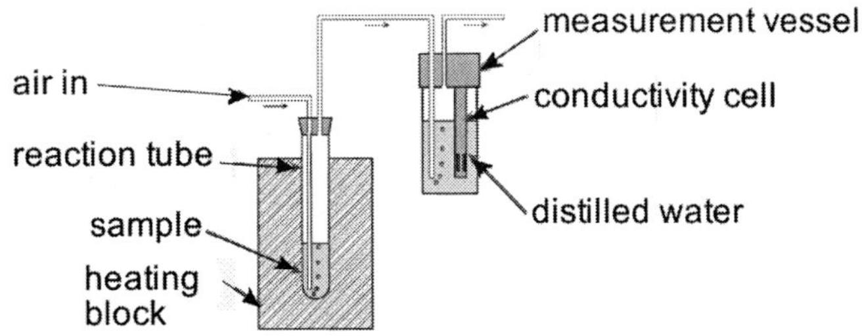

Figure 2. Schematic arrangement of rancimat test equipment.

3. Results and Discussion

3.1. Jatropha curcas biodiesel
The analysis of biodiesel was performed on the parameter of FAME content, kinematic viscosity at 40 oC, acid value, iodine number, water content, flash point, density at,15 oC and oxidation stability, 110 oC. The parameter that analyzed was fulfills the requirement on EN 14214 -2006 standard except for oxidation stability which has value 1.37 hours and far from the standard of 6 hours. Thus, jatropha curcas biodiesel is vulnerable to oxidation process. The high value on Iodine number in Jatropha curcas biodiesel is one of the suspect that responsible to oxidation process in biodiesel, because compound with double bond on biodiesel easy to react with oxygen. Jatropha curcas biodiesel start to oxidized and decompose after 1.37 hours of reaction. The result of oxidation reaction is a compound that catalyzed increasing of the acid value on biodiesel. While increasing in acid value on biodiesel the other parameter such viscosity, cetane number, caloric value and kinematic viscosity will change, the water content on biodiesel also increase for the reason that oxidation reaction also produce water.

3.2. Rate of oxidation reaction
One of the methods to determine stability of biodiesel due to oxidation process is biodiesel rancimat test method also known as accelerated reaction test because the reaction was performed in a temperature higher than ambient temperature. Latest European Standard for biodiesel (EN 14214-2012) protocol shown that temperature requirement for oxidation stability test by rancimat equipment is 110 °C and dry air flow rate 10 L/h, thus oxygen was available on the excess amount compare to the amount of unsaturated fatty acid methyl ester and assuring that oxidation process depend only on unsaturated compound on biodiesel. The idea of accelerated test is to find the shorter time to determine parameter of oxidation kinetics from which the data could be extrapolated to the ambient temperature condition. However, in the case of antioxidant activity test, high temperature experiment could decrease activity of antioxidant itself. Tert Butyl Hydroquinone (TBHQ) in soybean biodiesel could maintain oxidative stability in 544 hours on the 45 °C, but when the temperature was raises to 98 °C, the oxidative stability will decrease to 28 hours [10]. This could make sense if previous experiments could have different

IOP Conf. Series: Materials Science and Engineering **835** (2020) 012038 doi:10.1088/1757-899X/835/1/012038

result, even with the same materials and also the same antioxidant. Reaction between gas phase and liquid phase always involve mass transfer and chemical reaction. The overall reaction will determine by the slower step. To analyze which step is the slower, quasy steady state reaction model could be applied. In this experiment, biodiesel with and without antioxidant were heated in biodiesel rancimat test equipment at 110 °C. In a certain time, iodine number of the sample was analyzed. The value of iodine number was shown the amount of double bond on biodiesel, which has correlation to the amount of biodiesel that decompose by oxidation reaction. Fig. 3 and fig. 4 show the result of the analysis of iodine number for biodiesel with and without antioxidant.

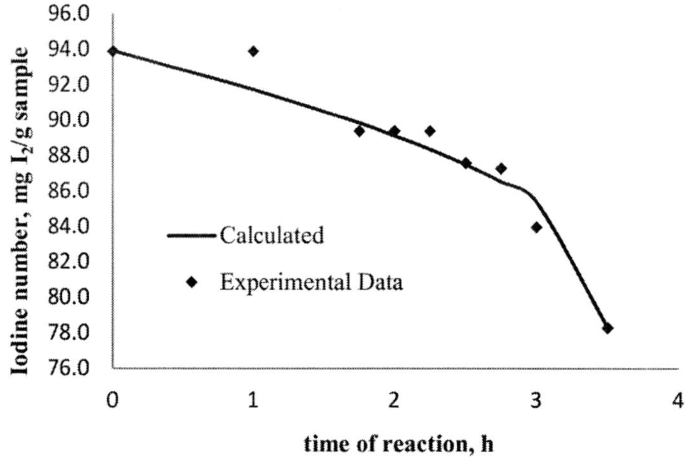

Figure 3. Correlations between Iodine number of biodiesel without antioxidant and reaction time on accelerated reaction test, 110 °C

As shown in Fig.3, iodine number of biodiesel without antioxidant will start to decrease in periods of 1 hour to 2 hour after reaction was started. This mean that oxidation reaction to the double bond of the compound in biodiesel started in a period 1 to 2 hour after reaction was initialized. The results of iodine number analysis also agree with analysis of stability on biodiesel by rancimat test method which has result 1.37 hour on induction period. Fitting of the experiment data with the developed equation (18), (20) and (21) give the result as follows

α = 10,29466; k* = 0,03644; k = 0,02130; Average of deviation is = 3.04%

$\frac{P_B}{H}$ = 0,021107

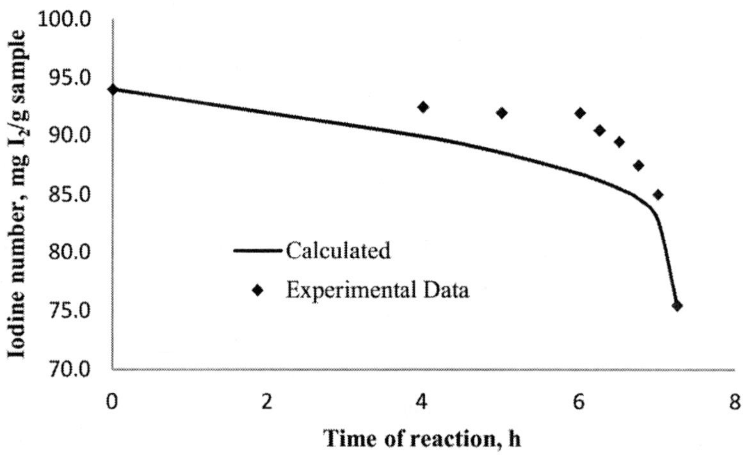

Figure 4. Correlations between Iodine number of biodiesel with pyrogallol 120 ppm and reaction time on accelerated reaction test, 110 °C

From Fig. 4, Iodine number slowly decrease until 6 hours of accelerated test then sharply decrease the value of iodine number after 6 hours. This result agree with the experiment that Jatropha curcas biodiesel contain 120 ppm pyrogallol has induction period time 6.02 hours, which is fulfill the requirement of biodiesel standard according to EN 14214-2003 and SNI 04-7182-2006. For the newer European Standard (EN 14214-2012) the amount of pyrogallol that should introduced will be higher because EN 14214 requirement is 8 hours for the induction period time.
Fitting of the data give the result for equations (20) and (21) as follows
α = 10.29466; k* = 0.04357; k = 0.05977; Average of deviation is = 0.77%,

$\frac{P_B}{H} = 0.166183$

By comparing the value of chemical reaction constant (k) and mass transfer constant (H) it was concluded that overall reaction was determine by chemical reaction step.

3.3. *Jatropha curcas biodiesel*

The The value of activation energy and frequency factor for the oxidation process on jatropha curcas biodiesel with 120 ppm pyrogallol was determine by re arrange equation

$$C_A = C_{A0}.e^{-kt}$$

$$\frac{\ln\left(\frac{C_{Ao}}{C_A}\right)}{t} = k \quad (22)$$

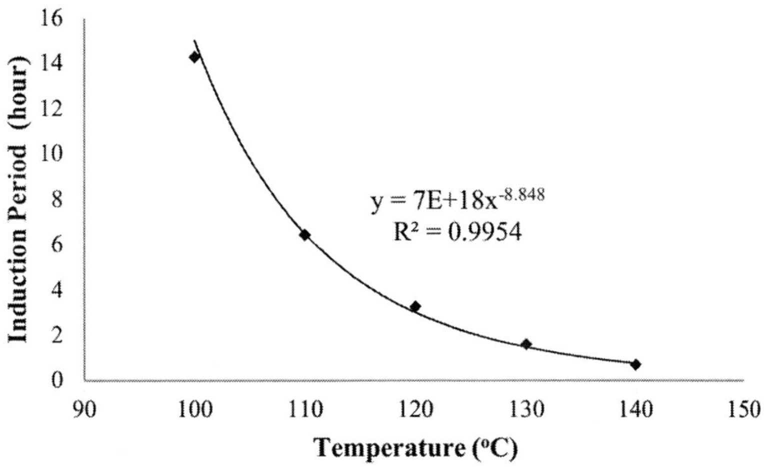

Figure 5. Induction Period value on biodiesel with pyrogallol 120 ppm

On the Induction Period (IP) point, the double bond on biodiesel was start oxidized to acidic volatile compound that absorbed on the measured vessel and increasing conductivity of the distilled water in a vessel. In this experiment, the measurement of iodine number in the IP point is 90.8 mg I_2/100 gram of sample, while initial iodine number is 94 mg I_2/100 gram of sample. For the same sample, the ratio of iodine number on the IP point (C_A) to initial iodine number value (C_{AO}) is constant, thus a set of data could have derived as shown in Table 1.

Table 1. Calculation of reaction constant (k) base on time of reaction

Time of reaction, t (hour)	Temperature of reaction T (K)	C_A	C_{AO}	$\ln(C_{AO}/C_A)$	1/t	k	ln(k)	1/T
0.72	413	90.8	94	0.03464	1.38889	0.04810	3.03437	0.00242
1.38	403	90.8	94	0.03464	0.72464	0.02510	3.68496	0.00248
2.81	393	90.8	94	0.03464	0.35587	0.01233	4.39606	0.00254
6.07	383	90.8	94	0.03464	0.16474	0.00571	5.16623	0.00261
14.1	373	90.8	94	0.03464	0.07092	0.00246	6.00905	0.00268
35.81	363	90.8	94	0.03464	0.02793	0.00097	6.94110	0.00275

ln k = ln A –(E/RT) (23)

Correlation between lk k and 1/T was shown in Fig. 6.

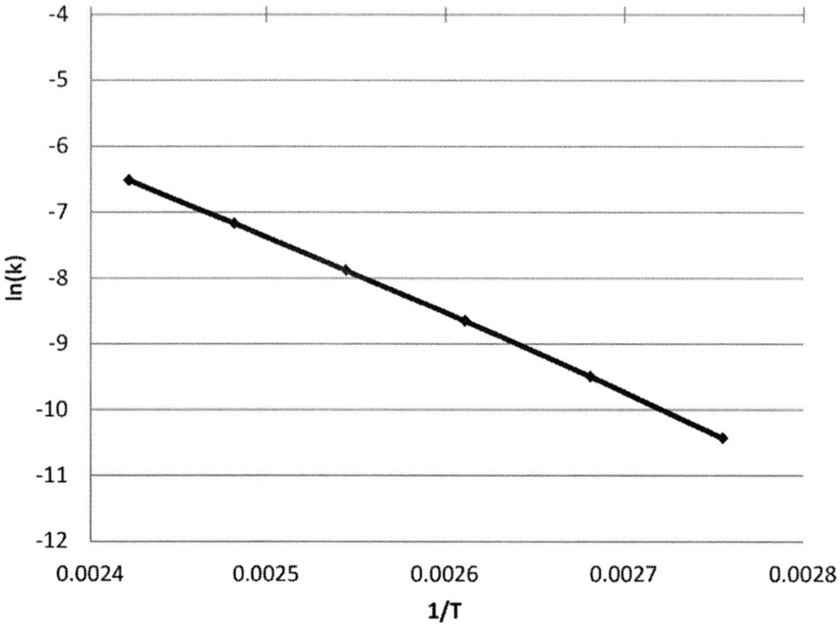

Figure 6. Correlation between 1/T and ln k in the oxidation process on jatropha curcas biodiesel with pyrogallol 120 ppm

The value of
A = exp (25.353) = 1.025 x 10^{11}/ hour
E = 11705 x 8.314 = 97315.37 J/mol

Base on the value of activation energy and pre exponential Arrhenius equation, it could be estimated that introducing 120 ppm pyrogallol on the jatropha curcas biodiesel and stored in 30°C Iodine number will decrease 3.4% during 2.3 years' storage. However, this calculation base on the oxidation reaction only without any other source that could interfere the result, such as moisture, light and metal that in some research is significantly to changes the rate of reaction [23]. A study on the thermal behavior of Jatropha curcas biodiesel [24] shown that weight loss in a range of 280-400 °C is about 96% on both nitrogen and air medium. Thermogravimetric (TGA) analyzer for this study was operating under heating rate 10 °C/min and constant flow of nitrogen and air at a rate (20±0.5) ml/min. Furthermore, the study also shown that activation energy on biodiesel weight loss in nitrogen medium is smaller (17.81 – 20.70 kJ/mol) than in air medium (22.69 – 24.71 kJ.mol). Base on the definition of activation energy in which minimum energy required to cross the barriers for initiating a chemical reaction, the rate of thermal decomposition on jatropha biodiesel is faster in nitrogen medium than in air medium. Comparing the result of thermal decomposition to the oxidation decomposition, it was shown that activation energy on oxidation decomposition is much higher with the value is about 97.35 kJ/mol. Thus rate of thermal decomposition is higher than oxidation decomposition on a temperature above flash point of Jatropha curcas biodiesel (175 °C).

The frequency factor of a reaction expressed by 'A', is show the degree of collisions of a reaction per hour. The value of frequency factor on jatropha curcas biodiesel oxidation reaction is 1.025 x 10^{11}/ hour = 28.47 x 10^{6}/ sec. Comparing to thermal decomposition of Jatropha curcas biodiesel in which has frequency factor 10.44 – 12.20 /sec [24], it shown that oxidation reaction has higher collision, however this collision does not propagate to the chemical reaction between oxygen with biodiesel. It seems that higher collision frequency is more between oxygen to antioxidant in which in this case is pyrogallol. As shown in a study by Fattah et al [25] which is concluded that Pyrogallol is found to be the most effective antioxidant to improve the oxidation stability in case of almost all biodiesels reviewed.

4. Conclusion

Oxidation reaction on biodiesel could determine by measuring iodine number changes, because oxygen will react with double bond on unsaturated fatty acid compound in biodiesel. Rate of oxidation reaction on jatropha curcas biodiesel determined by chemical reaction step compared to mass transfer step with or without antioxidant. In a presence of 120 ppm pyrogallol, the value of activation energy and frequency factor on Arrhenius constant is A = 1.025 E+11/ hour (28.47 x 10^6/ sec) and E = 97315.37 J/mol. Extrapolation of the data for the ambient temperature condition 30 °C shown that during 2.3 years the number of iodine number for jatropha curcas biodiesel will decrease 3.4%. However, this result should be considered that the calculation does not count on the effect of metal, light and microbiology process to the decomposition by oxidation process and total decomposition that might be also take a place on the Jatropha curcas biodiesel.

Acknowledgement

Author team would like to thank to Ministry of Technology and Higher Education Republic of Indonesia for the research grant Hibah Penelitian Terapan 2017.

References

[1] Gao, C., Zhai, Y., Ding, Y., Wu, Q., Application of sweet sorghum for biodiesel production by heterotrophic microalga Chlorella protothecoides, Applied Energy, 87 (2010) 756–761.
[2] Costa, J.F., Almeida, M.F., Ferraz, M.C.M.A., Dias, J.M., Biodiesel production using oil from fish canning industry wastes, Energy Conversion and Management, 74 (2013) 17–23
[3] Marchetti, J.M. and Errazu, A.F., Esterification of free fatty acids using sulfuric acid as catalyst in the presence of triglycerides, Biomass and Bioenergy, 32 (2008) 892 – 895.
[4] Arbenz A., Averous L., Synthesis and Characterization of Innovative Materials Based on Triglycerides from Microalgae, 8th Workshop on Fats and Oils as Renewable Feedstock for the Chemical Industry March 29-31, 2015 Karlsruhe, Germany.
[5] Brennan L., Owende P., Biofuels from microalgae—A review of technologies for production, processing, and extractions of biofuels and co-products Renewable and Sustainable Energy Reviews, Volume 14, Issue 2, February 2010, Pages 557–577
[6] Canakci M, Van Gerpen J., Biodiesel production from Oils and Fats with high Free Fatty Acids, 2001 American Society of Agricultural Engineers ISSN 0001–2351.
[7] Chisti Y., Research review paper Biodiesel from microalgae, Biotechnology Advances 25 (2007) 294–306
[8] Waynick J.A., Characterization of Biodiesel Oxidation and Oxidation Products, Task 1 Results. SwRI® Project No. 08-10721, August 2005
[9] Knothe, G., Dependence of biodiesel fuel properties on the structure of fatty acid alkyl esters, Fuel Processing Technology, 86 (2005) 1059– 1070.
[10] Shahidi, F., Bailey's Industrial Oil and Fat Product, John Wiley & Son's, 2005
[11] Berthiaume D., Study of the Rancimat Test Method in Measuring the Oxidation Stability of Biodiesel Ester and Blends, NRCan project, November 2006
[12] Fattah I.M.R., Masjuki H.H., Kalam M.A.,.Hazrat M.A., Masum B.M., Imtenan S., Ashraful A.M. , Effect of antioxidants on oxidation stability of biodiesel derived from vegetable and animal based feedstocks, Renewable and Sustainable Energy Reviews 30 (2014) 356–370
[13] Focke W.W., Westhuizen I,. Grobler A.B.L.,. Nshoane K.T., Reddy J.K., Luyt A.S., The effect of synthetic antioxidants on the oxidative stability of biodiesel Fuel 94 (2012) 227–233
[14] Jain S., Sharma M.P., Stability of biodiesel and its blends: A review, Renewable and Sustainable Energy Reviews 14 (2010) 667–678
[15] Jain S., Sharma M.P., Oxidation stability of blends of Jatropha biodiesel with diesel, Fuel 90 (2011) 3014–3020.

[16] Fattah I.M.R., Masjuki H.H., Kalam M.A.,.Hazrat M.A., Masum B.M., Imtenan S., Ashraful A.M. , Effect of antioxidants on oxidation stability of biodiesel derived from vegetable and animal based feedstocks, Renewable and Sustainable Energy Reviews 30 (2014) 356–370

[17] Sarin A., Arora R., Singh N.P., Sharma M., Malhotra R.K., Influence of metal contaminants on oxidation stability of Jatropha biodiesel, Energy 34 (2009) 1271–1275

[18] Leung , D.Y.C., Koo, B.C.P., and Guo, Y., Degradation of biodiesel under different storage conditions, Bioresource Technology, 97 (2006) 250–256

[19] Andras F., Sigurd S., The influence of tocopherols on the oxidation stability of methyl esters. Journal of the American Oil Chemists Society 2007; 84(6):579–85.Dunn, R.O., Effect of antioxidants on the oxidative stability of methyl soyate (biodiesel), Fuel Processing Technology, 86 (2005) 1071– 1085.

[20] Levenspiel O., Chemical Reaction Engineering, Mc Graw Hill, 1999

[21] Supriyono, Sulistyo H, Almeida M.F., Dias J.M., Influence of synthetic antioxidants on the oxidation stability of biodiesel produced from acid raw Jatropha curcas oil, Fuel Processing Technology, 132 (2015), pp 133–138.

[22] Saluja R.K., Kumar V., Sham R., Stability of biodiesel – A review, Renewable and Sustainable Energy Reviews 62(2016)866–881

[23] Chouhan A.P.S., Singh N, Sarma A.K., A comparative analysis of kinetic parameters from TGDTA of Jatropha curcas oil, biodiesel, petroleum diesel and B50 using different methods , Fuel 109 (2013) 217–224

[24] Fattah I.M,R,, Masjuki H.H.,Kalam M.A., Hazrat M.A., Masum B.M., Imtenan S., Ashraful A.M., Effect of antioxidants on oxidations stability of biodiesel derived from vegetable and animal based feeds tocks, Renewable and Sustainable Energy Reviews 30(2014)356–370

The Multi Agent System for Job Recommendation

Meilany Nonsi Tentua*[1,2], Azhari Azhari[2], Aina Musdholifah[2]

[1]Teknik Informatika, Fakultas Teknik, Universitas PGRI Yogyakarta, Indonesia
[2]Department of Computer Science and Electronics, FMIPA UGM, Yogyakarta, Indonesia

meilany@upy.ac.id, arisn@ugm.ac.id, aina_m@ugm.ac.id

Abstract. The number of available job portals causes abundant information. Therefore, a system of recommendations is needed by job seekers to find jobs that fit their profile. Offering job vacancies on job portals are changing every time because there are always additional job search data and job opening data. Multi-agents system is a technology that can be used to handle information changes.This article proposed a recommendation system is expected to help job seekers to get jobs in accordance with the field of science they have. The system will monitor what work is offered by online job portals. From the results of observations, the job content offered will be used as a reference so that if new content is entered, the agent will automatically provide input to the job seeker. Based on the results of the implementation of the recommendation system using a multi-agent system can provide search results that are in accordance with what is inputted by the user based on the profile they have. These search results can recommend jobs that match the job seeker profile.

1. Introduction

Today, there are many portals that offer online job openings (Job Portal) in Indonesia, such as Indeed.com, Jobindo.com, karir.com, loker.id and kompaskarir.com. Job portal provides job search features based on position, company name, job location and salary amount. The online job portal application allows job seekers and recruiters to be directly connected [1].

The number of available job portals causes abundant information. This often causes job seekers to get confused and eventually take jobs that are not in accordance with their profile. Therefore, a system of recommendations is needed by job seekers to find jobs that fit their profile. The recommendation system also helps people to find interesting items by utilizing previous user interactions [2]. In addition, the recommendation system has a successful use in e-commerce applications to handle problems related to information overload efficiently [3]..

Offering job vacancies on job portals are changing every time because there are always additional job search data and job opening data. So that for job recommendation systems must contain technology that can adapt to changes in the environment in the system. In addition, the recommendation system must also be able to gather information from several existing job portals. Multi-agents system is a technology that can be used to handle information changes. An agent is a computer system, which is in a certain environment, which is capable of performing autonomous actions that are flexible, in order to meet the design goals [4]. Multi-agent system has been widely applied in several applications such as job search agencies[5], health care [6] , biomedical [7], and so on. In this paper, a system of recommendations for finding a job using a multi-agent system will be proposed. This job recommendation system platform uses agent technology, its JADE.

2. Literature Review

2.1. Recommendation System

The recommendation system is an information filtering system that handles the problem of information overload [8] [8] by filtering fragments of information from a large number of data dynamically information generated according to user preferences, interests, or observed behavior about items [9] [9].

Several methods have been applied by previous researchers about recommendation systems such as profile matching [10] [11], collaborative filtering [12] [13] [14], clustering [15] , decision tree [16], [16], classification [17], machine learning [18], content base [19], hybrid [20]

Basically, the methods in the recommendation system can be divided into three categories: content-based, collaborative filtering and hybrid models. The recommendation system based on content uses items that interact with users in the past, then calculates similarity scores between pairs of items and ranks new recommendations accordingly [21]. The collaborative filtering approach suggests items to certain users, who interact with other users who have positive interactions [12]. Hybrid method is a method which combines both techniques [22].

2.2. Multi-Agent System

Multi-agents system is a technology that can be used to handle information changes. An agent is a computer system, which is in a certain environment, which is capable of performing autonomous actions that are flexible, in order to meet the design goals [4].

Several characteristics of multi agent systems can be used for the purpose of this study

- Autonomy, that is, agents can carry out independent and indirect tasks by users, other agents or by the environment.
- Intelligence, Reasoning, and Learning: Every agent must have a minimum standard to be called an agent, namely intelligence (intelligence). In the concept of intelligence, there are three components that must be possessed: an internal knowledge base, the ability to reason based on the knowledge base that is owned, and the ability to learn to complete in a changing environment.
- Mobility and Stationery: specifically for mobile agents, he must have the ability to represent the characteristics he has is mobility. In contrast to stationary agents. However, it still must have the ability to send messages and communicate with other agents.
- Delegation: In accordance with the appearance and we have discussed in the resolution section, the agent moves in a move task that is ordered by the user. This phenomenon of delegation (delegation) is the main characteristic of a program called an agent.
- Reactivity: Another characteristic agent is the ability to be able to quickly with the help of changing information in the environment (enviornment). Environments that can include: other agents, users, outside information, etc.
- Proactive and Target Oriented: Proactive nature is arguably a continuation of reactivity. Agents are not only required to change the environment, but also must take step by step what to take. For this reason, the agent must be designed to have a clear goal (goal), and to always be oriented towards the goals it is carrying out (goal oriented).
- Communication and Coordination Ability: Agents have the ability to communicate with users and other agents. To be able to coordinate with other agents in carrying out tasks, standard languages are needed to communicate, such as Knowledge Queries and Language Manipulation (KQML) and Knowledge Interchange Format (KIF).

The use of multi-agents on the recommendation system has been done a lot before. A multi-agent is used in the TV recommendation system based on the user's desire to select thousands of TV programs broadcasted by hundreds of TV stations [23] [24]. News recommendations that use a multi-agent framework by utilizing Social Media Preferences [25].

3. Methodology

Based on the explanation above to overcome the problem, the researcher proposes a multi-agent system for job recommendations that are in accordance with the field of job seekers. An agent is a computer system that is in an environment and has the ability to act autonomously in the environmental situation in accordance with the objectives designed [4].

The system is used to recommend work in accordance with the field of science that has been taken by job seekers. The system will monitor what work is offered by online job portals. From the results of observations, the job content offered will be used as a reference so that if new content is entered, the agent will automatically provide input to the job seeker. This proposed system is expected to help job seekers to get jobs in accordance with the field of science they have.

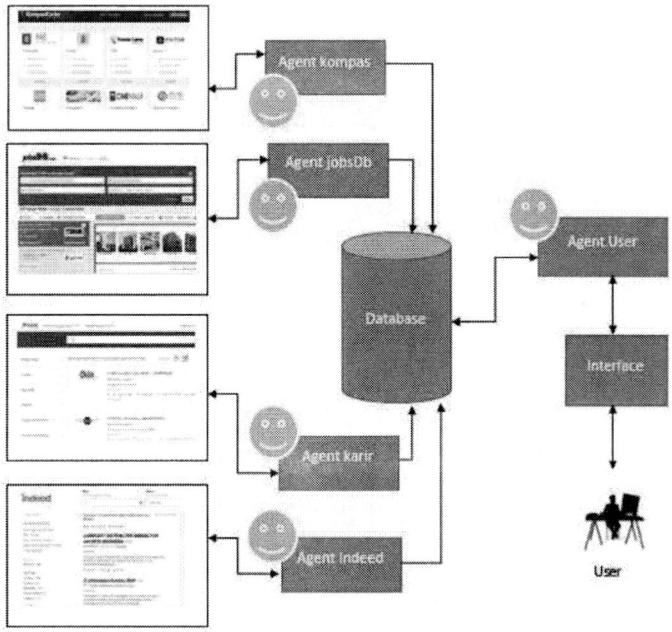

Figure 1 . Agent Work Flow Chart

Figure 1. Shows agent workflows in this job recommendation system. On the basis of this system the agent is grouped into two criteria based on its function, namely:

- information gathering agent
  This agent serves to collect information from various job portals currently available, such as indeed.com, jobsdb.com, karir.com, etc.
- agent user
  This agent serves to search for job openings that have been collected in the database based on user profiles

Any changes that occur in the environment will also result in changes to the agent's actions. The environment used is an online job portal on the internet. If there is a change in the portal, it will give an impact on the system agent. This environment will be modeled into a database that will be updated based on updates that occur in online portal jobs.

4. Implementation

The implementation of the job recommendation system based on the design consists of two interfaces, an interface for the user and an interface for displaying search results.

IOP Conf. Series: Materials Science and Engineering **835** (2020) 012039 doi:10.1088/1757-899X/835/1/012039

4.1. a. User interface
The user will enter the name, status, and profile. After that when the user clicks ok, the agent will search for the appropriate profile in the database.

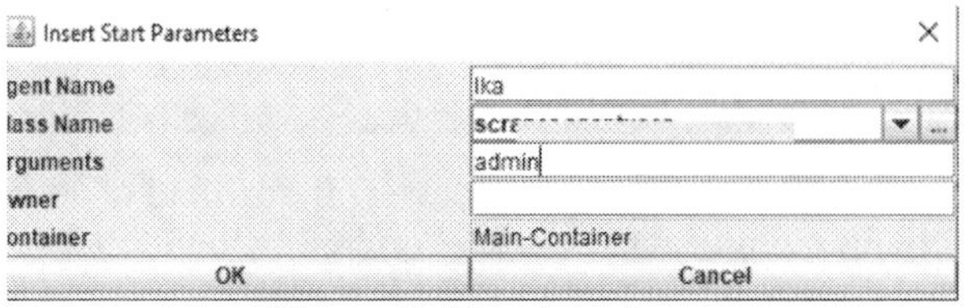

Figure 2. User interface

4.2. b. Interface results
The results of the search request performed by the user will be displayed on the results interface. The results displayed are the name of the appropriate job and the name of the company offering the job.

```
Nama : PT Surya Akapindo Semarang +13 lokasi
Pekerjaan : Staff Administrasi Pembukuan

Nama : SG Yogyakarta Yogyakarta
Pekerjaan : Admin

Nama : PT. Central Prima Kelola - Central Park Mall Jakarta
Pekerjaan : Admin Engineering / Data Entry

Nama : PT Indoreksa Lokamandiri Jakarta
Pekerjaan : Service Engineer Admin

Nama : PT. Asuransi Sinar Mas Yogyakarta Yogyakarta
Pekerjaan : Admin Klerikal

Nama : PT Sumber Daya Bergengsi Indonesia +20 lokasi
Pekerjaan : Staff Administrasi
```

Figure 3. Interface Results

5. Discussion
The online job portal application allows job seekers and recruiters to be directly connected. The number of available job portals causes abundant information. This often causes job seekers to get confused and eventually take jobs that are not by their profile. Utilization of the multi-agent system in the job recommendation system created can gather all job information offered by all online job portals. The same work from several portals will be sorted and only displayed once. So the user will not be offered repeatedly. In the system that has been created, data is collected from several online job portals such as, indeed.com, jobsdb.com, karir.com. Job data collection on the online job portal is done by the agent and then collected in one table. The results of data collection job can be seen in Figure 4.

id	name_job	company
30	ADMIN OFFICER	Angkasa Pura Supports Semarang
31	administrasi pembayaran	PT PELANGI FORTUNA GLOBAL Salatiga
32	Admin	SG Yogyakarta Yogyakarta
33	Recruitment Officer (RO)	PT Ponoeljobs Indonesia Jakarta
34	Staf Analis	SG Yogyakarta Yogyakarta
35	MAGANG DATA ENTRY	PT PELAYARAN TEMPURAN EMAS 4 ulasan Surabaya
36	Staff Administrasi Pembukuan	PT Surya Akapindo Semarang +13 lokasi
37	Cleanning Service OB Office Boy Asisten Masak	PT Olala Trading International Jakarta
38	Data Entry Team	PT Surya Akapindo Jakarta +13 lokasi
39	Salesman/Salesgirl Otomotif (Surabaya dan Sekitarn...	PT Mobilkamu Group Indonesia Sidoarjo
40	Sales Executive (Mojokerto)	PT Mobilkamu Group Indonesia Mojokerto
41	Marketing Otomotif	PT Mobilkamu Group Indonesia Gresik
42	Marketing Mobil Baru All Brand	PT Mobilkamu Group Indonesia Mojokerto

Figure 4. Collecting data

Each online job portal has a different presentation in providing information, even from the same company. So that in this system each online job portal has one agent to retrieve information. Retrieval of information tailored to the needs of the system, so that it can be put together in the table. By utilizing the characteristics of the autonomous agent, the agent is assigned to retrieve job information every time there is a change in the online job portal

On the user side, the system will provide information about what work is by user requests. An agent is assigned to search for information based on the suitability of the user in the collection of work information that has been collected in the table. Furthermore, the system will provide information to the user automatically or job recommendations based on changes that occur in the online job portal.

6. Conclusion

.The online job portal application allows job seekers and recruiters to be directly connected. The number of available job portals causes abundant information. This often causes job seekers to get confused and eventually take jobs that are not by their profile. Utilization of the multi-agent system in the job recommendation system created can gather all job information offered by all online job portals. The same work from several portals will be sorted and only displayed once. So the user will not be offered repeatedly. Based on the results of the implementation of the recommendation system using a multi-agent system can provide search results that are in accordance with what is inputted by the user based on the profile they have. These search results can recommend jobs that match the job seeker profile.

References

[1] S. Arunkumar, S. Gautam, and A. Behera, "Multifunctional Online Job Portal," Journal of Network Communications and Emerging Technologies (JNCET) vol. 8, no. 4, p. 4, 2018.

[2] K. C. Kara, S. Esen, N. Kahyalar, A. A. Karakaş, and T. Aytekın, "Design and implemenatation of a job recommender system," in *2017 International Conference on Computer Science and Engineering (UBMK)*, 2017, pp. 729–733.

[3] S. T. Al-Otaibi, "A survey of job recommender systems," *Int. J. Phys. Sci.*, vol. 7, no. 29, Jul. 2012.

[4] Wooldridge, "An Introduction to Multi Agent Systems." [Online]. Available: https://www.pdfdrive.com/wooldridge-an-introduction-to-multi-agent-systems-d5107234.html. [Accessed: 21-May-2019].

[5] X. Qiu, N. Zhou, and X. Wang, "Design of Labour Agency Platform Based on Agent Technology of JADE," in *Computer and Computing Technologies in Agriculture V*, vol. 368, D. Li and Y. Chen, Eds. Berlin, Heidelberg: Springer Berlin Heidelberg, 2012, pp. 41–45.

[6] R. Morejon Novales, "A Multi-Agent Approach To Data Mining Processes: Applications To Health Care," Mestre Em Informática, Pontifícia Universidade Católica Do Rio De Janeiro, Rio de Janeiro, Brazil, 2018.

[7] M. Camara, O. Bonham-Carter, and J. Jumadinova, "A multi-agent system with reinforcement learning agents for biomedical text mining," in *Proceedings of the 6th ACM Conference on Bioinformatics, Computational Biology and Health Informatics - BCB '15*, Atlanta, Georgia, 2015, pp. 634–643.

[8] J. A. Konstan and J. Riedl, "Recommender systems: from algorithms to user experience," *User Model. User-Adapt. Interact.*, vol. 22, no. 1, pp. 101–123, Apr. 2012.

[9] F. O. Isinkaye, Y. O. Folajimi, and B. A. Ojokoh, "Recommendation systems: Principles, methods and evaluation," *Egypt. Inform. J.*, vol. 16, no. 3, pp. 261–273, Nov. 2015.

[10] A. Gupta and D. Garg, "Applying data mining techniques in job recommender system for considering candidate job preferences," in *2014 International Conference on Advances in Computing, Communications and Informatics (ICACCI)*, 2014, pp. 1458–1465.

[11] V. Desai, D. Bahl, S. Vibhandik, and I. Fatma, "Implementation of an Automated Job Recommendation System Based on Candidate Profiles," *Int. Res. J. Eng. Technol. IRJET*, vol. 04, no. 05, p. 4, May 2017.
[12] C. Fry, "A Comparison of Collaborative Filtering Algorithms for Job Recommendation Using Apache Mahout," *Proj. Present. Fac. Calif. State Polytech. Univ. Pomona*, p. 101, 2016.
[13] W. Shalaby *et al.*, "Help Me Find a Job: A Graph-based Approach for Job Recommendation at Scale," *ArXiv180100377 Cs*, Dec. 2017.
[14] R. G. Belsare and D. V. M. Deshmukh, "Employment Recommendation System using Matching, Collaborative Filtering and Content Based Recommendation," *Int. J. Comput. Appl. Technol. Res.*, vol. 7, no. 6, pp. 215–220, Jun. 2018.
[15] I. Editor, "Survey on Recommendation System using Data Mining and Clustering Techniques."
[16] A. Utku, H. Karacan, O. Yildiz, and M. A. Akcayol, "Implementation of a New Recommendation System Based on Decision Tree Using Implicit Relevance Feedback," *JSW*, vol. 10, pp. 1367–1374, 2015.
[17] G. Özcan and Ş. G. Öğüdücü, "Applying Classifications Techniques in Job Recommendation System for Matching of Candidates and Advertisements," *Int. J. Intell. Comput. Res.*, vol. 8, no. 1, p. 9, 2017.
[18] M. Nilashi, "An Overview of Data Mining Techniques in Recommender Systems," *J. Soft Comput. Decis. Support Syst.*, vol. 3, no. 6, pp. 16–44, Oct. 2016.
[19] A. Elsafty, M. Riedl, and C. Biemann, "Document-based Recommender System for Job Postings using Dense Representations," in *Proceedings of the 2018 Conference of the North American Chapter of the Association for Computational Linguistics: Human Language Technologies, Volume 3 (Industry Papers)*, New Orleans - Louisiana, 2018, pp. 216–224.
[20] J. Séguéla and G. Saporta, "A hybrid recommender system to predict online job offer performance," p. 15.
[21] N. D. Almalis, G. A. Tsihrintzis, and N. Karagiannis, "A content based approach for recommending personnel for job positions," in *IISA 2014, The 5th International Conference on Information, Intelligence, Systems and Applications*, 2014, pp. 45–49.
[22] R. Burke, "Hybrid Web Recommender Systems," in *The Adaptive Web: Methods and Strategies of Web Personalization*, P. Brusilovsky, A. Kobsa, and W. Nejdl, Eds. Berlin, Heidelberg: Springer Berlin Heidelberg, 2007, pp. 377–408.
[23] K. Kurapati, S. Gutta, D. Schaffer, J. Martino, and J. Zimmerman, "A Multi-Agent TV Recommender," Adaptive Systems Department, Philips Research Briarcliff 345 Scarborough Rd. Briarcliff Manor, NY 10510, USA p. 8.
[24] Y. Blanco-Fernandez, J. J. Pazos-Arias, A. Gil-Solla, M. Ramos-Cabrer, B. Barragans-Martinez, and M. Lopez-Nores, "A Multi-Agent Open Architecture for a TV Recommender System: A Case Study Using a Bayesian Strategy," in *IEEE Sixth International Symposium on Multimedia Software Engineering*, Miami, FL, USA, 2004, pp. 178–185.
[25] M. A. Ghalib Ahmed Tahir, "Personalized News Recommendation based on Multi-agent framework using Social Media Preferences - IEEE Conference Publication." [Online]. Available: https://ieeexplore.ieee.org/document/8538403. [Accessed: 24-May-2019].

Designing Android Based Education Game Aksara Jawa Using Shuffle Random Algorithm

Nugroho Dwi Saputro[1], Triana Romadhani[2], Febrian Murti Dewanto[3]

[123]Faculty of Engineering and Informatics, Universitas PGRI Semarang, Jl. Sidodadi-Timur No.24 Semarang, Central Java 50232, Indonesia

nugputra@upgris.ac.id

Abstract. The lack of appropriate supporting media in learning at the age of children makes the young generation now less familiar and still having difficulty learning Javanese script (Aksara Jawa called in indonesia). For this reason, we need appropriate media in the introduction of Javanese script, one of which is through interesting educational games and can provide information about Javanese script. Games are forms of application that can be used as learning media where the process can be done with the concept of learning while playing for children. In this study the algorithm used was the Shuffle Random Algorithm. With this randomization algorithm, it is expected to avoid repetition of the problem so that the learning process can run effectively. Furthermore, this application was tested using the Black Box, White Box and User Acceptance Test. With the Black Box and White Box testing get a percentage of 100% while with the User Acceptance Test this application has an average of 86%, so it can be concluded that this educational game is feasible to be used as a learning medium for Javanese script.

1. Introduction

The majority of Javanese literacy learning is delivered using the lecture method. The use of instructional media is very minimal, there is still a lack of attractive, interactive and modern Javanese language learning media that are able to attract students' interest in learning Javanese script. To motivate students, it is necessary to apply the development of instructional media that can attract interest and foster enthusiasm of students in learning Javanese script or *aksara Jawa*. With an interesting learning process, material that is considered difficult can be delivered more easily, can be well received by students [1]. With the development of technology, Javanese script or *aksara Jawa* learning can be applied in the form of educational games. Multiple choice questions can be made randomly, or question can change. The random method is carried out using the Random Shuffle Algorithm which is applied to educational games in Javanese script. There are various shuffling algorithms to provide randomization techniques on the problem so that the questions that come out will be different and can be generated without repetition or duplication. Random shuffle algorithm is randomization of the index order of a record or array [2]. the author will create an Android-based Java alphabet learning application which in a multiple-choice problem applies the random shuffle algorithm. Later this application is intended for students in grades 3 through 6 with learning material that has been adapted to the curriculum.

2. Methods

2.1. Unified Modeling Language

In the development of the system the author uses the prototyping method. This method performs system development using an approach to create programs quickly and gradually so that it can be evaluated directly by the user. This prototyping method has the advantage that the user can find out the suitability of the resulting application without having to wait until the application is implemented. This type of research and development was chosen because it is very appropriate to develop an educational game. The prototyping method has several stages, namely the analysis phase, the system design stage, the prototyping evaluation stage, the implementation phase and the system testing stage [3]. The system design that was built to present the process modeling is to use the Unified Modeling Language (UML). The design stages in the Javanese script educational game design use use case diagrams, sequence diagrams and activity diagrams [4].

Use Case design of the Javanese alphabet educational game illustrates the user interface between the user and the application. Use Case diagram of Javanese alphabet educational game as follows:

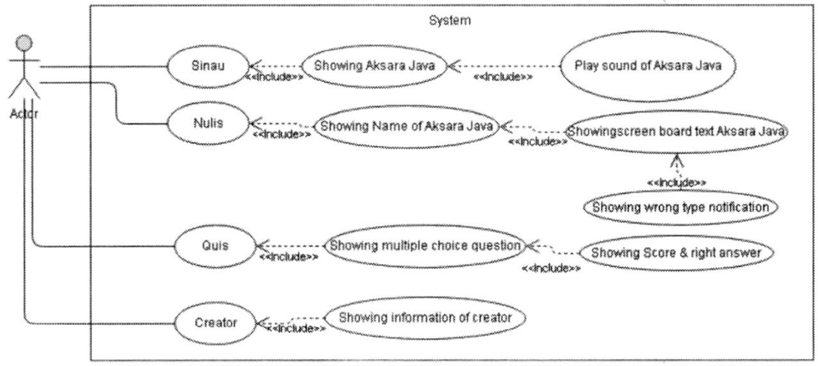

Figure 1. Use case

Class diagrams or class diagrams illustrate the structure of the system in terms of defining the classes created to build the system. Class diagram shows the relationship between classes in the system how they collaborate with each other in achieving goals. In this application the class structure helps in the visualization of a system and is the most widely used diagram type. Class diagram shows the relationship between classes and detailed explanation of each class that contains objects and attributes.

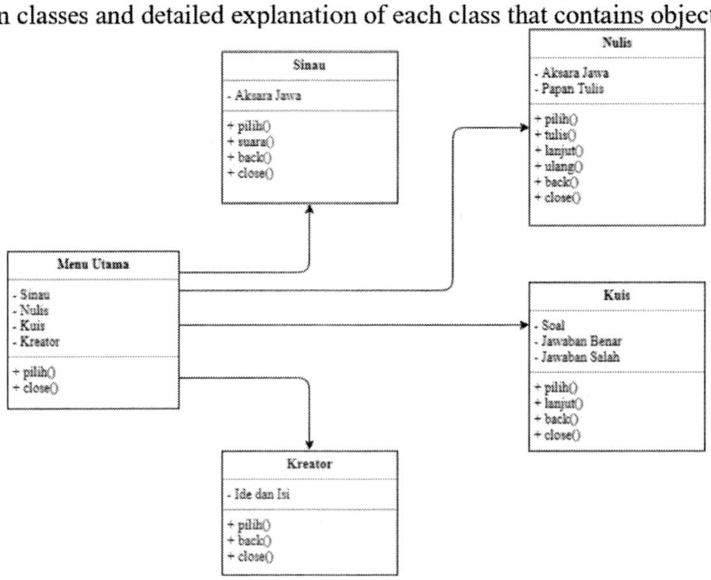

Figure 2. Class diagram

IOP Conf. Series: Materials Science and Engineering **835** (2020) 012040 doi:10.1088/1757-899X/835/1/012040

2.2. Random shuffle algorithm

Random shuffle algorithm is randomization of the index order of a record or array. This randomization is like someone shuffling cards, all cards are shuffled so that the order will be shuffled. Another example for example A is array 5 x 1, A = [1 2 3 4 5] then the random shuffle process will randomize the index array from array A to A1 = [5 1 3 2 4] or become another array arrangement. In the programming language the random shuffle function can not only randomize numbers, but can also randomize string arrays or a mixture of strings and numbers [5].

Another example is if we have 10 numbers in the list, with random techniques, we will get random numbers from these 10 numbers. To get numbers randomly from the list. Shuffle is a method of the Collections class. The shuffle method is used to randomize the contents of a list, so when we call data on a List with a certain No Index, the data that will appear is random. Change the Statement in the getData () method to be like this:

```
void getData(ArrayList<Integer> data){
Collections.shuffle(data);
System.out.println("Got number: "+data.get(0));

}
```

```
        // setelah di masukkan, acak
        Collections.shuffle(quiz);
        // tampilkan soal yang sudah di acak
        tampilkanSoal(soalTampil);
    }


    //DATA SOAL
     // Mengambil data dari arrayList. ArrayList.get(posisi berapa)
     //arrayList berawal dari 0, sama seperti array[]

    public void tampilkanSoal(int index){
        gambarSoal.setImageResource(quiz.get(index).getGambar());
        soal.setText(quiz.get(index).getSoal());
```

Figure 3 Code for randoming question

Figure 4 User interface *Aksara Java* aplication

The user interface in the *aksara Java* educational game application illustrates the structure, appearance of image placement and placement of buttons that the user uses to interact with the software. Design and build Javanese alphabet educational game using Android Studio for application programming, then use Corel Draw for making background images and icon images.

3. Results and discussion

3.1. User Acceptance Test

An Android-based Java Script Educational Game with randomized Shuffle Algorithm has gone through the User Acceptance Test process which was conducted on 4 respondents taken from users and principals with 3 testing areas, namely: usefulness, Ease of Use, and format.

3.1.1. Usefulness

The benefits of the educational game of Javanese script are taken in terms of productivity, performance, effectiveness, and usability aspects. From this field, there are two questions, namely in terms of productivity and effectiveness, the percentage value is only 82.9%, which has not reached a score of 85%, because there are still a number of respondents who still feel that the Javanese alphabet education game is not very productive and less effective. In terms of productivity the educational game of Javanese script still cannot replace the conventional Javanese script completely, and for the effectiveness of this game it has not been tested, and must be tested further.

Table 1. User acceptance test for usefulness

Respondent	**Question & Score**			
	1	2	3	4
1	3	3	3	4
2	3	3	3	4
3	3	3	4	4
4	3	3	3	4
Sum	12	12	13	16
percentage	75%	75%	81,25%	100%

3.1.2. Ease of Use

The use field for User Acceptance Test there are 4 questions that will be answered by 4 respondents with 4 weight choices in each question. The field of use has two analyses whose percentage value has reached 87.5%. Respondents felt that this Javanese alphabet educational game made it easier for users to learn Javanese alphabet, because the Javanese alphabet educational game made users to hear, remember, memorize and write. This educational game also has features to correct answers to wrong questions, there are sound effects and values that have been displayed.

Table 2. User acceptance test for ease of use

	Question & Score			
Respondent	1	2	3	4
1	4	4	4	3
2	3	4	3	3
3	3	4	3	3
4	3	4	4	4
Sum	13	16	14	13
Percentage	81,25%	100%	87,5%	81,25%

3.1.3. Form

The form field for testing Java Javanese educational game contains 2 questions namely about the presentation of output and information content on Javanese literary educational games. Analysis of the User Acceptance Test testing from the form field gets satisfactory results from the two questions given to the six respondents. Based on the User Acceptance Test conducted on 4 respondents taken from Users and Principals it can be concluded that the Javanese script educational game has an average percentage of 87.5%, then the Javanese alphabet educational game is satisfying in the field of expediency, ease of ease and form.

3.2. Black Box Testing
Based on the results of the questionnaire suitability of navigation and content, it can be seen that the assessment of the appearance of each component of the Javanese alphabet educational game in accordance with the function and appearance generated has been adjusted to the user, as well as in terms of content or material that is deemed sufficient, because the educational game is only in use for children 3-6 Elementary School. Testing of Javanese alphabet educational game using the Black Box Testing method, which gets 100% calculation results from three people in accordance with their respective test fields as well as one random user. So, the Javanese alphabet educational game meets the design goals, so this educational game is fit for use.

3.3. White Box Testing
Based on the white box test results from the calculation results of the cyclomatic complexity of the random shuffle algorithm produced is 2. Because the value is less than 10, it means that it is included in an algorithm that is not complex and meets software engineering criteria. Based on these results it can be concluded that this application passed the white box test.

4. Conclusion
Android-based Javanese alphabet educational game application using the Random Shuffle Algorithm as a learning media development is feasible to use. By testing Black Box and White Box, the percentage is 100%. Based on the User Acceptance Test this application has a good average of 86%.

References
[1] S. R. D. Rachman, "Perancangan aplikasi," *Semin. Nas. Teknol. Inf. Dan Multimed. 2015*, vol. 2013, no. Sentika, pp. 1–6, 2015
[2] S. Wijayanti, A. Nurhuda, and R. Andrea, "Edugame ' Etam-Tainment ' Pembelajaran Bahasa Kutai dengan Shuffle Random dan Agen Cerdas," vol. 7, no. 3, pp. 302–307, 2018.
[3] K. ARIZA, "Pembuatan Aplikasi Informasi Tagihan Listrik Berbasis Android," *Fak. Ilmu Komput.*, p. 3, 2014.
[4] A. B. Mutiara, A. Rifky, A. Muslim, and O. Teddy, "Testing Implementasi Website Rekam Medis Elektronik Opeltgunasys Dengan Metode Acceptance ...," *Semin. Ilm. Nas. Komput. dan Sist. Intelijen 2014)*, vol. 8, no. October, pp. 1–7, 2014.
[5] S. Widya, C. Dharma, and J. M. Y. No, "Appliying Shuffle Random Algorithm On Magic Time Edugame Based On," pp. 203–208, 2017.

IOP Conf. Series: Materials Science and Engineering **835** (2020) 012041 doi:10.1088/1757-899X/835/1/012041

Electronic state of sulfide-based alkali-ion conducting solid-state electrolytes applied to all- solid-state secondary batteries

Yoshiyuki Kowada[1], Yasukazu Akaki[1], Akitoshi Hayashi[2], Masahiro Tatsumisago[2]

[1] Hyogo University of Teacher Education, Shimokume 942-1, Kato-shi, Hyogo 673-1494, Japan

[2] Department of Applied Chemistry, Osaka Prefecture University, Gakuen-cho 1-1, Nakaku, Sakai 599-8531, Japan

novita@upgris.ac.id

Abstract. All-solid-state batteries are expected as the next generation secondary batteries for their higher energy density, in-flammable properties, and so on. In order to develop these batteries, there are several problems to improve. One of them is the lower ionic conductivities of the solid-state electrolyte. In order to improve the ionic conductivity, electronic states of the sulfide-based alkali-ion conducting solid-state electrolyte were calculated by the DV-Xα cluster method. The cluster models were constructed by the coordination number reported by experimental methods and the bond length estimated from the ionic radii of each ion. The movement of the cation was simulated by several model clusters with different positions of the moving ion. The relationship between ionic conductivity and the differential total bond overlap population (DBOP) of the moving ion was discussed for the sulfide-based solid-state electrolytes in the several different systems. In any cases, the smaller change of DBOP of the moving cations played an important role for the fast ion movement in the solid-state electrolytes. This bonding state of the moving cations is one of the characteristics of the electronic state in the sulfide-based solid-state electrolytes.

1. Introduction

Recently, solid-state electrolytes are very attractive because of their application for all-solid-state batteries. Those batteries are expected as the next generation secondary batteries for their higher energy density, in-flammable properties, and so on. There are, however, several serious problems in developing commercially available battery cells. One of the importants to improve is the ionic conductivities of the solid-state electrolyte. In the case of the lithium ion secondary batteries, there were significant improvement of the ionic conductivity in the sulfide-based lithium ion solid-state electrolytes. That is, LGPS ($Li_{10}GeP_2S_{12}$) series of superionic conductors were developed and showed incredibly high ionic conductivity ($25 mScm^{-1}$ at R.T.).[1,2] Therefore, the all-solid-state lithium ion batteries will be appeared in the next several years. On the other hand, there are the other problems in the lithium ion secondary batteries, that is, their high-cost, high-environmental impact and the low abundance ratio of its natural resources. Thus, the sodium all-solid-state batteries would be expected alternatives for the next generation low-cost secondary batteries. In the sodium ion batteries, the lower ionic conductivity of the solid-state electrolytes is still a serious problem to apply to the all-solid-state secondary batteries. In this study, electronic states of the sulfide-based alkali ion conducting solid-state

electrolytes were calculated by the DV-Xα cluster method, which is one of the first-principle density functional method.
The relationship between the ionic conductivity and the bonding nature of the mobile ions in the alkali ion solid-state electrolytes was discussed.

2. Calculation Methods

In this study, the DV-Xα cluster method was used to obtain the electronic state of the solid-state electrolytes. This method is one of the first principles density functional methods and is a linear combination of atomic orbital (LCAO) molecular orbital method, in which the exchange potential, V_{XC}, is described by:

$$Vxc(r\uparrow) = \begin{bmatrix} -3 & \frac{1}{3} \\ -3\alpha & \rho\uparrow(r) \end{bmatrix}$$

where ρ is the electron density of the cluster and α is a constant. [3] α was fixed at 0.7 in the present work. The numerical basis functions were used in the method. As the results, *H* and *S* matrices were evaluated by the weighted sum of each value on the sample points distributed in the three-dimensional space.

$$S_{"\#} = \sum_{*} \omega(r_*)\phi^*_{"}(r_*)\phi_{"}(r_*) \quad (2)$$

$$H_{"\#} = \sum_{*} \omega(r_*)\phi^*(r_*)h(r_*)\phi_{"}(r_*) \quad (3)$$

where ϕi is the basis functions, h is the Hamiltonian, ω is the weight of each sample point. This method has several characteristics, such as the same accuracy for all elements in the periodic table, the shorter calculation time by using numerical integration than the common Hartree-Fock method, the accurate energy and transition intensity of the electronic transition spectrum obtained by the slater transition method.

We have used Mulliken population analysis to obtain the bonding state of the moving alkali ion. The ionicity and covalency of the chemical bonding of the alkali ion are discussed by using net charge and bond overlap population. The net charge is the measure of the ionic interaction between mobile alkali ion and the surrounding ions. The bond overlap population is the measure of the covalent interaction between the mobile alkali ion and the other ions.

Fig. 1 shows the schematic diagram of the model cluster for the sulfide-based solid-state electrolyte. This figure shows the $M_{16}X_4S_{16}$ (M=Li, Na; X=B, Si, P, Sn, Sb etc.) cluster, that is one of the simplest models of M_2S-XS_2 solid-state electrolytes. This model cluster was constructed from the structure of the Li ion superionic conducting glasses. The coordination number of sulfide ions to the alkali ion and the distances between the alkali ion and the first neighbor ions were reported about several kinds of the sulfide materials. [4-6] As based on these studies, we have adapted tetrahedral structure and estimated the bond length from the ionic radius of each ion reported by Shannon, [7] to calculate electronic state near alkali ions. In this model, a M ion (M=Li, Na) is located in the center and four sulfide ions are coordinated. Each sulfide ion is included in a XS_4 unit (X=B, Si, P, Sn, Sb etc.), which has regular tetrahedron structure. The outside sulfide ions of XS_4 units are terminated by M ions. Near the center tetrahedral unit, three M ions are located on the outside to keep the cluster neutral. The movement of the center M ion was simulated by five clusters with different position of the ion. The XS_2 content was changed with exchange X in XS_4 units with the other elements. In this case, the neutrality of the model clusters was adjusted by the number of the outside M ions. Then the contents of XS_2 in the clusters could become 25, 50, 75, and 100 mol% in the present work. To compare the theoretical results with the experimental ones, the bond overlap population of the composition corresponding to the experimental samples was obtained by interpolation of the results of these four compositions.

As previously reported, the covalency of bonding nature between the moving cations and surrounding ions were very important to realize the fast movement of the ions.[9-14] We investigated the total bond overlap population to estimate total of covalent interactions of moving cations and it was suggested that the cations, which could move easily, had small change of the total bond overlap population with the movement. In the present work we would like to introduce differential total bond overlap population (DBOP), to evaluate the relationship between the covalent interaction and the fast movement of the moving Li ions. That is obtained by subtracting the minimum from the maximum of the total bond overlap population with the movement of the lithium ion through the path. In this study the electronic state of the moving lithium ions were calculated in the many kinds of the sulfide-based solid-state electrolytes to compare with the ionic conductivity.

3. Results and discussion

We have already reported about the electronic state of the sulfide-based Li ion superionic conducting glasses in the several systems [15], and suggested that the covalent interaction between the mobile Li ion and the surrounding ions near conduction path was very important to understand the incredibly high ionic conductivity. In previous reports, the change of the covalent interaction between the mobile Li ion and the surrounding ions showed clear relationship with the ionic conductivities in the systems Li_2S-SiS_4-Al_2S_3 and Li_2S-SiS_4-P_2S_5,

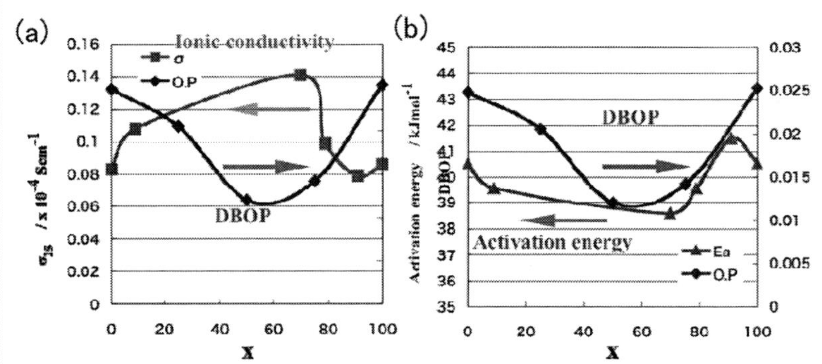

Figure 2 (a) The composition dependence of ionic conductivity and DBOP, (b) The composition dependence of the E_a of the ionic conductivity and DBOP in the $60Li_2S \cdot 40(xPS_{2.5} \cdot (100\text{-}x)BS_{1.5})$ glasses

that is, the smaller change of the covalent bonding of the mobile Li ion made the higher ionic conductivity of the superionic conducting glasses. This tendency of the superionic conductivity is also observed in the other systems, like Li_2S-B_2S_3- P_2S_5 shown in Fig. 2. In this system, there is a maximum in the ionic conductivity and a minimum in the activation energy of the ionic conductivity with the content of P_2S_5. The composition dependence of the ionic conductivity is much complicated than the previous reported systems. However, the composition dependence of the DBOP is corresponding to that of the activation energy of the ionic conductivity and has the opposite tendency toward the ionic conductivity. Thus, the result suggested that the covalent interaction of the mobile Li ion is very important to understand the high ionic conductivity in many kinds of the sulfide-based superionic conducting glasses. Fig. 3 shows the ionic conductivity at 25°C and activation energy for conduction as a function of the DBOP in the $60Li_2S \cdot 40(xPS_{2.5} \cdot (100\text{-}x)BS_{1.5})$ glasses. The evaluated correlation coefficient

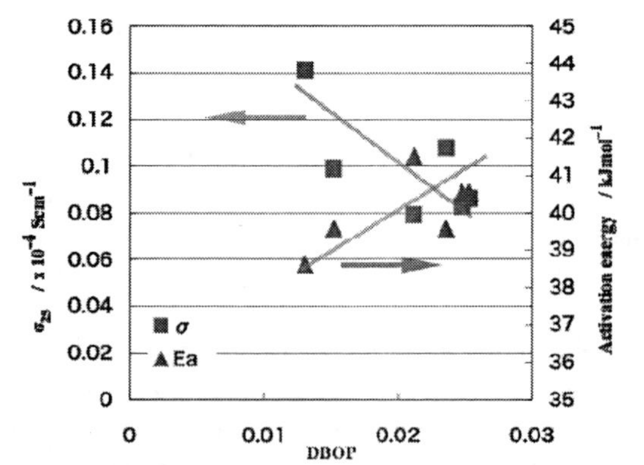

Fig. 3 Relationship between the ionic conductivity, Ea and DBOP in the $60Li_2S \cdot 40(xPS_{2.5} \cdot (100\text{-}x)BS_{1.5})$ glasses

between the ionic conductivity and DBOP is -0.717 and that between the activation energy and DBOP is 0.656. They are slightly smaller compared with the previously reported other systems. However, we can say that there is a good correlation between the DBOP and the ionic conductivity, though the variations of the ionic conductivity and the activation energy are very complex in this system.

The variation of DBOP in the cluster models of the Li ion superionic conducting glasses showed the good agreement with that of the activation energy of the ionic conductivities. The next, we would try to calculate DBOP in the Na ion solid-state electrolytes. Table 1 shows the DBOP and the change of

Table 1 DBOP and differential of NC with the movement of the Na ion in the model clusters $Li_{12}P_4S_{16}$ and $Na_{12}P_4S_{16}$

	DBOP	Differential of NC
Li12P4S16	0.023	0.015
Na12P4S16	0.072	0.029

Table 2 DBOP and differential of NC with the movement of the Na ion in the typical Na ion solid-state electrolytes

	DBOP	Differential of NC	σ / mScm^{-1}	
Na12P4S16	0.072	0.029	3.39	*t*-Na_3PS_4
Na16Sn4S16	0.052	0.014	3.0	$Na_{11}Sn_2PS_{12}$
Na12Sb4S16	0.048	0.01	3.0	Na_3SbS_4

the net charge of the mobile cations between the initial and final positions in the $M_{12}P_4S_{16}$ cluster model. In the case of the $Li_{12}P_4S_{16}$, which is the model for the typical Li ion conductor, the DBOP was 0.023, that is very small value described previously. In the $Na_{12}P_4S_{16}$ cluster, the DBOP was 0.072 with the movement of the Na ion from initial to final positions. This value was somewhat larger than that inthe $Li_{12}P_4S_{16}$ but it still small compared with the other Li ion conductors. This result suggests that the $M_{12}P_4S_{16}$ model cluster is useful to discuss the ionic conductivity of the Na ion solid-state electrolytes. Table 2 shows the experimentally obtained ionic conductivities in the several kinds of the typical Na ion conductors and the DBOP of the corresponding model clusters. In the Na ion conductors which shows the higher ionic conductivity than 3.0 mScm^{-1}, the DBOP also became small like the Li ion superionic conducting glasses.

DBOP had good correlations with experimental ionic conductivity and activation energy of sulfide-based solid-state electrolytes. We could expect to design new solid-state electrolytes by using this analysis method. These figures show calculated DBOP in the other combination of many kinds of elements, for example, fig. 4 (a) shows the results of 4th periodic elements, and fig. 4 (b) shows those of 5th periodic elements. DBOP becomes small and the minimums are observed in the several systems, such as Germanium, Gallium in 4th periodic elements, Tin, Antimony, Indium in 5th periodic elements. In the fact, the new solid-state electrolyte, Li4SnS4 was reported in 2012, that is corresponding to our calculation results. That is, it could be expected that new solid-state electrolytes would be found by using the chemical bonding analysis of the mobile alkali ions.

4. Conclusion

Electronic states of moving alkali ions in the sulfide-based solid-state electrolytes were calculated by the DV-Xα cluster method. The differential of the total bond overlap population (DBOP) was obtained from the model clusters constructed with no strict structural parameters by experimental measurements. The DBOP had good correlations with the ionic conductivity and activation energy in the system Li_2S-P_2S_5-B_2S_3. In addition, DBOP of the Na ion in the model clusters for the Na ion solid-state electrolytes had the similar tendency as the Li ion conductors. That is, DBOP showed relatively small values in the model clusters for the typical Na ion conducting solid-state electrolytes. This result suggests that the covalent interaction of the moving alkali ions is very important to understand the fast movement of the ions in the sulfide-based solid-state electrolytes. Furthermore, composition dependence of DBOP in the many kinds of systems Li_2S-P_2S5-M_xS_y was calculated and the results suggested that the DBOP estimated by the DV-Xα cluster method could be applied to design new sulfide-based superionic conducting glasses.

References

[1] Kamaya, N. et al. Nature Materials, 10, 682-686 (2011)
[2] Kato, Y. et al. Nature Energy, 1, 16030(2016)
[3] H. Adachi, M. Tsukada, C. Satoko, J. Phys. Soc. Japan 45, 875 (1978).
[4] C. Estournès, A. P. Owens, M. Ménétrier, A. Levasseur, K. J. Rao, and S. R. Elliott, J. Non-Cryst. Solids 171 (1994) 80-86.
[5] A. Pradel, G. Taillades, M. Ribes, and H. Eckert, J. Non-Cryst. Solids 188 (1995) 75-86.
[6] R. Prasada Rao, M. Seshasayee, J. Non-Cryst. Solids 352 (2006) 3310-3314.
[7] R. D. Shannon, Acta Crystallogr. A32 (1976) 751.
[8] K. Momma and F. Izumi, J. Appl. Crystallogr., 44, 1272-1276 (2011).
[9] Y. Kowada, H. Adachi, and T. Minami, J. Ceram. Soc. Jpn., 101, 1330-4 (1993).
[10] Y. Kowada, H. Adachi, M. Tatsumisago, T. Minami, J. Non-Cryst. Solids, 232-234, 497-501(1998).
[11] N. Imanaka, S. Tamura, G. Adachi, Y. Kowada, Solid State Ionics 130 (2000) 179-182.
[12] Y. Kowada, Y. Yamada, M. Tatsumisago, T. Minami, and H. Adachi, Solid State Ionics 136-137 (2000) 393-397.
[13] R. Araki, A. Hayashi, Y. Kowada, M. Tatsumisago, T. Minami, J. Non-Cryst. Solids (2001), 288(1- 3), 1-7.
[14] Y. Kowada, M. Okamoto, I. Tanaka, H. Adachi, M. Tatsumisago, T. Minami, J. Non-Cryst. Solids 345&346 (2004) 489-493
[15] Y. Kowada, M. Tatsumisago, T. Minami and H. Adachi, Advances in Quantum Chemistry 54 (2008) pp. 255-270

The Effect Of Chloride Acid (Hcl) Concentration And Fermentation On Bioethanol Levels From Breadfruit (*Artocarpus Artilis*) Substrate

Nasrul Rofiah Hidayati

Chemical Engineering Departement, Faculty of Engineering, Universitas PGRI MADIUN

nasrul.rofiah@unipma.ac.id

Abstract. At present, the public demand for fuel oil (BBM) is increasing inversely proportional to its availability. One of the efforts to reduce people's dependence on BBM is to utilize alternative vegetable-based energy such as bioethanol. Bioethanol is a fermented product that can be made from a substrate containing carbohydrates. Breadfruit (*Artocarpus artilis*) is one of the agricultural products that have a fairly high starch content of 89% and is not included as a staple food source in Indonesia. Utilization of breadfruit is still not optimal because breadfruit is only used as a snack or made into flour. This study aimed to make bioethanol with breadfruit raw material (*Artocarpus artilis*). The method of this study consisted of two stages, the first step was the hydrolysis of breadfruit flour with hydrochloric acid (HCL) catalyst. The concentrations of HCL used in this study were 1%, 1.5%, and 2%. The second stage of fermentation using *Saccharomyces cerevisiae* was with 72 hours fermentation time, 120 hours and 168 hours. From the results of the research, it was found that the highest levels of bioethanol to the treatment of HCL 2% concentration and fermentation duration of 168 hours was 17.6%.

Keywords: bioethanol, breadfruit, HCL concentration

1. Introduction

Indonesia is still very dependent on fossil-based fuels as an energy source. Petroleum is a non-renewable energy source. Nevertheless, it is important to realize that the amount of petroleum reserves is decreasing every period. The government of Indonesia responded to the scarcity of petroleum energy in the future by issuing Presidential Regulation No. 5 of 2006 concerning energy policy. Chapter II Article 2 of PP No.5 of 2006 states that the fulfillment of energy consumption in Indonesia is targeted to come from biofuels, which amounts to 5% of total energy consumption. Biofuel is a renewable energy development based on vegetable or often called Biofuel (BBN). One example of biofuel based fuel is bioethanol.

Bioethanol (bioethanol) is ethanol (ethyl alcohol) whose production process uses natural raw materials and biological processes, in contrast to synthetic ethanol obtained from chemical synthesis of hydrocarbon compounds. The characteristics of bioethanol include flammable, soluble in water, biodegradable, not carcinogenic, and if there is pollution it does not have a significant environmental impact (Kusnadi, 2009). Ethanol is an organic compound of the primary alcohol group. The physical and chemical properties of ethanol depend on the hydroxyl group (Rizani in Rahmawati, 2010). The physical properties of ethanol can be seen in Table 2.1. the following:

Table .1. Physical properties of ethanol

Mr	46,07 g/mol
Freezing point	-114,10C
Normal boiling point	78,320C
Denture at 20^0C	0,7893 g/ml
Water solubility 20^0C	Very soluble
Viscosity at 20^0C	1,17 cP
Specific heat, 20^0C	0,579 cal/g^0
Combustion heat, 25^0C	7092,1 cal/g
Evaporation heat $78{,}32^0$C	200,6 cal/g

Source: Rizani, 2010.

Plants that have the potential for starch content can have potential as raw materials for ethanol production, such as breadfruit (*Artocarpus artilis*). The availability of breadfruit (*Artocarpus artilis*) in Indonesia is quite large and the starch content is quite high at 89%. In each fruiting season, the breadfruit commodity harvest reaches four to twenty tons/hectare.

For the people of Indonesia, the use of breadfruit (*Artocarpus artilis*) is still not optimal because it is only used as a snack. Based on the above background, breadfruit (*Artocarpus artilis*) was used as raw material for bioethanol. Making bioethanol from breadfruit substrate was done by hydrolyzing cellulose in breadfruit using hydrochloric acid (HCL) and fermentation using *Saccharomyces cerevisiae.*

2. Method

The test material used in the study was 3 kg of breadfruit which was obtained from Ngariboyo District, Magetan Regency. Breadfruit fruit was peeled, cut into pieces and dried, then mashed into breadfruit flour. Each sample of breadfruit flour as much as 40 grams per hydrolyzed using hydrochloric acid (HCL) with a concentration of 1%, 1.5% and 2% for 24 hours. After 24 hours each treatment was added 10% *Saccharomyces cerevisiae* with 72 hours, 120 hours and 168 hours fermentation. The following was the ethanol fermentation process carried out by the bacterium *Saccharomyces cerevisiae* (Lud Waluyo, 2004: 159).

$$\underset{\text{Starch}}{(C_6H_{10}O_5)n} + \underset{\text{Water}}{nH_2O} \xrightarrow{HCL} \underset{\text{Glucose}}{(C_6H_{12}O_6)}$$

After getting the results of glucose filtrate from the hydrolysis process, the glucose results was fermented to produce bioethanol with the help of *Saccharomyces cerevisiae* microbes with the following reaction equation:

$$\underset{\text{Glucose}}{C_6H_{12}O_6} \xrightarrow{S.\ cereviceae} \underset{\text{Ethanol}}{2C_2H_5OH} + 2CO_2$$

According to Hambali et al. (2007: 40), bioethanol production includes three series of processes, namely preparation of raw materials, fermentation and purification. After fermenting, the sample was distilled and ethanol levels were measured using the specific gravity method.

IOP Conf. Series: Materials Science and Engineering **835** (2020) 012042 doi:10.1088/1757-899X/835/1/012042

3. Results And Discussion

The process of hydrolysis is the process of changing cellulosic molecules into glucose which was carried out with the help of a hydrochloric acid catalyst (HCL). HCL concentrations in this study were 1%, 1.5% and 2% with 24 hour hydrolysis time. Cellulose that had been converted into glucose would be converted into ethanol in the fermentation process by *Saccharomyces cerevisiae*. The fermentation time was 72 hours, 120 hours and 168 hours. Based on the research, the following results are obtained:

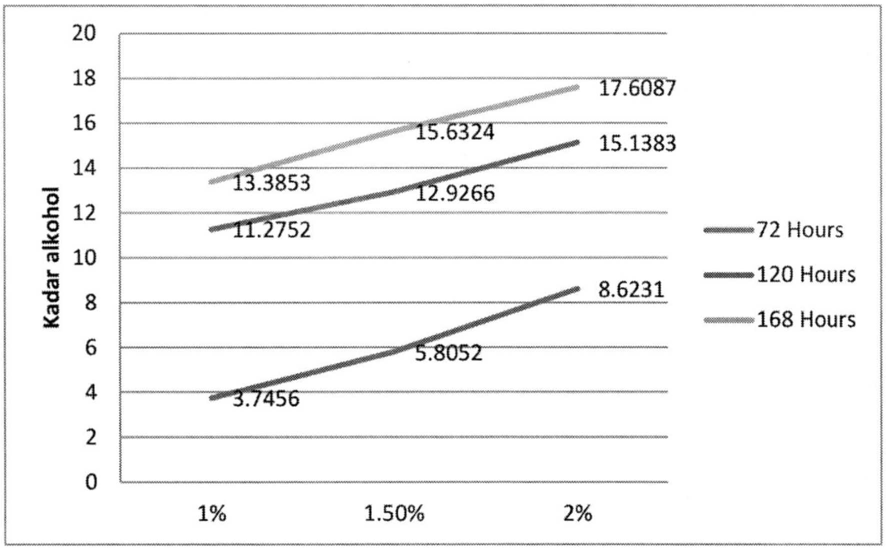

Figure 1. The graph between bioethanol levels and fermentation time

Based on Figure 1 above, it can be seen that the levels of bioethanol produced are increasingly increasing at ever-increasing HCL concentrations of 1%, 1.5% and 2% with a fermentation time of 72 hours, 120 hours, and 168 hours. This was because the reaction speed increases with increasing concentration acid used. So the higher the concentration of HCL used, the faster the hydrolysis reaction occurs and the starch molecules decomposed into glucose increases. Increased glucose levels will increase the level of ethanol produced.

The graph above shows the most optimum time for the results of the highest ethanol content occurred in the fermentation of 168 hours at a concentration of 2% HCL of 17.6087 and the results of the smallest ethanol content indicated at 72 hours fermentation at a concentration of 1% HCL of 3.7456. The high or the low concentration of ethanol produced in the fermentation process was influenced by various factors such as pH, glucose concentration as a substrate, starter concentration, and fermentation temperature. In addition, in general cellulose hydrolysis is more difficult when compared to polysaccharide hydrolysis, this is caused by complex interfacial heterogeneous hydrolysis processes that are influenced by several factors including the structure and composition of lignocellulosic adsorbs and desorption of cellulose enzymes, such as the presence of enzyme inhibitors such as cellobiose and glucose (Nobel, 1990 in Devi and Kumar 2012).

4. Conclusion

Based on the results of the study, it can be concluded that there is an effect of the concentration of Chloride Acid (HCL) and fermentation time on the levels of bioethanol from the breadfruit substrate (*Artocarpus artilis*). The best ethanol content was given by 2% HCL concentration and 168 hours fermentation time with 17.6% ethanol content.

References

[1] Daud. M, Saffi.W, dan Syamsu.K, (2012). Biokonversi Bahan Berlignoselulosa Menjadi Bioetanol Menggunakan *Aspergillus niger dan Saccaromyses cereviciae. JurnalPerennial,*8(2):43-51(http://journal.unhas.ac.id/indek.php/perennial)

[2] Devi and Kumar. (2012). Production, Optimization and Partial Purification of cellulose by *Aspergillus niger* fermented With Paper And Timber Sawmill Industrial Wastes. *Journal of Microbiology and Biotechnology Research,* 2(1):120-128

[3] Hambali E, S. Mujdalipah, A.H.Tambunan,A.W.Pattiwiri dan R. Hendroko , 2008 . Teknologi Bioenergi . PT. Agromedia Pustaka . Jakarta .

[4] Rahmawati, A. 2010. Pemanfaatan Limbah Kulit Ubi Kayu (Manihot Utilissima Pohl.) dan Kulit Nanas (Ananas Comosus L.) pada Produksi Bioetanol Menggunakan *Aspergillus niger.* Skripsi tidak diterbitkan. Surakarta: Jurusan Biologi Fakultas Matematika Dan Ilmu Pengetahuan Alam Universitas Sebelas Maret.

[5] Waluyo,Lud. 2007.*Mikrobiologi Umum.* Malang: UMM Press

Interaction Analysis Application of Arduino Industrial Automation Trainer Based on Project to Improve Cognitive Ability and The Bodily-Kinesthetic Ability

S. Kartikawati[1], S. Nita[2]
[1]Program Studi Pendidikan Teknik Elektro, FKIP, Universitas PGRI Madiun
[2]Program Studi Teknik Informatika, FT, Universitas PGRI Madiun

[1]kartikawati@unipma.ac.id

Abstract. This research aims to determine the interaction/influence of the application of Arduino Industrial Automation trainer based on projects to enhance cognitive abilities and The Bodily-Kinesthetic abilities. college student This research is in the form of quantitative descriptive research. The sample used in this study is one class taken using Saturated Sampling techniques. The research class was treated by using a project-based Arduino Industrial Automation trainer. The independent variables in this study are project-based Arduino Industrial Automation trainer, while the dependent variable is cognitive abilities and The Bodily-Kinesthetic abilities.. There are 3 kinds of analysis techniques, namely instrument analysis techniques, prerequisite analysis techniques, and hypothesis analysis techniques. The measured research data are cognitive ability data and The Bodily-Kinesthetic ability data. The analysis of the hypothesis test of this study used the T-Test with a significance level of 5%. T-Test Criteria are as follows, Ho is accepted / H1 is rejected if T count is smaller than T table and Ho is rejected / H1 is accepted if T count is greater than T table. From the results of the instrumenT-test, namely the validation test, there are 75% questions stated valid. For the prerequisite test for the normality of cognitive abilities and The Bodily-Kinesthetic abilities on the pretest and posttest obtained L count < L table means the sample is normally distributed. For the prerequisite test for homogeneity of cognitive abilities and The Bodily-Kinesthetic abilities obtained F count < F table means the sample is homogeneous. The results of the T-test for the first dependent variable cognitive ability obtained T count = 24.41 and T table = 1.697, because T count is greater than T table then Ho1 is rejected, meaning that there is an interaction/influence of the implementation of the Arduino Industrial Automation trainer based on projects in improving cognitive abilities. While the results of the T-test for the dependent variable The Bodily-Kinesthetic ability obtained T count = 18.33 with T table = 1.697, because T count is greater than T table then Ho2 is rejected, it means that there is an interaction/influence of the application of the Arduino Industrial Automation trainer based on the project in increasing The Bodily-Kinesthetic abilities. So that it can be concluded that there is an interaction/influence of the application of the Arduino Industrial Automation trainer based on projects in improving cognitive abilities and The Bodily-Kinesthetic abilities.

Keywords: Industrial Automation Trainer, Arduino, Project, Cognitive, The Bodily-Kinesthetic.

1. Preliminary

In the era of the industrial revolution both 4.0 and 5.0 human resources are required to be able to possess and master several abilities including the ability to innovate, the ability to solve problems, the ability to create something new, the ability to think critically and other abilities in dealing with the problems of science and technology development. Therefore the learning process is needed by an educator or educated to be able to have the knowledge and ability to create something useful for the development of science and technology.

With the ability of educators and educators, it is expected that the learning process will produce better products or outcomes so that this will have an impact on the achievement of the expected learning objectives. One of the learning innovations is to use interesting learning media and produce outcomes that can improve cognitive abilities and other measurable abilities. The use of appropriate learning media must provide stimulation for educated students in the ongoing learning process.

In the learning process in the classroom, it appears that learning has used instructor learning media with an experimental system and the results are still not optimal, because educated just use the media so educated still have difficulty in understanding the concept of electromechanical material. Where the media in question is an industrial automation media trainer that leads to the field of system assembly in the industrial world. This media is a device/system automation media. Utilization of control systems such as computers that are used to control industrial machines and process controls to replace human labor operators.

With this, the research in this study supports the improvement of the ability and work or educated product, so the learning process that is implemented must be truly capacity-based and product-based. One of the abilities needed is cognitive ability and The Bodily-Kinesthetic ability. To be able to measure the value of cognitive abilities and The The Bodily-Kinesthetic abilities in the learning process required learning media based on projects. To improve cognitive abilities and educated The The Bodily-Kinesthetic abilities, the use of project-based learning media can appropriately be optimized to be able to improve cognitive abilities as well as educated The Bodily-Kinesthetic abilities.

Educated ability can function as a reaction to the learning process carried out during learning activities, especially those relating to the assessment of cognitive domains and The Bodily-Kinesthetic educated abilities. In this study, to measure cognitive abilities and The Bodily-Kinesthetic abilities, industrial automation media trainers are used using project-based Arduino.

2. Method

2.1. Population

The population in this study was 6 (six) semester students totaling 31 students as an experimental class. The sampling procedure in this study uses the Saturated Sampling technique which means that in this study all students have the same opportunity or rights to be able to become a sample in the study.

2.2. Research design

In this study using an experimental research approach to the design of the Control Group pretest-posttest. The sampling procedure in this study used a purposive sampling technique. The sample of the research class was given treatment using an industrial automation media trainer using a project-based Arduino system. After the learning process in the experimental class is carried out, then calculated the average value of achievement (gain) to see changes in values. The independent variable in this study is the application of industrial automation media trainers using the project-based Arduino system. While the dependent variable is cognitive ability and The Bodily-Kinesthetic ability of students.

2.3. Data collection technique

Data collection techniques with the test method that is on the instrumenT-test aspects of cognitive aspects and aspects of The Bodily-Kinesthetic ability. Taking the test scores using the instrumenT-test in the form of a written test for aspects of cognitive abilities and oral tests for aspects of The Bodily-Kinesthetic aspects.

2.4. Data analysis technique

2.4.1. Analysis of Research Prerequisites

Data analysis of test results used to measure cognitive aspects and The Bodily-Kinesthetic aspects of students obtained from the pretest and posttest in the form of written test questions and oral tests. Student pretest and posttest data were tested for normality and homogeneity as a prerequisite test using Liliffort. The results of the prerequisite test are used as a basis for further testing using the T-test on the cognitive abilities and The Bodily-Kinesthetic abilities of students.

2.4.2. Research Hypothesis Analysis

Hypothesis analysis using The T-test. The hypothesis proposed is that there are interactions/effects of the application of industrial automation media trainers using the project-based Arduino system on the cognitive abilities and The Bodily-Kinesthetic abilities of students. The decision-making criteria are as follows:

a) If the significance value is smaller than 0,05 then H_0 rejected /H_1 be accepted.

b) If the significance value is greater than 0,05 then H_0 be accepted /H_1 rejected.

3. Results and Discussion

3.1. Results

3.1.1. Student Cognitive Ability Before Treatment (Pretest)

Data on the results of evaluations of students shows that the average value of learning outcomes is still low 50.64. Pretest results from 31 students, the highest score is 60, the lowest score is 35. There are 26 students with fewer categories, and 7 students with enough categories. The data shows that the students' cognitive abilities are not yet optimal.

Table 1. Pretest Cognitive Ability Score

No	Class Intervals	Number of Student (f)
1	31-35	2
2	36-40	4
3	41-45	3
4	46-50	8
5	51-55	7
6	56-60	7
	Total	31

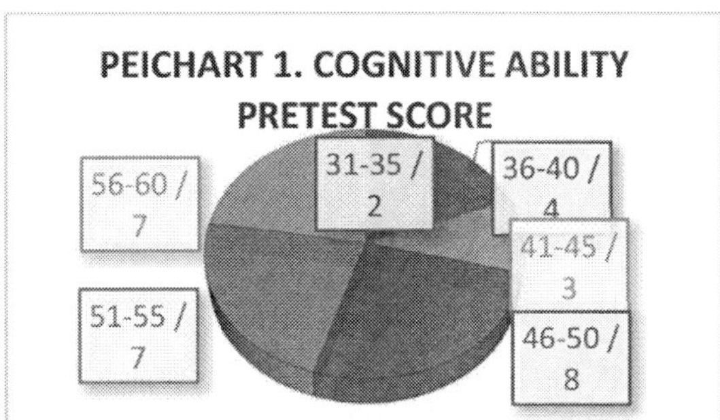

3.1.2. Cognitive Ability of Students After Treatment (Postest)

Data on the results of evaluations of students after being treated in the process of teaching and learning using industrial automation media trainers using Arduino based on projects shows that the average value of cognitive abilities there is an increase from 50.64 increased to 83.54. This shows that there is a significant increase in the average value

Table 2. Posttest Cognitive Ability Score

No	Class Intervals	Number of Student (f)
1	66-70	2
2	71-75	4
3	76-80	3
4	81-85	10
5	86-90	5
6	91-95	8
	Total	31

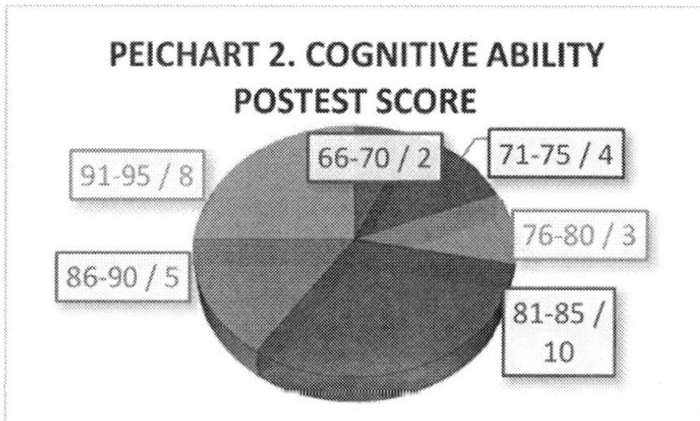

3.1.3. The Bodily-Kinesthetic Ability Before Treatment (Pretest)

Data from the evaluation of students' creative abilities shows that the average value of The Bodily-Kinesthetic abilities is still lacking at 50.67.

Table 3. Pretest Score The Bodily-Kinesthetic Ability

No	Class Intervals	Number of Student (f)
1	43-45	2
2	46-48	4
3	49-51	3
4	52-54	8
5	55-57	7
6	58-61	7
	Total	31

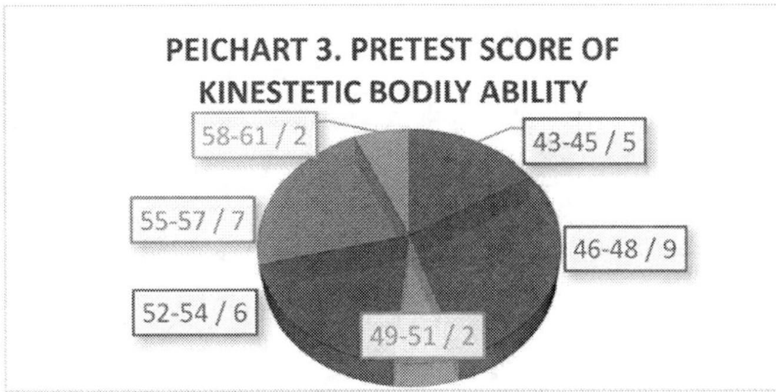

3.1.4. The Bodily-Kinesthetic Ability After Treatment (Postest)

Data on the results of evaluations of the The Bodily-Kinesthetic abilities of students after being treated in the teaching and learning process shows that the average value of the The Bodily-Kinesthetic abilities of students there is an increase from 50.67 to 73.87, This shows that there is a significant increase in the average value.

Table 4. Postest Ability Score BodilyKinestetik

No	Class Intervals	Number of Student (f)
1	63-65	2
2	66-68	4
3	69-71	3
4	72-74	8
5	75-77	7
6	78-80	7
	Total	31

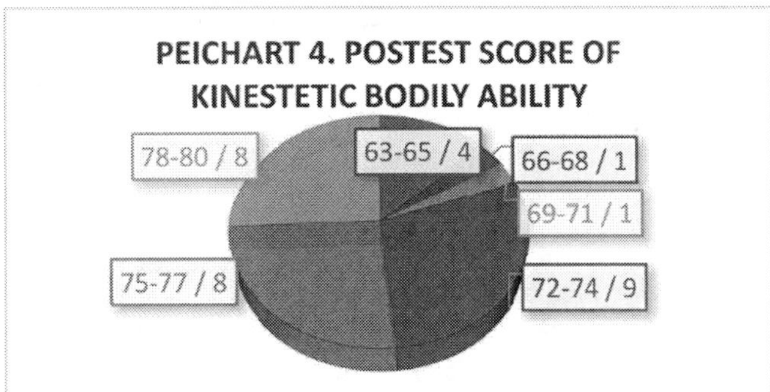

3.2. Analysis results

3.2.1. Test for Normality and Homogeneity of Cognitive Abilities

From the results of the pretest normality test students obtained an L count of 0.144 with an L table of 0.161 which indicates that the L count is smaller than the L table, which means that the sample is normally distributed. Whereas for the posttest obtained L count of 0.155 and L table of 0.161 then the L count is smaller than The L table, meaning that the sample is normally distributed. From the pretest and posttest homogeneity test obtained an F count of 1.364 with an F table of 1,840. This shows that the F count is smaller than the F table, meaning that the sample is homogeneous.

Table 5. Cognitive Ability Normality Test Results

Tes	L_{count}	L_{table}	Test Dcision	conclusion
Pretes	0,144	$L_{(0,05;22)} = 0,161$	H_0 Be accepted	Normal Distributd
Postes	0.155	$L_{(0,05;22)} = 0,161$	H_0 Be accepted	Normal Distributd

Table 6. Homogeneity Test Cognitive Ability Analysis Results

F_{count}	F_{table}	Criteria	Test Decision
1.364	1,840	$F_{hitung} \leq F_{table}$ H_o Be accepted	Homogeneous Samples

3.2.2. Test for Normality and Homogeneity in The Bodily-Kinesthetic Capabilities

From the pretest normality test results obtained by students L count of 0.159 with L table of 0.161 which shows that the L count is smaller than The L table, meaning that the sample is normally distributed. Whereas for The posttest obtained L count is 0.130 and L table is 0.161 then The L count is smaller than The L table, meaning that the sample is normally distributed. From the pretest and posttest homogeneity test obtained an F count of 0.944 with an F table of 1,840. This shows that the F count is smaller than the F table, meaning that the sample is homogeneous.

Table 7. Results of The Bodily-Kinesthetic Ability Normality Test Results

Tes	L_{count}	L_{table}	Test Decision	Conclusion
Pretest	0,159	$L_{(0,05;22)} = 0,161$	H_0 Be accepted	Normal Distributed
Postest	0.130	$L_{(0,05;22)} = 0,161$	H_0 Be accepted	Normal Distributed

Table 8. Homogeneity Ability Test Results Analysis of The Bodily-Kinesthetic

F_{count}	F_{table}	Criteria	Test Decision
0.944	1,840	$F_{hitung} \leq F_{table}$	H_o Be accepted Homogeneous Samples

3.2.3. T-test for Cognitive Ability

From the T-test results of students' cognitive abilities obtained T count of 24.41 with a T table of 1.697 which shows that T count is greater than T table, it means that there is an interaction/influence between the application of industrial automation media trainers by using project-based Arduino on the cognitive abilities of students.

Table 9. Results of T Cognitive Ability Test Results

T_{count}	T_{table}	Criteria	Test Decision
24.41	1,697	$T_{table} \leq T_{count}$	H_o Be accepted

3.2.4. T-test for The Bodily-Kinesthetic Abilities

From the results of the T-test the The Bodily-Kinesthetic ability of students obtained T count of 18.33 with a T table of 1.697 which shows that the T count is greater than the T table, this means that there is an interaction/influence between the application of industrial automation media trainers and project-based Arduino on The Bodily-Kinesthetic capabilities.

Table 10. Results of T The Bodily-Kinesthetic Ability Test Results

T_{count}	T_{table}	Criteria	Test Decision
18.33	1,697	$T_{table} \leq T_{hitung}$	H_o Be accepted

3.3. *Discussion*
From the results of the study and the results of data analysis showed that the T-test with a T count greater than the T table with a significance level of 5% indicates that the null hypothesis (H_0) was rejected or hypothesis one (H_1) is accepted, meaning that the application of industrial automation media trainers using the project-based Arduino system has a positive impact in improving students' cognitive abilities and The Bodily-Kinesthetic abilities. This is evident in the application of industrial automation media trainers using the project-based Arduino system that can provide new knowledge to students that we do not stop at one area of expertise but we must master at the same time three areas of expertise namely mechanics, informatics, and electronics that work together in building a media model.

This shows that students are required to be able to apply the knowledge and also must be able to innovate to create media that can stimulate active learning processes. It can also stimulate the ability to be creative, well structured and active in solving problems related to the projects given by lecturers. It can be seen that the teaching and learning process in the class shows the enthusiasm of students in participating in the teaching and learning process and the students seem to be more capable in terms of knowledge and more able to be creative in solving problems encountered, and students have the ability to develop creations well through projects that are completed.

While the media used in the teaching and learning process has a positive function so that it has an impact on improving cognitive abilities and The The Bodily-Kinesthetic abilities. It can be seen that with this project-based learning, students in designing a product or making a model of learning media can improve the ability of knowledge and can automatically improve abilities related to motor movements of the brain by controlling body organs such as the ability to use the body easily and dexterously (The Bodily-Kinesthetic skill).

Apart from the results of the T-test which showed the interaction between the independent variable and the dependent variable, it was seen that the average value of the experimental class had higher cognitive abilities and The Bodily-Kinesthetic abilities compared to before treatment. This shows that the application of industrial automation media trainer by using a project-based Arduino system in improving cognitive abilities and The Bodily-Kinesthetic abilities of students works well. Therefore, the application of industrial automation media trainers using the project-based Arduino system can make students more concentrated, easier, more pleasant, and can explore combinations or interact with various scientific fields, and can develop The Bodily-Kinesthetic abilities possessed by each student through work or projects that must be completed properly. From the results of this study it can be seen that projet-based learning is very influential in improving cognitive abilities and The The Bodily-Kinesthetic abilities of students.

4. Conclusion
Based on the results of The research analysis shows that of the two dependent variables obtained T count> T table it means that there is a significant interaction/influence between the use of industrial automation media trainers by using a project-based Arduino system on the cognitive abilities and The Bodily-Kinesthetic abilities of students.

References
[1] Arikunto, S. (2010). Prosedur Penelitian Suatu Pendekatan Praktik. Jakarta: Rineke Cipta.
[2] Arikunto, S. (2013). Prosedur Penelitan. Jakarta: PT Rineka Cipta.
[3] Aqib, Z. (2013). Model-Model, Media, dan Strategi Pembelajaran Kontekstual (Inovatif), Bandung, Yrama Widya.
[4] Adi Purwanto (2016), Pengembangan Trainer Arduino untuk Standar Kompetensi Memprogram Peralatan Sistem Pengendali Elektronik yang Berkaitan Akses I/O Berbantu Mikroprosesor dan Mikrokontroler di SMKN 1 Nanggulan

[5] Cahya Alim Wijaya, J. S. (2017), Judul: Pengembangan Media Pembelajaran Pengendali Kecepatan dan Soft Starting Motor Listrik Berbasis Arduino pada Mata Kuliah Penggunaan dan Pengaturan Motor. *JUPITER*, 2(2), 15-20
[6] Eko Prianto, K. I. (2013). Desain Sistem Kendali Kecepatan Dan Counter Putaran Berbasis Teknologi Otomasi Pada Industri Kecil Dan Menengah. *Simposium Nasional Rapi Xii*, 1412-9612.
[7] Hamzah & Nina. (2010). Teknologi Komunikasi & Informasi Pembelajaran. Jakarta: PT Bumi Aksara.
[8] Sundayana. (2015). Statistika Penelitian Pendidikan. Bandung: PT Alfabeta.
Teknologi Informasi. 2005 http://id.wikipedia.org/wiki/Teknologi_informasi

Analysis Of The Impact Narrative Algorithm Method, Pseudocode And Flowchart Towards Students Understanding Of The Programming Algorithm Courses

Sekreningsih Nita[1], Sulistyaning Kartikawati[2]

[1]Informatics Engineering, Faculty of Engineering, University PGRI Madiun
[2]Electrical Engineering education, Faculty of FKIP University PGRI Madiun

nita@unipma.ac.id

Abstract.The definition of an algorithm is to describe a problem into a sequence of systematic steps in solving problems. Arranging the algorithm in sequence from beginning to end so that it can solve the problem. The results of the implementation of the algorithm are programs. So the purpose of making an algorithm is to facilitate the making of the program. There are 2 types of algorithms, 1) Text (narration and pseudocode), 2) Visual / image (flowchart). Comparison between the two, students understand it which is higher between the text or visual model. This study aims to see how much infuence the value of students understands the algorithm in text or in an image. In conclusion, students are easier to understand the types of algorithms in an image. The test uses the Paired Sample T test, with a 5% significance level obtained value = 0.05. This means that the t count is greater than t table of 2.06390 and P Value (0,06211 > 0,05) or T-stat smaller than T-critical two tile, it means H1 can be accepted or T-count < T-table (-1,95597<2,06390), so H0 can be accepted.

Keywords: Algorithm, Paired Sample T test

1. Introduction

Algorithms and Programming is one of the basic courses in the field of informatics engineering that must be taken by students in semester one. The algorithm is very important because it is used as a basic foundation in learning programming for beginners. Therefore, algorithmic is said to be the heart of informatics, (kujangcode.com, 2017). Many branches of computer science are referred to in algorithm terminology. But don't assume that algorithms are always identical to computer science, (dictio.id). For example, the steps in a food recipe are an algorithm, starting from reading one by one the manufacturing steps and then working on the process (taking action) as read. In general, those who work on the process are called processors. Processors can be humans, computers, robots, electronic devices, etc. Implementing an algorithm means working on the steps written in the algorithm, (Saniman and M. Fatoni, 2008)/ In order for an algorithm to be implemented on a computer the algorithm must be converted into a programming language notation so that it can be called a program. The program is written in one of the programming languages, the activity of writing the program is called programming. The person who writes the program is called a programmer. Therefore algorithms with programming cannot be separated, interrelated with each other, (ilmukomputer.org, 2009). In this study, the authors took a sample of the first semester students of class 1A, amounting to 25 students.

The author made observations while doing face-to-face learning in the class, especially when the algorithm and programming material. Based on the results of direct observations it can be concluded that the level of student reasoning varies in the material of the algorithm and programming. Some are easy to accept the theory and practice of algorithm; some are hard to understand. The main topic of algorithm theory states that there are 2 methods to understanding algorithms, in a text and an image. This research was conducted with the aim to see how much influence the value of students understanding the algorithm in text or in images.

2. Theoretical Review

According to Gun Gun Maulana (2017), an algorithm is a collection of commands to solve a problem. Commands can be translated in stages starting from beginning to end, while the problem can be anything, provided that it has the initial conditions that must be met before running an algorithm.

Whereas in wordpress.com written by (andikafisma, 2015), said that the algorithm is a sequence of steps in solving problems that are arranged systematically and logically. The keyword in the algorithm is the word "logical" and must be determined the value of true or false. There is a difference between algorithm and program. An algorithm is a step sequence specification for solving a problem, whereas a program is a collection of computer statements. Some experts give the formula that: Program = algorithm + language (data structure). The relationship between data structures and algorithms is closely related to the program. Making a good algorithm without choosing the right data structure (language) will make the program less good, and vice versa. The algorithm created has many advantages, namely: (1) Writing the algorithm does not depend on any programming language, (2) Algorithm notation can be translated into various programming languages, (3) Whatever programming language is used, the output will be the same because the algorithm is the same, (andikafisma, 2015).

In general algorithm notation is grouped into 3 forms, namely: (a) Descriptive / natural / verbal, (b) pseudocode, and (c). flow chart. The verbal way and pseudocode are more on forms of writing / code, while the flowchart is more on graphic forms (Procceding, Bambang S., 2013).

Following is the description of the three algorithm notations:

a. Descriptive (Natural), means that the algorithm is written in the form of sentences of human language such as Indonesian or English. This descriptive notation is suitable for short algorithms, for long notations are less effective to use.
   Example :
   The algorithm for buying clothes
   1. Enter the clothing store
   2. Look at the size, repeat if the size does not fit , find another size
   3. Look at the other clothes
   4. Check the price, is the price suitable?
      a. if yes, buy
      b. f not, go to step 3
   5. Finished, go home.
b. Pseudocode, derived from the words pseudo and code. Pseudo means imitation and code means code in the form of instructions written in computer language or programming language code. Pseudocode is usually used to describe the logical sequence of a program regardless of how the programming language.
   Example :
   Calculate the Area of a Rectangle
   Formula : lw = long (l) * wide(w)
   Calculate the Area of a Rectangle

1) input (l)	OR	1) input ← l
2) input (w)		2) input ← w
3) lw = l * w		3) lw ← l * w
4) output (lw)		4) output ← lw

c. Flow Chart (Flowchart), is a chart that illustrates the sequence of activities or program steps from beginning to end by using symbols or pictures that represent the functions of program steps and flow lines that show the sequence of symbols to be worked on.
Example:

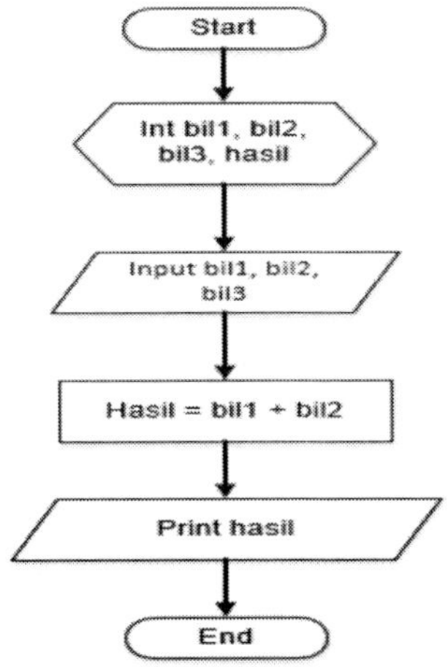

Figure 1. Flowchart Program

Taken from pradigtaagus writing on blogspot.com which was uploaded on August 16, 2019. Following in the table below contains the symbols commonly used for Flowcharts (Programs).

Table 1. Program Flowchart Symbols.

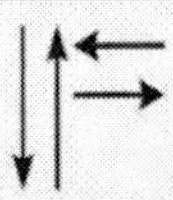	**Flow Direction symbol** Yaitu simbol yang digunakan untuk menghubungkan antara simbol yang satu dengan simbol yang lain. Simbol ini disebut juga connecting line.	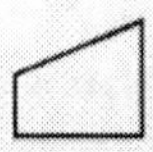	**Simbol Manual Input** Simbol untuk pemasukan data secara manual on-line keyboard
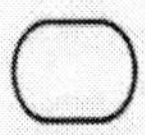	**Terminator Symbol** Yaitu simbol untuk permulaan (start) atau akhir (stop) dari suatu kegiatan	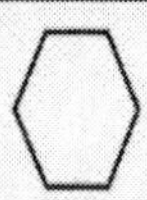	**Simbol Preparation** Simbol untuk mempersiapkan penyimpanan yang akan digunakan sebagai tempat pengolahan di dalam storage.
	**Connector Symbol** Yaitu simbol untuk keluar - masuk atau penyambungan proses dalam lembar / halaman yang sama.	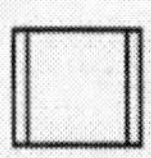	**Simbol Predefine Proses** Simbol untuk pelaksanaan suatu bagian (sub-program)/prosedure
	**Connector Symbol** Yaitu simbol untuk keluar - masuk atau penyambungan proses pada lembar / halaman yang berbeda.	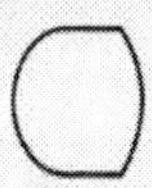	**Simbol Display** Simbol yang menyatakan peralatan output yang digunakan yaitu layar, plotter, printer dan sebagainya.
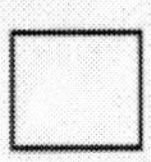	**Processing Symbol** Simbol yang menunjukkan pengolahan yang dilakukan oleh komputer	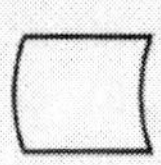	**Simbol disk and On-line Storage** Simbol yang menyatakan input yang berasal dari disk atau disimpan ke disk.
	**Simbol Manual Operation** Simbol yang menunjukkan pengolahan yang tidak dilakukan oleh computer		**Simbol magnetik tape Unit** Simbol yang menyatakan input berasal dari pita magnetik atau output disimpan ke pita magnetik
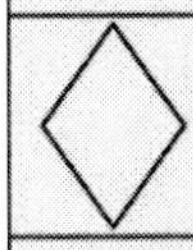	**Simbol Decision** Simbol pemilihan proses berdasarkan kondisi yang ada.	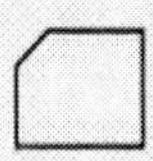	**Simbol Punch Card** Simbol yang menyatakan bahwa input berasal dari kartu atau output ditulis ke kartu
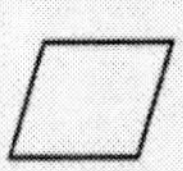	**Simbol Input-Output** Simbol yang menyatakan proses input dan output tanpa tergantung dengan jenis peralatannya		**Simbol Dokumen** Simbol yang menyatakan input berasal dari dokumen dalam bentuk kertas atau output dicetak ke kertas.

3. Methodology

The testing method used by the author to be able to declare valid results using the T-Paired Sample T-test. This test is used to determine the presence or absence of differences in the average effect between two groups of samples that are paired (related) with the same object, (duwiconsultan, 2011).

Paired T Test is a comparative test conducted on a paired sample. This test is similar to the independent t test, but is more specific for paired or related samples. Therefore it is also called the pairing t test. The author conducted a paired t test with the help of Ms. application. Excel, (statistika.com, 2013).

Indications for use of this test include:
a. Used to compare the mean of a paired sample
b. Paired samples are a group of samples with the same subjects but undergo two different treatments or measurements.
Formula T Paired :

$$t = \frac{\bar{x}_d - \mu_0}{s_d/\sqrt{n}}$$

OR : t = *d*/(*sd*/Akar(*n*)), is :
t = t count
d = difference 2 / average, or (x_d-μ_0)
sd = standart deviasi of difference
n = Total Sample.

Decision of Paired Hypothesis :
a. – table > -t count it or t count > t tabel or Absolut t count > Absolut t table:
There Are Significant Differences or Ho Rejected.
b. – table < -t count or t count < t table or Absolut t count < Absolut t table:
There Are Significant Differences or Ho Rejected.

4. Results and Discussion

Sample data for algorithm values using descriptive (text) and image (image) methods as in the table below
a. Manual calculation: Sample data (N = 25):

Table 2. Data sample

no	pre	post	selisih
1	75	80	-5
2	77	70	7
3	66	68	-2
4	70	80	-10
5	66	79	-13
6	77	75	2
7	80	70	10
8	77	80	-3
9	79	70	9
10	69	68	1
11	68	80	-12
12	77	79	-2
13	68	79	-11
14	68	80	-12
15	68	80	-12
16	70	80	-10

no	pre	post	selisih
17	75	79	-4
18	70	80	-10
19	68	79	-11
20	76	60	16
21	76	70	6
22	75	77	-2
23	75	80	-5
24	68	68	0
25	77	80	-3

The manual testing process obtained the following results:

Table 3. Test Results with: Paired T-test

25	N	Sample
24		DF = N-1
0.05		Critical Limit
2.0638986		value T table on DF and critical limit
72.6		average value / Post-test average
75.64		Average value/ Pre-test average
-3.04		the average value of the two pre-post variables
7.9972912		standard deviation of the difference in measurement
-1.900644		t Count
No Significant Difference		
H0 accepted		

b. The process of testing with Paired t-test using the help of Ms. Excel as follows:

Table 4. The Testing Process with Ms. Excel

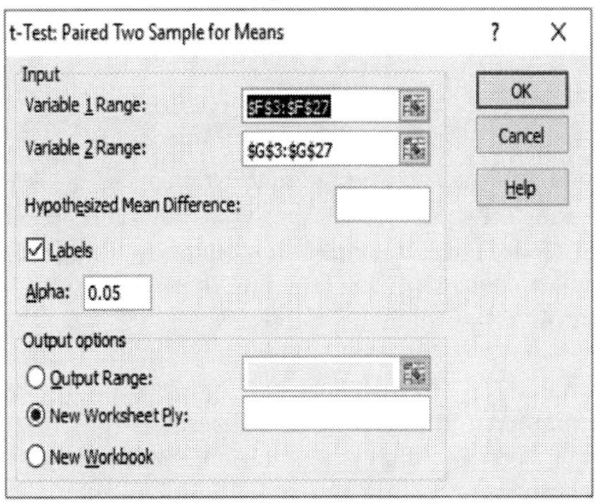

The final results of the Paired Sample T-test calculation are shown in the table below:

Table 5. Test Results with Ms. Excel

t-Test: Paired Two Sample for Means

	Variable 1	*Variable 2*
Mean	72.60000	75.72000
Variance	20.25000	32.04333
Observations	25.00000	25.00000
Pearson Correlation	(0.22213)	
Hypothesized Mean Difference	-	
df	24.00000	
t Stat	(1.95597)	
P(T<=t) one-tail	0.03110	
t Critical one-tail	1.71088	
P(T<=t) two-tail	0.06221	
t Critical two-tail	2.06390	

The testing steps are as follows:

1) Determine the Hypothesis. Ho: there is no difference between the average algol value with descriptive methods and images / symbols. Ha: there is a difference between the average algol value with descriptive methods and images / symbols.
2) Determine the level of significance. The test uses a two-tailed test with a significance level of a = 5%. Means we take the risk of making a wrong decision to reject the correct hypothesis as much as 5% (significance 5% or 0.05)
3) Determine t-count. From the table we get the value of t-count (-1.95597)
4) Define t-table. T-table is sought at a = 5%: 2 = 2.5% (2-tailed test) with degrees of freedom (df) n-1 or (25-1 = 24). With 2-sided testing (significant = 0.025) results obtained for t table of (2,06390)
5) Testing Criteria Ho is accepted if -t table> t arithmetic, Ho is rejected if t arithmetic> T table
6) Based on probability: Ho is accepted if P value> 0.05 and Ho is rejected if P value <0.05, then by comparing t-count with t-table and the probability is obtained –t count > -t table (-1,95597> 2, 06390) and P value (0.06221> 0.05) then Ho is accepted.
7) Conclusion, because the value of -t count> -t table (-1.95977 > 2.06390) and P value (0.06221> 0.05), Ho is accepted, meaning that there are differences in students' understanding of the algorithmic method in the picture with the method algorithm in text. By looking at the Mean value in the Paired Sample test of (72,600 and 75,720) obtained a positive p value of (0.06221) means that the average student is easier to understand the algorithm using the drawing method than the black algorithm in text.

5. Conclusion

Based on the results of calculations of an average of 25 students of class 1A, it can be seen that the T-stat is smaller than the T-critical two tile which means that H1 can be accepted. The algol method in the picture can facilitate student understanding, or T-count <T-table (-1.9555597 <2.06390), then H0 can be accepted. So it can be concluded that the algorithmic method with images gives more value than the value for the descriptive algorithm method (text)

References

[1] Algoritma dan Pemrograman Dengan C++. diperoleh 13 Agustus 2015, from http: // ilmukomputer.org / 2009 / 03 / 11 / algoritma-dan-pemrograman-dengan-c/
[2] Algoritma dan Pemrograman, diperoleh 10 Agustus 2015, from https : // andikafisma.wordpress.com/algoritma-dan-pemrograman/
[3] Gun Gun,M., 2017, *Pembelajaran Dasar Algoritma dan Pemograman menggunakan El-Goritma Berbasis Web*, Jurnal Teknik Mesin (JTM): Vol. 06, ISSN 2549 – 2888
[4] http: //duwiconsultant.blogspot.com/2011 /11/paired-samples-t-test.html
[5] https://kujangcode.com/blog/2017/03 /02 /jantung-ilmu-informatika-algoritma/ 3

[6] Bambang Sumarno HM, 2013, Jurusan Pendidikan Matematika FMIPA UNY, Proc, *Seminar Nasional Matematika dan Pendidikan Matematika FMIPA UNY Yogyakarta*, MT–16, ISBN : 978–979–16353–9–4. bambang@uny.ac.id
[7] Saniman dan Muhammad Fathoni ,2008, *Pengantar Algoritma dan Pemrograman*, Jurnal Saintikom, Vol. 4/ No.1/ Januari
[8] Donald E. Knuth, 1973, " *The Art of Computer Programming*", Second Edition Vol-I " *Volume 1*, second edition, 1973, xiii+634pp, ISBN 0-201-03809-9.
[9] Donald E. Knuth, 1997, " *Fundamental Algorithms* " , Third Edition ,Reading,Massachusetts: Addison-Wesley, xx+650pp. ISBN 0-201-89683-4
[10] Donald E. Knuth, 1997, "*Seminumerical Algorithms*", Third Edition, Reading, Massachusetts: Addison-Wesley, xiv+762pp. ISBN 0-201-89684-2

IOP Conf. Series: Materials Science and Engineering **835** (2020) 012045 doi:10.1088/1757-899X/835/1/012045

Color Based Segmentation Using Fuzzy C Means For Bamboo As EnvironmentallyFriendly Material

Khoiriya Latifah, Joko Siswanto, Bambang Supriyadi, Abdul Rochim

[1,2,3]Faculty of Engineering and Informatics, Universitas PGRI Semarang, Jl. Sidodadi-Timur No.24 Semarang, Central Java 50232, Indonesia
[4]Faculty of Engineering and Informatics, Universitas Islam Sultan Agung Semarang

khoiriyalatifah@upgris.ac.id, jokosiswanto@upgris.ac.id, b_4mbs@yahoo.co.id, abdulrochim@unissula.ac.id

Abstract. Bamboo is an environmentally friendly material. Many benefits of bamboo that we can get. Indonesia as a bamboo-producing country needs easy techniques to make good classification of bamboo. Bamboo is composed of fibers and fiber adhesives. There are various kinds of bamboo in Indonesia. This study uses digital image processing with fuzzy c means based segmentation to identify bamboo. Segmentation is an important thing in image processing. By using fuzzy c means in segmentation in this study obtained good segmentation results. This study uses 4 types of bamboo, namely Javanese bamboo, Ori bamboo and Petung bamboo and Wulung bamboo. There are 40 images as training images and 12 test images. The results of segmentation show that fuzzy c means produces good segmentation with the number of iterations between 20-23 and time ranging from 0.11 to 0.15. The accuracy of this test reaches 80%.

1. Introduction

Bamboo is an environmentally friendly material. Indonesia is the third bamboo producing country in the world after China and India. So that Indonesia has good potential to develop bamboo into an environmentally friendly material for example to produce bamboo fiber which is a natural fiber. There are various kinds of bamboo that grow in Indonesia so we need an easy and inexpensive way to classify bamboo. One of them is using digital image processing methods. In image processing an important step is segmentation. The most important part of image processing is image segmentation. [1.]. Segmentation accuracy determines the likelihood of success or failure of the computerized image analysis procedure. [2.] Segmentation is an image processing step by dividing the image into a number of regions or objects. The level of division depends on the problem being solved so that the development of segmentation techniques is carried out to obtain good and quality segmentation results for a particular image. Because the image quality is influenced by several factors such as lighting, degraded images or blurred colors, and images that contain noise, each image requires a different segmentation technique. Fuzzy clustering as a good method to be used in grouping spatial data and image analysis (Laboratory of Data analysis and Pattern Recognition) [5]. Fuzzy C-Means is a data grouping technique where the existence of each data point in a cluster group is determined by the degree of membership and each object can be a member of several clusters. Fuzzy C-Means is simple, easy to implement, has the ability to group large data, is more robust against outlier data and produces good segmentation [6]. Based on the provisions of the cluster validation that the values closer to one have better cluster quality. [7].

Fuzzy c-mean clustering is an iterative algorithm to find final groups of large data sets such as images so that it will take more time to implement. An improved fuzzy c-mean algorithm that takes less time in finding clusters and is used in image segmentation. [8]. Experimental results on segmentation of synthetic and real images demonstrate that the proposed Fuzzy C Menas algorithm is effective and robust [9]. FCM has been used with some success in image segmentation in general [10]. From the above background, in this study to segment bamboo writer's image using the fuzzy c means method. The purpose of this research is to implement the Fuzzy C Means algorithm to segment bamboo images as an environmentally friendly material.

2. Methods

The stages in this research include

1. Data collection
   Data collection is the stage where the author collects data by conducting library studies to collect data about bamboo fibers, the benefits of bamboo and image processing methods for images that have unstructured patterns or natural patterns, image segmentation methods
2. Data Acquisition
   Data acquisition is the process of getting data in the form of bamboo images. The tool used is the Canon D 60 camera. The process of shooting bamboo is done in the morning at 10-11 am where the sun is in bright conditions there is no cloud and no significant wind.
3. Pre Processing
   Initial processing is carried out to obtain sharper images and convert into HSV images of color spaces that represent colors as seen by the human eye. H comes from the word "hue", S comes from "saturation", L comes from the word "luminance", I comes from the word "intensity", and V comes from "value".

To get the values of H, S, V based on R, G, and B, there are several ways.
Acharya & Ray [2005]. The formulas used are as follows:

$$r = \frac{R}{(R+G+B)}, g = \frac{G}{(R+G+B)}, b = \frac{B}{(R+G+B)}$$

$$V = \max(r, g, b)$$

$$S = \begin{cases} 0, & \text{jika } V = 0 \\ 1 - \frac{\min(r,g,b)}{V}, & V > 0 \end{cases}$$

$$H = \begin{cases} 0, & jika\ S = 0 \\ \frac{60*(g-b)}{S*V}, & jika\ V = r \\ 60 * \left[2 + \frac{b-r}{S*V}\right], & jika\ V = g \\ 60 * \left[4 + \frac{r-g}{S*V}\right], & jika\ V = b \end{cases}$$

$$H = \mathbf{H + 360\ jika\ H < 0}$$

4. Fuzzy C Means Segmentation
This stage serves to segment Fuzzy C Means based on bambo imagery. In this study the case of grouping in cluster analysis, the correlation matrix used is the similarity matrix of the objects to be grouped. The principle is that the higher the value of similarity between objects with one another, the value of observations between objects has many similarities (meaning it is possible to become one group).

2. Result and Discussion

Based on laboratory testing of fiber, bamboo which has fiber content, fiber length, inter-fiber adhesive, namely lignin and the diameter of bamboo fiber, the higher the bamboo stems will be stronger. So that bamboo has a larger fiber, including a good type of bamboo. This research will segment the image with the fuzzy c means method. The data used in this study were 32 test data and there were 7 training data on each bamboo. Researchers used four types of bamboo that are often used in Indonesia, especially in the city of Semarang, including Java Bamboo, Ori Bamboo, Petung Bambu and Bambu Wulung.

Table 1. Training image and test image

No	Jenis Bambu	Jumlah Citra Pelatihan	Jumlah Citra Pengujian	Format
1	Jawa	32	7	. jpg
2	Ori	32	7	. jpg
3	Petung	32	7	. jpg
4	Wulung	32	7	. jpg

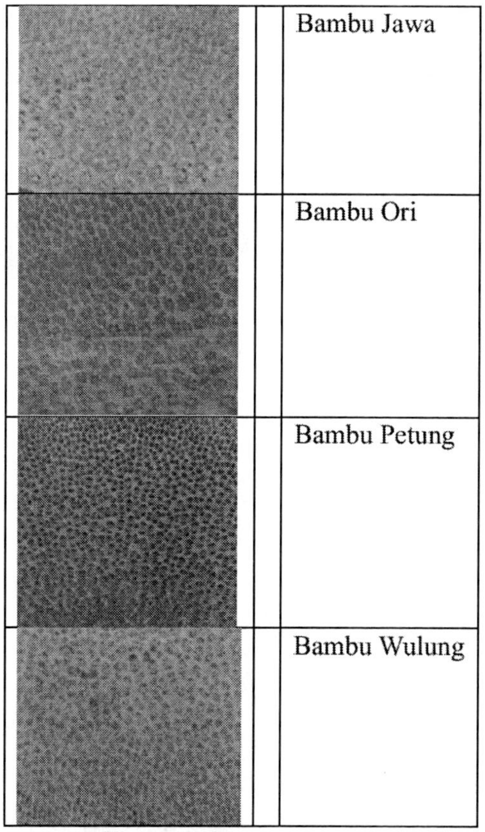

Figure 2. Bamboo Image

In this research, to find out the size of bamboo fiber, the writer uses the Fuzzy C Means segmentation method.

1. In fuzzy clustering, each point has a degree of belonging to clusters, as in fuzzy logic, rather than belonging completely to just one cluster
   uk (x): the degree of being in the kth cluster

$$u_k(x) = \frac{1}{\sum_j \left(\frac{d(\mathrm{center}_k, x)}{d(\mathrm{center}_j, x)}\right)^{2/(m-1)}}.$$

$$\text{center}_k = \frac{\sum_x u_k(x)^m x}{\sum_x u_k(x)^m}.$$

For m = 2, this is equivalent to normalising the coefficient linearly to make their sum When m is close to 1, algorithm is similar to k-means

X_l – sample feature vectors
V_i - vector of cluster centroid
M - number of sample feature vectors
N - dimension of the sample feature vectors
K - number of cluster centroids

2. Fuzzy membership

$$R_{lj}(t) = \frac{1/d_{lj}^2(t)}{\sum_{m=1}^{K}(1/d_{lm}^2(t))}, \; 1 \le l \le M, \; 1 \le j \le K.$$

3. Cluster centroid initialization
   first frame : random select
   otherwise : prediction from previous frame

4. Cluster centroid update

$$V_{ji}(t+1) = V_{ji}(t) + \frac{\sum_{l=1}^{M} R_{lj}(t) \cdot w_l \cdot (X_{li} - V_{ji}(t))}{\sum_{l=1}^{M} R_{lj}(t) \cdot w_l}$$
$$1 \le i \le N, \; 1 \le j \le K$$

5. Dynamic Growing of Centroids
   - entering and leaving regions are manually defined
   - Creation
   - we find a subset of samples where the Euclidean distance between each of these samples and its associated cluster centroid j exceeds a threshold Φ_j
   - Erasure
     The position of cluster centroid j is within a leaving region
     The number of the samples corresponding to cluster centroid j is too small to represent the smallest object in the scene.
6. Modeling of Cluster Centroids
   There may be objects which correspond to two or more cluster centroids in one frame
   For two centroid trajectories exist over the same sequence of frames, if the differences between the centroids in each frame are approximately constant and small, two trajectories are merged
   The results of the Fuzzy C Means segmentation for bamboo Wulung are as follows:

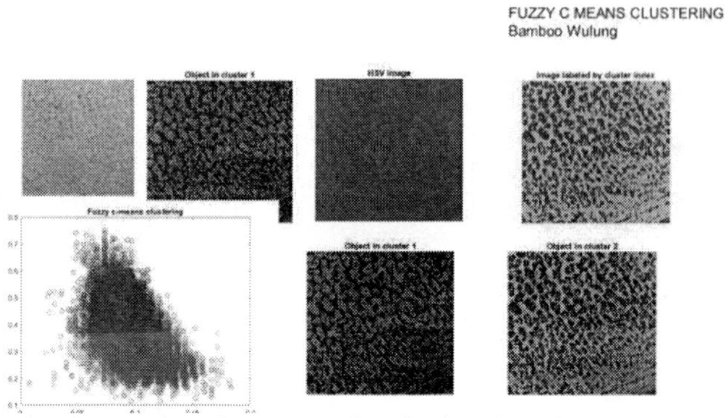

Figure 3. The results of segmentation for bamboo Petung are as follows

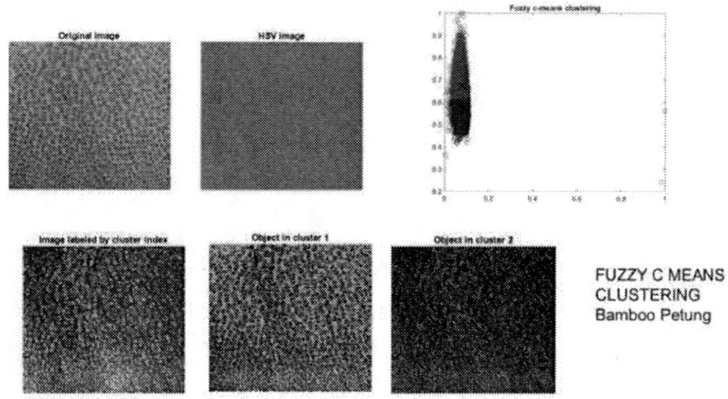

Figure 4. The results of segmentation for bamboo Jawa are as follows

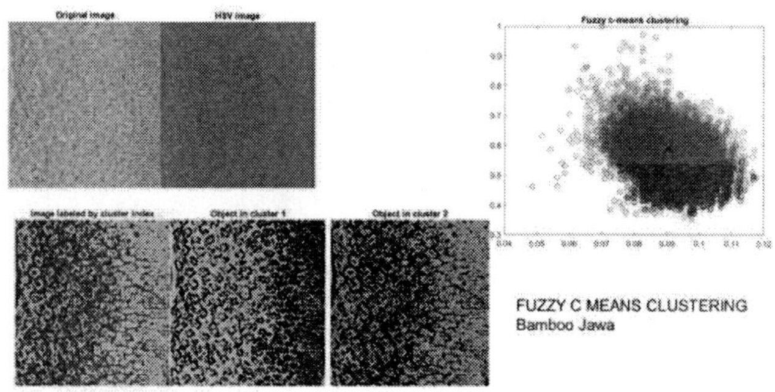

Figure 5. The results of segmentation for bamboo Ori are as follows

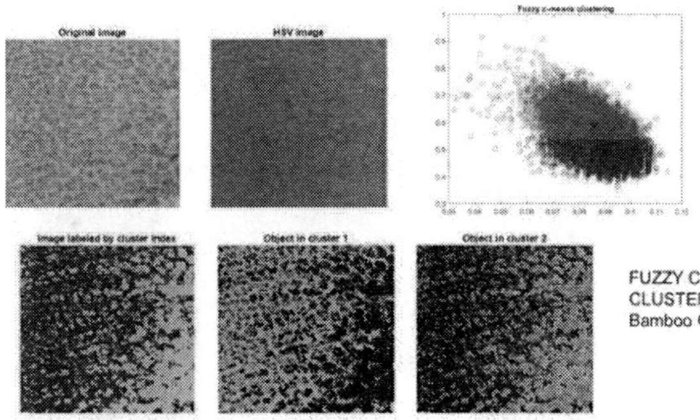

FUZZY C MEANS CLUSTERING
Bamboo Ori

Fuzzy C Means method produces good segmentation, the results of image segmentation clearly visible difference between which fiber and which fiber adhesive. This Experimental results show that the FCM method is effective and robust. Fuzzy C Menas segmented around 20-23 iterations with a duration of about 0.11-0.15. As for more clearly in the following table 1:

Tabel 2 IterasiCount Fuzzy C Means pada citra bambu

No	Bamboo	Iteration Count	Time	Partition Matriks
1	Wulung	23	0,15	137.080677
2	Petung	23	0,17	124.043284
3	Ori	22	0,12	232.093546
4	Jawa	20	0,11	134.880044

3. Conclusion

Experimental results show that the FCM method is effective and more robust in image segmentation. Future work will focus on adaptively deciding the penalized parameters of this algorithm as well as compensating for the intensity of homogeneity while segmenting the image data.

Acknowledgment

Thank you to the Fiber Laboratory of the Ministry of Industry of the Republic of Indonesia in the Field of Research Facilities and Standardization of the Pulp and Paper Center and thank you to the computer laboratory of the University of PGRi Semarang and the Laboratory of Material Unissula.

Reference

[1]. Reem Alzu'bi, Medical Image Segmentationvia Optimized K-Means, International Conference on Current Trends in Computer, Electrical, Electronics and Communication (ICCTCEEC-2017), 978-1-5386-3243-7/17/$31.00 ©2017 IEEE]

[2]. Eko Prasetyo, Pengolahan Citra Digital dan Aplikasinya Menggunakan Matlab, 2011, Yogyakarta : Andi Offset

[3]. Darma Putra, Pengolahan Citra Digital, 2010, Yogyakarta : Andi Offset.

[4]. Amiya Halder, Soumajit Pramanik. Dynamic Image Segmentation using Fuzzy C-Means based Genetic Algorithm. International Journal of Computer Application. 2011; 2(6)]

[5]. Klawonn dand Höppner, (2001), "What is Fuzzy about Fuzzy Clustering? Understanding and Improving the Concept of the Fuzzier". Science Journal, http://public.rz.fh-wolfenbuettel.de/klawonn.]

[6]. I Made Budi Adnyana, Implementasi Algoritma Fuzzy C Means dan Stastitical Region Merging Pada Segmentasi Citra, EKSPLORA INFORMATIKA, Vol. 6, No. 2, Maret 2017 hal 179-189].

[7]. Nur Indah Selviana1, Mustakim2, Analisis Perbandingan K-Means dan Fuzzy C-Means untuk Pemetaan Motivasi Balajar Mahasiswa, Seminar Nasional Teknologi Informasi, Komunikasi dan Industri (SNTIKI) 8, ISSN : 2085-9902, Pekanbaru, 9 November 2016. Hal 95-105

[8]. Hind Rustum Mohammed, Improved Fuzzy C-Mean Algorithm for Image Segmentation, (IJARAI) International Journal of Advanced Research in Artificial Intelligence, Vol. 5, No.6, 2016, pp 7-10, www.ijarai.thesai.org]

[9]. Yong Yang, IMAGE SEGMENTATION BY FUZZY C-MEANS CLUSTERING ALGORITHM WITH A NOVEL PENALTY TERM, Computing and Informatics, Vol. 26, 2007, 17–31.

[10]. S. M. Yammary AA, Farag,andS.Hsu,"Afuzzyhyperspectralclassifier for automatic target recognition (ATR) systems," Pattern Recogn. Lett., vol. 20, pp. 1431–1438, 1999.]

[11]. Acharya, T.; Ray, A.K. 2005; *Image Processing Principles and Applications*; New Jersey: John Wiley & Sons, Inc.

IOP Conf. Series: Materials Science and Engineering **835** (2020) 012046 doi:10.1088/1757-899X/835/1/012046

The Effect Of Decomposer Microorganism Additions On The Natural Hydroponic Technology (Nht) Systems Of Pakchoy Growth

Ary Susatyo Nugroho*, Endah Rita Susatya Dewi, Muhammad Anas Dzakiy

Program Studi Pendidikan Biologi FPMIPATI, Universitas PGRI Semarang, Jl. Sidodadi-Timur No. 24 Semarang, Central Java 50232, Indonesia

*arysusatyon@gmail.com

Abstract. Nowadays there is a tendency of the people, especially the middle class and above, to choose vegetables with high quality and safe for body. Therefore, farmers are required to always innovate in terms of vegetable cultivation, so that the crops can compete and be accepted by the community. This study aims to determine the effect of the addition of decomposer microorganisms on the Natural Hydroponic Technology (NHT) system on the growth of Pakchoy plants. The study was conducted on May to July 2019 in the agricultural land of Ngesrep Balong Village, Limbangan District, Kendal Regency, Central Java. This study uses a completely randomized design (CRD) with one treatment factor, the type of decomposer microorganisms. The research data was taken through several measurement techniques and direct observation of objects. Data were taken three times on the 14th day, 21st day, and 28th day after planting. Data were then analyzed statistically with ANOVA (Analysys of Variance) at 95% confidence level and continued with LSD test. The analysis shows that the addition of decomposer microorganisms to the Natural Hydroponic Technology (NHT) system can increase the growth rate of Pakchoy plants. Decomposer microorganisms with trademarks ST (Sukses Tani) and EM-4 (Effective Microorganism-4) produce the same growth rate.
Keywords: hydroponics, microorganisms, decomposers, Pakchoy

1. Introduction

Natural Hydroponic Technology (NHT) is one of the developments in hydroponic technology that aims to answer the farmers' needs in improving agricurtural bussiness. NHT is a hydroponic technology that uses the flow of natural springs, land slopes, and the decomposition of natural materials [1]. In the NHT model, water needs of cultivated plants are obtained from natural springs such as the rivers that are in higher places. The water can flow by itself towards hydroponic installations located in lower places caused by gravity. In addition, the nutrient requirements are obtained from the decomposition of organic material connected to the flow of water [2].

The source of organic material that will be broken down into nutrients for cultivation plants on the NHT system can be in the form of organic material placed in the composter tub. In addition, the NHT system can use fish pool wastewater that consists of fish food waste and feces as a source of plant nutrients. In the NHT, conditions composter or fish pools and hydroponic systems are made similar to the conditions of natural ecosystems [3].

IOP Conf. Series: Materials Science and Engineering **835** (2020) 012046	doi:10.1088/1757-899X/835/1/012046

The organic material decomposition process will occur naturally in every ecosystem, but the decomposition variations that depend on the amount and type of organic material, the number and type of decomposer microorganisms, and the environmental conditions in which the decomposition takes place. In the NHT system, a high decomposition rate is needed because it affects the availability of cultivated plants. If decomposition rate is low, it will cause nutrient deficiencies for the growth of cultivated plants.

To increase the rate of decomposition of organic matter on the NHT system, there are several factors that must be considered, such as the use of organic material which should be easily composed in sufficient quantities, the addition of decomposer microorganisms as decomposition agents for organic matter, and the regulating environmental conditions so that the performance of decomposer microorganisms can take place optimally.

The research aims to find out the effect of the addition decomposer microorganism on *Natural Hydroponic Technology* (NHT) system with water sources and organic matter in the form of fish pool water waste on the growth of pakchoy plants. The researcher hopes this research will be beneficial for the improvement of the NHT system in increasing vegetable production, and the development of science in general.

2. Methods

2.1. Time and Place of the Research
The research was conducted in Ngesrep Balong Village, Limbangan, Kendal Regency, Central Java. It was helm on June until August 2019 in the green house on the farmland.

2.2. Research Design
This reseach used Complete Randomized Design (CRD) with one treatment factor that is the addition of decomposer microorganisms with different trademarks. Organic material sources used were from the waste water of the tilapia pool which is flowed into the planting tub. Tilapia used are 5 cm in size with a stocking density of 100 fish/m^2 surface area. The addition of decomposer microorganisms treatment is carried out in a filter tub containing bioball. The treatment is as follows:

P0 : Without adding decomposer microorganism (as the control)
P1 : Adding decomposer microorganism with a trademark "Sukses Tani"
P2 : Adding decomposer microorganism with a trademark "EM-4"

Each treatment above is done 3 times.

2.3. Research Variables
a. Independent Variable : The research independent variable is adding decomposer microorganism
b. Dependent Variable : The research dependent variable is the growth of pakchoy plant production. The growth and production variables that is observed are as follows:
   1) The Height of Plants
   2) The number of leaves
   3) The Wet Weight of Plants

2.4. Method of Data Collection.
The research data needed is taken from the measurement techniques and direct observation of objects. The measurement and production of plants are taken by measuring, counting, and weighing objects according to the observed variables. A water quality data as supporting data is taken after pool water waste passes through a filter bath that has been given the addition of decomposer microorganisms. Data collection on the growth and production of plants, as well as water quality data is carried out on day-28 after planting.

2.5. Data Analysis
The quantitative data obtained will be analyzed statistically with ANOVA (Analysys of Variance) with a 95% confidence level and continued with the LSD test (Least Significant Difference). Besides using Anova, the data will be analyzed qualitatively.

3. Results and discussion

The quality of pool water waste as a source of pakchoy plant nutrients is measured after pool water waste passes through the filter tub which has been added with decomposer microorganisms. The quality of pool water waste consist of ammonia, nitrites, and nitrates in can be seen in Table 1.

Table 1. Ammonia, Nitrite, and Water Nitrate levels in the NHT System (Day-28 after planting) with the Addition of Different Decomposer Microorganisms

Treatment	Ammonia (mg/L)	Nitrit (mg/L)	Nitrat (mg/L)
Control	0,25	15,78	42,2
Sukses Tani	0,26	10,68	45,2
EM-4	0,25	13,98	44,8

Ammonia in pool water waste comes from the decomposition of organic material in the form of metabolic waste and leftover food that is not consumed. In general, the ammonia released by fish through the osmoregulation process ranges from 80-90%. Meanwhile, the ammonia released through faeces ranges from 10-20%. In addition, ammonia in water comes in two forms, namely NH_4^+ and NH_3 ions. Unionized free ammonia (NH_3) is toxic to aquatic organisms [4].

Ammonia waste level in fish pool after passing through the filter is relatively low, ranging from 0.25 to 0.26 mg/L. The low ammonia level is caused by the nitrification process when pool water waste has passed through the filter. In the process of nitrification, ammonia is converted to nitrite by the bacterium Nitrosomonas. The addition of probiotics containing Nitrosomonas can reduce the concentration of ammonia in fish pool water. [5]. Pool water waste which was given several types of probiotics can reduce ammonia levels up to 0.188-0.267 mg/L. Low ammonia levels indicate that the process of converting ammonia to nitrite runs smoothly [6].

Nitrite is the result of oxidation of ammonia through the process of nitrification by the bacterium Nitrosomonas. Nitrites are toxic to organisms and cannot be used by plants [7]. Nitrite level in pool water waste that has passed through the filter is very high, which is around 10.68-15.78 mg/L. The high level of nitrite is caused by the accumulation of ammonia in pool water waste successfully remodeled by the bacteria Nitrosomonas into nitrite [8].

Nitrite will be oxidized by bacteria Nitrobacter to nitrate immediately in the process of nitrification. In this research, the highest nitrite level was in the control treatment, followed by treatment with the addition of EM-4 trademark decomposer microorganisms and the nitrite level from the addition of Sukses Tani trademark decomposer microorganisms was lower. Meanwhile, the highest nitrate level was found in the treatment with the addition of Sukses Tani decomposer microorganism. After finding the highest level nitrate, the addition of EM-4 trademark decomposer microorganisms was added and the lowest was in the control treatment.

The level difference between nitrite and nitrate is caused by the results of the control treatment, the conversion of nitrites to nitrates in the nitrification process is only carried out by bacteria that are naturally present without any additions so that the process of converting nitrites to nitrates is not optimal. Meanwhile, converting nitrites to nitrates can be increased by using the treatment with the addition of EM-4 and Sukses Tani decomposer microorganisms.

Pool water waste which is given several types of probiotics can reduce nitrite levels and increase nitrate levels. The nitrification process will cause a reduction of ammonia levels in water, while nitrate levels will increase. Nitrification process fails, the ammonia or nitrite levels in the water will increase.

The quality of pool water waste analyzed is based on phosphate levels, dissolved oxygen (DO), and biochemical oxygen demand (BOD). Based on the results of the study were obtained that levels of phosphate, DO, and BOD of pool water waste that had passed through the filter tube in all treatments were in normal conditions. The detail about levels of phosphate, DO, and BOD of pool water waste is presented in Table 2.

Table 2. Phosphate, DO, and BOD levels of the water in the NHT System (Day-28 after planting) with the Addition of Different Decomposer Microorganisms

Treatment	Phosphat (mg/L)	DO (mg/L)	BOD (mg/L)
Controll	7,68	5,2	3,5
Sukses Tani	8,34	5,6	3,9
EM-4	8,58	6,6	4,4

The process of decomposition of organic material in pool water waste will provide nutrients for cultivated plants indirectly. After that, the nutrients will be absorbed and utilized by plants to grow and thrive. the plant growth can be measured by increasing plant height, number of leaves, and wet weight of plants. The height plant growth, number of leaves, and heavy wet plants can be used as an indicator of nutrient installation by plants. The detail of pakchoy plant growth data in this study can be seen in Table 3.

Table 3. The Height Plants, the Number of Leaved, and Wet Weight Pakchoy of each treatment in day 28 after planting

Treatments	Height (cm)	The Number of Leaves	Wet Weight (gram)
Control	21,30[a]	18,33[a]	93,07[a]
Sukses Tani	27,47[b]	18,67[a]	120,77[b]
EM-4	26,50[b]	18,67[a]	116,50[b]

Note: the numbers accompanied by the same notation in the same column show no significant difference at the 95% test level.

The results of the study showed that there is a difference between the control treatment and the treatment of the addition of decomposer microorganisms to pakchoy growth. In the control treatment, pakchoy plants showed the lowest growth. Meanwhile, the treatment with the addition of decomposer microorganisms of the trademark Sukses Tani and EM-4 showed the same growth and higher than the control treatment.

Pakchoy can be given the addition of decomposer microorganisms treatments can grow better than pakchoy that was not given the addition of decomposer microorganisms. It shows that the availability of pakchoy nutrients with the addition of decomposer microorganisms is available in sufficient quantities for its growth. Meanwhile, pakchoy growth is not optimal because of the availability of nutrients in pakchoy which is not given additional decomposer microorganisms is insufficient.

The availability of nutrients for pakchoy plants caused by wastewater from Tilapia fish culture contains a lot of organic materials, including leftovers that are not consumed, fish dung, and additional food waste in the form of green leaves. This organic material contains nutrients for plant growth. Fish dung contains ammonia while the remaining food additives in the form of green leaves contain lots of carbohydrates and minerals [9]. In fish pool water wastes there are many humus and feed residues that contain a lot of N, P, and K and other minerals in quite high amounts. These elements are nutrients that are needed for plant growth. Besides that, these nutrients are nutrients that are not yet available to plants because they are still in the form of organic compounds, so they cannot be absorbed by plants [10].

Organic materials in fish pool water waste must be decomposed to produce inorganic nutrients available to plants. The decomposition of organic material will produce macro nutrients such as NH_4^+, NO_3^-, PO_4, K, Ca, Mg, and SO4, as well as other micronutrients. Amino which is a waste of leftover food and fish metabolism results will undergo nitrification. In this section, the rest of the fish's food and feces are converted by the Nitrosomonas bacteria into nitrites. Furthermore, nitrite will be oxidized by bacteria Nitrobacter to produce nitrate as nutrients in plants [11].

In NHT system, a high rate of decomposition of organic material is needed because the availability of nutrients for cultivatied plants is very dependent on it. If the decomposition rate is low, the crop will lack nutrients. Meanwhile, If the decomposition rate is high, the nutrient requirements of the cultivated plant will be fulfilled.

The process of decomposition of organic material in pool water waste has basically taken place naturally. Besides that, the number of species and the number of individual decomposer microorganisms in the natural control treatment is insufficient to maximally decompose organic matter in pool water waste. Thus the available nutrients needed for cultivating plants cannot be fulfilled at all. That is evidenced by the growth and production of cultivated plants that are not optimal.

During treatment with the addition of decomposing microorganisms from Sukses Tani and EM-4 trademarks, the rate of decomposition of organic matter is higher because the number of species and the number of individual decomposer microorganisms is greater than control treatments. It affects nutrients that are needed by cultivated plants are also available in sufficient quantities. It is proven by the optimal growth and production of cultivated plants. The addition of probiotics in fish pools can increase nitrate levels which are essential nutrients needed in large quantities for plant growth and production.

4. Conclusion

Based on the result of the research, it can be concluded that the addition of decomposers microorganisms with the Natural Hydroponic Technology (NHT) system, water sources, and organic matter in the form of fish pool water waste can increase the growth and production of pakchoy plants. The researcher hopes this research could be useful for improving the NHT system in increasing vegetable production, and developing science in general.

Acknowledgement

We sincerely thank the Ministry of Research, Technology and Higher Education, and LPPM of PGRI Semarang University for facilitating this research.

References

[1] Nugroho, AS., ERS Dewi., R. Rosyida dan MA. Dzakiy. 2017. Pemanfaatan Aliran Sungai Dalam Pengembangan *Natural Hydroponic Technology* (NHT). *Prosiding Seminar Nasional Simbiosis II*. Universitas PGRI Madiun.

[2] Nugroho, A.S. 2018. Penerapan Konsep-Konsep Ekologik dalam Biopreneur Berbasais Sumberdaya Hayati untuk Mendukung SDGs. *Prosiding Seminar Nasional Sains dan Entepreneur IV*. Universitas PGRI Semarang.

[3] Nugroho, A.S. 2018. *Produksi Tanaman Sayuran dengan Natural Hydroponic Technology. Monograf.* LPPM Universitas PGRI Semarang. Semarang.

[4] Effendi, H. 2003. Telaah Kualitas Air Bagi Pengelolaan Sumberdaya dan Lingkungan Perairan. Kanisius. Yogyakarta.

[5] Lisna dan Insulistyowati. 2015. Potensi Mikroba Probiotik_FM dalam Meningkatkan Kualitas Air Kolam dan Laju Pertumbuhan Benih Ikan Lele Dumbo (*Clarias gariepinus*). Fakultas Peternakan. Universitas Jambi. 8 hal.

[6] Pratama, W.D., Prayogo dan A. Manan. 2017. Effect Addition of Different Probiotik in Aquaponic System towards Water Quality in Aquaculture Catfih (*Calarias sp.*). *Journal of Aquaculture Science*. Vol 1(1) : 27-35.

[7] Djokosetiyanto, D., A. Sunarma dan Widanarni. 2006. Perubahan Amonia (NH_3-N), Nitrit (NO_2-N) dan Nitrat (NO_3-N) pada Media Pemeliharaan Ikan Nila Merah (*Oreochomis sp.*) di dalam Sistem Resirkulasi. Jurnal Akakultur Indonesia. 5(1) : 13-20.

[8] Thesiana, L. dan A. Pamungkas. 2015. Performance Test of *Recirculating Aquaculture System* (RAS) Technology to Water Quality Condition in Spriny Lobster *Panulirus homarus* Nursery.

Jurnal Kelautan Indonesia. Vol. 10, No. 2 : 65-73.

[9] Nelson, R. 2009. Methods of Hydroponic Production. *Aquaponics Journal*. Montello. USA. http://www.aquaponicsjournal.com. diakses tanggal 22 Januari 2016.

[10] Diver S. 2006. Aquaponics – integration of hydroponics with aquaculture. *ATTRA - National Sustainable Agriculture Information Service*. Diakses dari https://attra.ncat.org/ diakses tanggal 22 Januari 2016.

[11] Wahab, M. A., Jellali, S., and Jedidi, N., (2010), Ammonium biosorption onto sawdust: FTIR analysis, kinetics and adsorption isotherms modeling, *Bioresource Technology*, 101(14): 5070-5075.

IOP Conf. Series: Materials Science and Engineering **835** (2020) 012047 doi:10.1088/1757-899X/835/1/012047

Damarjati Application for Development of Javanese Language Learning in High Schools

Suyitno[1], Ngatmini[2], Setyoningsih Wibowo[3]

[1]Faculty of Science Education, Universitas PGRI Semarang, Jl. Sidodadi-Timur No.24 Semarang, Central Java 50232, Indonesia
[2]Faculty of Language and Art, Universitas PGRI Semarang, Jl. Sidodadi-Timur No.24 Semarang, Central Java 50232, Indonesia
[3]Faculty of Enginering and Informatics, Universitas PGRI Semarang, Jl. Sidodadi-Timur No.24 Semarang, Central Java 50232, Indonesia

suyitno@upgris.ac.id

Abstract. The background of this study is the difficulty of Javanese high school language teachers in implementing the 2013 Local Content Java Language Curriculum. The problem discussed is how the application of the Damarjati application in the development of HOTS-based Javanese language learning in high school. The objective to be achieved is to describe the standard operating procedures of the Damarjati application in the development of HOTS-based Javanese language learning in high school. The object of this research is the application of the Damarjati application in the development of HOTS-based Javanese language learning with a population of Javanese language teachers and high school students in Central Java. The phenomenological approach applied in this research is carried out by developing products in the learning process with the Content Management System (CMS) format. Based on in-depth studies it was concluded that: 1) the Damarjati application was opened through damarjati.upgris.ac.id .; choose classes and semesters; choose themes and sub-themes, and provide corrections and comments; and 2) the application of the Damarjati application can develop HOTS-based Javanese language learning in high schools through the *nggatekake* menu, *tetakon*, *naliti*, *nyoba*, *ngatonake*, and *gladhen*. Our gratitude goes to the In the 7th International Conference on DV-X melalui Method Committee, through Ms. Mega Novita for her willingness to give our team the opportunity to join this highly qualified international conference.

1. Introduction

The background of this study is the difficulty of Javanese high school language teachers in implementing the 2013 Local Content Java Language Curriculum. This was identified in the Facilitation of Curriculum Development and Deepening of Local Content Materials in the Javanese Language of Central Java High School, at Siliwangi Hotel, Semarang, 15-18 May 2017. In its implementation, the 2013 Curriculum was aimed at developing XXI century skills, namely learning that integrates literacy skills, skills knowledge, skills and attitudes, and mastery of technology. Learning activities are developed by reducing *Lower Order Thinking Skills* (LOTS) and add on *Higher Order Thinking Skills* (HOTS). (Dirbin SMA, 2017: 2-4). This is in accordance with the Strategic

Plan for UPGRIS Research, in the Roadmap for Superior and Identity HR Research. In that field it is planned that 2017-2019 can implement technology in the form of prototype models and produce web-assisted HOTS model products, the Damarjati application is developed. The problem of this research is how the application of Damarjati application in the development of HOTS-based Javanese language learning in high school. The goal to be achieved is to describe the standard operational procedures of the Damarjati application in the development of HOTS-based Javanese language learning in high school.

The XXI Century is a global era and spread to all aspects of human life, including the world of education. Therefore, learning Javanese must lead to it. Means of communication in this global era are dominated by internet networks, one of which is the use of the World Wide Web or often called the web. The web is the largest collection of documents stored on various servers connected to a network (Rianawaty, 2011: 1). Documents used on the web are developed in a format Hypertext Markup Language (HTML). Through this format it is possible to link from one document to another. Learning that utilizes the web as it is commonly referred to as *Web Based Learning* or *E-learning.* The use of the web in this study was chosen by the type of web enhanced course, which utilizes the internet to support the improvement of the quality of learning in the classroom (Rianawaty, 2011: 3-4). The time needed in this learning model is 5-10 months divided into 5 stages. Stage 1, 3, and 5 are implemented remotely using the web, while stages 2 and 4 are carried out conventionally by face to face (Idayoce 2014: 1). Permendikbud Nomor 22/2006 contains that learning aims to provide the ability to think logically, analytically, systematically, critically, and creatively, and be able to work together with students. These objectives are in accordance with the main principle *Higher Order Thinking Skills* (HOTS). Itasari (2012: 1) explain that *Higher Order Thinking Skills* (HOTS) in the Indonesian language is interpreted as a high-level thinking ability. HOTS is one of the approaches in learning that teaches students to think critically, logically, reflectively, metacognitively, and creatively. Such thinking skills can arise if students are confronted with problems they have not encountered before. Referring to Krulik and Rudnick's opinion, Itasari mentioned that the characteristics of thinking skills consist of memorizing (recall thinking), basic (basic thinking), critical (critical thinking), and creative (creative thinking). Critical thinking according to Romlah (in Itasari, 2012: 2) is a disciplined thinking controlled by consciousness. This creative thinking will produce something complex. Complexity is seen by the activities to unite ideas, create new ideas, and determine their effectiveness.

2. Methods

2.1. Object of research

The object of this research is the application of the Damarjati application in the development of HOTS-based Javanese language learning. The study population was Javanese language teachers and high school students in Central Java who were randomly selected.

2.2. Research stages

This research applies a phenomenological approach. In this case the research activities carried out by developing products in the learning process with the content format Management System (CMS) Damarjati application for high school students. The stages in conducting this research are: 1) goal setting and problem formulation; 2) research design and data collection techniques; 3) conducting research: data collection and analysis; 4) formulation of conclusions; and 5) reporting (Moustakas, 1994 in Creswell, 2013: 105-149), like the following picture.

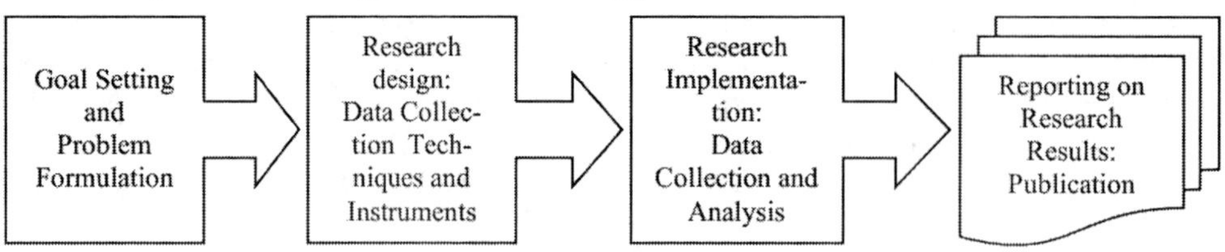

IOP Conf. Series: Materials Science and Engineering **835** (2020) 012047 doi:10.1088/1757-899X/835/1/012047

Figure 1. Phases of Phenomenology Research

3. Results and Discussion

Through research it can be stated that the standard operational procedures of the Damarjati application include four steps, namely: finding a web address, finding class and semester information, entering the material menu, entering the submateri menu.

3.1. *Damarjati application address*

The first step to enter the Damarjati web address can be found by address **damarjati.upgris.ac.id.** This address will open the main page that loads: the front page (figure 2.1); menu list (figure 2.2); class and semester (figure 2.3); material and submissions (figure 2.4); and the comments page (figure 2.5) like the picture below.

Figure 2.1. Figure 2.2. Figure 2.3. Figure 2.4. Figure 2.5.

Figure 2. The main page of Damarjati

3.2. *Class and semester*

This page contains class and semester information. Through this page, users can choose class X semester 1, class X semester 2, class XI semester 1, class XI semester 2, class XII semester 1, and class XII semester 2. In addition to selecting classes and semesters, this page can be viewed core competency (KI) and basic competencies (KD) which is developed in each class and semester chosen.

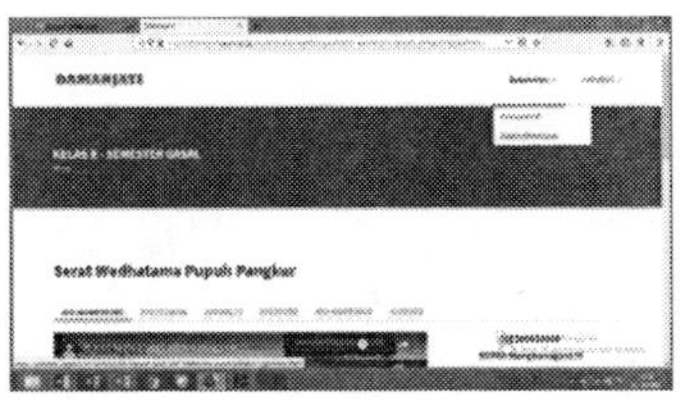

Figure 3. The class and semester page

3.3. *Material and submissions*

On this page the material to be discussed has been determined, for example, *Serat Wedhatama Pupuh Pangkur*. Then you can choose the submersion to be studied more specifically through the menu: *ayo nggatekake* 'observing', *ayo tetakon* 'questioning', *ayo naliti* 'associating', *ayo nyoba* 'experimenting', *ayo ngatonake* 'networking/communicating', and *gladhen* 'exercises'.

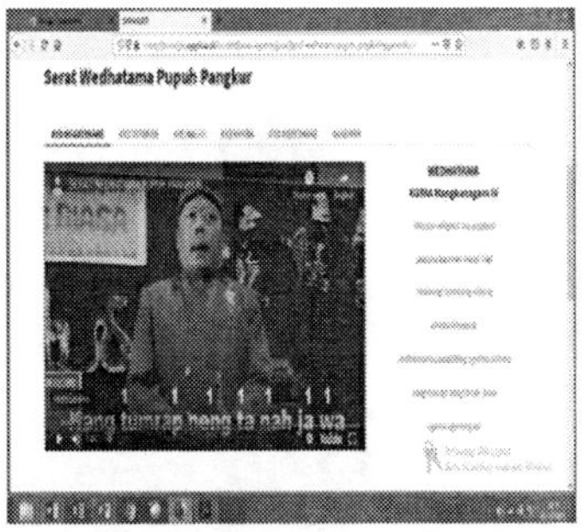

Figure 4.1. The observing page

Figure 4.1 shows the display of the submission *ayo nggatekake*. In this submission, a video is presented in the 'verse' of the Pangkur song, which is equipped with notation and moving poetry. This is intended to guide the observer in observing the songs and poems of the Pangkur song.

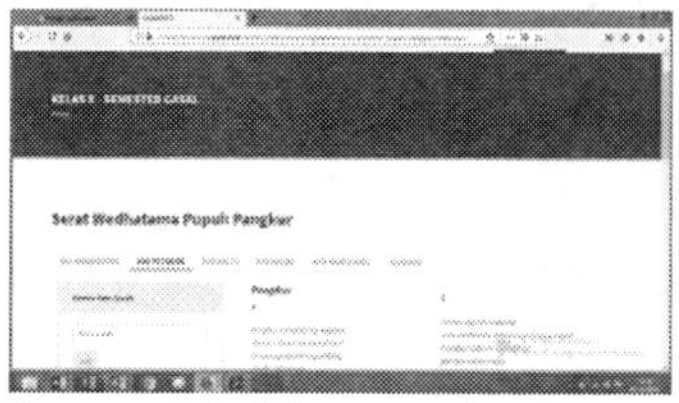

Figure 4.2. The questioning page

The next sub-topic is *ayo tetakon* 'questioning'. This page shows a page like Figure 4.2. In this submission the user can ask all the words contained in Pangkur song. If a word is written in the "kata susah" column and then clicks the "cari" column, then an answer will automatically appear: lime clay root word, its morphological process, and the meaning or meaning of a word.

On this *ayo naliti* 'associating' submission, users observe several verses of Pangkur song. In each verse of the Pangkur song the first line number is written down to the seventh line; each line will be written with a syllable fragment which is *guru wilangan*; then the last syllable vowel of each line will be displayed differently from other letters. This was to mark the *guru lagu*. By observing several verses of a song, a characteristic of a Macapat song will be found.

Figure 4.3. The associating page

Based on the characteristics that have been deduced from the above association process, students are instructed to compose songs of Pangkur using their own language. If students experience difficulties, the teacher can guide by choosing certain themes. After students succeed in composing songs, in this case Pangkur song, the students' work is uploaded into the upload column provided (Figure 4.4). The upload can be seen and commented on by anyone who opens the Damarjati application.

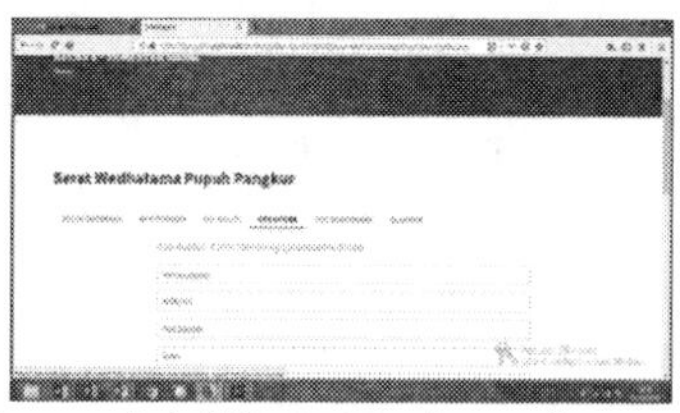

Figure 4.4. The experimenting page

Figure 4.5. The networking page

The next step, the song that has been composed by the students was sung by themselves and video. After that, the video is uploaded using the URL format facility and uploaded to YouTube, as shown in Figure 4.5. Similar to the previous step, the video can be seen and commented on by everyone who opens the Damarjati application. Students who compose songs and upload videos can read comments, both related to the writing and the video. Based on the various comments of readers, viewers, and listeners of the song earlier, students can make improvements.

These steps (1-5) can also be applied to *Sinom*, *Pocung*, *Gambuh*, *Kinanthi*, *Dhandhanggula*, and other songs.

The execises page, like Figure 4.6 contains questions to test students' understanding of the material being studied. Student identity is recorded accurately. Thus, the teacher can conduct a comprehensive evaluation. Understanding ability is assessed based on answers to questions; writing ability and critical attitude can be assessed based on the songs created by students; and communication skills or expressions can be assessed through video recordings of students communicating their work.

Figure 4.6. The exercises page

3.4. The comments page

As noted above, the comments page (Figure 5) is provided for all users to comment on student uploads. All written and video student work; both the song material, discourse fiction, discourse nonfiction, customs, and Javanese script, commented on by application users through this page.

Figure 5. The comments page

4. Conclusion

Based on the above research it can be concluded that the standard operational procedures of the Damarjati application include: 1) opening the address through **damarjati.upgris.ac.id.;** 2) choose classes and semesters; 3) choose a theme and sub-theme with the menu *ayo nggatekake* 'observing', *ayo tetakon* 'questioning', *ayo naliti* 'associating', *ayo nyoba* 'experimenting', *ayo ngatonake* 'networking/communicating', and *gladhen* 'exercises'; and 4) users make corrections and comments. Through the application of the Damarjati application, it can achieve research objectives in developing HOTS-based Javanese language learning in high schools, which can be achieved through the Damarjati menu facility: *nggatekake*, *tetakon*, *naliti*, *nyoba*, *ngatonake*, and *gladhen*.

IOP Conf. Series: Materials Science and Engineering **835** (2020) 012047 doi:10.1088/1757-899X/835/1/012047

Acknowledgement
This research cannot be successful and published without constructive contributions from several parties. Our gratitude goes to the In the 7th International Conference on DV-Xα Method committee, through Mrs. Mega Novita for her willingness to give our team the opportunity to join this highly qualified international conference. Thanks also to the team members, Mrs. Ning, Mrs. Mimin, Tyas, and Yusuf who were always united in planning, implementing, and compiling the results of this research. Hopefully this small work is of great benefit for the development of Javanese language learning.

References
[1] Creswell, John W. 2015. *Penelitian Kualitatif & Desain Riset: Memilih Di antara Lima Pendekatan, Edisi ke-3.* Terjemahan *Qualitative Inquiry & Research Design: Choosing Among Five Approaches, Third Edition.* Yogyakarta: Pustaka Pelajar.
[2] Dinas Pendidikan Provinsi Jawa Tengah. 2015. *Kurikulum 2013 Muatan Lokal Bahasa Jawa SMA/SMALB/SMK/MA/MAK Provinsi Jawa Tengah.* Semarang.
[3] Direktorat Pembinaan SMA. 2017. *Implementasi Pengembangan Kecakapan Abbad 21 dalam Perencanaan Pelaksanaan Pembelajaran (RPP).* Jakarta: Dirjen Dikdasmen-Kemendikud.
[4] Idayoce.2014. *Model Pembelajaran Berbasis web (e-Learning).* www.idayoce.com. Diunduh 19 Agustus 2018.
[5] Ismawati, Esti. 2012. *Perencanaan Pengajaran Bahasa.* Yogyakarta: Ombak.
[6] Itasari, Yuliana. 2012. *Higher Order Thinking Skill (HOTS).* http://yuliannaitasari.blogspot.com/2012/12/higher-order-thinking-... Diunduh 22 Agustus 2018.
[7] Isodonk. 2016. *Pembelajaran dan Penilaian HOTS Kurikulum 2013.* https://isodonk.blogspot.com/2017/03/pembelajaran-dan-penilaian-... Diunduh 22 Agustus 2018.
[8] Mahsun. 2014. *Teks Dalam Pembelajaran Bahasa Indonesia Kurikulum 2013.* Jakarta: PT Rajagrafindo Persada.
[9] Moeliono, Anton M (ed.). 1990. *Kamus Besar Bahasa Indonesia, Cetakan ketiga.* Jakarta: Balai Pustaka.
[10] Nuh, Mohammad. 2013. *Pengembangan Kurikulum 2013.* Jakarta: Kementerian Pendidikan dan Kebudayaan. Dicetak oleh UPGRIS Press.
[11] Rianawati, Ida. 2011. *Pembelajaran Berbasis Web (Web Based Learning) atau E-Learning.* http://idarianawaty.blogspot.com/2011/02/ pembelajaran-berbasis-... Diunduh 19 Agustus 2018.

Design of Product Monitoring System Using Internet of Things Technology for Smart Manufacturing

Marti Widya Sari[1], Herianto[2], IGB Budi Dharma[3], A.E. Tontowi[4]

[1,2,3,4]Mechanical and Industrial Engineering Department, Universitas Gadjah Mada Yogyakarta, Indonesia
[1]Informatics Department, Universitas PGRI Yogyakarta, Yogyakarta, Indonesia

[1]marti.widya.sari@mail.ugm.ac.id

Abstract. The application system in the industrial era 4.0 is currently growing rapidly. One of them is the implementation of smart manufacturing in the industrial sector. In smart manufacturing, production processes can be monitored automatically and real time, starting from material selection, then the production process to the planned product. In this study discussed the design of a monitoring system on smart manufacturing based on internet of things technology. Smart technology is implemented on material scans automatically based on color sensors, then from the material it will also be known that with these materials produce certain products and require any material. In addition, the system will also automatically detect through the initial material to find out which processes will be passed. The methods used include literature studies, system requirements analysis, system design and system testing. The results of this study are in the form of a hardware system design and display of production process monitoring system applications. In the system that is made also a notification is displayed if a bottle neck occurs in the production process.

1. Introduction

Information and communication technology is highly needed for modern manufacturing nowadays. The Forth Industrial Revolution (Industry 4.0) integrates all current technologies such as wireless network, sensors, Internet of Things (IoT), control and robotics, artificial intelligence and information management system to create a Cyber Physical System (CPS) and smart factories [1],[2]. Industry 4.0 refers to a number of automated systems equipped with automatic data exchange and manufacturing technology capabilities [3],[4],[5]. These systems exert a transformative effect on economic and social processes [5]. Recently, Industry 4.0 concepts have attracted considerable attention in advanced manufacturing countries as a means of streamlining production. The widespread application of CPS in manufacturing environments renders manufacturing systems increasingly smart for more effective production process [6][7]. To advance research on the implementation of Industry 4.0, this study discusses smart manufacturing system for Industry 4.0 [8], [9]. The IoT is considered to be a modern manufacturing concept under Industry 4.0 and has adopted recent advances, such as cutting-edge information technology (IT) infrastructure for data acquisition and sharing, which greatly influence the performance of a manufacturing system [10]. The real-time data collection and sharing are based on key technologies such as radio frequency identification (RFID) and wireless communication standards [11][12]. By using RFID technology, physical manufacturing flows such as the movements of materials and associated information flows such as the visibility and traceability of various manufacturing operations can be seamlessly integrated [3], [13],[14]. Several real-life cases of IoT-enabled manufacturing have been reported [15]. In addition to IoT-based manufacturing, it needs to be supported by flexible manufacturing for a more effective and efficient production process. Flexible manufacturing allows for the addition of hardware, software changes and production equipment functions [16], [17].

2. Manufacturing

2.1 Smart Manufacturing

Figure 1 presents a framework of Industry 4.0 smart manufacturing systems, including smart design, smart machining, smart monitoring, smart control, smart scheduling, and industrial applications, which are the focus of this work. The vertical axis shows issues in another dimension of Industry 4.0 ranging from sensor and actuator deployment to data collection, data analysis, and decision making [6],[18]. In Industry 4.0, data gathering and analysis are the main sources of the smartness of activities shown on the horizontal axis.

2.2 Smart Monitoring and Control

Monitoring is an important aspect in the operation, maintenance, and optimal scheduling of Industry 4.0 manufacturing systems [19]. The widespread deployment of various sensors has made smart monitoring possible [20],[21]. For example, data on various manufacturing objects, such as temperature, electricity consumption, vibrations, and speed, can be obtained in real time. Smart monitoring provides not only a graphical visualization of these data but also alerts when abnormality occurs in machines or tools. Internet of things are key technologies that enable smart monitoring in Industry 4.0 smart manufacturing systems [2], [4].

Production behavior and procedures can then be carried out automatically and effectively because of the well established structures and services. With the aid of data input mechanisms, the output resolutions are fed back to the parties involved in different ways [5], [18].

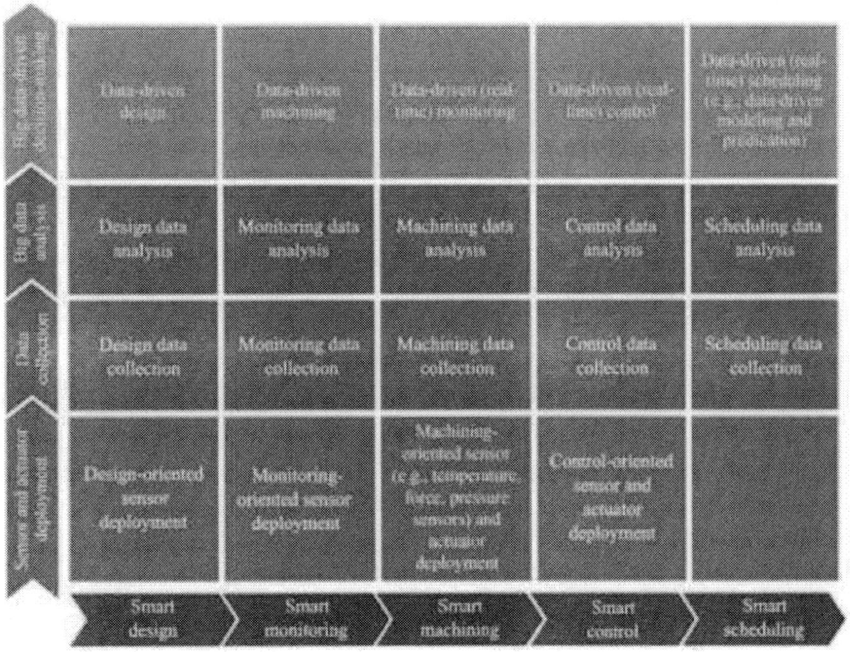

Fig 1. Conceptual framework of Industry 4.0 smart manufacturing systems [6]

3. Research Method

Based on some of the studies above, there is still not much research that discusses the production process monitoring system in the smart manufacturing based internet of things technology, especially those using material selection at the beginning of the process and flexible manufacturing. in this research discussed about the design of a production process monitoring system in real time that can display activities on the production floor starting from material detection, work in process needs of each material, detection of bottle neck up to the final product. The research methods used include the following: literature study, system requirements analysis for hardware and software needed, system design and system testing. Literature studies are conducted to get complete information and knowledge about the theme of the research that will be conducted.

3.1 System Analysis

Before designing a monitoring system, a system requirements analysis was carried out, both hardware and software requirements. Hardware requirements include colour sensors, PIR sensors, RFID and RFID tags, Arduino Mega, ethernet shields, routers and Raspberry Pi mini PCs. The hardware system scheme is shown in the Figure 2 below. Colour sensors are used to detect initial material based on material colours. Colour sensors are used because the product will be made using material of a certain color. RFID is used to store material data to make it easier to detect during the production process. Routers are used to connect to the internet, because the monitoring system is built on internet of things. Then the Raspberry Pi mini PC is used to control the system as a whole, including the smart material detection process. Furthermore, for software systems using PHP for display on the web and databases to support the overall system data availability. The monitoring system is web-based because it can be implemented on a multi-platform or compatible basis for all operating systems so that it is more flexible. Then for storing the database using MySQL.

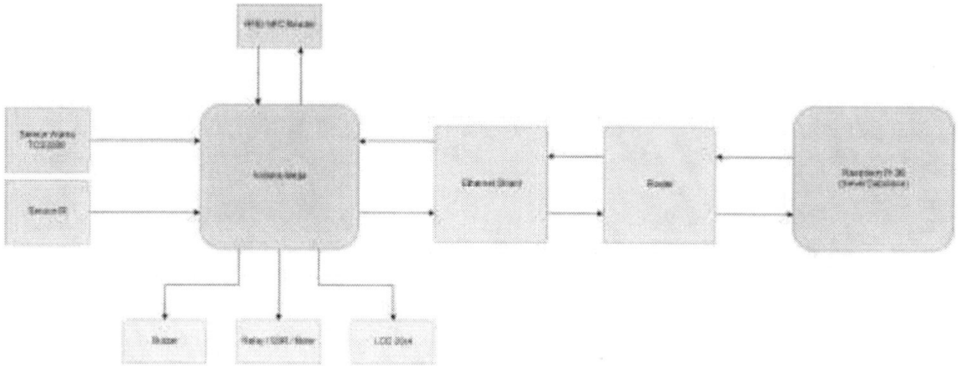

Fig. 2 Hardware system scheme

3.2 System Design

The design of a monitoring system is generally explained in the following Figure 3. In designing this monitoring system it starts with scanning the initial material through a sensing process. The sensor used in this system is a colour sensor, because the product to be made is related to the colour of the material. Then the material will pass Work in Process (WIP) 1 and 2 for the production process.

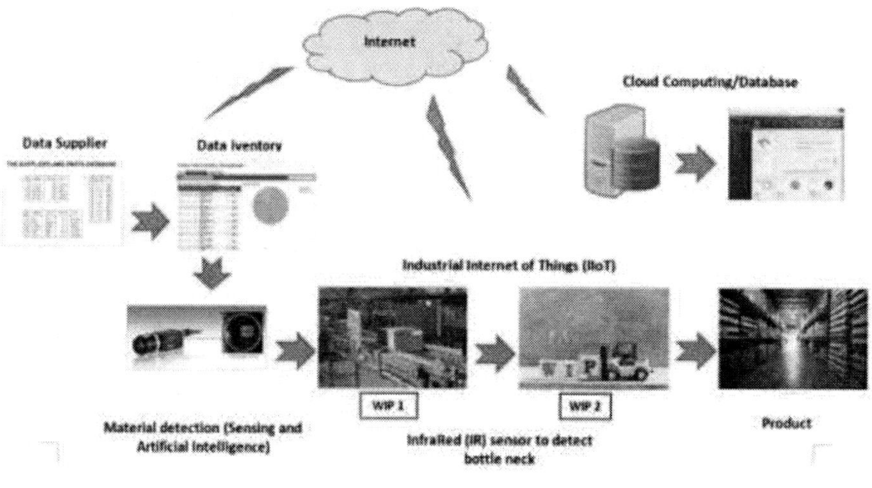

Fig. 3 Design of monitoring system

Between WIP 1 and WIP 2, several PIR sensors were installed to detect the bottle neck. A bottle neck can occur if there is a queue that exceeds the capacity between WIP 1 and WIP 2. The monitoring system also displays notification of bottle neck during the production process. Furthermore, the material that has been processed in WIP 1 and WIP 2 will enter the final stage, namely the product form. All production processes databases are stored in cloud computing based.

4. Results and Discussion

4.1 Results

The following picture is the result of designing a product monitoring system application on smart manufacturing. On the dashboard display of monitoring application, there is material data, in this case there are 3 materials to support the product to be made. The process monitoring shows data on material that is running and being processed. Then in this system there is also information on the final product that will be produced. The results of the design of the monitoring system are shown in Figure 4 below.

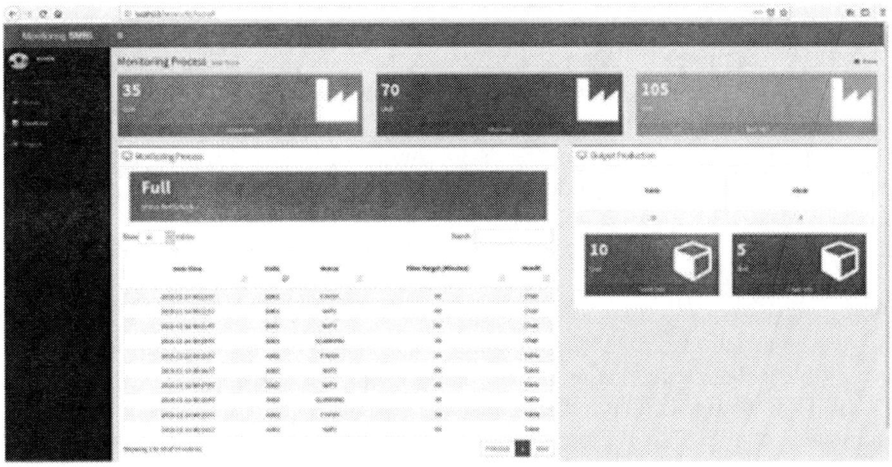

Fig 4. Display of monitoring system dashboard

4.2 Discussion

In the system view, Figure 4 contains some information that can be used as an analysis material for the results of monitoring system testing. On the system display dashboard, on the top side, there is any material information provided and how many, the numbers will change according to the production process that requires available material. In this system three types of material are provided to make a product. Furthermore, on the middle side there is a column to display the status and notification of bottle neck during the production process. Status bottle neck will appear if there is damage or there is a queue in the production process between WIP 1 and WIP 2. The monitoring system also shows the target time needed for a production process from the beginning of material selection to the final product. On the right side there is Output Production data, which shows the number and types of products that have been produced.

5. Conclusion

The conclusion from the results of this study is that the system designed can be used to monitor the production process in smart manufacturing, from the material selection process, the production process to the final product. This monitoring system also displays information on the production process, bottle neck and WIP requirements for each product. Material selection in this system uses color sensors to make a particular product based on material colors. As a propose for further research can use image processing for material selection at the beginning of the production process, so that the material detection process can more details. Then, measurements can also be made and analysis of time allocation for the overall production process.

References

[1] F. Zezulka, P. Marcon, and Z. Bradac, "Systems for Industry 4 . 0 and the IIoT Communication Systems for Industry 4 . 0 and the IIoT Communication Systems for IIoT Communication Systems for Industry Industry and the the Communication Systems for Industry the," *IFAC-PapersOnLine*, vol. 51, no. 6, pp. 150–155.

[2] X. Li, D. Li, J. Wan, and A. V Vasilakos, "A review of industrial wireless networks in the context of Industry 4.0," pp. 23–41, 2017.

[3] H. Lu and C. Weng, "Technological Forecasting & Social Change Smart manufacturing technology , market maturity analysis and technology roadmap in the computer and electronic product manufacturing industry," *Technol. Forecast. Soc. Chang.*, vol. 133, no. March, pp. 85–94, 2018.

[4] E. Mueller and X. C. Ralph, "Challenges and Requirements for the Application of Industry 4 . 0 : A Special Insight with the Usage of Cyber-Physical System," *Chinese J. Mech. Eng.*, vol. 30, no. 5, pp. 1050–1057, 2017.

[5] J. Astrid, S. Martínez, M. Pérez, J. Antonio, M. Saucedo, T. Eloy, S. Fierro, and P. Vasant, "Industry 4 . 0 framework for management and operations : a review," *J. Ambient Intell. Humaniz. Comput.*, vol. 9, no. 3, pp. 789–801, 2018.

[6] P. Zheng, H. Wang, Z. Sang, R. Y. Zhong, Y. Liu, C. Liu, and K. Mubarok, "Smart manufacturing systems for Industry 4 . 0 : Conceptual framework , scenarios , and future perspectives," vol. 13, no. 2, pp. 137–150, 2018.

[7] M. Y. Afanasev, Y. V Fedosov, A. A. Krylova, and S. A. Shorokhov, "Modular Industrial Equipment in Cyber-Physical Production System : Architecture and Integration," pp. 3–9, 1960.

[8] E. Ndip-agbor, J. Cao, and K. Ehmann, "Towards smart manufacturing process selection in Cyber-Physical Systems," *Manuf. Lett.*, vol. 17, pp. 1–5, 2018.

[9] K. Ian, I. K. Craig, E. Olivier, O. Ian, K. Ian, and K. Craig, "ScienceDirect Perspective for equipment automation Perspective for equipment automation Perspective for equipment automation Perspective for equipment automation process industries process industries process industries process industries in in in in," 2017.

[10] R. Y. Zhong, X. Xu, E. Klotz, and S. T. Newman, "Intelligent Manufacturing in the Context of Industry 4 . 0 : A Review," *Engineering*, vol. 3, no. 5, pp. 616–630, 2017.

[11] M. Gaynor and J. Waterman, "Design framework for sensors and RFID tags with healthcare applications," *Heal. Policy Technol.*, vol. 5, no. 4, pp. 357–369, 2016.

[12] R. Y. Zhong, Q. Y. Dai, T. Qu, G. J. Hu, and G. Q. Huang, "Robotics and Computer-Integrated Manufacturing RFID-enabled real-time manufacturing execution system for mass-customization production," vol. 29, pp. 283–292, 2013.

[13] S. S. Kamble, A. Gunasekaran, and S. A. Gawankar, "Sustainable Industry 4 . 0 framework : A systematic literature review identifying the current trends and future perspectives," *Process Saf. Environ. Prot.*, vol. 117, pp. 408–425, 2018.

[14] S. Lu, C. Xu, R. Y. Zhong, and L. Wang, "Computers & Industrial Engineering A passive RFID tag-based locating and navigating approach for automated guided vehicle," *Comput. Ind. Eng.*, no. xxxx, pp. 0–1, 2018.

[15] P. Stief, J. Dantan, A. Etienne, and A. Siadat, "ScienceDirect ScienceDirect Architecture and development of an Industrial Internet France of Things framework for realizing services in Industrial Product Service Systems A new methodology to analyze the functional and physical architecture of existing products for an a , assembly oriented product family identification," *Procedia CIRP*, vol. 72, pp. 880–885, 2018.

[16] M. Bortolini, F. G. Galizia, and C. Mora, "Reconfigurable manufacturing systems : Literature review and research trend," *J. Manuf. Syst.*, vol. 49, no. September, pp. 93–106, 2018.

[17] A. Santana, P. Afonso, A. Zanin, and R. Wernke, "Scie enceDir rect Scie enceDir rect ScienceDirect ScienceDirect Smar rt chang eable ma anufactu uring sys stems Smar rt chang eable ma anufactu uring sys stems Costing models for capacity optimization in Industry 4 . 0 : Trade-off between used capacity and operational efficiency," *Procedia Manuf.*, vol. 28, pp. 3–9, 2018.

[18] J. Ignacio, R. Molano, J. Manuel, C. Lovelle, C. Enrique, M. J. Javier, R. Granados, and R. González, "Metamodel for integration of Internet of Things , Social Networks , the Cloud and Industry 4 . 0," *J. Ambient Intell. Humaniz. Comput.*, vol. 9, no. 3, pp. 709–723, 2018.

[19] H. Boyes, B. Hallaq, J. Cunningham, and T. Watson, "The industrial internet of things (IIoT): An analysis framework," *Comput. Ind.*, vol. 101, pp. 1–12, Oct. 2018.
[20] A. Moktadir, S. Mithun, S. Kusi-sarpong, and A. Ali, "Assessing challenges for implementing Industry 4 . 0 : Implications for process safety and environmental protection," *Process Saf. Environ. Prot.*, vol. 117, pp. 730–741, 2018.
[21] M. W. Sari, P. W. Ciptadi, and R. H. Hardyanto, "Study of Smart Campus Development Using Internet of Things Technology," *IOP Conf. Ser. Mater. Sci. Eng.*, vol. 190, no. 1, p. 012032, 2017.

IOP Conf. Series: Materials Science and Engineering **835** (2020) 012049 doi:10.1088/1757-899X/835/1/012049

Designing Enterprise Architecture in Koperasi Karyawan using TOGAF Architecture Development

Noora Qotrun Nada[1], Setyoningsih Wibowo[2], Mega Novita[3]

[1,2,3] Faculty of Engineering and Informatics, Universitas PGRI Semarang
Jl. Sidodadi-Timur No.24 Semarang, Central Java 50232, Indonesia

noora@upgris.ac.id

Abstract. The Open Group shows that Enterprise Architecture (EA) can produce an IT blueprint that can be used to determine business, technology, and information to support the achievement of an organization's vision and mission. EA can also be seen as a model, a set of principles and methods used to produce the actual organizational structure of a company with business processes used including business operations, business planning, information technology, and information systems. EA is also in the form of models, graphics and narratives that will explain the scope of a company's business processes and design. Koperasi Karyawan TEKAD do not yet have information systems and operations are driven by manual system. The purpose of this study was to design an EA, specifically in Koperasi Karyawan using the TOGAF framework. In other words, this architecture was created by emphasizing the relationship between IT and business. The EA model is intended to maximize business alignment with IT in all situations, but depends on the nature of the business and the capacity in business and IT management in Koperasi Karyawan. This EA design produces a blueprint that can be used so that service quality and performance can be improved.

1. Introduction

The current condition is that Koperasi TEKAD has not used an information system to support its business. Based on interviews with administrators, information systems (IS) and information technology (IT) should have started to be built considering that their members are getting bigger. The administrators and managers also concurrently lecturers and university employees, often the affairs of Koperasi become unmanaged because it requires a long time. Information systems and information technology are expected to provide solutions to this problem. Koperasi as an organization can be seen as an enterprise, and to support the vision and mission of an organization requires an Enterprise Architecture (EA) plan.

Information systems and information technology not only provide increased efficiency and effectiveness of organizational performance, but also have become a major empowering organization to carry out business processes and achieve the business goals of the organization by: enabling new ways of managing the organization, increasing productivity and performance, developing new businesses and provide competitive advantage [1].

The Open Group suggests that EA can produce an IT blueprint that can be used to determine business, technology and information to support the achievement of the vision and mission of an organization [2]. EA can also be seen as a model, a set of principles and methods used to produce an enterprise organizational structure that is true to the business processes that are used include business operations, business planning, information technology and information systems. EA also takes the form of models, graphics and narration that will explain the scope of business processes and the design of an enterprise [3].

The purpose of this study is to design an EA, specifically in Koperasi Tekad using the TOGAF framework. In other words, this architecture was created by emphasizing the relationship between IT and business. The EA model is intended to maximize business alignment with IT in all situations, but depends on the nature of business and capacity in business as well as IT management in Koperasi.

2. Method

This research is a research and development (Research and Development) because this research aims to produce or develop a product in the form of a blueprint instead of research that finds theory. Because the output produced from this research is a product in the form of Blueprint Corporate Architecture in Koperasi TEKAD Semarang, the design uses the TOGAF ADM framework. The first stage as an introduction is to determine management requirements in full by the framework. At this stage, researchers conduct information searches and analyze needs by conducting interviews with Tekad Koperasi stakeholders. Stakeholders are manager and members of Koperasi. This aim is to determinate targets and desired steps with the new system.

The next step is to identify the problem. This stage is carried out by conducting interviews with stakeholders in this case the management and members of Koperasi Karyawan TEKAD. Then a meeting was held to determine the requirements. This is adjusted to the goals of the organization. Next, design the business architecture by showing the current architecture (baseline) and the proposed architecture. During this stage, a gap analysis is also carried out between the baseline and the target. Data architecture and applications are carried out at a later stage. And finally the architectural design of technology in this case in the form of network infrastructure and hardware design. The resulting blueprint is an amalgamation of business architecture design, data, applications, and technology.

3. Results and discussion

The Open Group Architecture Framework (TOGAF) is a framework and method that is widely accepted in the development of enterprise architecture. The Open Group Architecture Framework (TOGAF) provides a detailed method of how to create, run and implement enterprise architecture called the Architecture Development Method (ADM) [3].

ADM consists of several stages needed to build an EA, the ADM stages as shown in Figure 1, which is a flexible method that can search for various techniques in modeling that have been used in a design, because this method can follow developments as well during the design run.
The Preliminary Phase describes the preparation and initiation activities required to prepare to meet the business directive for a newenter prise architecture. Based on observations and analyses conducted on all activities, there are several problems experienced in Koperasi Karyawan

a. Savings and Loans takes a long process. The submission process goes through several stages and takes a long time. Deposit and billing information not yet available. Information and reporting is done once during the Annual Member Meeting.
b. Reporting on trades is still manual and offline system. Trade is still very dependent on shopkeepers.
c. Access to information to members is still limited. The goods and services provided by cooperatives are still not widely available to members. So far, it is still through the WA group.

d. Includes policies and procedures that apply. Policies, procedures, goods, and services have not been widely used by members.

The solution to be able to overcome the problems in Tekad Employee Cooperatives is to implement an Koperasi Information System consisting of Savings and Loans, information and financial reporting.

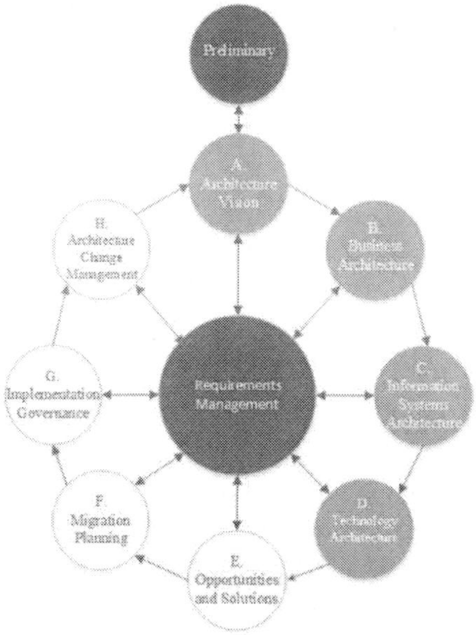

Figure 1. TOGAF ADM cycle

Architecture Vision describes the initial phase of an architecture development cycle. It includes information about defining the scope, identifying the stakeholders. Koperasi TEKAD is growing bigger. This is indicated by the increasing number of members and also the investment and the money in it. The purpose and objective of this cooperative is to work in the field of Savings and Loans, Trade and Services. To achieve these goals and objectives the cooperative carries out business activities including savings and loan, food or snack sales, graduation photo services and toga rental. Four types of general architecture are part of the EA, including Business Architecture, Data Architecture, Applications Architecture, and Technology Architecture. Information Systems architecture is a combination of data architecture and application [3]. Figure 1 illustrates the vision and real principles regarding the development of EA, the principle is used as a benchmark to assess the success of EA development in an organization.

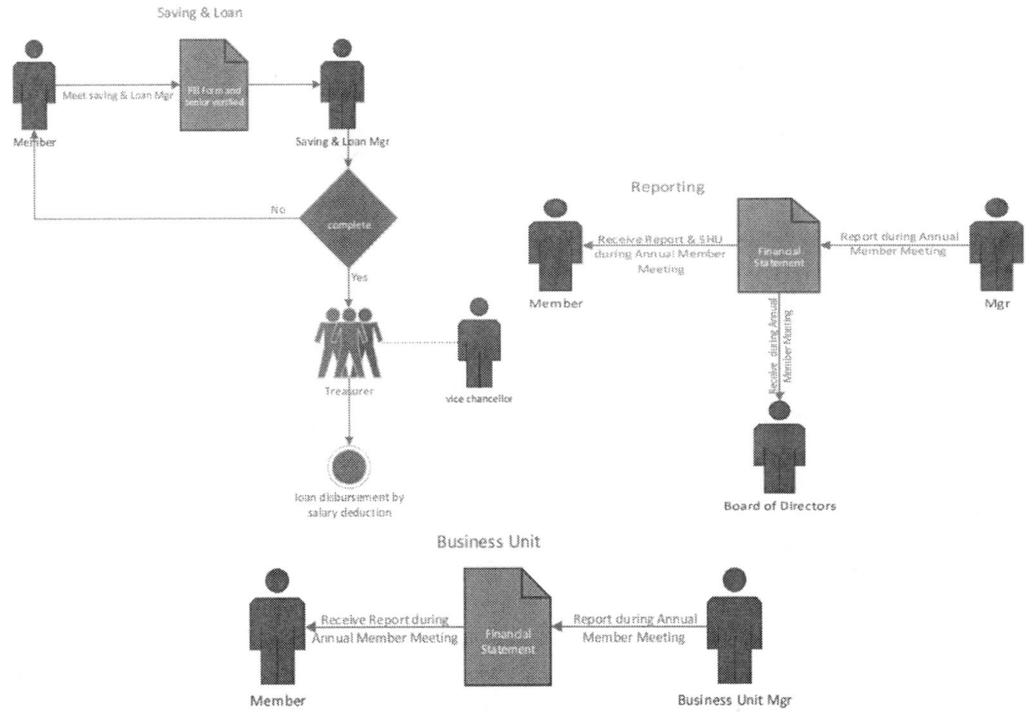

Figure 2. Baseline Business Process Koperasi TEKAD

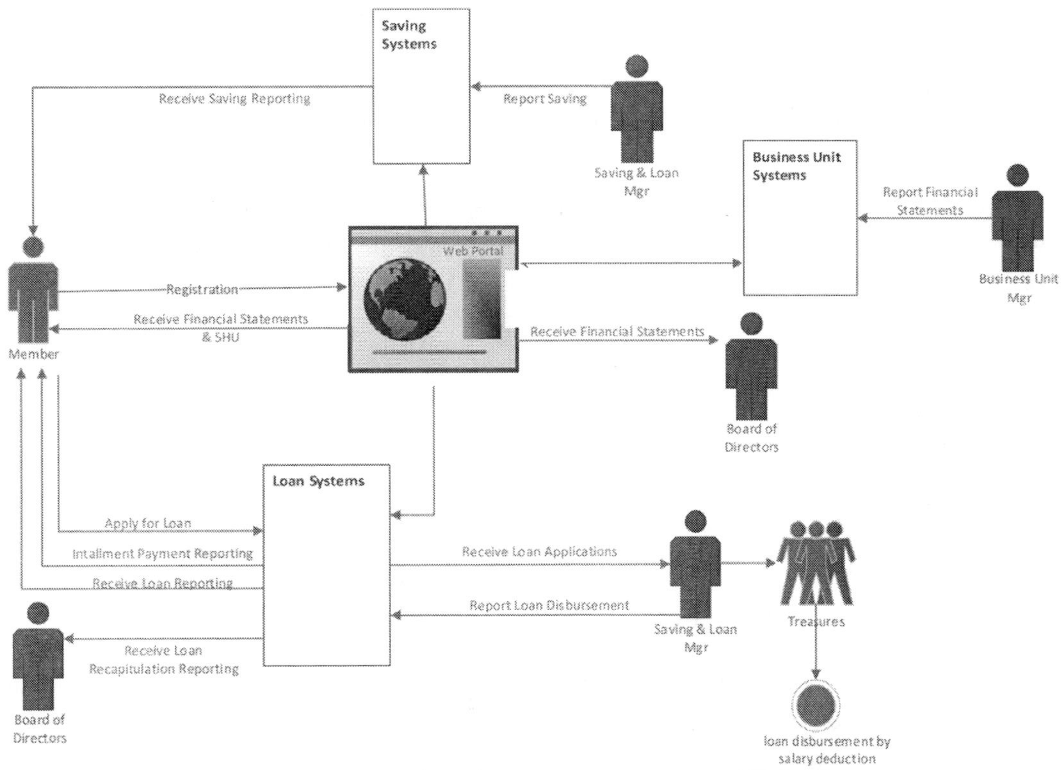

Figure 3. Proposed Business Processes Koperasi TEKAD

TOGAF also has four types of architecture that can support the overall EA as follows:
1. Business Architecture
Business Architecture describes the development of a Business Architecture to support an agreed Architecture Vision. Modeling business processes in business architecture has a purpose, to provide a clear picture of the current state of the organization. The form of business process analysis in cooperatives for operational functions is illustrated by two things, the baseline or current conditions in Figure 2 and the proposed business processes in Figure 3.
2. Data Architecture
This architecture describes the structure of data assets as well as organizational data management resources logically and physically. The data architecture component provides a structure for detailed documentation of information for organizations. Currently, Koperasi Karyawan TEKAD does not have a data architecture yet, so this stage will be proposed to design a data architecture so that it can be applied to support the business processes. Part of data architecture in this definition is made by identifying prospective class candidates.
3. Application Architecture
An architecture that provides blueprints for use on individual application systems that are applied. Application architecture is the action that decides the development and application of solutions or services that are being made for the organization. The design for application architecture is a combination of processes that use parts of the overall application and business models, information and technology that can design an expected business application.
4. Technology Architecture
The architecture needed to help spread business, data and application services to develop software and hardware capabilities. The technology architecture can include IT infrastructure, middleware, networks, communications, processes, and standards. Technology architecture is an approach to explain the structure and relationship of technology companies now and in the future to make the value of technology. Implementation of Governance phase and Change Management phase in this research are not discussed, will be continued in another research.

4. Conclusion
Based on the discussion, it can be concluded that Koperasi Karyawan TEKAD currently does not have an information system to manage Koperasi operations. So in this study, the authors designed an enterprise architecture so that the systems designed can be integrated, and the design of the enterprise architecture uses the TOGAF ADM framework that starts from Architecture Vision phase to Architecture Technology phase which produces blueprint designs to be able to assist every activity in Koperasi Karyawan TEKAD.

Acknowledgments
Support from Lembaga Penelitian dan Pengabdian Kepada Masyarakat (LPPM) Universitas PGRI Semarang is gratefully acknowledged.

References
[1] M. G. Kasenda, E. Nugroho, S. Sulistyo, M. T. Informasi, F. Teknik, and J. Bulaksumur, "Sektor Publik Menggunakan Kerangka the Open Group Architecture Framework (Togaf)," *Semin. Nas. Sist. Inf. Indones. 22 Sept. 2014*, no. September, pp. 51–58, 2014.
[2] T. O. Group, TOGAF Version 9, San Fransisco: The Open Group, 2009.
[3] Kridanto Surendro, "Pengembangan Rencana Induk Sistem Informasi, Informatika, Bandung.," p. 2009, 2009.
[4] R. Yuliana and B. Rahardjo, "Designing an agile enterprise architecture for mining company by using TOGAF framework," *Proc. 2016 4th Int. Conf. Cyber IT Serv. Manag. CITSM 2016*, 2016.
[5] Mukhfid. (2017). Perencanaan Arsitektur Sistem Informasi Enterprise Koperasi dengan TOGAF ADM Framework. *Vol 8 No 1 (2017): Gema Wiralodra.*

[6] A. A. Arman, G. Ruhman, and R. Hurriyati, "Information Systems Architecture and Technology: Proceedings of 39th International Conference on Information Systems Architecture and Technology – ISAT 2018," vol. 852, 2019.
[7] R. Alm and M. Wißotzki, "TOGAF adaption for small and medium enterprises," *Lect. Notes Bus. Inf. Process.*, vol. 160, pp. 112–123, 2013.
[8] L. Urbaczewski and S. Mrdalj, "A Comparison of Enterprise Architecture Frameworks," *Issues Inf. Syst.*, vol. 7, no. 2, pp. 18–23, 2006.

IOP Conf. Series: Materials Science and Engineering **835** (2020) 012050 doi:10.1088/1757-899X/835/1/012050

Geometry and Stability of Small Gold Cluster Ions by Graph Theory and Hückel model

Rika Sekine, Takashi Kasai, Eri Furuta, and Masaya Okakura

Department of Chemistry, Faculty of Science, 836 Ohya, Suruga-ku, Shizuoka, 422-8529 Japan

sekine.rika@shizuoka.ac.jp

Abstract. Structure and stability of the gold cluster ions of which skeleton are synthesized as a complex were analyzed using the Hückel method based on graph theory. Hückel Energy (HE) and Topological Resonance Energy (TRE) were determined for neutral, positive ion, and negative ion clusters, where all the isomers of the gold cluster up to octamer were considered. Since some graphically designed isomers include bonds that cannot be realized in three-dimensional space, the screening was carried out by a molecular force field calculation with LAMMPS (lammps.sandia.gov/.). Among the isomers thus obtained, both HE and TRE were most stable when the tetramer was Au_4^{2+} with a tetrahedral structure, and with the hexamer, Au_6^{2+} with two tetrahedrons sharing one side. The complexes with these structures have actually been synthesized. On the other hand, the synthesis example of the most stable cluster for octamer Au_8 has not been reported.

1. Introduction

The purpose of this study is to clarify the geometry characteristics and stability of gold clusters using graph theory and Hückel model. The reason why the Hückel method was chosen is that it is easy to perform calculations. Table 1 shows the number of isomers of clusters of several sizes [1]. The number of isomers increases rapidly as the number of atoms increases, and the number of isomers exceeds 10,000 for octamers. There are too many isomers to obtain precise result.

Table 1. Number of connected simple graphs (graphical isomers) for M_n, n=4-9.

n	isomers	n	isomers
4	6	7	853
5	21	8	11,117
6	112	9	261,080

Although the calculation is relatively simple, the Hückel method is not considered bad for Na, Cu, and Au clusters that have one 1s electron in the outermost shell. Figure 1 shows the most stable structures of M_3 – M_8 clusters where M=Na [2], Cu [3] and Au[4] obtained by *ab-initio* or DFT calculation. The stuructures obtained by the Hückel method are the same as those for Na clusters, except for octamer. For Cu and Au clusters, the structures are almost the same with a few exceptions. From this comparison, it was considered a good approximation to discuss cluster stability using the Hückel method.

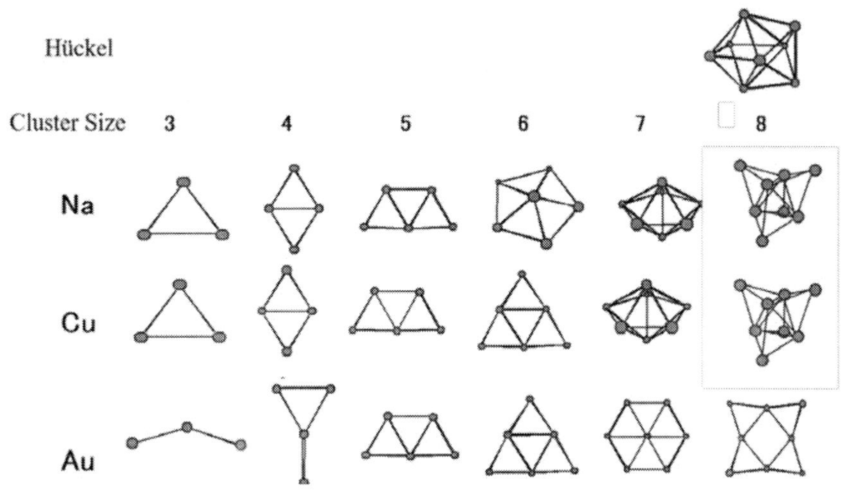

Figure 1. Most stable structures for M_3-M_8 clusters for Na [2], Cu [3], and Au [4]. The most stable octamers obtained by Hückel method [1] is also shown.

Various gold complexes have been synthesized so far [5-9]. The variety of isomers is abundant and ideal for comparison. This is the reason why we chose gold clusters in order to elucidate determining factor of the stability of metal cluster isomers.

2. Computational Methods

2.1. *Adjacent Matrices*

First, we create a list of all isomeric graphs by creating code in Mathematica or C ++ [10,11]. The graph is represented by an adjacency matrix. Once we get adjacency matrix, Hückel Energy (HE) is automatically obtained by calculating the eigenvalues of the adjacency matrix. In this study, Topological Resonance Energy (TRE), that is a kind of aromatic stabilization energy, was also calculated. The procedure for obtaining TRE is quite complex, but it can be also obtained from the adjacency matrix. For a detailed description of TRE, see the review [12].

2.2. *Elimination of Impossible Structures*

Next, we will examine whether the stability of the complex actually synthesized can be explained from the values of HE and TRE. It is necessary to remove impossible isomers before examination. Some graphically designed isomers contain bonds that are not possible in 3D space. For example, it is impossible for K5 complete graph, where every atom is bonded to every other atom, to set all bond lengths are equal. We screened the isomers by molecular mechanics (MM) calculation with LAMMPS [13]. The screening procedure for removing impossible isomers is as follows. The atom is placed at the position corresponding to the given number. Bonds are created based on the adjacency matrix. Then the MM calculation is carried out, where

1) the bond potential is expressed using the Morse function,
2) set the angle function to be most stable at 180 degrees with harmonic oscillator approximation,
3) a repulsive force works between atoms which are not bonded.

The 3D-structure obtained in this way is defined as the most stable one corresponding to a given adjacency matrix (graph-structure). For the 3D-structures, we examined the distances of all atom pairs [14]. The judgement based on the distance between atoms is carried out as follows. If the bond atom pair distance is 0.9 r_0 to 1.1 r_0, where r_0 is the equilibrium internuclear distance, and if the distance between unbonded atom pairs is more than 1.1 r_0 simultaneously, this structure is permitted. A slightly relaxed condition, 0.85 r_0 and 1.15 r_0, is also used. After this screening, one pentamer, 13 hexamer, 207 heptamer and 5,079 octamer isomers are excluded, where the criteria are 0.9 r_0 and 1.1r_0. Figure2 shows the screening results for octamer. Isomers that may exist in more severe conditions are shown in green. Clusters that cannot exist are shown in red. Isomers that can exist under relaxed conditions are shown in blue. The vertical axis is the value obtained by MM calculation of LAMMPS, and it shows that the stability increases as the value decreases. We can see that no octamer isomer with more than 20 bonds can exist.

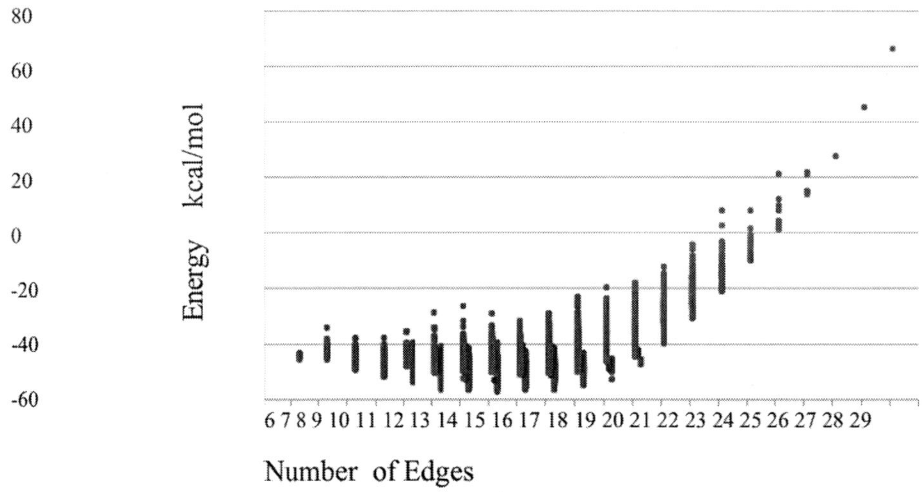

Figure 2. Possible and impossible octamer isomers screened by LAMMPS. The bond atom pair distance r is compared with the equilibrium internuclear distance r_0.

Green: possible, 0.9 $r_0 < r <$ 1.1 r_0. Blue: maybe, 0.85 $r_0 < r <$ 0.9 r_0, 1.1 $r_0 < r <$ 1.15 r_0. Red: impossible, $r <$ 0.85 r_0, 1.15 $r_0 < r$. The distance between unbonded atom pair was also examined, see the text.

3. Results and discussion

3.1. *HE and TRE*

We calculated not only neutral but also positive and negative ion clusters and compared HE and TRE values. Figures 3 and 4 show the mean values of HE and TRE, respectively, as a function of number of bonds in different charges for octamers. As shown in Fig. 3, the HE increases monotonously only when the charge is +6, and the others show the shape of a bowl rising to the right. Also, the smaller the charge, the more slowly it rises. It can be seen that the reversal of the HE occurs due to the increase in the number of bonds, and the charges +2, +4 are larger than the neutral and charge -2. Such a tendency was also seen in the tetramer, pentamer, hexamer and heptamer. As an overall trend, it can be said that HE increases as the charge becomes more positive and the number of bonds increases.

As shown in Fig. 4, the mean value of TRE takes a positive when the charge is +6, +4 or +2, and takes negative when the charge is neutral or negative. When examining each isomer, most of the TREs with positive values were +4 and +2.

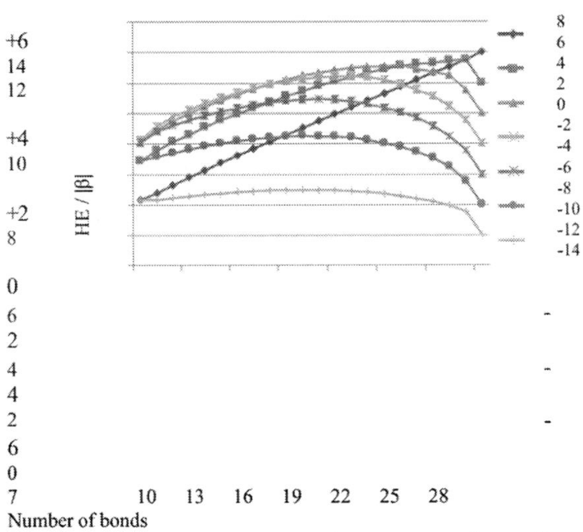

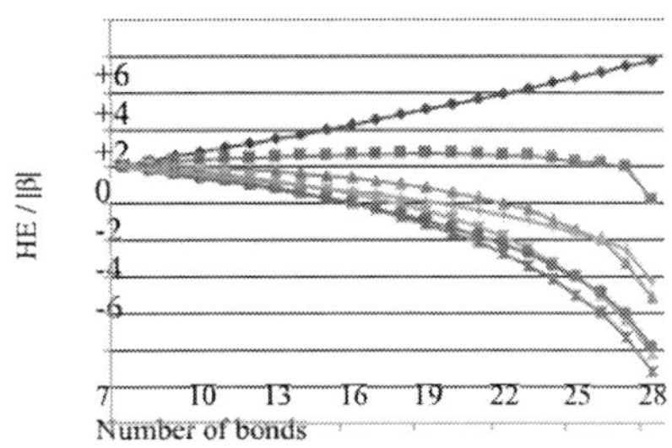

Figure 3. The mean HE values vs. number of bonds for octamer isomers with +6, +4, +2 (positive charge), 0 (neutral), and -2, -4, -6 (negative charge). (left)

Figure 4. The mean TRE values vs. number of bonds for octamer isomers with +6, +4, +2 (positive charge), 0 (neutral), and -2, -4, -6 (negative charge). (Right)

3.2. Comparison with synthesized clusters

Table 2 summarizes the HE and TRE values of isomers that have the same structure and charge as the synthesized complexes. Table 2 also shows how large HE and TRE are among the same charge isomers. When ranking, we excluded structures that could not be realized in 3D. Figure 5 shows skeletal structure of synthesized gold cluster. In Fig. 5, the most stable octamer isomers are also shown.

Table 2. HE and TRE values for the synthesized Au complexes. The alphabets and the numbers in the "complex" column correspond to that in Fig. 5 and the references, respectively.

luster Size	bonds	charge	complex	HE/ |β|	Rank -ing	TRE/ |β|	Rank -ing
4	6	+2	$\{[^tBu]_3PAu]_4\}^{2+}$ (a) [5]	6.00	1	1.33	1
6	11	+2	$[Au_6(PPH_3)_6]^{2+}$ (b) [6]	9.66	1	0.76	7
6	10	+2	$[Au_6(dppp)_4]^{2+}$ (c) [7]	9.12	6	0.66	10
7	16	+1	$[Au_7((PPH_3)_7]^{+}$ (d) [7]	11.87	1	-0.79	489
7	13	+3	$[Au_7(dppp)_4]^{3+}$ (e) [8]	11.13	7	1.53	8
8	12	+2	$[Au_8(PMes_3)_6]^{2+}$ (f) [7]	12.36	790	1.29	2
8	16	+2	$[Au_8((PPH_3)_7]^{2+}$ (g) [7]	13.60	21	0.70	147
8	16	+2	$[Au_8(dppp)_4]^{2+}$ (h) [7]	12.47	646	-0.47	4441
8	16	+2	$[Au_8(PPh_3)_8]^{2+}$ (i) [9]	12.52	580	-0.36	3970
8	19	+2	Not synthesized (j)	13.87	1	-0.19	3047
8	13	+2	Not synthesized (k)	13.09	142	1.31	1
8	15	+4	$[Au_8(dppp)_4Cl_2]^{2+}$ (l) [7]	12.17	88	2.00	45

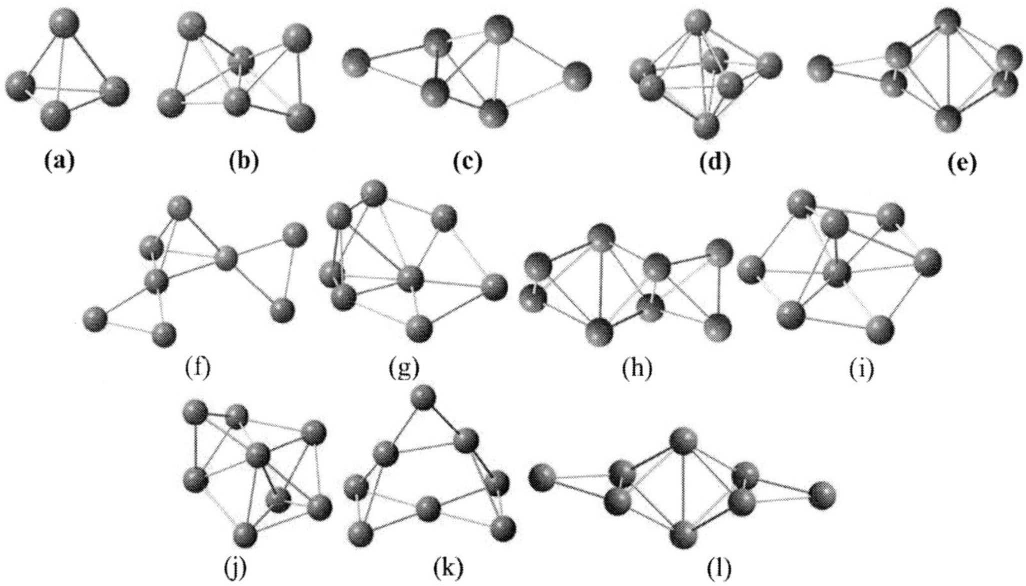

Figure 5. Skeletal structure of synthesized gold cluster. (a) tetramer. (b) and (c) hexamer. (d) and (e) heptamer. (f) – (i) octamer (+2). The most stable octamer (+2) isomer (j) in HE (k) in TRE. (l) octamer (+4). See also Table2.

The most stable tetramer is shown in Fig.5 (a), which is tetrahedron, +2 ion isomer. Here both HE and TRE are the largest among the tetramer isomers. A complex with this isomer as a skeleton was synthesized in 1993 by Zeller *et al.* [5]. No other synthesis examples of gold tetramer were found.

Two types of Au complexes with hexamer skeletons are synthesized, shown in Fig.5 (b) and (c). The structure (b) consists of two tetrahedrons with sharing one edge, +2 ion clusters. Here HE is the largest and TRE is the 7th in all the hexamers. This complex was synthesized by Briant *et al.* in 1993 [6]. The complex (c) is recently synthesized [7]. Both HE (the 6th) and TRE (the 10th) shows that this isomer is relatively stable. As shown in Fig. 5 (d) and (e), two types again of Au complexes with a heptamer skeleton are synthesized [7,8]. As to isomer (d), HE is the largest. The isomer (e) has also large HE value. Figure 5 (f) – (i) shows synthesized complexes with octamer skelton, which have all +2 charge [7,9]. The TRE for isomer (f) and the HE for isomer (g) are relatively large but that is not the case for isomer (h) and (i). Furthermore, the complex with the most stable octamer isomer in HE (j) or that in TRE (k) has not been synthesized yet. Tsukuda states based on the shell model that the Au^{2+} ion is a six-electron system and should be stable when it has a flat structure [15]. This may be the reason why isomers (h) and (i) are synthesized.

There is only one example of complex where the charge for the octamer skeleton is +4, as shown in (l). The HE and TRE values for this isomer is not small. It suggests that other complexes will be synthesized from the stability evaluated by the HE and TRE values.

4. Conclusion

For tetramer, hexamer and heptamer clusters, the gold complexes of which skelton is the most stable isomer , based on HE, has been synthesized. The skeleton isomers of the synthesized complexes are also stable from the TRE perspective.

However, for octamers, there has been no report that the most stable complex in HE with the skeletal isomer has been synthesized. Many kinds of complexes with octamer skelton have been synthesized, and their HE and TRE are not necessarily large. Other factors such as shell structure may affect the ease of synthesis of the complex.

References

[1] Y. Wang, T. George, D. M. Lindsay, and A. C. Beri, "The Hückel model for small metal clusters. II. Orbital energies, shell structures, ionization potentials, and extrapolation to the bulk limit", *J. Chem Phys.*, vol. 86, no.6, pp.3493-3499, 1987.

[2] I. A. Solov'yov, A. V. Solov'yov, and Walter Greineret, "Structure and properties of small sodium clusters," *Phys. Rev. A - At. Mol. Opt. Phys.*, vol. 65, no. 5, pp. 532031–5320319, 2002.

[3] D. Die1, B-X Zheng, L.-Q Zhao1, Q-W Zhu1 Z.-Q. Zhao, "Insights into the structural, electronic and magnetic properties of V-doped copper clusters: comparison with pure copper clusters," *Sci. Rep.*, vol. 6, no. 1, p. 31978, Aug. 2016. DOI: 10.1038/srep31978

[4] L. Rincon, A. Hasmy, M. Marquez, C. Gonzalez, J. Wu, S. Pan, X. Zhu, Z. Cai, P. Zhang, and C. Zhang, "A perturbatively corrected tight-binding method with hybridization: Application to gold nanoparticles", *Chem., Phys. Lett.*, vol. 503, no. 1–3, pp. 171-175, 2011, doi: 10.1016/j.cplett.2010.12.075

[5] E. Zeller, H. Beruda, and H. Schmidbaur, "Tetrahedral gold cluster $[Au_4]^{2+}$: crystal structure of $\{[(tert\text{-}Bu)_3PAu]_4\}^{2+}(BF_4^-)_2 \cdot 2CHCl_3$, *Inorg. Chem.*, vol. 32, no. 15, pp. 3203-3204, 1993, doi: 10.1021/ic00067a002.

[6] C. E. Briant, K. P. Hall, D. M. P. Mingos and A. C. Wheeler, "Synthesis and structural characterisation of hexakis(triphenyl phosphine)hexagold(2+) nitrate, $[Au_6(PPh_3)_6][NO_3]_2$, and related clusters with edgesharing bitetrahedral geometries", *J. Chem. Soc., Dalton Trans.*, No. 3, pp. 687-692, 1986, doi: 10.1039/DT9860000687.

[7] K. Konishi, "Phosphine-coordinated pure-gold clusters: Diverse geometrical structures and unique optical properties/responses," *Structure and Bonding*, vol. 161. Springer Verlag, pp. 49–86, 10-Apr- 2014.

[8] Y. Shichibu, M. Zhang, Y. Kamei, and K. Konishi, "$[Au_7]^{3+}$: A missing link in the four-electron gold cluster family," *J. Am. Chem. Soc.*, vol. 136, no. 37, pp. 12892–12895, 2014. doi: 10.1021/ja508005x.

[9] B. S. Gutrath, U. Englert, Y. Wang, U. Simon, "Molecular and Electronic Structure of the Cluster $[Au_8(PPh_3)_8](NO_3)_2$," *Eur. J. Inorg. Chem.*, vol. 2016, no. 7, pp. 975–981, Mar. 2016.. doi: 10.1002/ejic.201300148.

[10] R. Sekine, A. Murono, M. Okakura, Y. Kobayashi, Y. Nakagami, "Analysis for the stability of all sodium isomers by graph theory and molecular mechanics", *J. Comput。Chem. Jpn.*, vol. 17, no. 3, pp. 117-119, 2018, doi: 10.2477/jccj.2018-0026. (in japanese)

[11] R. Sekine, A. Murono, Y. Kobayashim and Y. Nakagami, "Graph theoretical study on metal clusters (4)", *Bull. Soc. DV-X*□, vol. 31, pp.103-106, 2018. (in japanese)

[12] J. I. Aihara, "Graph theory of aromatic stabilization," *Bulletin of the Chemical Society of Japan*, vol. 89, no. 12., pp. 1425–1454, 2016

[13] S. Plimpton, A. Thompson, S. Moore, A. Kohlmeyer, and R. Berger, "LAMMPS Molecular Dynamics Simulator", https://lammps.sandia.gov/.

[14] R. Sekine, M. Okakura, and Y. Nakagami, "Screening of achievable Na clusters by molecular mechanics method", *Bull. Soc. DV-X*□, vol. 31, pp. 107-110, 2018. (in japanese)

[15] T. Tsukuda, "Structures of chemically modified superatoms toward creation of artificial atoms and quasi-molecules in nanoscale", *Mol. Sci.*, vol. 13, A0108, 2019. doi: 10.3175/molsci.13.A0108. (in japanese).

IOP Conf. Series: Materials Science and Engineering **835** (2020) 012051 doi:10.1088/1757-899X/835/1/012051

Extraction Sentiment Analysis Using naive Bayes Algorithm and Reducing Noise Word applied in Indonesian Language

Aris Tri Jaka Harjanta[1], Bambang Agus Herlambang[2]

[1,2] Faculty of Engineering and Informatics, Universitas PGRI Semarang, Jl. Sidodadi-Timur No.24 Semarang, Central Java 50232, Indonesia

aristrijaka@upgris.ac.id

Abstract. Sentiment Analysis is now very important and very useful in machine learning technology where a contextual mining of text to identify and extract subjective information in the source, and in helping to understand social sentiment from comments In general, sentiment analysis can be classified into three broad categories namely sentiment positive and negative. One method of machine learning is the Deep Belief Network (DBN). DBN which is included in the Deep Learning method, is by stacking several algorithms with several extraction features that utilize all resources optimally. This research has two points. First, it aims to classify positive, negative, and neutral sentiments for the test data. The following experiments provide a system of sentiment analysis through the naive Bayes algorithm to calculate sentiment and to improve accuracy by reducing noise in words applied in Indonesian language. From this research, a good level of accuracy can be obtained for extending sentiment using 10-Cross Validation resulting in an accuracy rate of 78.33% with an increase of 6.66% for systems that do not use stopwords, which means reducting noise words and the use of the naïve bayes classification method can be used to determine analysis sentiment on Indonesian language text.

1. Introduction

Nowadays digital data is getting bigger and bigger, and with the current data we are able to determine several things related to the acceleration of decision making, both in the field of trade, government, health and education. According to the internetwordstats survey 4,241,972,790 est. population for Asia in 2019 - Area: 39,365,000 sq km 2,275,469,859 Internet users and 53.6% penetration rate as of July 31/19

Country	**INDONESIA**
population	269,536,482
Area	1,904,443 sq km
Capital City	Jakarta - 9,751,937 Pop.
GNI (per capita)	$3,540 (2017)
Internet users in June/2019	171,260,000
penetration	63.5%

Based on the data above, the number of internet users is very large, and one of the interactions between internet users is in the form of comments written in a post, both written and video. Comments written in a post are representations of users who interact with the owner to show sentiment. So that an extraction from the post in the form of tadai text is very much needed considering the amount of data or the number of comments and limitations on the ability to read humans, this requires a machine learning system capable of extracting sentiments in the comment text in Indonesian language.

2. Research Methods

2.1. Process Documents

Tokenization of words(1), We use Tokenization of words to convert sentences into words, the output of this tokenization is words that can be converted into data frames that make it easier for a learning machine to process understanding of the meaning of a word. Tokenization of words is a crucial step in turning words into numerical data for the next process.Transform case(2) the use of transformcase is to change all attributes to be the same and the same type so that it will be very easy when doing matching in a text search without constraints of case type differences either uppercase or lowercase. In this case, we change the case to the lowercase for all processed text.Filter Stopwords(3) to delete words that are not needed or unwanted in this project is to filter words using the stopwords library filter, where with a library we can add and subtract words that are not needed in a way that is more flexible and can be placed in a database separate from the system.

2.2. Cross Validation

Cross Validation is a nested operator. Here the operator has two sub processes: Training subprocess and Testing subprocess. The training subprocesses are used to train the system with data that has been labeled. The trained model is then applied to the Testing subprocess. Model performance is measured during the ExampleSet Input Testing phase partitioned into a set of parts k of the same size. From the subset, one subset is retained as a test data set (ie input from the Testing subprocess). Evaluation of the performance of the model on an independent test set results in a good estimate of performance on an invisible data set. This also shows if 'overfitting' occurs. This means that the model represents test data very well, but does not generalize very well to new data.

2.3. Training with Naïve Bayes

In general, the Bayes theorem is stated as follows:

$$P(A|B) = \frac{P(B|A)P(A)}{P(B)} \quad (1)$$

In this notation P (A | B) means the probability of event A if B occurs and P (B | A) chance of event B if A occurs.

2.4. Testing

Cross-validation is one of the statistical methods used to evaluate the performance of models or algorithms where the data is separated into two subsets namely learning process data and validation data. Models or algorithms are trained by a subset of learning and validated by a validation subset. Furthermore the selection of CV types can be based on the size of the dataset. Usually the K-fold CV is used because it can reduce computing time while maintaining the accuracy of the estimate.

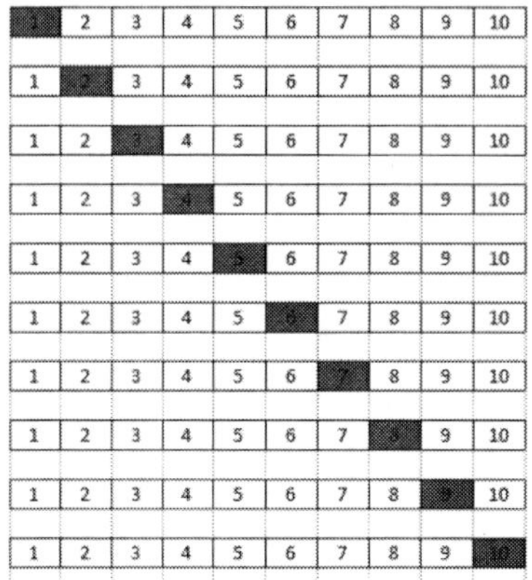

Figure 1. Figure 10-Cross Validation Schema.

The 10-fold CV is one of the recommended K-fold CVs for the selection of the best model because it tends to provide an estimate of accuracy that is less biased compared to the usual CV, leave-one-out CV and bootstrap. In 10 fold CVs, the data is divided into 10 folds of approximately the same size, so we have 10 data subsets to evaluate the performance of the model or algorithm. For each of the 10 data subsets

3. Result

3.1. Read the document

Read the document from comma separated value text to get all documents and then set the role operator functioned in arranging parts of the document that can read and group data into columns, so that it can be identified which columns will be processed as what, for example columns id is an integer which means that the column only contains the index of the data, then the column label contains the binominal type data which is the label of each data record, and the text column contains sentiment data in the form of polynominal type text clusters, so all roles must be determined in accordance with its function correctly in order to obtain the right data.

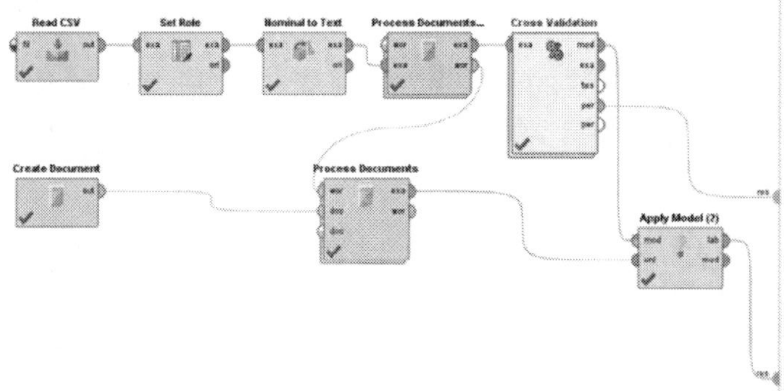

Figure 2. Design process

Each process is connected and can be seen in the picture above, all processes are interconnected with one another in terms of the output of an operator being input for the next operator, and testing of sample documents that are entered and interacted into the system that has been trained to be in check into the operating system.

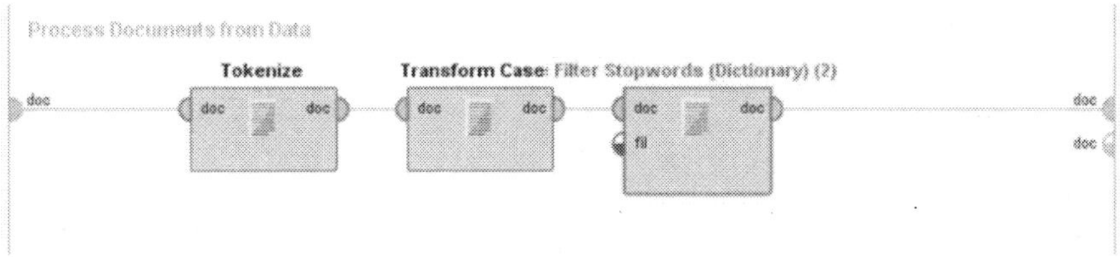

Figure 3. Preprocessing text

In Figure 3 is a preprocessing text system where the incoming text is carried out in the initial management by applying the Tokenization operation and then transformed to the lower case and then by filtering the stopword dictionary operator by reducing unnecessary words. The results of preprocessing this text will be input for further processing

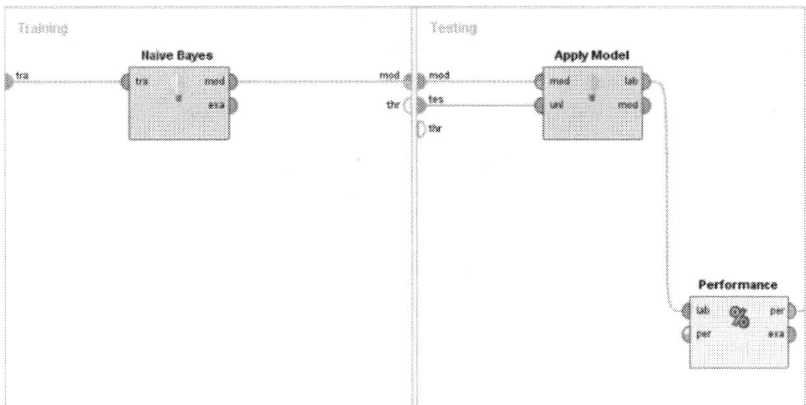

Figure 4. Training and Testing Process

The application of Naïve Bayes algorithm when training data is to be able to provide data to determine the type of sentiment in accordance with the results of the training, and testing by applying a model that has been designed and applied from the start. By measuring performace to get accuracy through 10-Cross validation schemas as a result. The results obtained from this system are as can be seen in the table confusion matrix below

Table 2. Confusion matrix without stopword dictionary

	True negative	True positive	Class precision
Pred. negative	24	11	68.57%
Pred. positive	6	19	76.00%
Class recall	80.00%	63.33	

Based Confusion Matrix on table 2 it can be concluded while the resulting vector performave is accuracy: accuracy: 71.67% +/- 19.33% (micro average: 71.67%)

Table 3. Confusion matrix with stopword dictionary

	True negative	True positive	Class precision
Pred. negative	25	8	75.75%
Pred. positive	5	22	81.48%
Class recall	83.33%	73.33	

Based on table 3 it can be concluded while the resulting vector performave is accuracy: 78.33% +/- 20.86% (micro average: 78.33%).

4. Conclusion

Based on sentiment testing experiments in Indonesian using the naïve Bayes algorithm, it can be seen in the confusion matrix table above that there are two results, with soppword and without stopword, from the result without stopword yields an accuracy of 71.67% while with stopword dictionary produces an accuracy of 78.33% in this increased by 6.66% which means a significant increase. Technically it can be concluded that word reduction can improve accuracy in sentiment extraction in an Indonesian language text.

Reference lists

[1] M. Barmer, Principles of data mining. 2007.
[2] J. Jotheeswaran, R. Loganathan, and M. S. B, "Feature Reduction using Principal Component Analysis for Opinion Mining," vol. 3, no. 5, pp. 118–121, 2012.
[3] C. Rohrdantz, M. C. Hao, U. Dayal, L. Haug, H. P. Labs, and D. A. Keim, "Feature-Based Visual Sentiment Analysis of Text Document Streams," vol. 3, no. 2, 2012.
[4] W. Zhang and F. Gao, "An Improvement to Naive Bayes for Text Classification," Procedia Eng., vol. 15, pp. 2160–2164, Jan. 2011.
[5] H. Jeong, D. Shin, and J. Choi, "FEROM: Feature Extraction and Refinement for Opinion Mining," ETRI J., vol. 33, no. 5, pp. 720–730, Oct. 2011.
[6] M. Yusuf and D. D. Santika, "ANALISIS SENTIMEN PADA DOKUMEN BERBAHASA INDONESIA DENGAN PENDEKATAN SUPPORT VECTOR MACHINE," 2011.
[7] B. Liu, "Sentiment Analysis : A Multi-Faceted Problem," no. 1, 2010.
[8] V. Roth and T. Lange, "Feature Selection in Clustering Problems," 2000.
[9] J. Read, "Using emoticons to reduce dependency in machine learning techniques for sentiment classification," Proc. ACL Student Res. Work., 2005.
[10] F. Sebastiani, "Machine learning in automated text categorization," ACM Comput. Surv., vol. 34, no. 1, pp. 1–47, Mar. 2002.
[11] A. P. Windarto *et al.*, "Analysis of the K-Means Algorithm on Clean Water Customers Based on the Province," *J. Phys. Conf. Ser.*, vol. 1255, no. 1, 2019, doi: 10.1088/1742-6596/1255/1/012001.
[12] Budiharjo, T. Soemartono, A. P. Windarto, and T. Herawan, "Predicting tuition fee payment problem using backpropagation neural network model," *Int. J. Adv. Sci. Technol.*, vol. 120, pp. 85–96, 2018, doi: 10.14257/ijast.2018.120.07.
[13] Sumijan, A. P. Windarto, A. Muhammad, and Budiharjo, "Implementation of Neural Networks in Predicting the Understanding Level of Students Subject," *Int. J. Softw. Eng. Its Appl.*, vol. 10, no. 10, pp. 189–204, 2016.
[14] Budiharjo, T. Soemartono, A. P. Windarto, and T. Herawan, "Predicting School Participation in Indonesia using Back-Propagation Algorithm Model," *Int. J. Control Autom.*, vol. 11, no. 11, pp. 57–68, 2018.

Temperature and Humidity Observation System in ATmega8 Microcontroller-Based Homes

Abdul Latif*, Kuat Supriyadi

Department of Electrical Engineering, [Universitas Islam Sultan Agung], Semarang Indonesia

*Correspondent Email: 4bdul@gmail.com

Abstract. In our daily life it is sometimes necessary to measure the temperature and humidity of the environment around us. Many kinds of sensors for measuring temperature or humidity, such as LM35, DS18S20 or DHT11. To measure the temperature and humidity of the environment around us, we can choose a sensor that is quite cheap, namely DHT11. DHT11 has a calibrated digital output. This sensor consists of resistive type humidity measuring components and temperature measurements via NTC and is connected with 8 bit uC so that it gives good results, sufficient response speed, has good resistance to interference and is quite cheap in price. The interface used is a single write serial interface which is quite fast and easy. Small sensor size, low power consumption requirements and able to transmit its output within a distance of 20 meters. The disadvantage of this sensor is the lack of accuracy, besides the temperature measurement range is only 0 to 50 degrees Celsius but with a fairly cheap price this sensor can be an alternative to be used in temperature and humidity measurements that do not really need good accuracy. For example measurement of room temperature, measurement of temperature and humidity of incubators and others. If the sensor distance to the uC is less than 20 meters, then a 5K pull up resistor must be installed on the data pin. Meanwhile, if the distance is more than 20 meters, then the size of the pull up resistor should be adjusted. The power supply needed by DHT11 ranges from 3.5V to 5V. Access to the sensor is only allowed more than 1 second after the first power supply is supplied. It is also necessary to add a 100nF capacitor between the VCC and GND pins for the power supply filter.

1. Introduction

Various types of technology have been created by many people to be able to facilitate their daily activities in carrying out their work. As one of the developing technologies is technology in the field of temperature measurement [1] [2] - [4] and humidity. Temperature gauges [5] - [9] and humidity are very much needed in certain cases. For example, in a storage warehouse it is very important to pay attention to the temperature and humidity of the warehouse room to store goods properly, in the computer server room also requires a certain temperature so that the server can still work well, as well as in a house, so that residents in the house can monitor and control the temperature inside. Departing from this, the researchers wanted to make the application of DHT11 [10], [11], [11] - [14] as an ATmega8-based temperature and humidity sensor, DHT11 sensor as a temperature and humidity sensor, LCD as a display, and a minimum system ATmega8 microcontroller as the center. The results show the ATmega8 Microcontroller [16] - [18] has an input in the form of a temperature sensor, this sensor will detect the temperature and humidity inside the house and display it on the LCD.

2. Methodology

A. Power Supply

In general, the term power supply means a rectifier filter system, the circuit of which converts the AC voltage from the PLN source voltage to a pure DC voltage. The basic components used in power supply circuits are transformers, rectifiers, stabilizers.

B. DHT11 Sensor

The DHT11 sensor is a sensor with a digital signal calibration that is able to provide temperature and humidity information. This sensor is classified as a component that has a very good level of stability, moreover coupled with the ability of the ATmega8 microcontroller [19]. The highest quality products, fast response readings, and anti-interference trays, at an affordable price. DHT11 [14], [20], [21] has a very accurate calibration feature. This calibration coefficient is stored in the OTP program memory, so that when an internal sensor detects a temperature [22] or humidity, this module reads the sensor's coefficient. Its small size, with signal transmission of up to 20 meters, makes this product suitable for many temperature and humidity measurement applications. For shapes and legs DHT11 [23] - [26] can be seen in figure 1.

Figure 1. DHT11

C. Microcontroller ATmega8

The ATmega8 microcontroller [27] has 28 pins for the PDIP model. The names of the pins on this microcontroller include: • VCC for positive power supply voltage. • GND for negative power supply voltage. • PortB (PB0 - PB5) as an Input / Output port and also has other capabilities. • PortC (PC0 - PC6) as Input / Output ports for ATmega8 and ADC ports. • PortD (PD0 - PD7) as an Input / Output port and also Program Filling in the AVR Microcontroller To program in the AVR microcontroller, Atmel has provided special software that can be downloaded from the official website of Atmel. The software is AVR Studio [27]. This software uses assembly language as its intermediary language. Besides AVR Studio, there are several third party software that can be used to create programs on AVR. This third-party software uses high-level programming languages such as C, Java, or Basic. To transfer from the computer to the chip, you can use several methods such as using a JTAG cable or using an atmospheric vehicle registration, or you can also use a USB downloader.

D. LCD (Liquid Crystal Display)

LCD (Liquid Crystal Display) is a type of display media that uses liquid crystal as the main viewer [28]. LCD has been used in various fields such as electronic devices such as televisions, calculators, or computer screens. In this research, the LCD application used is dot matrix LCD with 2x16 characters. The LCD is very function as a viewer which will later be used to display the temperature and humidity sensor DHT11

E. How it Works

When the power supply is powered by an electric current in the form of AC current, the AC current is converted to a DC current in the power supply. Then transformed into the MCU (Microcontroller unit) [29], the DHT11 sensor sends a signal into the MCU and the results are displayed on the LCD.

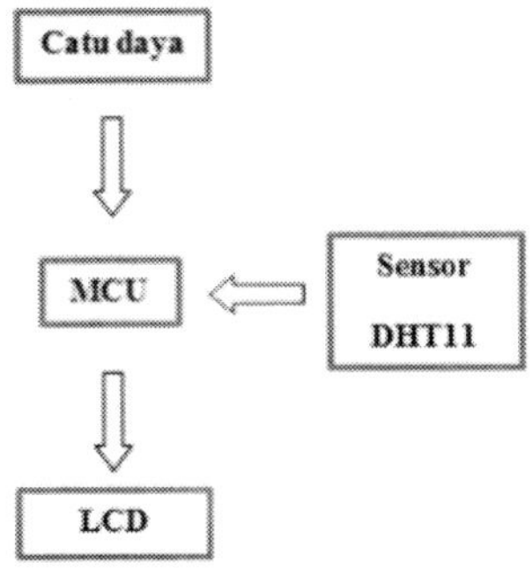

Figure 2. Tool block diagram.

The input and output addresses of the microcontroller are shown in table 1 below.

Table 2. List Of Microcontroller Inputs And Outputs

No	Alamat Output	Jenis Alat
1.	Port D4 – Port D6	LCD
2.	Port D3	Sensor DHT11

F. Electronic Design

1) Power Supply Circuit

The power supply circuit provides a voltage supply to the controller. The power supply circuit gets a voltage source from PLN of 220 VAC. The 220 VAC voltage is then reduced to 12 VAC via a voltage-reducing transformer. The 12V AC voltage is directed by the bridge diode to the DC voltage. The output of this bridge diode then enters the polar capacitor whose function is to stabilize the voltage. The 100nF capacitor serves to get rid of noise at the voltage. The following is a series of drawings, layout and component layout of the power supply.

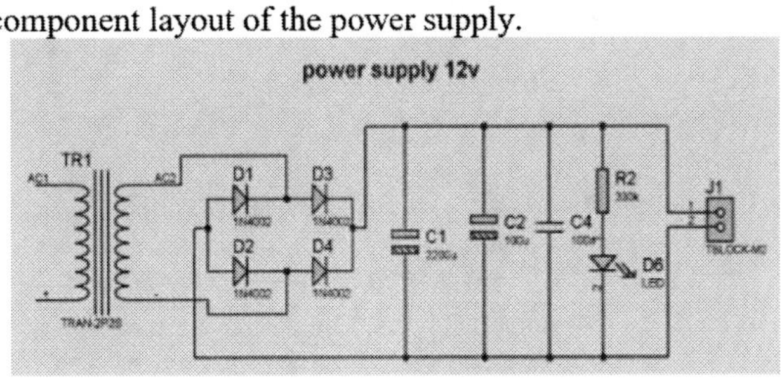

Figure 3. Power Supply

2) Minimum System Series

This microcontroller circuit is the main controller for the system that has been made. Basically this microcontroller circuit is a minimum system of ATmega8 microcontrollers.

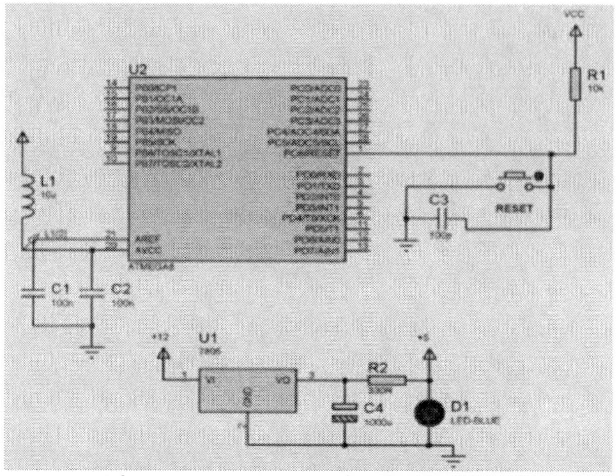

Figure 4. ATmega8 minimum system

G. Overall Series

The whole circuit is a combination of a series of power supplies and a microcontroller whose results are displayed by the LCD.

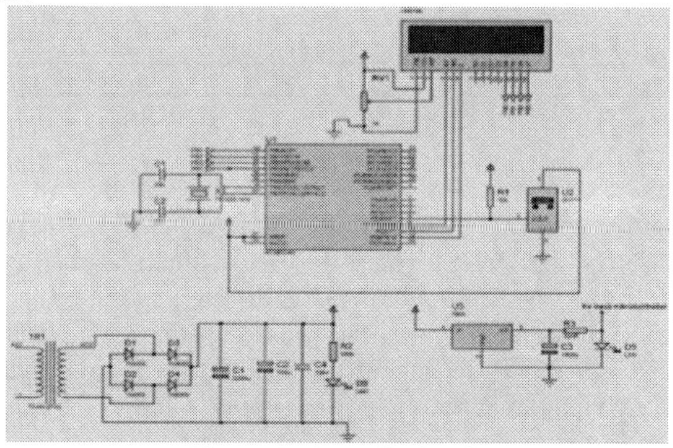

Figure 5. Overall circuit

H. Measurement Analysis

The measurement this time was carried out in a 3x3m room with the window open and the weather conditions bright, the following are the results of the measurement table:

Table 3. MEASUREMENT TOOLS

No	Sensor DHT11		Termometer	Hygrometer
	suhu	kelembaban	suhu	kelembaban
1	25 °C	95% RH	24 °C	1
2	26 °C	93% RH	26 °C	2
3	26 °C	95%RH	25 °C	3

From the results of the measurements of the two samples above between props with a temperature thermometer and hygrometer, there is no significant difference in temperature and humidity.

3. Conclusion

The incubator analyzer works properly after measurement using comparative equipment. From the results of the trial by comparing the values produced by the TA module with a comparison tool, the results were quite significant. At humidity the error value was below 2% and was still on the tolerance threshold ± 5%, where the biggest error in the measurement value was 37 ° C which was 1.56% and the smallest error at the measurement of 35 ° C is 0.75%.

References

[1] T. P. Tunggal, A. W. Apriandi, J. E. Poetro, E. T.Helmy, and F. Waseel, "Prototype of Hand Dryer with Ultraviolet Light Using ATMega8," J. Robot. Control, vol. 1, no. 1, pp. 7–10, 2020.

[2] Iswanto, P. Megantoro, and D. V. Senzas, "Calibrator for Temperature Measurement Device with Raspberry Pi-Based Interface," Int. J. Innov. Technol. Explor. Eng., vol. 8, no. 12, pp. 4862–4866, 2019.

[3] Iswanto, J. Syaftriadi, A. Nur, N. Chamim, R. O. Wiyagi, and R. Syahputra, "LED and Servo Motor Control Via Bluetooth Based on Android Applications," Int. J. Recent Technol. Eng., vol. 8, no. 2, pp. 6227–6231, Jul. 2019.

[4] I. Iswanto, W. S. Agustiningsih, F. Mujaahid, R. Rohmansyah, and A. Budiman, "Accumulator Charging Control with Piezoelectric Based on Fuzzy Algorithm Scheduling," TELKOMNIKA (Telecommunication Comput. Electron. Control., vol. 16, no. 2, p. 635, Apr. 2018.

[5] Z. Dzulfikri, N. Nuryanti, and Y. Erdani, "Design and Implementation of Artificial Neural Networks to Predict Wind Directions on Controlling Yaw of Wind Turbine Prototype," J. Robot. Control, vol. 1, no. 1, pp. 20–26, 2020.

[6] Y. Kondratenko, V. Korobko, O. Korobko, and O. Gerasin, "Pulse-phase control system for temperature stabilization of thermoacoustic engine model driven by the waste heat energy," in 2015 IEEE 8th International Conference on Intelligent Data Acquisition and Advanced Computing Systems: Technology and Applications (IDAACS), 2015, vol. 1, no. September, pp. 58–61.

[7] M. Kassim, Cik Ku Haroswati Cik Ku Yahaya, and Mohd Nazri Ismail, "A prototype Of Web Based Temperature Monitoring system," in 2010 2nd International Conference on Education Technology and Computer, 2010, vol. 5, pp. V5-266-V5-270.

[8] A. Latif, K. Shankar, and P. T. Nguyen, "Legged Fire Fighter Robot Movement Using PID," J. Robot. Control, vol. 1, no. 1, pp. 15–19, 2020.

[9] Z. Liu, L. Wang, G. Xi, Z. Luo, and Y. Li, "Temperature and Humidity Sensor Location Optimization Based on BP Neural Network," in 2015 Fifth International Conference on Instrumentation and Measurement, Computer, Communication and Control (IMCCC), 2015, pp. 283–287.

[10] J. Du, J. Guo, D. Xu, and Q. Huang, "A remote monitoring system of temperature and humidity based on OneNet cloud service platform," in 2017 IEEE Electrical Design of Advanced Packaging and Systems Symposium (EDAPS), 2017, pp. 1–3.

[11] X. Li and Y. Yu, "A high accuracy temperature control system based on ARM9," in 2011 International Conference on Electrical and Control Engineering, 2011, pp. 23–26.

[12] R. Y. Adhitya et al., "Comparison methods of Fuzzy Logic Control and Feed Forward Neural Network in automatic operating temperature and humidity control system (Oyster Mushroom Farm House) using microcontroller," in 2016 International Symposium on Electronics and Smart Devices (ISESD), 2016, pp. 168–173.

[13] Bharathkumar V., Irshad SM, Gowtham S., and R. Geethamani, "Microcontroller based digital meter with alert system using GSM," in 2017 11th International Conference on Intelligent Systems and Control (ISCO), 2017, pp. 444–448.

[14] W. Jian, Z. Hanqun, X. Jiaxin, and S. Yuxiao, "Radio Frequency-Based Body Temperature and Walking Steps Monitor of Dairy Cow," in 2011 International Conference on Computer Distributed Control and Intelligent Environmental Monitoring, 2011, pp. 249–252.

[15] R. K. Kodali and K. S. Mahesh, "Low cost ambient monitoring using ESP8266," in 2016 2nd International Conference on Contemporary Computing and Informatics (IC3I), 2016, pp. 779–782.

[16] N. A. A. Rahman and A. B. Jambek, "Wireless sensor node design," in 2016 3rd International Conference on Electronic Design (ICED), 2016, pp. 332–336.

[17] P. Megantoro, A. Widjanarko, R. Rahim, K. Kunal, and A. Z. Arfianto, "The Design of Digital Liquid Density Meter Based on Arduino," J. Robot. Control, vol. 1, no. 1, pp. 1–6, 2020.

[18] Y. Sanjaya, A. Fauzi, D. Edikresnha, M. M. Munir, and Khairurrijal, "Air temperature regulation in a chamber for rotary forcespinning," in 2016 International Seminar on Sensors, Instrumentation, Measurement and Metrology (ISSIMM), 2016, pp. 28–31.

[19] I. Firmansyah, W. A. Sukarto, B. Hermanto, J. M. Asta, and H. Nugraha, "Wi-Fi based temperature monitoring system for thermal analysis with Kalman Filter implementation," Proceeding - 2016 Int. Conf. Comput. Control. Informatics its Appl. Recent Prog. Comput. Control. Informatics Data Sci. IC3INA 2016, pp. 1–5, 20160.

[20] R. Amri, N. L. Marpaung, E. Ervianto, and Nurhalim, "Design of firing detector system by peat land with woody peat types using it's heat characteristics," in 2017 5th International Conference on Instrumentation, Control, and Automation (ICA), 2017, pp. 130–134.

[21] N. H. Wijaya, A. G. Alvian, A. Z. Arfianto, J. E. Poetro, and F. Waseel, "Data Storage Based Heart and Body Temperature Measurement Device," J. Robot. Control, vol. 1, no. 1, pp. 11–14, 2020.

[22] S. Santra, U. Jana, S. Pal, and S. K. Ray, "Humidity Sensing of Zinc Oxide Nanorods Based Prototype Using Arduino Uno Microcontroller Platform," in 2018 IEEE SENSORS, 2018, pp. 1–4.

[23] M. Sharma, A. Kumar, and A. Bachhar, "I2P air purifier with air quality monitoring device," in 2017 2nd International Conference on Communication and Electronics Systems (ICCES), 2017, vol. 2018-Janua, no. Icces, pp. 478–481.

[24] P. Tapak and M. Csiba, "LoT Plant Watering," in 2018 16th International Conference on Emerging eLearning Technologies and Applications (ICETA), 2018, pp. 563–568.

[25] S. Gangopadhyay and M. K. Mondal, "A wireless framework for environmental monitoring and instant response alert," in 2016 International Conference on Microelectronics, Computing and Communications (MicroCom), 2016, pp. 1–6.

[26] N. H. Wijaya, Z. Oktavihandani, K. Kunal, E. T.Helmy, and P. T. Nguyen, "The Design of Tympani Thermometer Using Passive Infrared Sensor," J. Robot. Control, vol. 1, no. 1, pp. 27–30, 2020.

[27] V. R. Mutha, N. Kumar, and P. Pareek, "Real time standalone data acquisition system for environmental data," in 2016 IEEE 1st International Conference on Power Electronics, Intelligent Control and Energy Systems (ICPEICES), 2016, pp. 1–4.

[28] V. Raju, A. S. N. Varma, and Y. S. Raju, "An environmental pollution monitoring system using LORA," in 2017 International Conference on Energy, Communication, Data Analytics and Soft Computing (ICECDS), 2017, pp. 3521–3526.

[29] K. Kunal, A. Z. Arfianto, J. E. Poetro, F. Waseel, and R. A. Atmoko, "Accelerometer Implementation as Feedback on 5 Degree of Freedom Arm Robot," J. Robot. Control, vol. 1, no. 1, pp. 31–34, 2020.

IOP Conf. Series: Materials Science and Engineering **835** (2020) 012053 doi:10.1088/1757-899X/835/1/012053

Design and Development of Heart Rate Per Minutes Based on Atmega16 Microcontroller with Alarm Warning

Nur Hudha Wijaya[2*], Achmad Kholilul Mujib[1], Agung Budi Santoso[1], Kuat Supriyadi[2]

[1]Department of Electrical Engineering, [Universitas Muhammadiyah Semarang], Indonesia
[2]Department of Medical Electronics Technology, [Universitas Muhammadiyah Yogyakarta], Indonesia

*Corresponden Email hudha@umy.ac.id

Abstract. Sport is an activity to train one's body, not only physically but also spiritually, which is useful for maintaining and improving the quality of one's health. The simplest and most cost-effective exercise is jogging, which is usually done in the morning or evening. Sports activities will stimulate the heart to work extra in pumping blood and will facilitate blood in the body. In order to achieve the maximum benefits, it is really advisable to do sports activities in a regular and well-structured way, because if you exercise excessively, it will be very dangerous if the heart works beyond its normal limits. In this study, a heart rate meter was made which has the principle of controlling and monitoring heart rate per minute, with the hope to make it easier to calculate the heart rate while exercising and provide safety because this tool can inform the normal heart rate when exercising for users. This system is made by using the atmega16 microcontroller as the main module. This monitoring process is carried out by connecting the pulse sensor on the finger whose output is a numeric character on the LCD and the sound of a warning on the buzzer if the heart rate exceeds the normal limit. Based on the results of testing of 20 respondents, the device has been able to calculate heart rate per minute and the buzzer can sound as a warning alarm when detecting heart rate if the heart rate exceeds the normal limit. In addition, the tool designed has a percentage error of 0.72%.

1. Introduction

Sport is an activity to train one's body, not only physically but also spiritually, which is useful for maintaining and improving the quality of one's health. The simplest and most cost-effective exercise is jogging, which is usually done in the morning or evening. Sports activities will stimulate the heart to work extra in pumping blood and will expedite blood in the body. So that the benefits achieved are more leverage, it is really recommended to do sports activities with regular and well-structured way, because if you exercise excessively it will be very dangerous if the heart works beyond its normal limits. Heart rate per minute for normal heart in adults is 60-100 [1]–[4]. If an adult exceeds the normal per minute heart rate this will cause fatal heart disease. The research on heart rate detection that has been previously carried out is titled "Digital heart rate measuring device based on microcontroller ATMega8535". Where in the study heart rate measurements using sensors in the form of electronic circuits such as LEDs and light sensors. Therefore, the research "Design of a Heart Rate

Per Minute Microcontroller-Based Heartbeat Counter with Alarm Warning" was raised so that sports such as jogging can still be done safely and comfortably without worrying about excessive exercise with the presence of this heart rate meter, making it easier to watch - detect heart rate when you exercise. In this research, a device that can calculate heart rate per minute will be made to monitor the heart rate while exercising with a warning sound.

2. Literature Review

2.1 Heart

The heart is a vital organ and is the last defense for life other than the brain. The pulse in the heart cannot be controlled by humans. Cone heart is located in the thoracic cavity (chest) between the two lungs posterior to the sternum (breastbone) [2]. Normally adults beat about 60-100 times per minute. In children the baby beats faster, but emotional and hormonal activity can affect the heart rate so that it changes its beat [3] - [4].

2.2 Mikrokontroler ATMega16

AVR is an 8-bit RISC (Reduce Instruction Set Compute) microcontroller based on Harvard architecture made by the Atmel company [5]–[8]. AVR (Advanced Versatile RISC) or Alf and Veg's Risc Processor derived from the names of two Norwegian Institute of Technology (NTH) students, Alf-Egil Bogen and Vegard Wollan [5]. AVR has an advantage compared to other microcontrollers which has a faster program execution speed because most of the instructions are executed in 1 clock cycle, faster than other types of microcontrollers. AVR microcontroller [9]–[12] has complete features such as internal ADC, internal EEPROM, timer / counter, watchdog timer, PWM (Pulse Width Modulation), I / O port, serial communication, comparator, I2C and other enhancements. In general, the AVR microcontroller is divided into five types of families, namely TinyAVR, MegaAVR, XMEGA AVR, UC3 AVR32, and AVR32 AP7. The difference that is owned by the five types of AVR is a feature that is in each microcontroller, which has advantages and disadvantages of each. ATMega16 [13]–[16] is a MegaAVR type which is a low-power 8 bit CMOS microcontroller and has 32 general purpose registers [6]. Figure 1 is a configuration of the ATMega16 AVR pin [17]–[20].

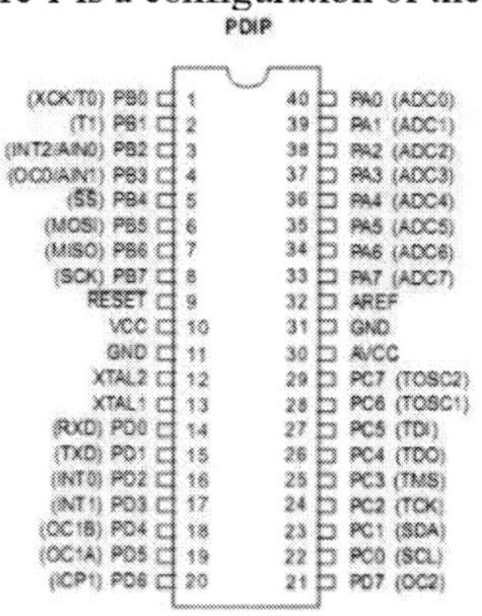

Figure 1. ATMega16 AVR pin configuration

2.3 Pulse Sensor

The pulse sensor works by utilizing light. When this sensor is placed on the surface of the skin, most of the light is absorbed or reflected by organs and tissues (skin, bones, muscles, blood), but some of the light will pass through body tissue that is quite thin [7]. Figure 2 is the physical form of the pulse sensor.

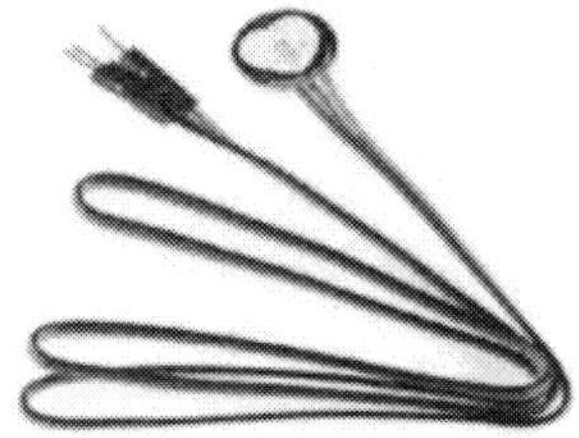

Figure 2. Pulse Sensor

2.4 Liquid Crystal Display (LCD)

Liquid Crystal Display (LCD) [2], [21]–[23] is a form of liquid crystal that will emulsify if given a voltage. The LCD used is a 16 X 2 LCD with each character formed by 8 pixel rows and 5 pixel columns [8]. Figure 3 is a physical form of LDC 16x2.

Figure 3. 16x2 LCD

2.5 Buzzer

Buzzer [24]–[27] is an electronic component that functions to convert electrical vibrations into sound vibrations. Basically the working principle of the buzzer is almost the same as the loudspeaker, so the buzzer also consists of a coil mounted on the diaphragm and then the coil is flowed so that it becomes electromagnetic, the coil will be attracted into or out, depending on the direction of the current and the magnetic polarity, because the coil is mounted on the diaphragm back and forth so that it makes air vibrate which will produce sound [9]. Figure 4 is the physical form of the buzzer.

Figure 4. Buzzer

3. Methodology

3.1 Literature Review

The stage of conducting a literature review begins by collecting various references in the form of books, journals, theses, online sources related to the final project Designing a Heart Rate Per Minute Counter Based on ATMega16 Microcontroller with a Warning Alarm.

3.2 Needs Analysis

This needs analysis includes hardware and software requirements.

3.3 Research Planning

Research planning includes: a. Hardware design Hardware design is done to get a picture of the design of electronic circuits and patterns of communication between existing hardware. b. Tool work system design The tool work system design phase is carried out after an overview of communication between hardware is obtained.

The working system of this tool includes the design of sensor sensitivity in responding to the heart rate of the input / output device communication with the microcontroller, as well as the algorithm applied to the microcontroller. c. Software design The programming language used in this study is the basic compiler language. The programming algorithm designed refers to the working system of the tool, where this program allows the microcontroller to be able to process signals from input / input devices and provide output / output device measures.

3.4 Integration
In the Integration phase, the results of the planning and needs analysis are processed to be made into a whole system. This stage is carried out in order to realize this tool into real form, by integrating system design, hardware and software.

3.5 Testing
This testing phase involves testing the hardware and software.

3.6 Implementation
This stage is the final stage after a series of testing of the device.

4. System Design
The design that will be carried out in this study includes the design of hardware (hardware) and software design (software). The design of a hardware device begins with designing a block diagram and the working principle of a tool, then continues to design a series of tools by combining the whole device into a controlled system.

4.1 Block Diagram and Tool Working Principle
a. Block diagram
Figure 5 is a block diagram of a minute-based heart rate counting device based on the ATMega16 microcontroller with a warning alarm.

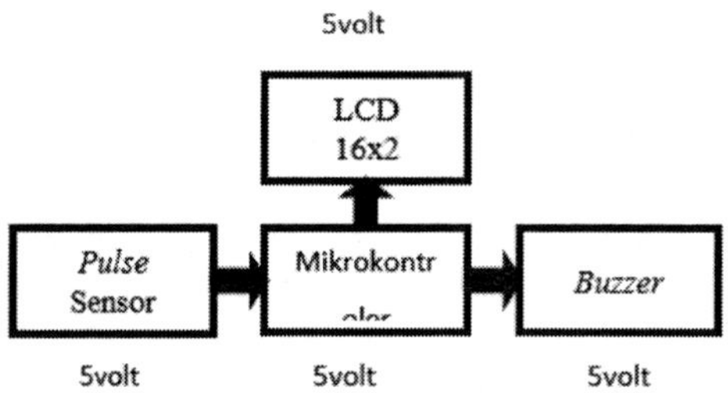

Figure 5. System Block Diagram

b. The Working Principle of the Tool
This tool will work based on the system as a whole and integrated from each series in which the principle of working the series of tools or systems are as follows: 1) Pulse sensors function to detect heart rate. When this sensor is placed on the surface of the skin, most of the light is absorbed or reflected by organs and tissues (skin, bones, muscles, blood), but most of the light will pass through the body tissue which is quite thin. Heart rate detected through the skin surface by the sensor functions as an input (input) and is still a voltage which is also an analog quantity to be sent to the microcontroller. Furthermore, the microcontroller processes the signal or data from the heart rate using the ADC facility to convert the output voltage into digital data that is recognized by the microcontroller. 2) Microcontroller processes the signal or data from the heart rate by counting the heart rate data continuously every minute, then get the heart rate data per minute. The heart rate data per minute is obtained by multiplying the heart rate calculation results for 15 seconds by 4, because 15 seconds is ¼ minutes. The heart rate data will be compared continuously with the normal heart rate. Furthermore, the microcontroller will send data or signals as output to the buzzer and LCD circuit. 3) This system is equipped with an output device (output) such as an LCD that functions as a medium for displaying heart rate data and a buzzer device that functions as a warning alarm sound when the heart rate data exceeds normal limits.

4.2 Hardware Design

Hardware (hardware) is the main and most important component in making this system [28]–[34]. The hardware design in this study include: a. The Minimum Circuit Design of the ATMega16 System The microcontroller used is the type of AVR ATMega16 which is a single chip that functions as a data processing center and device controller. ATMega16 was chosen because it is easy to get on the market, has a number of pins according to needs and a practical circuit because it is arranged in one module that is very easy to use, simple and does not take place. In order for the microcontroller to work, an additional circuit is needed for the microcontroller called the minimum system circuit, the function of this circuit is almost the same as the motherboard contained in the computer. In this study, the minimum system applied is quite simple, for hardware communication it can be directly connected to the pin in the port of the microcontroller. Figure 6 is the design of the ATMega16 minimum system schematic circuit.

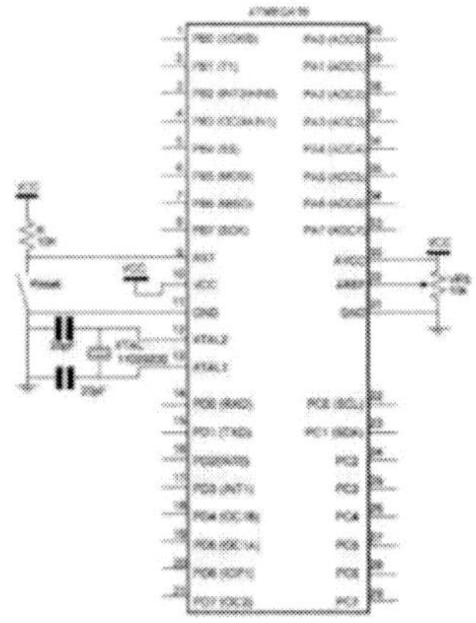

Figure 6. ATMega System Minimum Schematic Series16

The pin functions on the minimum ATmega16 microcontroller system can generally be used as I / O. The ports on ATMega16 are ports A, B, C and D, where each port consists of 0-7. Examples of ports consist of ports A.0 to portA.7, as well as other ports. In addition to general functions as I / O, each port has its own special function. There are facilities that serve to interrupt, serial communication, generate PWM pulses, and others. In this heart rate / heart counter study, the special function, namely ADC (Analog to Digital Converter), is available on the ATMega16 microcontroller, there are 8 ADCs which happen to be inport, not owned by other ports. Where the maximum ADC on the ATMega16 microcontroller is 10 bits, if it is minimized 1023. The use of ADC facilities in this study because the output of the pulse sensor is still a voltage that is still in the form of an analog quantity. While the microcontroller used can only process input in the form of logic 0 and 1 or can only recognize 0 volts and 5 volts. Therefore ADC is needed to convert the output voltage into digital data that is recognized by the microcontroller. b. LCD Circuit Design The LCD circuit planning aims to make the LCD as a display of the number of heart beats per minute fully controlled by the ATMega16 microcontroller. LCD used type 16x2 which means 16 columns and 2 rows. On the LCD there are 16 pins that will be connected to port B on the ATMega16 microcontroller. Figure 7 is the design of the LCD circuit.

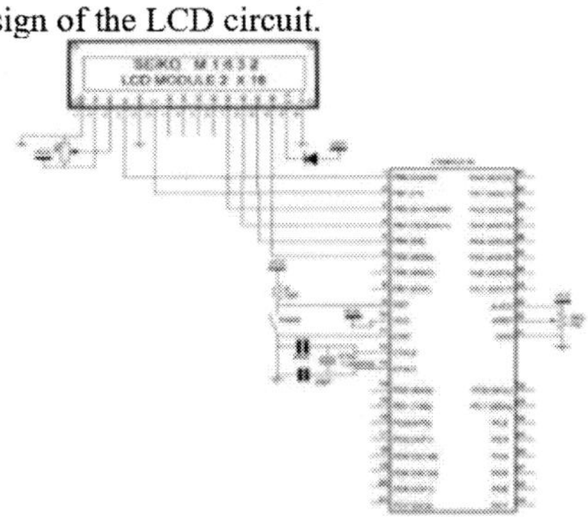

Figure 7. LCD schematic circuit

In LCD circuit design, RS pin is connected to PortB.0, pin E is connected to PortB.1, DB4 pin is connected to PortB.2, DB5 pin is connected to PortB.3, DB6 pin is connected to PortB.4, and DB7 pin is connected with PortB.5. so not all LCD pins are connected to the pins on the microcontroller, only the 6 pins above are needed to be connected to the microcontroller pin so that it can be controlled by the microcontroller ATMega16. c. Design of Buzzer Circuits The buzzer circuit functions as a warning in the form of sound, so that the buzzer circuit can be controlled by a microcontroller it needs to be connected to a specific pin on the microcontroller. Figure 8 is the design of the Buzzer series.

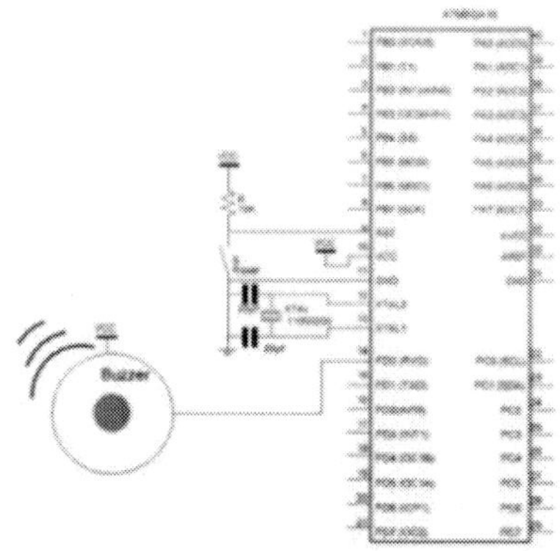

Figure 8. Buzzer Schematic Series

In the design of the buzzer circuit connected to portD.0 there is no specific reason for the selection of ports because for buzzer control such as the above circuit which has received a positive voltage (VCC) first so that the microcontroller only needs to provide negative voltage, such a circuit is called active low circuit. In controlling it, it only needs the I / O control function, so all ports on the microcontroller can control it. d. Pulse sensor design In this pulse sensor circuit planning it aims to control or process the sensor output using a special ADC facility on the ATMega16 microcontroller. Figure 9 is the design of the pulse sensor circuit.

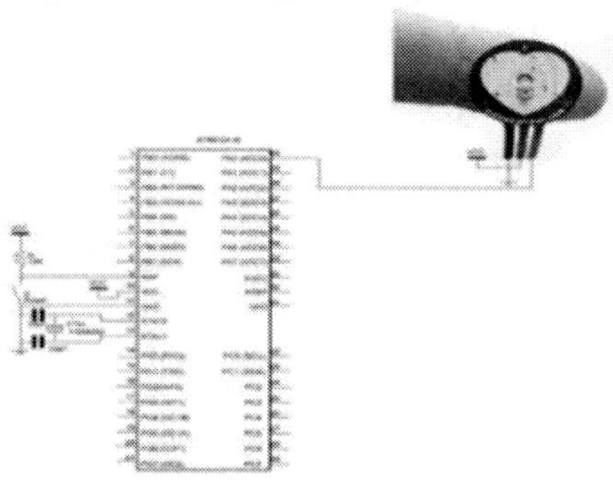

Figure 9. Pulse Sensor Schematic Circuit

In the design of a pulse sensor circuit connected to ADC 0 which happens to be located in PortA.0. ADC value will change according to the output of the sensor in the form of voltage. e. Overall Hardware Design. After designing the hardware, the next stage is the design phase for the entire hardware. Hardware such as pulse sensors, buzzers, and LCDs are connected to the microcontroller via predefined ports. At this stage, the entire hardware unit or set of tools is transformed into a prototype so that it can be directly simulated. Figure 10 is a whole schematic set of hardware.

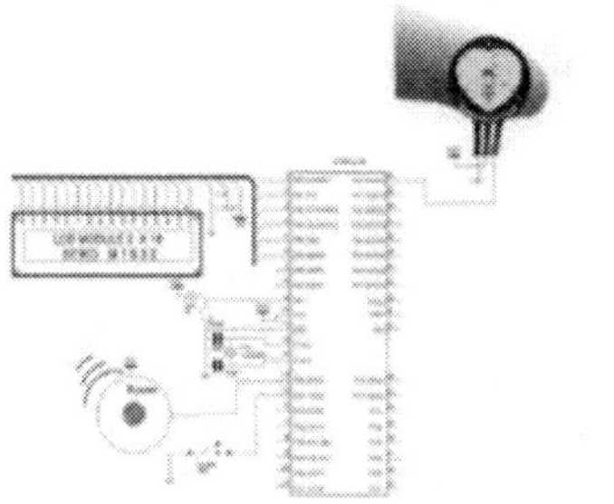

Figure 10. Overall Hardware Schematic Series

4.3 Software Design

Software is needed as a protocol between the microcontroller and other hardware components. Based on the concept in hardware design, the designed program is expected to be able to process heart rate data to the microcontroller or vice versa. a. The design of microcontroller software flow diagrams To facilitate the preparation of software on the microcontroller, it is necessary to make a flow diagram (flowcart). Figure 11 is a microcontroller software flow chart.

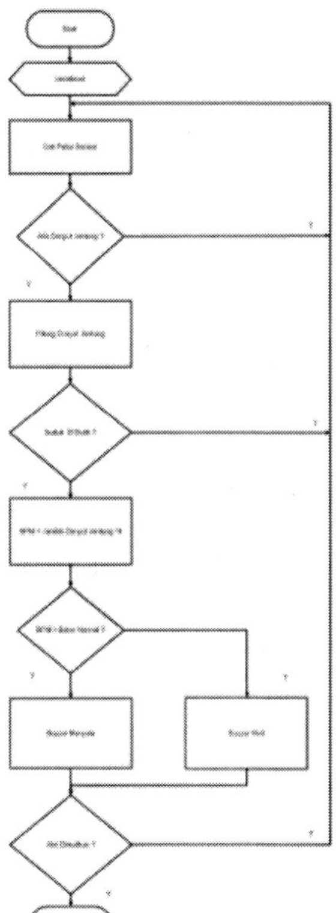

Figure 11. Flow Chart of Microcontroller Software

When the program starts, the initialization process begins with the registration of the microcontroller and the microcontroller pin configuration that will be used to control the components and input data from the sensor. Consists of configuring to activate the timer, ADC configuration, LCD pin configuration, buzzer configuration and start button configuration. Make the variables needed for processing values, the variables used are time variables, ADC values, heart counts, BPM, and normal limits. The workflow of this system will work when the start button is activated. Previously the on button is activated then the sensor is first mounted by the hand. Then when the start button is activated the sensor will work (counting the number of pulses). If a pulse is detected, the system will automatically calculate the heart rate for 15 seconds, then the timer will stop first and the Beat Per

Minute (BPM) calculation will be carried out. In this process BPM can be calculated by multiplying the heart rate calculation results for 15 seconds by 4, because 15 seconds is ¼ minutes. Then the process continues if the heart rate data per minute exceeds the normal limit, the buzzer will sound, if not the buzzer will turn off. The last process if the device is not turned off then the process to calculate the heart rate will be repeated again. If the device is turned off, the program is finished. b. Design of ATMega Microcontroller Programming Algorithm 16 The design of a program algorithm aims to determine the program flow before the program is written and incorporated into the microcontroller. Algorithm design makes it easier when writing programs and makes writing programs more directed. The program algorithm will define actions to be taken by the microcontroller such as receiving an input signal and giving an output signal to the hardware. This programming algorithm functions to define the variables used for program writing.
Making software algorithms using the AVR Basic Compiler program that functions to write program code and compile it into a hex file. The hex file that is generated after the compilation process will be downloaded into the microcontroller. So the microcontroller will work in accordance with the commands that exist in flash memory.

5. Result and Discussion

After the design and manufacture of tools and programs have been completed, a system test is performed to find out whether the tools that have been designed can run as planned. Tests carried out include testing hardware such as a minimum circuit ATMega16 system, LCD, Pulse sensor, and buzzer. The test for the incorporation of all components used in testing the overall system. The tests are carried out separately on each unit or series block and carried out in an integrated system for the overall testing.

5.1 Overall System Testing

This overall test involves the performance of all components in order to determine whether the device is able to work properly and can detect heart rates or not. Overall testing is done by assembling all components or tools used in an integrated manner. The whole system testing stages are as follows: a. Connects all hardware components such as minimum ATMega16 system, LCD, Pulse sensor, and Buzzer. b. Download the program into the ATMega16 microcontroller. This test aims to determine whether the heart rate calculation process in this system can function as expected. Figure 12 is a display of the overall test results. The monitoring process in the heart rate meter will work in accordance with the sensor response that is read and programmed in the microcontroller. Indicator of success on this tool is in the form of an output number of heartbeats that appear on the LCD.

Figure 12. Display of Overall Testing Results

In the device, if the sensor has responded then the system will run in accordance with the program and produce an output heart rate that will be displayed on the LCD as shown in figure 12. In addition, if the number of heartbeats exceeds the normal limit, the output will be a warning sound on the buzzer. In the picture 12 characters displayed on the LCD are the symbol t for time, N for normal BPM, pulse for input heart rate that is read on the pulse sensor at the time of measurement and BPM for the number of heartbeats detected within 1 minute.

5.2 Testing Analysis

From the overall results of testing the equipment that has been done, it can be explained that the heart rate calculator based on ATmega16 microcontroller with this warning alarm functions in accordance with the previous design. . Based on the test, it is known that the heart rate calculation tool per minute can work to respond to the heart rate per minute in accordance with the input read in the system and the number of heartbeats that are read according to a person's condition at the time of measurement. Each hardware component such as the minimum circuit ATMega16 system, Liquid Crystal Display (LCD), Pulse sensor, and buzzer can be connected between devices and can function and work well too.

Table 1. Tool Testing Results

Nama	BPM		*Buzzer*	Error (%)
	Alat	Manual		
Mila	104	104	berbunyi	0.00%
Deska	96	96	Tidak berbunyi	0.00%
Dona	108	108	berbunyi	0.00%
Dwi	128	124	berbunyi	3.23%
Edwar	124	124	Berbunyi	0.00%
Erny	112	108	Berbunyi	3.70%
Januar	88	88	Tidak berbunyi	0.00%
Louis	84	84	Tidak berbunyi	0.00%
Maria	120	120	Berbunyi	0.00%
Patrisia	104	104	Berbunyi	0.00%
Peter	104	104	Berbunyi	0.00%
Prita	108	104	Berbnyi	3.85%
Rauda	116	112	Berbunyi	3.57%
Riska	88	88	Tidak berbunyi	0.00%
Sella	140	140	Berbunyi	0.00%
Via	92	92	Tidak berbunyi	0.00%
Willy	104	104	Berbunyi	0.00%
Yatno	92	92	Tidak berbunyi	0.00%
Yudi	88	88	Tidak berbunyi	0.00%
Yuwono	116	116	Berbunyi	0.00%
Total *Error*				14,35 %
Rata-rata Persentase *Error*				0,72%

Based on the results of tests that have been carried out, the results of data consisting of 20 respondents indicate that each person has a different heart rate depending on conditions during the test such as in a state of calm or in a state of many activities. Factors that differentiate a person's heart rate are also due to a person's different health, different quality of exercise, age, and many more factors that can influence a person's heart rate. In a test consisting of 20 respondents, there were some people whose heart rate exceeded the normal limit, so the buzzer was active (beeping). In addition, it can be seen that the heart rate test results show a relatively small error and the results are almost stable between tool measurements with manual measurements. This shows that the measurement data experienced quite small errors. Comparison of measurement results with manual calculation tools (calculation of the pulse on the wrist) can be seen that the measurement results error of 0.72%. This can be caused by manual calculations performed on the wrist to allow smooth pulses to go undetected. Whereas the measurement of heart rate using this measuring instrument is carried out on the fingers with sensors mounted attached to the fingers. In the heart rate sensor testing that has been done, the resulting output in the form of data or heart rate signals that are more easily detected at the fingertips. The light on the sensor will be able to work effectively if placed on the tip of the finger because this section is thin which makes the blood flow in the body read properly. Proper placement with the location of the sensor on the surface of the fingertips can display data more constant and stable. Placement of the fingers is recommended by the use of this tool to see the results of testing carried out easily and precisely with a very small error. From the results of tests conducted, this tool can be used to calculate heart rates properly.

6. Conclusion

After testing and analyzing the heart rate measuring system per minute based on the ATmega16 microcontroller with this warning alarm, the conclusions can be drawn as follows: 1. The tool can calculate the human heart rate per minute with an average percentage error of 0.72 %. 2. Overall the tools made can work and function as expected. Heart rate measuring devices have given a warning with an alarm when the heart rate per minute exceeds the recommended normal limit of 60-100. 3. This heart monitoring device has a different heart rate input response according to the condition at the time of measurement which is read by the pulse sensor.

IOP Conf. Series: Materials Science and Engineering **835** (2020) 012053 doi:10.1088/1757-899X/835/1/012053

References

[1] N. H. Wijaya, A. G. Alvian, A. Z. Arfianto, J. E. Poetro, and F. Waseel, "Data Storage Based Heart and Body Temperature Measurement Device," J. Robot. Control, vol. 1, no. 1, pp. 11–14, 2020.

[2] D. Sarkar and A. Chowdhury, "Low cost and efficient ECG measurement system using PIC18F4550 microcontroller," in 2015 International Conference on Electronic Design, Computer Networks & Automated Verification (EDCAV), 2015, pp. 6–11.

[3] V. K. Singhal, V. Menezes, S. Chakravarthy, and M. Mehendale, "8.3 A 10.5μA/MHz at 16MHz single-cycle non-volatile memory access microcontroller with full state retention at 108nA in a 90nm process," in 2015 IEEE International Solid-State Circuits Conference - (ISSCC) Digest of Technical Papers, 2015, pp. 1–3.

[4] Iswanto, S. Suripto, F. Mujahid, K. T. Putra, N. P. Apriyanto, and Y. Apriani, "Energy Harvesting on Footsteps Using Piezoelectric based on Circuit LCT3588 and Boost up Converter," Int. J. Electr. Comput. Eng., vol. 8, no. 6, 2018.

[5] C. Cristea, A. Pasarica, G. Andruseac, B. Dionisie, and C. Rotariu, "A wireless ECG acquisition device for remote monitoring of heart rate and arrhythmia detection," in 2015 E-Health and Bioengineering Conference (EHB), 2015, pp. 1–4.

[6] K. Kunal, A. Z. Arfianto, J. E. Poetro, F. Waseel, and R. A. Atmoko, "Accelerometer Implementation as Feedback on 5 Degree of Freedom Arm Robot," J. Robot. Control, vol. 1, no. 1, pp. 31–34, 2020.

[7] M. Sejera et al., "Detection of arrhythmia while asleep through the use of an electrocardiogram integrated with a microcontroller and an alarming device," in 2017IEEE 9th International Conference on Humanoid, Nanotechnology, Information Technology, Communication and Control, Environment and Management (HNICEM), 2017, pp. 1–6.

[8] S. Lermthong, P. Suwanna, and S. Airphaiboon, "Bedside patient monitoring by NFC," in 2016 9th Biomedical Engineering International Conference (BMEiCON), 2016, pp. 1–4.

[9] S. Tjiharjadi and A. Fajar, "Human Heart Rate Detection Application," in 2017 International Conference on Soft Computing, Intelligent System and Information Technology (ICSIIT), 2017, pp. 167–172.

[10] P. Megantoro, A. Widjanarko, R. Rahim, K. Kunal, and A. Z. Arfianto, "The Design of Digital Liquid Density Meter Based on Arduino," J. Robot. Control, vol. 1, no. 1, pp. 1–6, 2020.

[11] T. K. Chakraborty, S. Rahman, and S. H. Rakib, "Simulation and implementation of microcontroller based gate drive circuit for three-phase MOSFET inverter," in 2017 International Conference on Energy, Communication, Data Analytics and Soft Computing (ICECDS), 2017, pp. 13–16.

[12] P. Adhikary et al., "A novel approach to find out QRS complex for ECG signal," in 2016 IEEE 7th Annual Information Technology, Electronics and Mobile Communication Conference (IEMCON), 2016, pp. 1–5.

[13] M. E. Karar and M. El-Brawany, "Embedded heart sounds and murmurs generator based on discrete wavelet transform," in 2016 Fourth International Japan-Egypt Conference on Electronics, Communications and Computers (JEC-ECC), 2016, pp. 34–37.

[14] T. P. Tunggal, A. W. Apriandi, J. E. Poetro, E. T.Helmy, and F. Waseel, "Prototype of Hand Dryer with Ultraviolet Light Using ATMega8," J. Robot. Control, vol. 1, no. 1, pp. 7–10, 2020.

[15] C. G. Bartone, L. Moore, and M. Kohli, "An e-textile antenna for body area network," in 2016 IEEE International Symposium on Antennas and Propagation (APSURSI), 2016, pp. 999–1000.

[16] S. Sangurmath and N. Daimiwal, "Application of photoplethysmography in blood flow measurement," in 2015 International Conference on Industrial Instrumentation and Control (ICIC), 2015, no. Icic, pp. 929–933.

[17] P. Bagchi et al., “Intelligent health diagnosis and patient guidance system,” in 2017 8th IEEE Annual Information Technology, Electronics and Mobile Communication Conference (IEMCON), 2017, pp. 269–272.
[18] A. Latif, K. Shankar, and P. T. Nguyen, “Legged Fire Fighter Robot Movement Using PID,” J. Robot. Control, vol. 1, no. 1, pp. 15–19, 2020.
[19] A. Najmurrokhman, R. Rahim, K. Kusnandar, B. Wibowo, D. E. Kurniawan, and Y. Sumaryat, “Design and Implementation of A Low Cost Heartbeat Meter using Vibration Sensor, ATmega 16 Microcontroller, and Android Based Smartphone,” in 2018 International Conference on Applied Engineering (ICAE), 2018, pp. 1–5.
[20] D. Murali, D. R. Rao, S. R. Rao, and M. Ananda, “Pulse Oximetry and IOT based Cardiac Monitoring Integrated Alert System,” in 2018 International Conference on Advances in Computing, Communications and Informatics (ICACCI), 2018, pp. 2237–2243.
[21] N. Akhter, S. Tharewal, H. Gite, and K. V Kale, “Microcontroller based RR-Interval measurement using PPG signals for Heart Rate Variability based biometric application,” in 2015 International Conference on Advances in Computing, Communications and Informatics (ICACCI), 2015, pp. 588–593.
[22] N. H. Wijaya, Z. Oktavihandani, K. Kunal, E. T.Helmy, and P. T. Nguyen, “The Design of Tympani Thermometer Using Passive Infrared Sensor,” J. Robot. Control, vol. 1, no. 1, pp. 27–30, 2020.
[23] R. Umega and M. A. Raja, “Design and implementation of livestock barn monitoring system,” in 2017 International Conference on Innovations in Green Energy and Healthcare Technologies (IGEHT), 2017, pp. 1–6.
[24] S. S. Thomas, A. Saraswat, A. Shashwat, and V. Bharti, “Sensing heart beat and body temperature digitally using Arduino,” in 2016 International Conference on Signal Processing, Communication, Power and Embedded System (SCOPES), 2016, pp. 1721–1724.
[25] Z. Dzulfikri, N. Nuryanti, and Y. Erdani, “Design and Implementation of Artificial Neural Networks to Predict Wind Directions on Controlling Yaw of Wind Turbine Prototype,” J. Robot. Control, vol. 1, no. 1, pp. 20–26, 2020.
[26] S. Sawidin, D. S. Pongoh, and A. A. S. Ramschie, “Design of Smart Home Control System Based on Android,” in 2018 International Conference on Applied Science and Technology (iCAST), 2018, pp. 165–170.
[27] V. Ahmed and S. A. Ladhake, “Design of Ultra Low Cost Cell Phone Based Embedded System for Irrigation,” in 2010 International Conference on Machine Vision and Human-machine Interface, 2010, pp. 718–721.
[28] A. Srinivasan, V. K. Gnanavel, Girideeshraj, P. Jain, and G. Rajat, “Remote Monitoring of Hazardous Gases in Industries (A Low Cost Device),” in 2018 Internat2018 International Conference on Computation of Power, Energy, Information and Communication (ICCPEIC)ional conference on computation of power, energy, Information and Communication (ICCPEIC), 2018, pp. 1–4.
[29] I. Iswanto, K. Purwanto, W. Hastuti, A. Prabowo, and M. Y. Mustar, “Smart Smoking Area based on Fuzzy Decision Tree Algorithm,” Int. J. Adv. Comput. Sci. Appl., vol. 10, no. 6, pp. 500–504, 2019.
[30] K. Purwanto, Iswanto, T. K. Hariadi, and M. Y. Muhtar, “Microcontroller-based RFID, GSM and GPS for motorcycle security system,” Int. J. Adv. Comput. Sci. Appl., vol. 10, no. 3, 2019.
[31] A. N. N. Chamim, M. E. Fawzi, I. Iswanto, R. O. Wiyagi, and R. Syahputra, “Control of Wheeled Robots with Bluetooth-Based Smartphones,” Int. J. Recent Technol. Eng., vol. 8, no. 2, pp. 6244–6247, Jul. 2019.
[32] A. N. N. Chamim, M. Heru Gustaman, N. M. Raharja, and I. Iswanto, “Uninterruptable Power Supply based on Switching Regulator and Modified Sine Wave,” Int. J. Electr. Comput. Eng., vol. 7, no. 3, p. 1161, Jun. 2017.
[33] A. N. N. Chamim, D. Ahmadi, and Iswanto, “Atmega16 Implementation As Indicators Of

Maximum Speed," Int. J. Appl. Eng. Res., vol. 11, no. 15, pp. 8432–8435, Jun. 2016.
[34] Iswanto, J. Syaftriadi, A. Nur, N. Chamim, R. O. Wiyagi, and R. Syahputra, "LED and Servo Motor Control Via Bluetooth Based on Android Applications," Int. J. Recent Technol. Eng., vol. 8, no. 2, pp. 6227–6231, Jul. 2019.

Decision Support System in Recommending Climbing Tourism Destinations with Profile Matching Method

Ni Luh Wiwik Sri Rahayu Ginantra[1], Gita Widi Bhawika[2], Ahmad Zamsuri[3], Fadjar Budianto[4], Achmad Daengs GS[4]

[1] STMIK STIKOM Indonesia, Denpasar - Indonesia
[2] Institut Teknologi Sepuluh Nopember, Surabaya - Indonesia
[3] Universitas Lancang Kuning, Pekanbaru - Indonesia
[4] Universitas 45 Surabaya, Surabaya - Indonesia

*wiwik@stiki-indonesia.ac.id

Abstract. Climbing is one of the extreme sports categories and recreational activities that are in great demand in the last decade, especially teenagers. Climbing offers a different sensation and challenge from other sports, besides this sport also offers a beautiful view when the climbers have reached the top of the mountain that is intended, of course, also improves fitness and health. The purpose of this study is to provide climbing tourism destination recommendations to groups of organizations and individuals who have a hobby or are the first to want to do this sport. Data collection in this study was carried out by giving questionnaires to 3 groups of nature lovers organizations, each organization was given 10 questionnaires. In this study, there are 5 alternatives used, including Mount Jaya Wijaya, Mount Ijen, Mount Kerinci, Mount Rinjani and Mount Bromo with 6 Criteria, including Beauty, Altitude, Location, Cost and Pathway. Based on calculations using the Profile Matching method, the results are obtained that Mount Rinjani is a recommended climbing tourism destination.

1. Introducing

Climbing is a trip that has the purpose of getting to the top of the mountain, climbing is also included in the category of extreme sports and recreational activities that are in great demand, especially young people. Climbing is also one of the new hobbies that are being loved lately, usually done by people who love adventure and natural organizations. A climber usually has a reason why he wants to be involved in this activity. For example, this sport has its own challenges that make it feel its own satisfaction because it can conquer the mountain, and can also be integrated with nature. However, in undergoing this sport it is necessary to pay attention to safety factors because usually, the road that will be traversed is quite difficult and steep, so information is needed before heading to the mountain to be climbed both the route, location distance, mountain height and so on [1]. In recent decades, the sport of climbing has received considerable attention from the community. This is because climbing sports for enthusiasts offer sensations and challenges that are different from other sports, besides this sport also offers a beautiful view of the climbers have reached the top of the mountain that is intended, of course, can also improve fitness and health [2]. However, the many mountains in Indonesia that can be used as a climbing tourist destination, make the fans of this sport confused. Therefore we need a system that can recommend climbing tourism destinations in Indonesia, as information for beginners and who are used to climbing. There are many branches in the field of computer science that are able to solve complex problems, this is evident from the many studies that have been done, such as research in the field of data mining [3]–[10], the field of artificial neural networks [11]–[24], and the field of decision support systems [25]–[32]. Based on this explanation, researchers used a decision support system to resolve the above problem. In this case, the researcher uses the Profile Matching method, because this method is one of the decision support system methods that works by comparing the GAP between Alternative values and criteria. There are several things that are known about GAP Analysis, one of which is a table of GAP weight values. In addition, this GAP analysis must also understand the concept of Priority Scale, because in making weights with a range of 0-5 based on the priority of each criterion.

Many previous studies discuss the profile matching method, including research propose a privacy-preserving multi-hop profile-matching protocol for Proximity-based mobile social networks (PMSNs) [33]. In this study, the profile matching method can be applied well. Subsequent research discusses the method of matching profiles to resolve the selection problem to a decision. This study uses 5 sample data, each of which has its own credit score. This study divides credit scores between 1 and three where number 1 becomes the lowest value (worst) while number 3 is the highest value (best) [34]. In this study, the profile matching method can also be applied well. Subsequent research Using the Profile Matching Method to Employee Position Movement. The position of employee mutations discussed in this study is in Tanggamus Regency. This research generates a ranking sequence of prospective employees who have good performance that has been selected, and the output of the application can help decision-makers (Decision makers) to choose an alternative task transfer or employee office mutation in Tanggamus Regency in order to increase employment and better performance better at a new workplace again. Based on this background, the research was conducted. It is expected that by using this Profile matching method, the results will be obtained in the form of recommendations for climbing tourism destinations to groups of organizations and individuals to be used as information before climbing activities.

2. Methodology

2.1. *Method of collecting data*

The data collection techniques in this study are:

a. Library Research (Library Research), which uses the library as a means to collect data and make books as references relating to research cases.
b. Field Research (Field Work Research) is research conducted directly using several techniques, namely as follows.
   1. Questionnaire
      The questionnaire is the process of collecting data or information by giving a question or statement to someone else. The author gives questionnaires to several members of the nature lovers community in the city of Denpasar.
   2. Interview
      The Interview is the process of collecting data face-to-face between the questioner and the questioner or answerer who aims to get the right information from a trusted resource person (nature lover community).

2.2. *Data analysis*

Data analysis in this study uses quantitative analysis. The data analysis technique used is the type of inferential statistics. Inferential statistics are statistics relating to data analysis (samples), then conclusions are generalized to the entire population and used to draw inference from sample to population.

2.3. *Research Location and Data Collection Time*

The location of the study was conducted in the city of Denpasar, with the time of data collection conducted from November 21 to December 4, 2019.

3. Results and Discussion

Profile Matching Method is one of the simplest methods in decision support systems by comparing the GAP between Alternative values and criteria. There are several things that are known about GAP Analysis, one of which is a table of GAP weight values. In addition, this GAP analysis must also understand the concept of Priority Scale, because in making weights with a range of 0-5 based on the priority of each criterion. The following is the weighting of the GAP value in the Profile Matching method as follows.

Table 1. GAP Value Weight

No	Difference (GAP)	Weight Value (Wj)	Explanation
1	0	6	There is no GAP (Competency as needed)
2	1	5,5	Individual competency over 1 levels
3	-1	5	Individual competencies lack 1 level
4	2	4,5	Individual competency over 2 levels
5	-2	4	Individual competencies lack 2 levels
6	3	3,5	Individual competency over 3 levels
7	-3	3	Individual competencies lack 3 levels
8	4	2,5	Individual competency over 4 levels
9	-4	2	Individual competencies lack 4 levels
10	5	1,5	Individual competency over 5 levels
11	-5	1	Individual competencies lack 5 levels

The weights and criteria in this study can be seen in the following table.

Table 2. Value Matching Criteria Profile Matching Method (GAP Analysis)

No	Criteria Name	Profile Criteria	Weight Value
1	Beauty (C1)	5	45% = 0.45
2	Height (C2)	4	25% = 0.25
3	Location (C3)	3	15% = 0.15
4	Cost (C4)	2	10% = 0.1
5	Pathway (C5)	1	5% = 0.05

Then based on the following survey respondents are the results of several respondents' assessment of the Alternative Profile of Mount Bromo (A1), Mount Kerinci (A2), Mount Ijen (A3), Mount Rinjani (A4) and Mount Jaya Wijaya (A5).

Table 3. Alternative Profile

No	Alternative	C1	C2	C3	C4	C5
1	Mount Bromo (A1)	5	5	2	2	2
2	Mount Kerinci (A2)	5	4	5	5	3
3	Mount Ijen (A3)	5	3	4	4	5
4	Mount Rinjani (A4)	5	4	3	3	3
5	Mount Jaya Wijaya (A5)	5	2	4	4	4

Then the GAP value is calculated between the subject's profile and the required profile. For more details can be seen in the following table.

Table 4. GAP Value between Alternative Profiles and Required Profiles

No	Criteria Name	A1	A2	A3	A4	A5	Profile Criteria	GAP A1	GAP A2	GAP A3	GAP A4	GAP A5
1	Beauty (C1)	5	5	5	5	5	5	0	0	0	0	0

No	Criteria Name	A1	A2	A3	A4	A5	Profile Criteria	GAP A1	GAP A2	GAP A3	GAP A4	GAP A5
2	Height (C2)	5	4	3	4	2	**4**	1	0	-1	0	-2
3	Location (C3)	2	5	4	3	4	**3**	-1	2	1	0	1
4	Cost (C4)	2	5	4	3	4	**2**	0	3	2	1	2
5	Pathway (C5)	2	3	5	3	4	**1**	1	2	4	2	3

Explanation :
GAP = Alternative Profile - Criteria profile

The next thing to do is calculate the GAP Mapping Value derived from the GAP analysis. For more details can be seen in the following table.

Table 5. GAP Mapping (Based on the GAP map table)

No	Criteria Name	A1	A2	A3	A4	A5	MAP GAP A1	MAP GAP A2	MAP GAP A3	MAP GAP A4	MAP GAP A5
1	Beauty (C1)	0	0	0	0	0	6	6	6	6	6
2	Height (C2)	1	0	-1	0	-2	5,5	6	5	6	4
3	Location (C3)	-1	2	1	0	1	5	4,5	5,5	6	5,5
4	Cost (C4)	0	3	2	1	2	6	3,5	4,5	5,5	4,5
5	Pathway (C5)	1	2	4	2	3	5,5	4,5	2,5	4,5	3,5

a. Calculating the Final Value
   To obtain the final grade, you can use the following formula:
   Final Value = (C1*45%) + (C2*25%) + (C3*15%) + (C4*10%) + (C5*5%)
   1. A1 = (6*45%) + (5,5*25%) + (5*15%) + (6*10%) + (5,5*5%) = 5,7
   2. A2 = (6*45%) + (6*25%) + (4,5*15%) + (3,5*10%) + (4,5*5%) = 5,45
   3. A3 = (6*45%) + (5*25%) + (5,5*15%) + (4,5*10%) + (2,5*5%) = 5,35
   4. A4 = (6*45%) + (6*25%) + (6*15%) + (5,5*10%) + (4,5*5%) = 5,875
   5. A5 = (6*45%) + (4*25%) + (5,5*15%) + (4,5*10%) + (3,5*5%) = 5,15
b. Ranking
   Based on the results of the calculation of the Final Value, the following is a table of the adoption of Alternative values.

Table 6. GAP Mapping (Based on GAP Map Table)

No	Alternative Name	Final Value	Explanation
1	Mount Bromo (A1)	5,7	Rank 2
2	Mount Kerinci (A2)	5,45	Rank 3
3	Mount Ijen (A3)	5,35	Rank 4
4	Mount Rinjani (A4)	5,875	Rank 1
5	Mount Jaya Wijaya (A5)	5,15	Rank 5

4. Conclusion

Based on the problems studied and testing that has been done, it can be concluded that the analysis of the Profile Matching method in recommending climbing tourism destinations, that Mount Rinjani which is Alternative 4 is recommended as the most important climbing tourism destination. Mount Rinjani is a mountain located on the island of Lombok, West Nusa Tenggara. The mountain which is the second-highest volcano in Indonesia with an altitude of 3,726 m above sea level and is located at latitude 8°25 'latitude and 116°28' east longitude. Followed by Mount Bromo in East Java and Mount Kerinci in Jambi province.

References

[1] Q. Hou, X. Meng, X. Huo, Y. Cheng, and J. Leng, "Effects of freeway climbing lane on crash frequency: Application of propensity scores and potential outcomes," *Physica A: Statistical Mechanics and its Applications*, vol. 517, pp. 246–256, 2019.

[2] D. Saul, G. Steinmetz, W. Lehmann, and A. F. Schilling, "Determinants for success in climbing: A systematic review," *Journal of Exercise Science and Fitness*, vol. 17, no. 3, pp. 91–100, 2019.

[3] R. Rahim *et al.*, "C4.5 Classification Data Mining for Inventory Control," *International Journal of Engineering & Technology*, vol. 7, pp. 68–72, 2018.

[4] I. Parlina *et al.*, "Naive Bayes Algorithm Analysis to Determine the Percentage Level of visitors the Most Dominant Zoo Visit by Age Category," *Journal of Physics: Conference Series*, vol. 1255, no. 1, 2019.

[5] I. S. Damanik, A. P. Windarto, A. Wanto, Poningsih, S. R. Andani, and W. Saputra, "Decision Tree Optimization in C4.5 Algorithm Using Genetic Algorithm," *Journal of Physics: Conference Series*, vol. 1255, no. 1, pp. 1–7, 2019.

[6] H. Siahaan, H. Mawengkang, S. Efendi, A. Wanto, and A. Perdana Windarto, "Application of Classification Method C4.5 on Selection of Exemplary Teachers," *Journal of Physics: Conference Series*, vol. 1235, no. 1, 2019.

[7] D. Hartama, A. Perdana Windarto, and A. Wanto, "The Application of Data Mining in Determining Patterns of Interest of High School Graduates," *Journal of Physics: Conference Series*, vol. 1339, no. 1, pp. 1–6, 2019.

[8] M. Widyastuti, A. G. Fepdiani Simanjuntak, D. Hartama, A. P. Windarto, and A. Wanto, "Classification Model C.45 on Determining the Quality of Custumer Service in Bank BTN Pematangsiantar Branch," *Journal of Physics: Conference Series*, vol. 1255, no. 012002, pp. 1–6, 2019.

[9] W. Katrina, H. J. Damanik, F. Parhusip, D. Hartama, A. P. Windarto, and A. Wanto, "C.45 Classification Rules Model for Determining Students Level of Understanding of the Subject," *Journal of Physics: Conference Series*, vol. 1255, no. 012005, pp. 1–7, 2019.

[10] S. Sudirman, A. P. Windarto, and A. Wanto, "Data Mining Tools | RapidMiner : K-Means Method on Clustering of Rice Crops by Province as Efforts to Stabilize Food Crops In Indonesia," *IOP Conference Series: Materials Science and Engineering*, vol. 420, no. 012089, pp. 1–8, 2018.

[11] E. Siregar, H. Mawengkang, E. B. Nababan, and A. Wanto, "Analysis of Backpropagation Method with Sigmoid Bipolar and Linear Function in Prediction of Population Growth," *Journal of Physics: Conference Series*, vol. 1255, no. 1, pp. 1–6, 2019.

[12] G. W. Bhawika *et al.*, "Implementation of ANN for Predicting the Percentage of Illiteracy in Indonesia by Age Group," *Journal of Physics: Conference Series*, vol. 1255, no. 1, pp. 1–6, 2019.

[13] A. Wanto *et al.*, "Analysis of the Accuracy Batch Training Method in Viewing Indonesian Fisheries Cultivation Company Development," *Journal of Physics: Conference Series*, vol. 1255, no. 1, pp. 1–6, 2019.

[14] I. S. Purba *et al.*, "Accuracy Level of Backpropagation Algorithm to Predict Livestock Population of Simalungun Regency in Indonesia Accuracy Level of Backpropagation Algorithm to Predict Livestock Population of Simalungun Regency in Indonesia," *Journal of Physics: Conference Series*, vol. 1255, no. 1, pp. 1–6, 2019.

[15] P. Parulian *et al.*, "Analysis of Sequential Order Incremental Methods in Predicting the Number of Victims Affected by Disasters," *Journal of Physics: Conference Series*, vol. 1255, no. 1, pp. 1–6, 2019.

[16] A. Wanto *et al.*, "Analysis of the Backpropagation Algorithm in Viewing Import Value Development Levels Based on Main Country of Origin," *Journal of Physics: Conference Series*, vol. 1255, no. 1, pp. 1–6, 2019.

[17] S. Setti, A. Wanto, M. Syafiq, A. Andriano, and B. K. Sihotang, "Analysis of Backpropagation

Algorithms in Predicting World Internet Users," *Journal of Physics: Conference Series*, vol. 1255, no. 1, pp. 1–6, 2019.

[18] T. Afriliansyah *et al.*, "Implementation of Bayesian Regulation Algorithm for Estimation of Production Index Level Micro and Small Industry," *Journal of Physics: Conference Series*, vol. 1255, no. 1, pp. 1–6, 2019.

[19] A. Wanto *et al.*, "Forecasting the Export and Import Volume of Crude Oil , Oil Products and Gas Using ANN," *Journal of Physics: Conference Series*, vol. 1255, no. 1, pp. 1–6, 2019.

[20] M. K. Z. Sormin, P. Sihombing, A. Amalia, A. Wanto, D. Hartama, and D. M. Chan, "Predictions of World Population Life Expectancy Using Cyclical Order Weight / Bias," *Journal of Physics: Conference Series*, vol. 1255, no. 1, pp. 1–6, 2019.

[21] W. Saputra, J. T. Hardinata, and A. Wanto, "Resilient method in determining the best architectural model for predicting open unemployment in Indonesia," *IOP Conference Series: Materials Science and Engineering*, vol. 725, no. 1, pp. 1–7, 2020.

[22] A. Wanto and J. T. Hardinata, "Estimations of Indonesian poor people as poverty reduction efforts facing industrial revolution 4 . 0," *IOP Conference Series: Materials Science and Engineering*, vol. 725, no. 1, pp. 1–8, 2020.

[23] A. Wanto *et al.*, "Model of Artificial Neural Networks in Predictions of Corn Productivity in an Effort to Overcome Imports in Indonesia," *Journal of Physics: Conference Series*, vol. 1339, no. 1, pp. 1–6, 2019.

[24] N. Nasution, A. Zamsuri, L. Lisnawita, and A. Wanto, "Polak-Ribiere updates analysis with binary and linear function in determining coffee exports in Indonesia," *IOP Conference Series: Materials Science and Engineering*, vol. 420, no. 012089, pp. 1–9, 2018.

[25] P. P. P. A. N. . F. I. R.H Zer, Masitha, A. P. Windarto, and A. Wanto, "Analysis of the ELECTRE Method on the Selection of Student Creativity Program Proposals," *Journal of Physics: Conference Series*, vol. 1255, no. 1, pp. 1–7, 2019.

[26] A. P. W. Budiharjo and A. Muhammad, "Comparison of Weighted Sum Model and Multi Attribute Decision Making Weighted Product Methods in Selecting the Best Elementary School in Indonesia," *International Journal of Software Engineering and Its Applications*, vol. 11, no. 4, pp. 69–90, 2017.

[27] D. R. Sari, N. Rofiqo, D. Hartama, A. P. Windarto, and A. Wanto, "Analysis of the Factors Causing Lazy Students to Study Using the ELECTRE II Algorithm," *Journal of Physics: Conference Series*, vol. 1255, no. 012007, pp. 1–6, 2019.

[28] K. Fatmawati *et al.*, "Analysis of Promothee II Method in the Selection of the Best Formula for Infants Under Three Years," *Journal of Physics: Conference Series*, vol. 1255, no. 1, pp. 1–7, 2019.

[29] P. Alkhairi, L. P. Purba, A. Eryzha, A. P. Windarto, and A. Wanto, "The Analysis of the ELECTREE II Algorithm in Determining the Doubts of the Community Doing Business Online," *Journal of Physics: Conference Series*, vol. 1255, no. 1, pp. 1–7, 2019.

[30] S. Sundari, Karmila, M. N. Fadli, D. Hartama, A. P. Windarto, and A. Wanto, "Decision Support System on Selection of Lecturer Research Grant Proposals using Preferences Selection Index," *Journal of Physics: Conference Series*, vol. 1255, no. 1, pp. 1–7, 2019.

[31] S. R. Ningsih, R. Wulansari, D. Hartama, A. P. Windarto, and A. Wanto, "Analysis of PROMETHEE II Method on Selection of Lecturer Community Service Grant Proposals," *Journal of Physics: Conference Series*, vol. 1255, no. 1, pp. 1–7, 2019.

[32] T. Imandasari, M. G. Sadewo, A. P. Windarto, A. Wanto, H. O. Lingga Wijaya, and R. Kurniawan, "Analysis of the Selection Factor of Online Transportation in the VIKOR Method in Pematangsiantar City," *Journal of Physics: Conference Series*, vol. 1255, no. 1, pp. 1–7, 2019.

[33] T. Susilowati, E. Y. Anggraeni, Fauzi, W. Andewi, Y. Handayani, and A. Maseleno, "Using Profile Matching Method to Employee Position Movement," *International Journal of Pure and Applied Mathematics*, vol. 118, no. 7, pp. 415–423, 2018.

[34] Z. Tharo and A. P. Utama Siahaan, "Profile Matching in Solving Rank Problem," *IOSR Journal of Electronics and Communication Engineering*, vol. 11, no. 05, pp. 73–76, 2016.

Conjugate Gradient Polak Ribiere In Improving Performance in Predicting Population Backpropagation

Medi Herman Tinambunan[1*], Erna Budhiarti Nababan[2], Benny Benyamin Nasution[2]

[1]Universitas Sumatera Utara, Medan - Indonesia
[2]Politeknik Negeri Medan, Medan - Indonesia

*medi.tinz93@gmail.com

Abstract. Perform predictive data in the form of time series required a correct method, one method is now often used is the propagation of this method is a method that is able to minimize the error value of the output of the predicted number of people, but still generate quite a lot of iteration that needs to be optimized by minimize iterations and use of time, then the use of conjugate gradient polak Ribiere are expected to minimize the use of time, the number of the epoch of the results of standard backpropagation.

1. Introducing

This time the prediction is very often done to determine the circumstances in the future, for example, a company wants to make predictions about earnings in the period of 10 years, a football coach to predict the strategies to be carried out by opponents that the team can do prevention at opponent's strategy. A good predictor that has a range of very slight difference with the data of fact, the less difference prediction data with the data facts, the better the accuracy of these predictions. This case discusses the prediction of the number of people with one method in which back propagation neural network. Backpropagation originally introduced by Rumelhart [1]. The algorithm will be improving the performance by using conjugate gradient. There are several types of Conjugate Gradient method, one of them is Polak Ribiere. In certain cases, Polak Ribiere to find a solution even though the starting point away from the point of minimum so as to converge more quickly than before. Thus in this research network training using the Conjugate Gradient Polak Ribiere on Backpropagation in the case study predicted the number of people in the province of North Sumatra. Conjugate Gradient Polak Ribiere on Backpropagation is one of the neural network methods that is widely used to predict. Moreover, there have been many studies using neural network methods [2]–[15].

Projected population is predicted the number of people based on a particular method by assuming births, deaths and migration. According to demographers, the prediction is generally used to predict the number of people for a period between census and after the census was held in a certain area, While the projection implies that the estimated number of people in the future who have not known the exact number, and the future value of an indication of the number of future residents if implemented in fertility, morality and certain immigration that may apply.

In previous studies, [16] conducted a study in predicting population density by using the binary sigmoid activation function and linear function. The forecast accuracy using a combination of these functions is 94%. The weakness of this study is that this study only uses binary sigmoid functions and linear functions without discussing the function of bipolar sigmoid at all. Therefore, by applying the method Conjugate Gradient Polak Ribiere expected to improve the accuracy of prediction of the number of people in North Sumatra province so that it can cope with a range difference prediction results with the original data is not too far away.

2. Research methods

This study uses data from the Central Bureau of Statistics of North Sumatra. The data used has a time span from 2011-2018 year, each data can be grouped based on the number of cities and districts in North Sumatra. Here is the data used;

Table 1. Data Population of Sumatra Utara (North Sumatra, Central Bureau of Statistics, 2019)

Years	Nias	Mandailing Natal	South Tapanuli	Central Tapanuli	...	Field	Binjai	Padangsidimpuan	Gunung Sitoli
2011	132605	408731	266282	314142	...	2117224	248456	193322	127382
2012	132860	410931	268095	318908	...	2122804	250252	198809	128337
2013	133388	413475	268824	324006	...	2123210	252263	204615	129403
2014	133388	413475	268824	324006	...	2123210	252263	204615	129403
2015	136115	430894	275098	350017	...	2210624	264687	209796	135995
2016	141403	435303	276889	356918	...	2229408	267901	212917	137693
2017	142110	439505	278587	363705	...	2247425	270926	216013	139281
2018	142840	443490	280283	370171	...	2264145	273892	218892	140927

The initial step in solving problems is to first make the selection data from the Central Statistics Agency website North Sumatra, then the data will be prepared to do the processing using standard backpropagation method and conjugate gradient backpropagation polak Ribiere. The next step the data will be normalized the data using a sigmoid function which is an asymptotic function did not reach 0 or 1 then needs to be normalized, Once the data is normalized will be made to the training network using a combination of inputs 3-5-1, 4-4-1, 5-3-1, 6-2-1, 2-6-1 this is done to get maximum results in the training process. In the case raised the authors used a combination of input 3-5-1 because it scores fewer errors and has better accuracy results. Once the data is normalized already done training the network and get the testing data that has been normalized earlier into a standard back propagation method This method is used because the method is capable of formulating knowledge and experience and is also flexible enough to change the rules of forecasts, Then, after the data is processed using backpropagation standards and get the next result of the same data is used to perform processing by using backpropagation optimized using the conjugate gradient polak Ribiere to get results in training close to the target and then do the testing, the results of the testing will be conducted comparison of the results so that it can be deduced whether by adding a conjugate gradient method to optimize the results polak Ribiere of accuracy compared with the results of standard propagation.

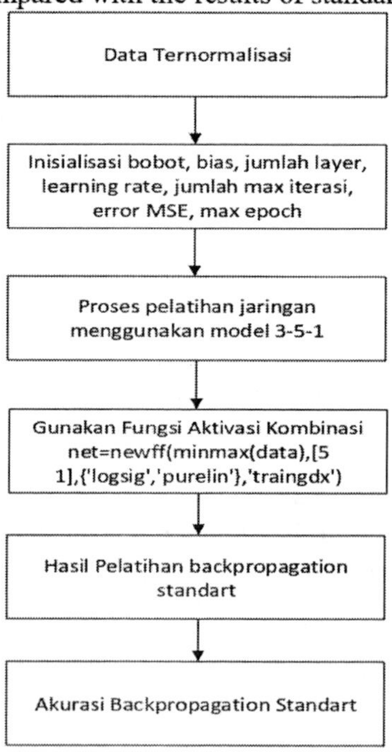

Figure 1. Block Diagram of Standard Backpropagation Methods

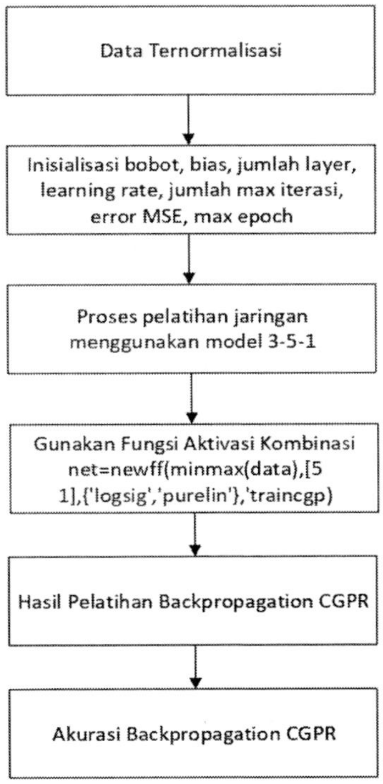

Figure 2. Block Diagram Backpropagation Conjugate Gradient Method Polak Ribiere

3. Results and Discussion

3.1. Standard Backpropagation Network

By using the standard back propagation method is carried out training of the data that has been normalized. Based on the data obtained epoch were 71 and working time 8 seconds performance results of the training of BP Standard are as follows:

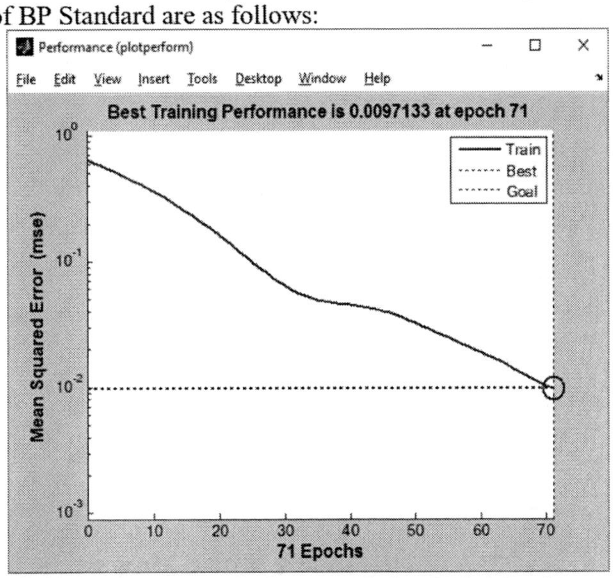

Figure 3. Training Performance Prediction Population Using Backpropagation Standart

From the results of the training showed that finding the best training patterns in the epoch that is large enough that 71 epoch and his best training performance MSE value is 0.0097133.

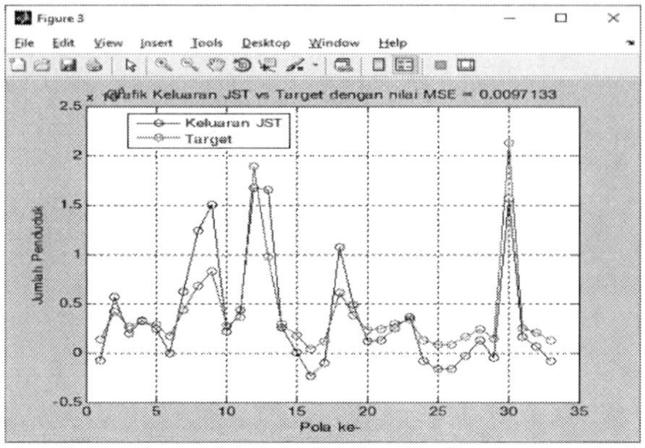

Figure 4. Comparison Chart View with Results Output Target Population Prediction Using Backpropagation Standart

From the results above shows that the output data of the neural network copies are still too far from the target so it needs to be maximized.

3.2. Network Conjugate Gradient Polak Ribiere

After training using standard backpropagation then done using conjugate gradient Backporpagation polak Ribiere with the data is normalized and the selection of training data and the data on the same target the MSE value obtained by 0.000054846 convergence in the epoch to 4. Here is a figure of the results of MSE:

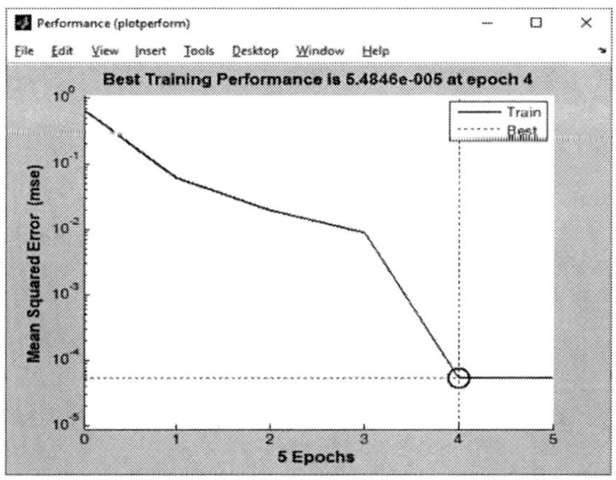

Figure 5. Backpropagation Training Performance Display CGPR in Predicting Population

From the picture above we can say that by using conjugate gradient obtained polak Ribiere 4 epoch and the time required in the process is only 1 second.

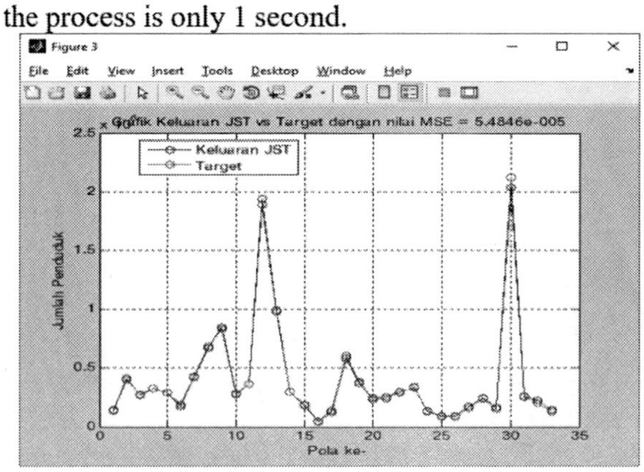

Figure 6. Graph View CGPR Backpropagation Training in Predicting Population

Based on the chart above shows that the output data of the neural network is almost close to the target or it could be said to have a range that is not too far from the targeted data.

3.3. Testing

Table 2. Results Projected Population North Sumatra

Target Data	Backpropagation Standard	Backpropagation CGPR
142 840	54 456	159 850
443 490	638 993	424 178
280 283	219 546	282 032
370 171	434 043	365 694
299 881	265173	299 627
182 673	13836	192 457
486 480	748 004	466 623
724 379	1319830	712 493
863 693	1540520	871 622
283 203	227 003	284 133
409 675	540 044	398 739
2155625	1647215	2111081
1035411	1705567	1048274
317 207	306 064	315 687
188 480	24560	199 074
48 119	222 524	52200
125 816	91 062	135 589
614 618	1093532	586 406
412 992	554 952	397 810
267 771	187 035	276 141
275 515	203 005	283 790
332 922	339 282	333 952
360 926	416 519	352 757
137 002	70 876	147 754
81 663	173 402	81 866
87 317	157 315	94 837
173 302	4942	184 695
253 500	158 710	258 967
162 581	24 626	174 240
2264145	1608996	2136018
273 892	203 247	278 524
218 892	84 525	229 321
140 927	64 278	152 256
MSE	0.0097133	0.000054846

4. Conclusion

After testing the results of research that discussed the problems of the conclusions obtained are as follows;

a. Backpropagation and conjugate gradient method polak Ribiere can be used to perform the prediction process.
b. Standard backpropagation method has poor accuracy in predicting the data changes.
c. The combination of the conjugate gradient backpropagation method polak Ribiere and using sigmoid activation function and Biner can improve performance in this case the time series data prediction accuracy significantly nearing the target data to the MSE value of 0.000054846.

References

[1] N. P. Sakinah, I. Cholissodin, and A. W. Widodo, "Prediksi Jumlah Permintaan Koran

Menggunakan Metode Jaringan Syaraf Tiruan Backpropagation," vol. 2, no. 7, pp. 2612–2618, 2018.

[2] E. Siregar, H. Mawengkang, E. B. Nababan, and A. Wanto, "Analysis of Backpropagation Method with Sigmoid Bipolar and Linear Function in Prediction of Population Growth," *Journal of Physics: Conference Series*, vol. 1255, no. 1, pp. 1–6, 2019.

[3] G. W. Bhawika *et al.*, "Implementation of ANN for Predicting the Percentage of Illiteracy in Indonesia by Age Group," *Journal of Physics: Conference Series*, vol. 1255, no. 1, pp. 1–6, 2019.

[4] W. Saputra, J. T. Hardinata, and A. Wanto, "Resilient method in determining the best architectural model for predicting open unemployment in Indonesia," *IOP Conference Series: Materials Science and Engineering*, vol. 725, no. 1, pp. 1–7, 2020.

[5] A. Wanto and J. T. Hardinata, "Estimations of Indonesian poor people as poverty reduction efforts facing industrial revolution 4 . 0," *IOP Conference Series: Materials Science and Engineering*, vol. 725, no. 1, pp. 1–8, 2020.

[6] A. Wanto *et al.*, "Model of Artificial Neural Networks in Predictions of Corn Productivity in an Effort to Overcome Imports in Indonesia," *Journal of Physics: Conference Series*, vol. 1339, no. 1, pp. 1–6, 2019.

[7] A. Wanto *et al.*, "Analysis of the Accuracy Batch Training Method in Viewing Indonesian Fisheries Cultivation Company Development," *Journal of Physics: Conference Series*, vol. 1255, no. 1, pp. 1–6, 2019.

[8] I. S. Purba *et al.*, "Accuracy Level of Backpropagation Algorithm to Predict Livestock Population of Simalungun Regency in Indonesia Accuracy Level of Backpropagation Algorithm to Predict Livestock Population of Simalungun Regency in Indonesia," *Journal of Physics: Conference Series*, vol. 1255, no. 1, pp. 1–6, 2019.

[9] P. Parulian *et al.*, "Analysis of Sequential Order Incremental Methods in Predicting the Number of Victims Affected by Disasters," *Journal of Physics: Conference Series*, vol. 1255, no. 1, pp. 1–6, 2019.

[10] A. Wanto *et al.*, "Analysis of the Backpropagation Algorithm in Viewing Import Value Development Levels Based on Main Country of Origin," *Journal of Physics: Conference Series*, vol. 1255, no. 1, pp. 1–6, 2019.

[11] S. Setti, A. Wanto, M. Syafiq, A. Andriano, and B. K. Sihotang, "Analysis of Backpropagation Algorithms in Predicting World Internet Users," *Journal of Physics: Conference Series*, vol. 1255, no. 1, pp. 1–6, 2019.

[12] T. Afriliansyah *et al.*, "Implementation of Bayesian Regulation Algorithm for Estimation of Production Index Level Micro and Small Industry," *Journal of Physics: Conference Series*, vol. 1255, no. 1, pp. 1–6, 2019.

[13] A. Wanto *et al.*, "Forecasting the Export and Import Volume of Crude Oil , Oil Products and Gas Using ANN," *Journal of Physics: Conference Series*, vol. 1255, no. 1, pp. 1–6, 2019.

[14] M. K. Z. Sormin, P. Sihombing, A. Amalia, A. Wanto, D. Hartama, and D. M. Chan, "Predictions of World Population Life Expectancy Using Cyclical Order Weight / Bias," *Journal of Physics: Conference Series*, vol. 1255, no. 1, pp. 1–6, 2019.

[15] N. Nasution, A. Zamsuri, L. Lisnawita, and A. Wanto, "Polak-Ribiere updates analysis with binary and linear function in determining coffee exports in Indonesia," *IOP Conference Series: Materials Science and Engineering*, vol. 420, no. 012089, pp. 1–9, 2018.

[16] A. Wanto, A. P. Windarto, D. Hartama, and I. Parlina, "Use of Binary Sigmoid Function And Linear Identity In Artificial Neural Networks For Forecasting Population Density," *International Journal Of Information System & Technology*, vol. 1, no. 1, pp. 43–54, 2017.

IOP Conf. Series: Materials Science and Engineering **835** (2020) 012056 doi:10.1088/1757-899X/835/1/012056

Effect of chitosan-zinc oxide coated in PVDF membrane; morphology and performance testing

A Febriasari[1], D Siswanta[2], N Riyanto[1], N Hidayat Aprilita[2], F Silvianti[3]

[1]Department of Chemical Engineering, Universitas Serang Raya, Banten, Indonesia
Email: arifinafebriasari@lppmunsera.ac.id or arifinafebriasari1725@gmail.com
[2]Universitas Gadjah Mada, Yogyakarta, Indonesia
[3]Department of Rubber and Plastic Processing Technology, Politeknik ATK Yogyakarta, Yogyakarta, Indonesia

Abstract. PVDF membrane modification has been done by coating it using chitosan which is mixed with zinc oxide. This study was conducted to determine the effect of chitosan-zinc oxide hybrid on PVDF membrane's hydrophobicity and performance. The coating has been processed by immersing PVDF membrane into a chitosan-zinc oxide hybrid solution with a ratio chitosan : zinc oxide of 1: 0 (M1), 6: 1 (M2), 4: 1 (M3), and 2: 1 (M4). Membrane morphology characterization was carried out by FE-SEM and FTIR. Water contact angle analysis and performance test was conducted to determine the effect of chitosan and zinc oxide on the hydrophobicity, permeability, percent rejection, and antifouling performance of PVDF membrane. The result of water contact angle analysis shows that the more zinc oxide was added to chitosan-zinc oxide hybrid, the hydrophobicity of membrane was decreased. Membrane performance test shows that the most stable membrane permeability to time is membrane M4, while the membrane with the most stable permeability to pressure is PVDF membrane without coating (membrane M). The best rejection performance ad flux recovery ratio of membranes is shown by membrane M4.

1. Introduction

Currently the development of membrane technology is still interested according to its application, whether for waste water separation, gas separation, biodiesel separation, and others. Poly-vinylidine fluoride (PVDF) is a polymer commonly used as a membrane base polymer. PVDF has the advantage that it is stable against chemicals, it also has good mechanical strength and thermal stability. PVDF membrane applications can be used both for ultrafiltration, microfiltration, membrane bioreactor, and gas separation. One of the weaknesses of pure PVDF membranes and membrane-based polymers in general is that it is easy to foul on the membrane surface which causes significant flux reduction in the compound transport process. The fouling is caused by the deposit of both chemical and microbiological substances on the membrane surface [1]. In this article modification on flat sheet PVDF membrane was carried out using chitosan and Zinc oxide to overcome the weakness of fouling on membranes [2]. Chitosan is a natural polymer which is the result of deacetylation of chitin which can be obtained from the skin of shrimp or crab. Chitosan has been widely used for the pharmaceutical industry, but currently many also examine the benefits of chitosan as an adsorbent for processing wastewater. Benavente (2008) examined chitosan as an adsorbent for metal ions, Oktari (2014) examined chitosan in the form of composite films modified with κ-Karaginan for the adsorption of methylene blue, and many more studies on chitosan with its application as adsorbents.Chitosan has the potential as a good adsorbent because it has chelating groups [3] . Goy et al. (2016) also reported the ability of chitosan in its activity as an antibacterial. With this consideration, chitosan is chosen as one of the substances used as coatings on PVDF membranes to prevent microbial buildup on the membrane surface [4]. Zinc oxide is a particle that has antimicrobial ability so that it can dispel bacteria that accumulate on the membrane surface. In addition, zinc oxide also has the ability to reduce the wetting effect that can occur on the membrane surface. Several types of nanoparticles have been studied that have potential in the process of degradation of organic matter and bacteria. Kumar et al. (2013) examined the potential of zinc oxide nanoparticles in the process of degradation of methyl orange dyes [5]. Sanna et al. (2016) also examined that the efficiency of degradation of organic dyes with zinc oxide nanoparticles was quite high in the sun [6]. And there are many more studies regarding the potential degradation of nanoparticles against organic matter. In addition to organic substances, Barnes et al. (2013) that titanium oxide and zinc oxide nanoparticles have the potential to degrade E. coli, S. aureus, P. aeruginosa, and B. subtilis bacteria [7]. It can be concluded that zinc oxide nanoparticles can have the potential to be applied in the waste treatment process. Research by combining polymeric membranes with nanoparticles is expected to not only stabilize polymer [4], the function of nanoparticles is also used as antifouling and anti microbacterial agents on membranes [8]

2. Materials and methods

2.1 Synthesis of zinc oxide.

A total of 50 grams of zinc acetate were dissolved in 100 mL of 0.5 M hydrochloric acid, then ammonium hydroxide 0.5 M was added. The pH of the solution was adjusted to pH 9 using concentrated ammonia. The solution was stirred with a hot plate at 85 °C for 5 hours. The precipitate formed was filtered and washed with distilled water. Then the sediment was dried in an oven at 130 °C and characterized by FE-SEM (JSM-7100F) [9]

2.2 Hybrid synthesis of chitosan-zinc oxide

A number of chitosan was dissolved in 2% acetic acid with a ratio of 1: 50, then stirred and heated at a temperature of 60 °C. After the chitosan solution is homogeneous, then mixed with zinc oxide with a weight ratio of chitosan and zinc oxide respectively; 2:1; 4:1; and 8:1. Stirring the solution is done at 1300 rpm for 30 minutes. Then the solution was taken 50 mL to be printed with Teflon mold and dried in an oven at 60 0C. After the membrane is formed then it is characterized. [2]

2.3 PVDF membrane coated with chitosan-zinc oxide

PVDF flat sheet membrane coated with chitosan-zinc oxide by immersion method. PVDF flat sheet was immersed in a mixture of chitosan-zinc oxide for 12 hours with a variety of chitosan-zinc oxide comparisons; 1: 0 (M1), 6:1 (M2), 4:1 (M3), and 2:1 (M4). Then the membrane was dried in an oven at 60 ° C. The membrane that has been dried was then morphologically characterized by FE-SEM (JSM-7100F), EDS mapping, FTIR, and water contact angle. PVDF membrane without layers (M) is also characterized as control.

2.4 Performance test with methylene blue

This application and performance test was conducted to determine the effect of application time on the flux value, the effect of pressure on the flux value, calculate the percent rejection of the membrane to the target compound, and calculate the flux recovery ratio (FRR) to determine the antifouling properties of the membrane. Calculation of percent rejection is done using the following formula

$$R\,(\%) = \left(1 - \frac{C_p}{C_f}\right) * 100 \qquad (1)$$

Where C_p and C_f are the concentration of the target compound in the permeate solution and the feed solution. Whereas to find out the FRR value, the following formula is used:

$$FRR\ (\%) = \left(\frac{J_{w2}}{J_{w1}}\right) * 100 \qquad (2)$$

Where J_{w1} is the value of membrane flux when the target compound is passed, and J_{w2} is the value of flux after the backwash membrane uses demineralized water. [10]

3. Result and Discussion

3.1 Morphological characterization

FE-SEM and EDS mapping identify morphological differences in each membrane M, M1, M2, M3, and M4 based on variations in membrane coating levels with chitosan and zinc oxide particles. Figure 1 shows the differences in PVDF membrane morphology before and after coated by chitosan (M and M1).

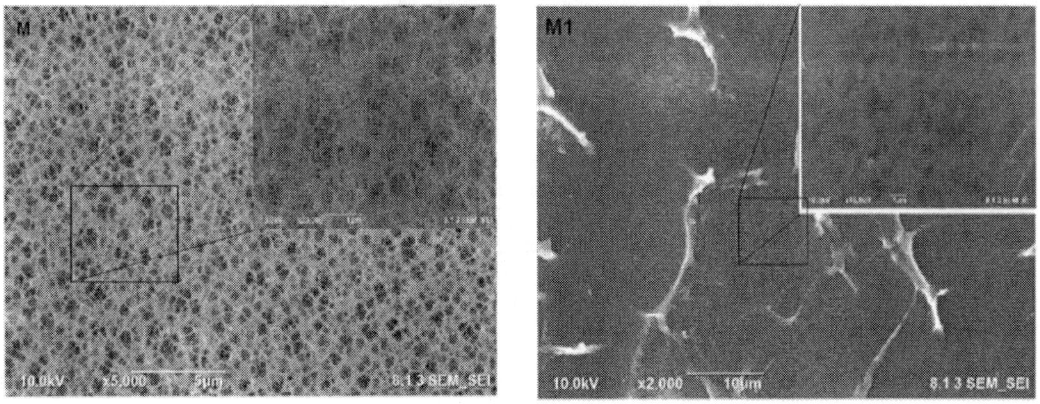

Figure 1. FE-SEM characterization results on M and M1 membranes

In Figure 1, membrane M pores appear more clearly than M1 membranes. M1 show that chitosan coats the pores of the PVDF membrane so that the PVDF membrane pore becomes faint. This will certainly affect the membrane filtration process. PVDF membrane has a pore size of 0.1 μ. After coating, the surface of the membrane becomes multilayer with the first layer is more dense than the second layer which is porous (anisotropic membrane)[11].

Coating membrane results with mixed matrix chitosan-zinc oxide can be seen in Figure 2 (red for the distribution of Zn elements and green for oxygen element distribution). SEM membrane images M2, M3, and M4 show different morphological structures in the EDS membrane M2, M3 and M4 mapping results showing the distribution map of zinc oxide particles in each membrane based on the ratio of chitosan and zinc oxide to the solution coated on the membrane surface. The distribution of Zn elements in the EDS map is proportional to the membrane surface roughness shown in the FESEM image. FESEM results on the M4 membrane show a kind of particle that is spread like a needle. It can be assumed that these particles are zinc oxide particles

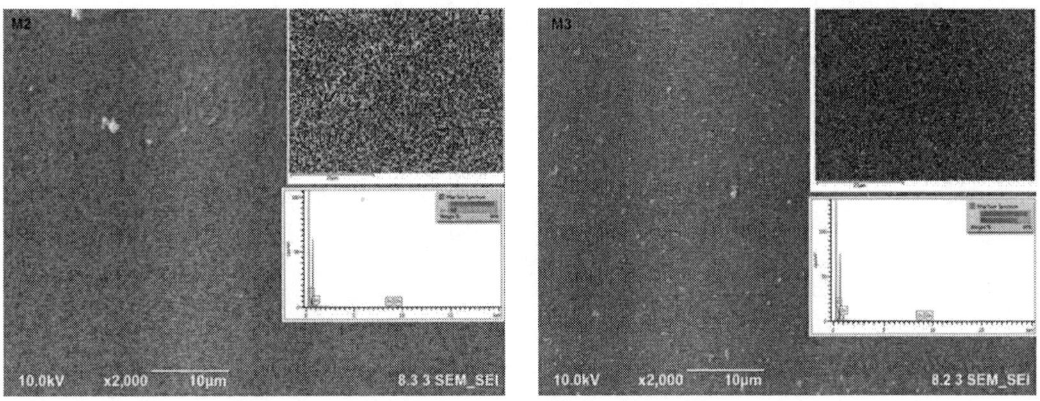

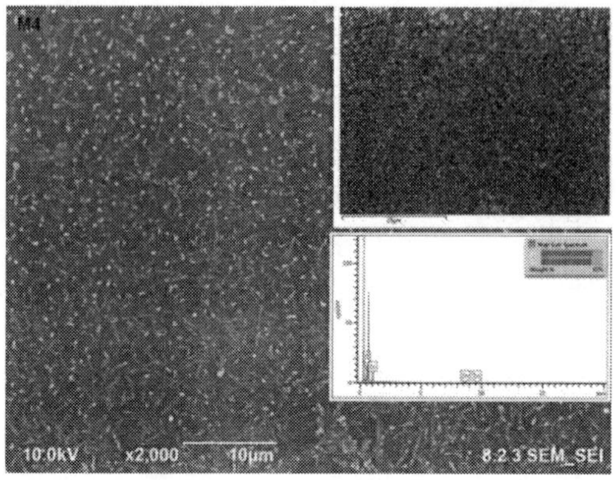

Figure 2. Results of FE-SEM and EDS mapping characterization for M2, M3 and M4 membranes. Red color represent Zn elements, green color represent oxygen elements.

Zinc oxide particles on the membrane are propagated and immobilized on the surface of the membrane with the aim that the antifouling function is obtained optimally. Zinc oxide on the membrane surface is expected to degrade deposited substances on the surface so that the backwash process can produce membranes that are free of substances that cause fouling and reduce membrane permeability. From the results of FESEM, we can conclude that the process of immobilizing zinc oxide particles on the membrane surface was successfully carried out. Mainly indicated by the M4 membrane in which zinc oxide particles are seen to spread evenly on the surface of the membrane.

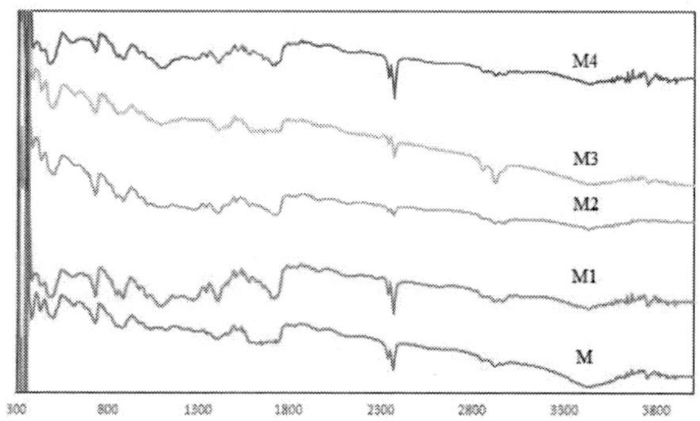

Figure 3. FTIR spectra of membranes

FTIR spectra of membranes M2, M3, and M4 can be seen in Figure 3. It can be seen that the peak at wave number 840 cm^{-1} shows the beta phase vibration of PVDF polymer, the number 2300-2400 cm^{-1} shows the vibration of C and N double bonds asymmetry in chitosan, wave numbers 2700-2978 show CH_2 bonds in chitosan [12][13]. The greater the intensity of the peak in the wave number between 300 and 400 cm^{-1} where in the wave number Zn element is detected [2]

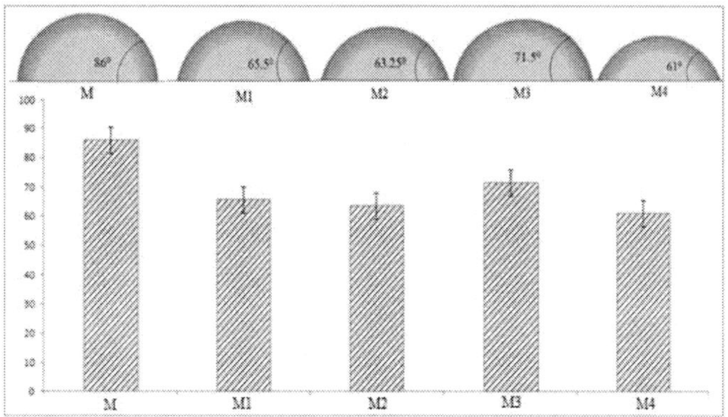

Figure 4. Water contact angle of membranes M, M1, M2, M3, M4

Water contact angle analysis on the membrane surface shows how much influence the coating of chitosan and zinc oxide has on the hydrophilicity properties of PVDF membrane surface. From Figure 4, it can be seen that the addition of chitosan (M1) can improve the hydrophilic properties of the membrane surface due to the chitosan properties that are easily bonded to water molecules. The addition of zinc oxide (M2, M3, and M4) further increases the surface hydrophilicity of PVDF membranes due to ionic properties in zinc oxide so that it is easy to bind to water. A slight increase in M4 is due to the uneven distribution of zinc oxide on the membrane surface [13][14]

3.2 Performance test of membranes

In this performance test, variations in pressure and time are taken to see the permeability capability of the membrane.

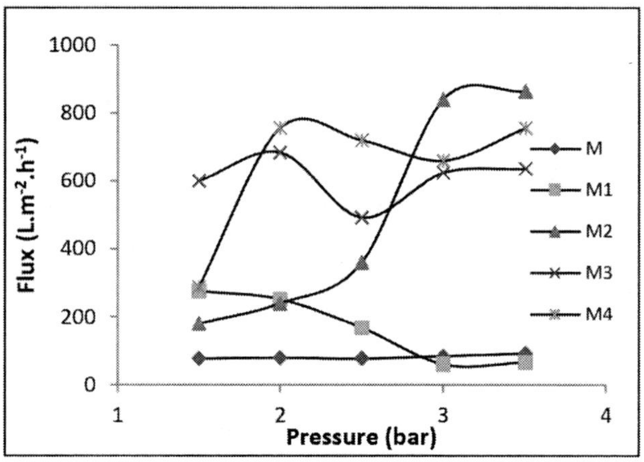

Figure 5. Membrane permeability test for flux value pressure variations

Graph on the figure 4 shows a fairly low and stable flux value of membrane M. The value of flux increases with the addition of chitosan and zinc oxide on membranes M1, M2, M3, and M4, this is because chitosan and zinc oxide make the surface of the membrane more hydrophilic, so that water is more easily absorbed to the membrane surface [15].

There are several factors that cause membrane flux values to go down or increase. The lower membrane flux indicates fouling caused by deposit of substances on the surface of the membrane. While the increase in membrane flux is usually caused due to membrane imbalance to the pressure applied.The increase in flux that occurs significantly in the M2 membrane is due to membrane instability to the pressure applied.

From the graph we can see that the pressure applied to the membrane actually affects the instability of chitosan and zinc oxide in membrane particles, therefore the application of the synthesis synthesis membrane should be carried out at a low pressure range of 1-2 bar.

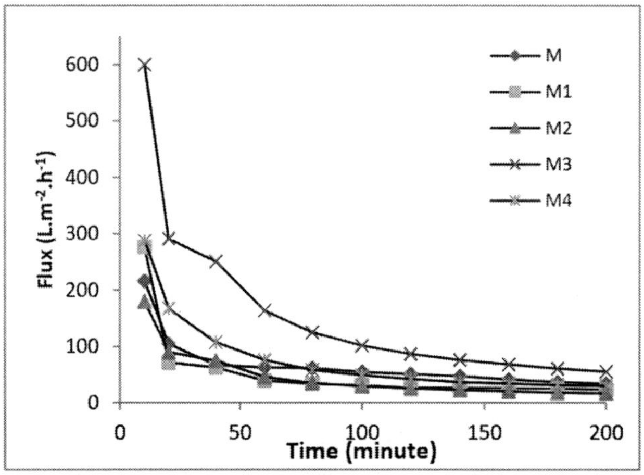

Figure 6. Membrane permeability test time

The permeability test is then carried out using 7 ppm methylene blue solution to determine the stability of membrane permeation over time. In this test, it was found that M3 membrane had the most significant decrease in flux compared to other membranes. While the membrane that appears to be stable with time is the M4 membrane.

As we discussed earlier, that membrane flux can be reduced due to fouling that occurs on the surface of the membrane. This is a normal condition as the phenomenon of fouling on the surface of the membrane cannot be avoided. Membrane stability over time can be influenced by several factors, including the appropriate operating pressure and antifouling performance on the membrane surface.

In this study we try to immobilize zinc oxide particles on the membrane surface in the hope that the membrane can reduce the percentage of fuling by itself so that it can reduce the time and water requirements for the backwash process.

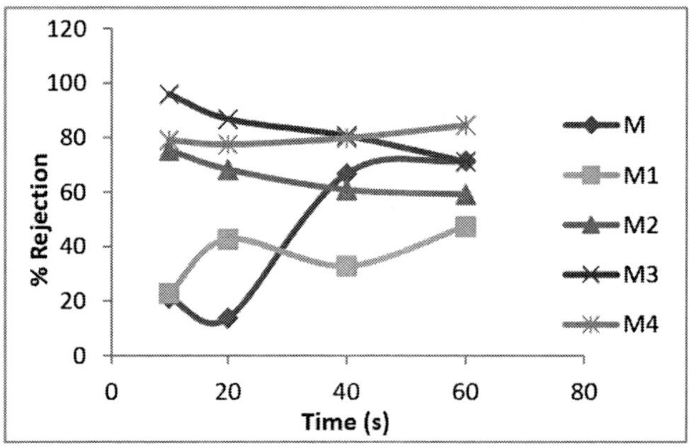

Figure 7. The percent rejection graph of membranes

The percent rejection graph in Figure 6 shows the performance of M, M1, M2, M3, and M4 membranes in rejecting methylene blue compounds with an initial concentration of 7 ppm. The highest percentage of rejection at the initial minute is owned by the M3 membrane, where the membrane is a PVDF membrane coated with chitosan-zinc oxide with a ratio of 4: 1. However, the percentage of rejection on this membrane decreased although not significant, this indicates that fouling affects the membrane rejection process against methylene blue compounds from the initial minute. While membrane M4 also has a high percentage of rejection compared to other membranes and tends to rise. This shows that at the initial minute, the rejection process can still increase even though the transport process has been going on for 1 hour. Excessive and inequality addition of zinc oxide to the coating process can also potentially damage the membrane because of its ability to degrade organic compounds. So it requires optimal levels of zinc oxide and the selection of polymers that are quite stable such as PVDF [16]

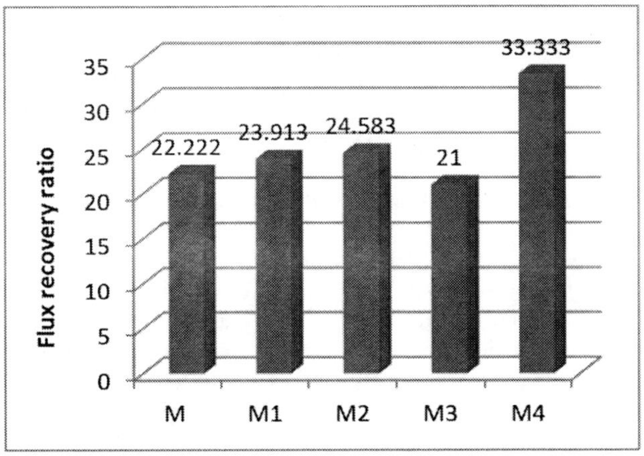

Figure 8. Flux recovery ratio of membranes

The level of flux recovery ratio (FRR) on the membrane shown in Graph 5.7 above identifies the antifouling performance of each membrane. Overall the membrane FRR value is below 50, which means that the membrane antifouling performance is still quite low. The effect of zinc oxide is seen in the increase in FRR values in M2 and M4 membranes. Zinc oxide is thought to reduce fouling on the surface of the membrane because of its ability to degrade pollutants deposited on the surface of the membrane [17]. However, this effect is not too significant.

This condition can be caused due to lack of reactivity of zinc oxide particles so that the degradation efficiency of zinc oxide particles against methylene blue is less optimal. Degradation particles such as zinc oxide will be more reactive if they are in smaller sizes such as nano size. Although it is feared that the more reactive zinc oxide particles will actually cause membrane instability itself because the membrane is made from organic polymers. Therefore the choice of the stability of the membrane material against the degrading agent itself is very necessary

4. Conclusions

PVDF membrane coating with chitosan-zinc oxide has been done by immersion method. The effect of chitosan-zinc oxide was identified by morphological characterization, water contact angle analysis, and performance test. The results of morphological characterization with SEM-EDS show that the spread of zinc oxide particles is most evenly distributed on the M4 membrane (membranes coated with chitosan-zinc oxide in a ratio of 2: 1). Analysis of water contact angle shows that more zinc oxide is added to the coating process, the higher the hydrophilicity of the membrane surface. Membrane performance test shows that the most stable membrane permeability to time is membrane M4, while the membrane with the most stable permeability to pressure is PVDF membrane without coating (membrane M). The best rejection performance of membranes is shown by membrane M4. While the antifouling performance of the coated membrane is not too significant compared with PVDF membrane (M) despite it has increased.

Acknowledgement

This research was funded by the Ministry of Research and Higher Education (DRPM RISTEKDIKTI) of Indonesia.

References

[1] S. Judd 2017 , "The status of membrane bioreactor technology," *Trends Biotechnol.*, vol. 26, no. 2, pp. 109–116, 2008.

[2] I. Munnawar *et al.* 2017, "Synergistic effect of Chitosan-Zinc Oxide Hybrid Nanoparticles on antibiofouling and water disinfection of mixed matrix polyethersulfone nanocomposite membranes," *Carbohydr. Polym.*, vol. 175, pp. 661–670.

[3] M. Benavente 2008, *Adsorption of metallic ions onto chitosan: equilibrium and kinetic studies*, no. May.

[4] H. Bai *et al.* 2015, "Multi-ion-crosslinked nanoparticles with pH-responsive characteristics for oral delivery of protein drugs," *J. Memb. Sci.*, vol. 12, no. 1, pp. 941–949.

[5] R. Kumar, G. Kumar, and A. Umar 2013, "ZnO nano-mushrooms for photocatalytic degradation of methyl orange," *Mater. Lett.*, vol. 97, pp. 100–103.

[6] V. Sanna, N. Pala, V. Alzari, D. Nuvoli, and M. Carcelli 2016, "ZnO nanoparticles with high degradation efficiency of organic dyes under sunlight irradiation," *Mater. Lett.*, vol. 162, pp. 257–260.

[7] R. J. Barnes, R. Molina, J. Xu, P. J. Dobson, and I. P. Thompson 2013, "Comparison of TiO2 and ZnO nanoparticles for photocatalytic degradation of methylene blue and the correlated inactivation of gram-positive and gram-negative bacteria," *J. Nanoparticle Res.*, vol. 15, no. 2.

[8] K. S. V Krishna Rao, P. Ramasubba Reddy, Y. I. Lee, and C. Kim 2012, "Synthesis and characterization of chitosan-PEG-Ag nanocomposites for antimicrobial application," *Carbohydr. Polym.*, vol. 87, no. 1, pp. 920–925.

[9] A. Gnanaprakasam, V. M. Sivakumar, P. L. Sivayogavalli, and M. Thirumarimurugan 2015, "Characterization of TiO2 and ZnO nanoparticles and their applications in photocatalytic degradation of azodyes," *Ecotoxicol. Environ. Saf.*, vol. 121, pp. 121–125.

[10] I. G. Wenten, P. T. P. Aryanti, and a N. Hakim 2016, "Scale-up Strategies for Membrane-Based Desalination Processes: A Review," *J. Membr. Sci. Res.*, vol. 2, pp. 42–58.

[11] M. N. Kajama, H. Shehu, E. Okon, and E. Gobina 2015, "Preparation and Characterization of Inorganic Membranes for Hydrocarbon Separation from N2 for Environmental Applications," *Energy Environ. Res.*, vol. 5, no. 1, pp. 110–120.

[12] J. Zahirifar, S. M. A. Moosavian, A. Hadi, P. Khadiv-Parsi, and J. Karimi-Sabet 2018, "Fabrication of a novel octadecylamine functionalized graphene oxide/PVDF dual-layer flat sheet membrane for desalination via air gap membrane distillation," *Desalination*, vol. 428, no. November 2017, pp. 227–239.

[13] H. M. Hegab, Y. Wimalasiri, M. Ginic-Markovic, and L. Zou 2015, "Improving the fouling resistance of brackish water membranes via surface modification with graphene oxide functionalized chitosan," *Desalination*, vol. 365, pp. 99–107.

[14] S. H. Woo, B. R. Min, and J. S. Lee 2017, "Change of surface morphology, permeate flux, surface roughness and water contact angle for membranes with similar physicochemical characteristics (except surface roughness) during microfiltration," *Sep. Purif. Technol.*, vol. 187, pp. 274–284.

[15] K. E. Geckeler and H. Court 1996, "Removal of Hazardous Substances from Water Using Ultrafiltration in Conjunction with Soluble Polymers," *Environ. Sci. Technol.*, vol. 30, no. 3, pp. 725–734.

[16] G. R. Xu, J. N. Wang, and C. J. Li 2013, "Polyamide nanofilm composite membranes (NCMs) supported by chitosan coated electrospun nanofibrous membranes: Preparation and separation performance research," *Desalination*, vol. 328, pp. 31–41.

[17] R. Lamba, A. Umar, S. K. Mehta, and S. Kumar 2015, "ZnO doped SnO 2 nanoparticles heterojunction photo-catalyst for environmental remediation," *J. Alloys Compd.*, vol. 653, pp. 327–333.

Text Mining Pre-Processing Using Gata Framework and RapidMiner for Indonesian Sentiment Analysis

S Kurniawan[1*], W Gata[1], D A Puspitawati[1], I K S Parthama[2], H Setiawan[3] and S Hartini[1]

[1]Sekolah Tinggi Manajemen Informatika dan Komputer Nusa Mandiri, Indonesia
[2]Universitas Pramita Indonesia, Indonesia
[3]Sekolah Tinggi Manajemen Informatika dan Komputer Bani Saleh, Indonesia

*kurniawan.sgt@gmail.com

Abstract. Research in the field of Text Mining in general still uses text in English, Arabic, China or others language, while for text in Indonesian is still very limited, so it requires good tools to help Indonesian researchers to conduct research in the field of text mining in Indonesian. Pre-processing is needed for text mining processes such as deleting notation '@', 'http' removal, Indonesian stopwords, normalizing acronym, slang words, emoticons, and Indonesian stemming. The GATA Framework Text Mining provided is one of the options for conducting text mining research in Indonesian and has been used by several researchers. There are several known data mining processing methods, including KKD, CRISP-DM, and SEMMA, all three of which are quite reliable methods. CRISP-DM which consists of; Bussiness Understanding, Data Understanding, Data Preparation, Modeling, Evaluation, and Deployment is a method that is quite widely used in research in the field of text mining which can be combined with text pre-processing. With so much research in the field of Text Mining in Indonesian, the need for pre-processing in Indonesian is very important. GATA Framework is an option for pre-processing devices that can be combined with Repidminer devices, as seen from the results of the excellent FUPRS.

1. Introduction

Research in the field of Text Mining in general still uses text in English, Arabic, China or others language, while for texts in Indonesian is still very limited, so it requires good tools to help Indonesian researchers to conduct research in the field of text mining in Indonesian. Pre-processing needed for text mining processes such as deleting '@' notation, 'http' removal, Indonesian stopwords, normalize acronym words, slang words, emoticons, and Indonesian stemming.

Many software can be used to do text mining, one of which is RapidMiner software. RapidMiner is one of the most widely used worldwide open source data mining solutions [1]. The use of RapidMiner in various studies related to data mining and text mining has been very good. In RapidMiner many menus are available and can be used for text mining, such as retrieving text data from Twitter, Facebook and pre-processing, before pre-processing using a classification algorithm commonly used for text mining such as Support Vector Machine (SVM) and Naïve Bayes (NB). Although the RapidMiner application has provided many menus, for the use of Indonesian Acronym, Indonesian stemming, Indonesian slang words, and others related to the Indonesian text pre-processing, but is still very limited and requires innovation.

Text mining or Natural Language Processing research, especially Indonesian Language Stemming, already exists in an application called Sastrawi using the PHP and Python programming languages that were built in 2016. In general, this device can be used but needs special expertise in installing it and there are still limitations if must be integrated with other software. One of the studies using the Sastrawi application is research conducted by Agastya and Artha which discusses the influence of Indonesian Language Stemmer [2].

In 2018, a personal named Windu Gata developed web-based applications for pre-processing such as the elimination of Indonesian stopwords, Indonesian stemming, Indonesian Acronym, Indonesian Slang and others intended to help students and other researchers in making research in the field of text-mining of Indonesian language. The application is built using a framework called the GATA Framework (http://www.gataframework.com). The application is an alternative in Indonesian pre-processing text, the application also provides an application program interface (API) feature for sending data from external applications. While the GATA framework is a framework based on the PHP programming language that was developed with the name MTG Framework in 2012 and changed its name to the GATA framework in 2017. GATA Framework has been able to overcome various external problems, namely usability, capability, response, security, existence, and reliability, as well as internal factors, namely ease of syntax or code that is easy to use and has used the Model View Controller (MVC) programming pattern [3]. Currently, the application device can process pre-processing in the form of a single form, or upload data in the form of Ms. Excell files with templates and web services.

In this study, the focus is on how the development and use of pre-processing text with GATA Framework Text Mining and RapidMiner in processing sentiment analysis in Indonesian. In addition, to evaluate the quality of the GATA Framework text mining software, this study uses the FURPS quality model method which consists of, Functionality, Use, Reality, Performance, and Support in the form of descriptive statistics. FURPS (Functionality, Usability, Reliability, Performance, Supportability) Quality Model is a model introduced by Rober Grady, where the development was carried out by IBM in Rational Software [3].

This study choose to use the CRISP-DM method in processing data, which consists of several stages, namely Bussiness Understanding, Data Understanding, Data Preparation, Modeling, Evaluation, and Deployment with a combination of text pre-processing using the GATA Framework pre-processing application combined with RapidMiner applications to process words in Indonesian, as shown in figure 1.

IOP Conf. Series: Materials Science and Engineering **835** (2020) 012057 doi:10.1088/1757-899X/835/1/012057

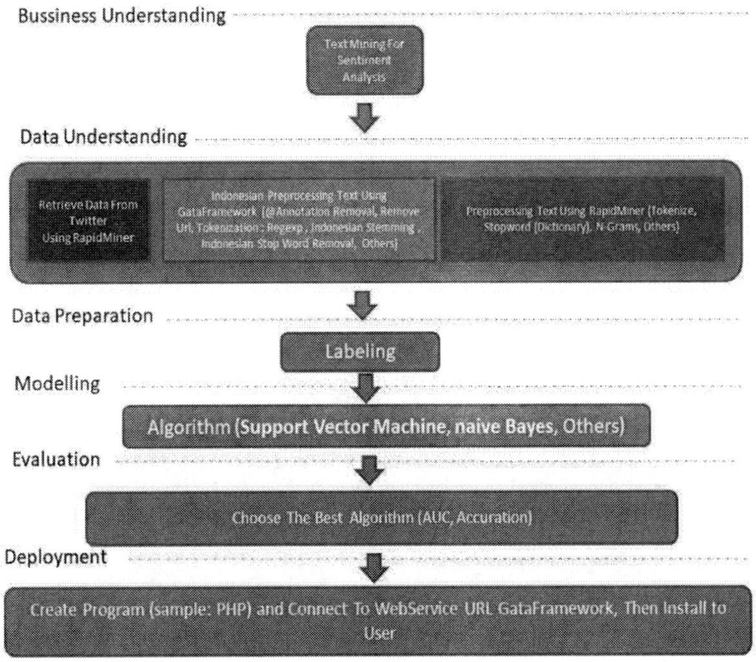

Figure 1. Combination of GATA Framework – Pre-processing and RapidMiner

2. Business Understanding

The use of data in the form of text comments or tweets from microblogging twitter is often used to process sentiments from each comment or tweet. The sentiment is an attitude, thought, or judgment prompted by feeling. Sentiment analysis, which is also known as opinion mining, studies people's sentiments towards certain entities. The Internet is a resourceful place with respect to sentiment information. From a user's perspective, people are able to post their own content through various social media, such as forums, micro-blogs, or online social networking sites [6].

Research related to sentiment analysis that has used text pre-processing, among others, as has been done by Nia Kusuma Wardhani for sentiment analysis in online news articles related to the coordinator of the maritime ministry using the Naïve Bayes classification algorithm and Support Vector Machine which was optimized with Particle Swarm Optimization by using Rapidminer. The results of the study obtained the Naive Bayes algorithm which was optimized with PSO having a high accuracy value and being a solution to the problem of sentiment analysis in online news articles [7].

Other research for sentiment analysis as conducted by Siswanto on research related to the classification of comment analysis on social media related to MotoGP, research uses the Naive Bayes classification algorithm and Support Vector Machine to classify comments on social media and then classify them as analysis of positive or negative sentiment with Rapidminer. The study resulted in accuracy for sentiment prediction of 95.50% for the Support Vector Machine algorithm [8].

There is also another study conducted by Tirana Noor Fatyanosa related to the comparison of the classification methods for sentiment analysis in Indonesian social media, in addition to using the pre-processing text also uses several classification methods compared to one another, the classification methods include; Summation, Average on Tweet, Average on Tweet with the objective score, Weighted Average, and Naïve Bayes method. The results of these experiments found that Naïve Bayes produced high precision, recall and accuracy values for neutral and positive sentiments, but it was not good for negative sentiment [9]. The overall research carried out related to sentiment analysis begins with the pre-processing phase of the text which then continues to use the classification method to obtain positive or negative results from the dataset used.

3. Data Understanding

When using Twitter social media, there are signs such as # (hashtag), @ (user), HTTP, and other signs that need to be removed so that text can be used better. While the use of pre-processing methods in Indonesian that can be used is the Indonesian acronym method, Indonesian stopwords, Indonesian Stemming, and others. Retrieving data from the Twitter microsite on RapidMiner can use the Search Twit operator as shown in Figure 2.

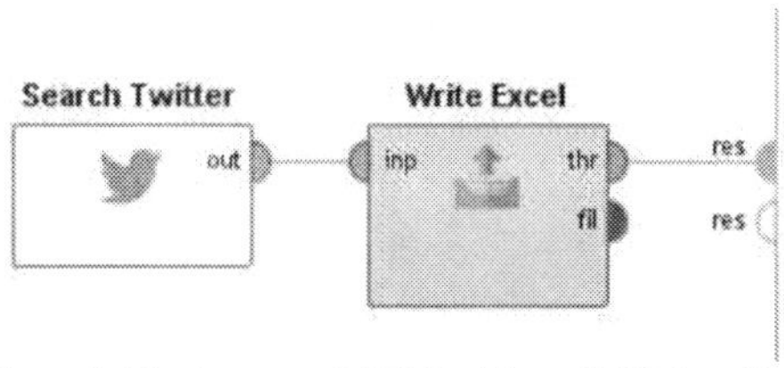

Figure 2. Search Twitter and Write Excell Using RapidMiner
Source: RapidMiner 9.2

In the RapidMiner application there is already a dictionary facility to change the acronym, and stopwords, but it is still limited to English, Chinese, and Arabic, while for Indonesian it is still not available. The Indonesian text processing stage for the removal of a hashtag, @, HTTP, Indonesian acronym, Indonesian Stopwords, and Indonesian stemming can use the GATA Framework application as shown in Figure 3. The technique commonly used to pre-processing Indonesian text is @annotation Removal, Remove URL, Tokenization: REGEXP, Transformation Not (Negative), Indonesian Stemming, and Indonesian Stopwords Removal. An explanation of each technique option in the GATA Framework text mining can be seen in table 1.

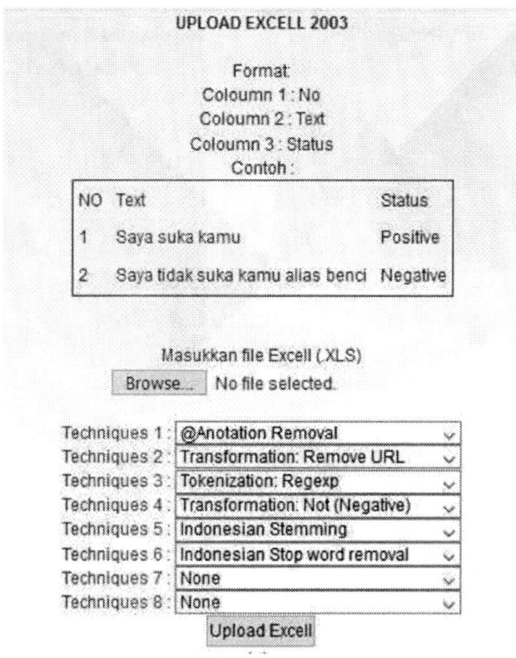

Figure 3. Indonesian Pre-processing Using GATA Framework
Source: htttp://www.gataframework.com/textmining

Table 1. Description techniques on Text mining – Pre-processing - GATA Framework

No	Techniques	Description
1	@ Annotation Removal	Remove the @ sign and description that is often used to greet or mark other accounts.
2.	Transformation: Remove URL	Remove the URL from the tweet used.–
3.	Tokenization: Regexp	Remove marks other than letters and also eliminate numbers.
4.	Transformation NOT	Connect words that have inverse meanings like "no", "no", etc.
5	Indonesian Stemming	Returning a word becomes a basic word.
6.	Indonesian Stopword removal	Remove words that have no meaning.

4. Data Preparing

At this stage, each data used will be labeled as needed, if you use the model sentiment then each comment can be labeled "Yes" or "No". The amount of data labeled "Yes" and "No" is recommended to have the same amount. If you have an unequal number, you can use the balancing method, namely SMOTE.

5. Modeling

After completing the data preparation stage, the next step is to do modeling using RapidMiner. In RapidMiner Pre-processing stages can be done and use sentiment algorithms such as SVM and NB algorithms. Some pre-processing features commonly used on RapidMiner, namely: Tokenize, Filter Stopwords (unnecessary words using a dictionary), Filter Token By Length, and Generate n-Grams (relationship of words to other words), as shown in figure 4 While the next step is the use of SVM and NB algorithms which are validated using 10-fold cross-validation and T-Test.

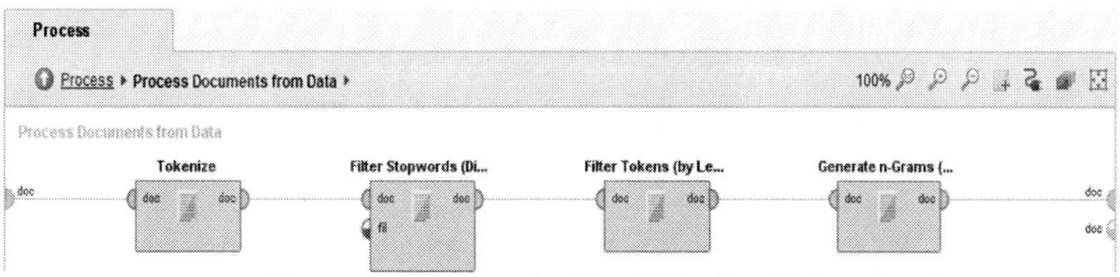

Figure 4. Pre-processing Text Using RapidMiner

The next step is to make the overall model of all operators use RapidMiner, namely: ReadExcel, SetRole, Nominal to Text, Process Document (figure 4), SMOTE, Weight by GINI Index, Multiply, Cross Validation for SVM, Cross Validation for NB, and T-Test, as shown in figure 5.

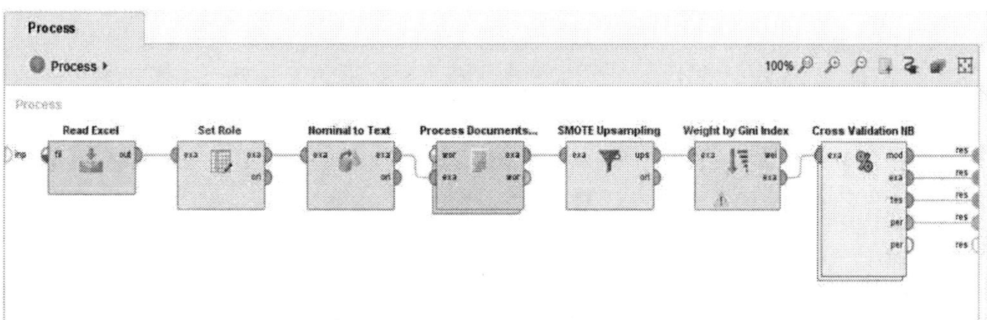

Figure 5. Example of Modeling Using RapidMiner With SVM and NB Algorithm
Source: RepidMiner 9.2

In the cross-validation stage, there are two columns, namely training and testing. The training column uses the SVM algorithm and NB algorithm can be seen as in Figure 6.

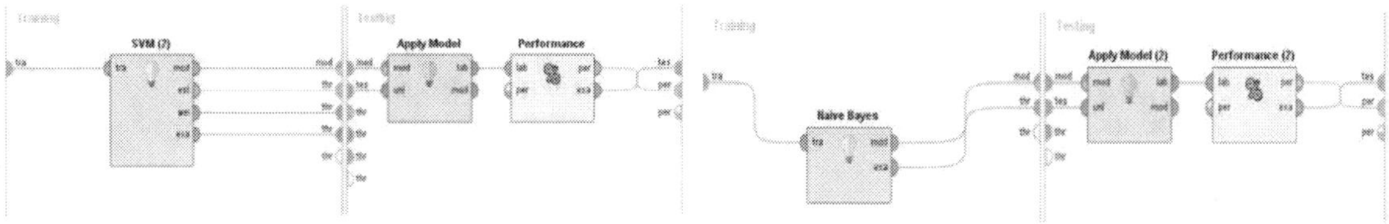

Figure 6. Example of Cross-Validation SVM and NB
Source: RapidMiner 9.2

6. Evaluation

At this stage, what must be done is to choose the best algorithm by looking at the value of accuracy produced by the RapidMiner application device. If you have got the best accuracy value from the algorithm model used and consider the accuracy to be good enough, then it will continue with the deployment stage.

In this study, the use of FURPS was used in the form of descriptive statistics to assess Text Mining - Pre-processing applications on the GATA framework. A total of 21 application users or researchers were given a questionnaire by giving questions about FURPS. The results of the questionnaire are explained in table 6, where the highest value of the questioner is 5 multiplied by the number of evaluators of 21, equal to 105. The scoring formula is the maximum number of divided values * 100.

Table 2. Results of the FURPS questionnaire

No	Aspek	Score					Result
		1	2	3	4	5	Value/max * 100
1	Functionality				13	8	87.62
2	Usability				15	6	85.71
3	Reliability			1	13	7	82.86
4	Performance		1	3	11	6	70.48
5	Supportability			1	13	7	82.86
				Average			**81.90**

From the results obtained the highest value is seen from the Functional parameter aspect which has a value of 87.62, while the lowest value is in the Performance parameter aspect. The average value of all aspects is 81.90, which is greater than 80 which means that the application GATA Framework text mining is very good to use.

7. Deployment

At this stage, we can use two models, use RapidMiner Server, or develop applications by choosing the programming language that suits your needs. In some studies that have been done, there are those who use VB.net and PHP at the deployment stage.

8. Conclusion

The amount of research in the field of text mining in Indonesian, the need for pre-processing in Indonesian is very important. GATA Framework is one of the recommended pre-processing device options, seen from the results of the excellent FUPRS. The device supports Data Understanding on the CRISP-DM method or other methods, namely '@' annotation Removal, Remove URL or 'http', Tokenization: REGEXP, Transformation Not (Negative), Indonesian Stemming, and Indonesian

Stopwords Removal. In addition, these tools can also be combined with RapidMiner tools, and have online process features and web services (API) that can be linked to other applications.

References

[1] Agastya, I. M. (2018). PENGARUH STEMMER BAHASA INDONESIA TERHADAP PEFORMA ANALISIS SENTIMEN TERJEMAHAN ULASAN FILM. *Jurnal Tekno Kompak*, 12(1), 18-23.

[2] Fahad Salmeen Al-Obthani, A. A. (2018). TOWARDS CUSTOMIZED SMART GOVERNMENT QUALITY MODEL. *International Journal of Software Engineering & Applications (IJSEA)*, 9(2), 41-50.

[3] M. Hofmann, R. K. (2014). *RapidMiner: Data Mining Use Cases and Business Analytics Applications*. Boca Ranton: Chapman and Hall/CRC.

[5] Nia Kusuma Wardhani, S. K. (2018). Sentiment Analysis Article News Coordinator Minister of Maritime Affairs Using Algorithm Naive Bayes and Support Vector Machine with Particle Swarm Optimization. *Journal of Theoretical and Applied Information Technology*, 96(24), 8365-8378.

[6] Nia Kusuma Wardhani, W. G. (2017). IMPLEMENTASI FRAMEWORK MTG DALAM PENGEMBANGAN CRUD, PEMBUATAN LAPORAN, GRAFIK MODEL DAN EKSPORT BERKAS MENGGUNAKAN BAHASA PEMROGRAMAN PHP. *International Journal of Human Capital Management (IJHCM)*, 1(02), 81-94.

[7] Tirana Noor Fatyanosa, F. A. (2017). *Classification Method Comparison on Indonesian Social Media Sentiment Analysis*. Paper presented at International Conference on Sustainable Information Engineering and Technology (SIET).

[8] Siswanto, Y. P. (2018). *Classification Analysis of MotoGP Comments on Media Social Twitter Using Algorithm Support Vector Machine and Naive Bayes*. Bali, Indonesia: International Conference on Applied Information Technology and Innovation (ICAITI).

[9] Umair Shafique, H. Q. (2014). A Comparative Study of Data Mining Process Models (KDD, CRISP-DM and SEMMA). *International Journal of Innovation and Scientific Research*, 12(1), 217-222.

[10] Xing Fang, J. Z. (2015). Sentiment Analysis Using Product Review Data. *Journal of Big Data*, 2(1), 2-14.

IOP Conf. Series: Materials Science and Engineering **835** (2020) 012058 doi:10.1088/1757-899X/835/1/012058

Application of Data Mining Technique using K-Medoids in the case of Export of Crude Petroleum Materials to the Destination Country

Fathur Rahman[1], Ihda Innar Ridho[1], M. Muflih[1], Sefto Pratama[1], Mokhamad Ramdhani Raharjo[1], Agus Perdana Windarto[2*]

[1] Universitas Islam Kalimantan Muhammad Arsyad Al Banjari Banjarmasin, Indonesia
[2] STIKOM Tunas Bangsa, Pematangsiantar, Indonesia

Email: *agus.perdana@amiktunasbangsa.ac.id

Abstract. The purpose of this research is to analyze and implement the data mining technique in the export of crude petroleum materials to the destination country. This is because Indonesia is a member of OPEC (Organization of the Petroleum Exporting Countries) which is one of the largest petroleum exporters in the world. It aims to obtain profits in the form of foreign exchange income obtained by the State. The data source was obtained from the Central Statistics Agency (BPS) with the website https://www.bps.go.id for data for 2017-2018. The calculation process The technique used is clustering with the K-Medoids algorithm. The calculation process is carried out using the help of Rapid Miner tools. The clusters used in this study are 2 namely: high cluster (C1) for export of crude oil materials and low cluster (C2) for crude oil materials. The results of the study stated that the high cluster (C1) consisted of 3 countries (Japan, Thailand and the United States) and the low cluster (C2) consisted of 6 countries (South Korea, Taiwan, China, Singapore, Malaysia and Singapore). It is hoped that the research results will be input and information for the government to rearrange policies in order to increase competitiveness, ensure business certainty and the sustainability of domestic industrial raw materials.

1. Introduction

The State of Indonesia is a member of OPEC (Organization of the Petroleum Exporting Countries) which is one of the largest petroleum exporters in the world. Indonesia is one of the exporting countries to developed and developing countries. The purpose of the exporter is to be able to make a profit. With this export, the government also receives income in the form of foreign exchange. The more exports, the greater the foreign exchange earned by the state. Goods exported by Indonesia consist of petroleum and natural gas (oil and gas) as well as non-oil / gas [1]. Petroleum is one of the commodities that are exported and imported in Indonesia because petroleum is one of the main energy that is widely used in almost every country. As a result of economic growth that continues to increase in other countries resulting in demand for oil continues to increase and make exporting activities run. Exports and imports play an important role in supporting a country's economy. In terms of oil demand, developed countries tend to run slowly even though the economy is going on. The main thing that causes the increase in demand, especially oil, is the increase in the population itself. Therefore it is necessary to review policies to increase competitiveness, ensure business certainty and the sustainability of domestic industrial raw materials specifically for the export of crude petroleum materials to destination countries. The policy can be in the form of a mapping of the destination country to export petroleum materials.

There are many computer science techniques that can solve these problems. The branches of computer science are Artificial Intelligence such as data mining [1], [2], Decision Support Systems (SPK) [3]–[6], expert systems [7], Artificial Neural Networks (ANN) [8]–[10], fuzzy logic [11] and others. In this research, data mining becomes a solution because data mining has an approach called clustering. Clustering The main concept emphasized is the search for cluster centers, where cluster centers are determined based on the minimum distance of each data in the cluster center. The data source uses data obtained from the Central Statistics Agency (https://www.bps.go.id). In this case the research focuses on the export of crude oil based on the destination country where the method used is K-Medoids. The purpose of the research is to apply K-Medoids in the export of crude petroleum materials to the destination country so that the results of the study can be input for the government of the country that has the highest priority in the export of crude oil based on the cluster that has been carried out.

2. Methodology

2.1 Data Mining

Data Mining is the process of discovering useful new correlations, patterns and trends by adding a large number of data repositories, using pattern recognition technologies such as static and mathematical techniques [2], [12].

2.2 Clustering Technique

Clustering is a process of grouping records, observations, which have the same object. Cluster technique consists of two methods: hierarchical clustering and non-hierarchical clustering [13], [14].

2.3 K-Medoids Method

This algorithm includes the Partitional Clustering algorithm and is one of the well-known methods. K-Medoids method is better than K-Means method in handling noise and outliers, because this method is not too influenced by outliers or other extreme values [13].

2.4 Research Method

In the application of data mining for the export of crude oil to the destination country, relevant data is needed about it. The data used in this study are the data of Crude Petroleum Exports to Destination Countries from 2017-2018 consisting of 9 countries namely Japan, South Korea, Taiwan, China, Thailand, Singapore, Malaysia, Australia and the United States taken from Central Statistics Agency with official website https://www.bps.go.id. then the data will be processed using the k-medoids method by taking the values of each attribute in the data to classify the crude oil export data.

3. Results and Discussion

In this clustering, the data that has been collected will be calculated in advance based on the data of crude oil export materials in the nine destination countries shown in the following table.

Table 1. Preliminary data on exports of Crude Petroleum Materials

Country of destination	2017	2018
Japan	1582,2	1315,4
South Korea	998,4	1001,7
Taiwanese	1082,0	908,0
China	1468,0	442,6
Thailand	2339,7	1819,9
Singapore	1536,0	959,8
Malaysia	1553,8	1148,2
Australia	1339,4	1099,4
United States of America	1 632,5	1 443,4

IOP Conf. Series: Materials Science and Engineering **835** (2020) 012058 doi:10.1088/1757-899X/835/1/012058

2.5 Centroid Data

In the application of the K-Medoids algorithm, the midpoint value can be determined randomly from the data obtained provided that the cluster created is 2, namely a high export cluster (C1) and a low export cluster (C2). Here are the initial centroids used:

Table 2. Centroid Data

	C1 (High Export Cluster)	C2 (Low Export Cluster)
United States of America	1632,5	1443,3
South Korea	998,4	1001,7

Here are the results of the first iteration calculation as shown in the following table:

Table 3. First Iteration Data

Country of destination	2017	2018	C1	C2	Euclidian Distance
Japan	1582,2	1315,4	16434,3	98991,49	16434,3
South Korea	998,4	1001,7	195733	0	0
Taiwanese	1082,0	908,0	287203,7	8863,29	8863,29
China	1468,0	442,6	1001765	313062,41	313062,41
Thailand	2339,7	1819,9	142459,5	670792,54	142459,45
Singapore	1536,0	959,8	233965,5	2293,21	2293,21
Malaysia	1553,8	1148,2	87221,74	22017,65	22017,65
Australia	1339,4	1099,4	118629,1	9886,29	9886,29
United States of America	1632,5	1443,4	0	195732,99	0

Based on table 3 that the cluster process can be done by centroid data in table 2. The cluster process is done by taking the shortest distance from each data that has been processed. From the initial data obtained the grouping in the first iteration as shown in the following table.

Table 4. Results of the First Iteration Grouping

Cluster 1	Cluster 2
Japan	South Korea
Thailand	Taiwanese
United States of America	China
	Singapore
	Malaysia
	Australia

The K-Medoids process stops if the deviation> 0, but if the deviation <0 then exchange the object with the data cluster to form a new set of objects as the medoid. The process of calculating the deviation by adding up the total new distance value - the total old distance that contains the distance of each object in each cluster with the new medoid. With the number of deviations shown in the following table.

Table 5. Total First Iteration Distance

C1	C2
16434,3	0
142459,5	8863,29
0	313062,4
	2293,21
	22017,65
	9886,29
158893,8	356122,9
515016,6	

Following are the results of data processing in the second iteration as shown in the following table:

Table 6. Second Iteration Data

Country of destination	2017	2018	C1	C2	Euclidian Distance
Japan	1582,2	1315,4	255277,8	761894,04	255277,75
South Korea	998,4	1001,7	670792,5	313062,41	313062,41
Taiwanese	1082,0	908,0	832819,3	216983,16	216983,16
China	1468,0	442,6	1897827	0	0
Thailand	2339,7	1819,9	0	1897826,99	0
Singapore	1536,0	959,8	740575,7	267563,84	267563,84
Malaysia	1553,8	1148,2	451966,8	497957,16	451966,79
Australia	1339,4	1099,4	520120,6	431514,84	431514,84
United States of America	1632,5	1443,4	142459,5	1001765,14	142459,45

The same thing is done as table 4, grouping based on Euclidian Distance as shown in the following table:

Table 7. Total Second Iteration Distance

C1	C2
255277,8	313062,4
0	216983,2
142459,5	0
	267563,8
	451966,8
	431514,8
397737,2	1681091
2078828,24	

From the sum of the new distance data - the old distance produces a deviation> 0, then the grouping process stops at the second iteration with the result of the intersection of 1563811.64.

Table 8. Standard Deviation Results

Standard Deviation = New Distence - Old Distance		
New Distance	Old Distance	Result
2078828,24	515016,6	1563811,64

2.6 Rapid Miner Implementation

Following is the data processing using k-medoids on Rapid Miner:

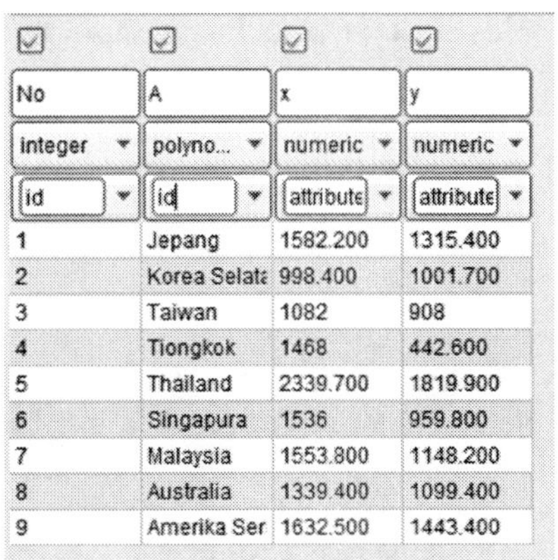

No	A	x	y
integer	polyno...	numeric	numeric
id	id	attribute	attribute
1	Jepang	1582.200	1315.400
2	Korea Selata	998.400	1001.700
3	Taiwan	1082	908
4	Tiongkok	1468	442.600
5	Thailand	2339.700	1819.900
6	Singapura	1536	959.800
7	Malaysia	1553.800	1148.200
8	Australia	1339.400	1099.400
9	Amerika Ser	1632.500	1443.400

Figure 1. Input Data

IOP Conf. Series: Materials Science and Engineering **835** (2020) 012058 doi:10.1088/1757-899X/835/1/012058

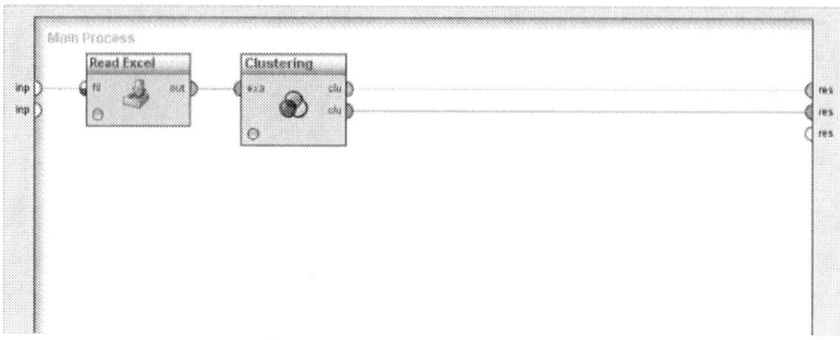

Figure 2. K-Medoid Clustering Modeling in Rapid Miner

By using k-medoids modeling like Figure 1, by initializing the number of clusters as much as 2 pieces, the results obtained with the cluster formed are 2, according to the definition of k value with the number of cluster_0 = 3 items and cluster_1 = 6 items.

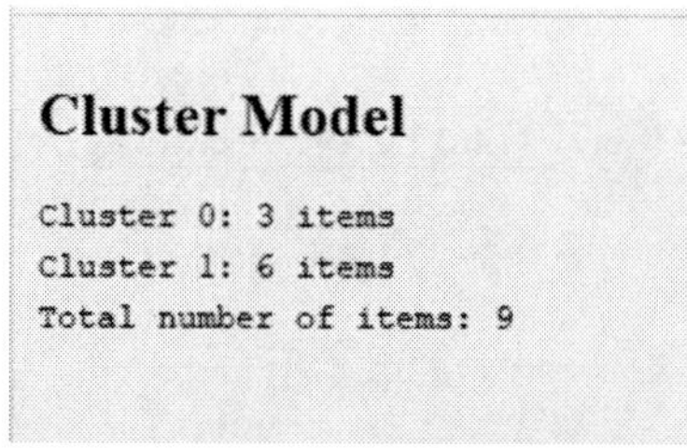

Figure 3. K-Medoids cluster results in the Rapid Miner implementation

The results of the analysis are then tested again using the existing Rapid Miner. By entering the analysis data into Rapid Miner. In Text View it will display the cluster model. Cluster models obtained from the results of testing of data using k-medoids clustering.

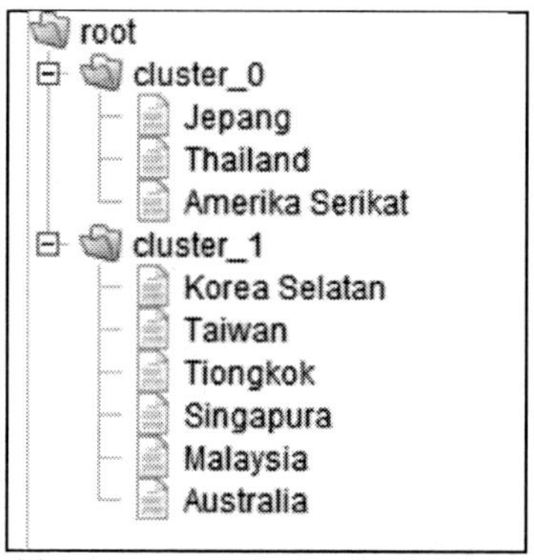

Figure 4. K-Medoids cluster results in the Rapid Miner implementation

Based on Figure 4, the results obtained from manual calculations and implementation with Rapid Miner are the same. The results of the cluster model consist of cluster 0 (high export clusters) namely 3 items (Japan, Thailand, United States), cluster 1 (low export clusters) which are 6 items (South Korea, Taiwan, China, Singapore, Malaysia, Australia). From a total of 9 items.

4. Conclusion

Based on the results of the study concluded the following conclusions:

1. The grouping of Crude Petroleum Material Export data has been successfully carried out to the Destination Country using the K-medoids algorithm with 2 clusters, namely for C1 (high exports) totaling 3 data: Japan, Thailand, United States and C2 (low exports) totaling 6 data namely: South Korea, Taiwan, China, Singapore, Malaysia, Australia.

2. With the existence of Rapid Miner software in this study, the accuracy of the data will be good enough for problems that occur related to the export of crude petroleum materials used as samples.

References

[1] W. Katrina, H. J. Damanik, F. Parhusip, D. Hartama, A. P. Windarto, and A. Wanto, "C.45 Classification Rules Model for Determining Students Level of Understanding of the Subject," *J. Phys. Conf. Ser.*, vol. 1255, no. 012005, pp. 1–7, 2019, doi: 10.1088/1742-6596/1255/1/012005.

[2] A. P. Windarto *et al.*, "Analysis of the K-Means Algorithm on Clean Water Customers Based on the Province," *J. Phys. Conf. Ser.*, vol. 1255, no. 1, 2019, doi: 10.1088/1742-6596/1255/1/012001.

[3] S. R. Ningsih, R. Wulansari, D. Hartama, A. P. Windarto, and A. Wanto, "Analysis of PROMETHEE II Method on Selection of Lecturer Community Service Grant Proposals," *J. Phys. Conf. Ser.*, vol. 1255, no. 1, pp. 1–7, 2019, doi: 10.1088/1742-6596/1255/1/012004.

[4] T. Imandasari, M. G. Sadewo, A. P. Windarto, A. Wanto, H. O. Lingga Wijaya, and R. Kurniawan, "Analysis of the Selection Factor of Online Transportation in the VIKOR Method in Pematangsiantar City," *J. Phys. Conf. Ser.*, vol. 1255, no. 012008, pp. 1–7, 2019, doi: 10.1088/1742-6596/1255/1/012008.

[5] K. Fatmawati *et al.*, "Analysis of Promothee II Method in the Selection of the Best Formula for Infants under Three Years," *J. Phys. Conf. Ser.*, vol. 1255, no. 1, 2019, doi: 10.1088/1742-6596/1255/1/012009.

[6] D. R. Sari, N. Rofiqo, D. Hartama, A. P. Windarto, and A. Wanto, "Analysis of the Factors Causing Lazy Students to Study Using the ELECTRE II Algorithm," *J. Phys. Conf. Ser.*, vol. 1255, no. 1, 2019, doi: 10.1088/1742-6596/1255/1/012007.

[7] Y. Malhotra, "Expert systems for knowledge management: crossing the chasm between information processing and sense making," *Expert Syst. Appl.*, vol. 20, no. 1, pp. 7–16, 2001, doi: 10.1016/S0957-4174(00)00045-2.

[8] Sumijan, A. P. Windarto, A. Muhammad, and Budiharjo, "Implementation of Neural Networks in Predicting the Understanding Level of Students Subject," *Int. J. Softw. Eng. Its Appl.*, vol. 10, no. 10, pp. 189–204, 2016.

[9] Budiharjo, T. Soemartono, A. P. Windarto, and T. Herawan, "Predicting tuition fee payment problem using backpropagation neural network model," *Int. J. Adv. Sci. Technol.*, vol. 120, pp. 85–96, 2018, doi: 10.14257/ijast.2018.120.07.

[10] Budiharjo, T. Soemartono, A. P. Windarto, and T. Herawan, "Predicting School Participation in Indonesia using Back-Propagation Algorithm Model," *Int. J. Control Autom.*, vol. 11, no. 11, pp. 57–68, 2018.

[11] A. Mulyanto and Abdul Haris, "Penerapan Metode Fuzzy Tsukamoto Untuk Menentukan Jumlah Jam Overtime Pada Produksi Barang di PT Asahi Best Base Indonesia (ABBI) Bekasi," *J. Inform. SIMANTIK Vol.1,* vol. 1, no. 1, pp. 1–11, 2016.

[12] D. Hartama, A. Perdana Windarto, and A. Wanto, "The Application of Data Mining in Determining Patterns of Interest of High School Graduates," *J. Phys. Conf. Ser.*, vol. 1339, no. 1, 2019, doi: 10.1088/1742-6596/1339/1/012042.

[13] B. Supriyadi, A. P. Windarto, T. Soemartono, and Mungad, "Classification of natural disaster prone areas in Indonesia using K-means," *Int. J. Grid Distrib. Comput.*, vol. 11, no. 8, pp. 87–98, 2018.

[14] A. P. Windarto, "Implementation of Data Mining on Rice Imports by Major Country of Origin Using Algorithm Using K-Means Clustering Method," *Int. J. Artif. Intell. Res.*, vol. 1, no. 2, pp. 26–33, 2017.

IOP Conf. Series: Materials Science and Engineering **835** (2020) 012059 doi:10.1088/1757-899X/835/1/012059

Application of MCDM using PROMETHEE II Technique in the Case of Social Media Selection for Online Businesses.

I Gede Iwan Sudipa[1], Cici Astria[2], Khairunnissa Fanny Irnanda[2], Agus Perdana Windarto[2]*, Nelly Khairani Daulay[3], Wildan Suharso[4], Harma Oktafia Lingga Wijaya[5]

[1] STMIK STIKOM Indonesia
[2] STIKOM Tunas Bangsa, Pematangsiantar, North Sumatra, Indonesia
[3] Universitas Bina Insan, Indonesia
[4] Universitas Muhammadiyah Malang, Indonesia
[5] STMIK Musirawas Lubuklinggau, Indonesia

Email: agus.perdana@amiktunasbangsa.ac.id *

Abstract. Internet media has become one of the means of product promotion that has very good prospects today. The research aims to recommend the right social media for online businesses. The data collection method was conducted by interview and questionnaire to 300 respondents who conducted online business in Pematangsiantar city by random sampling. Based on the results of interviews and questionnaires obtained assessment criteria namely security (C1), application features (C2), community (C3), ease of access (C4) and response speed (C5). The alternatives used in this research are Facebook (A1), Instagram (A2), Line (A3) and WhatsApp (A4). The settlement method applied is a decision support system with the PROMETHEE II algorithm. The results of the algorithm show that the right alternative for doing online business is Facebook (A1) with a net flow of 0.25 and followed by WhatsApp (A4) with a net flow of −0.1. The results of the study are expected to provide recommendations in conducting online business.

1. Introduction

The growing trend of the use of social media among the public and the development of internet technology, certainly brings quite encouraging opportunities for business people in general, this condition is supported by the advancement of mobile devices that are increasingly being used by the community. more creative among business people to take advantage of advances in cellular technology in running their business online. The development of the use of social media in Indonesia is growing rapidly. According to Nielsen's research, the growth rate of internet usage in Indonesia reached 26%. Indonesians spend 1.5 hours a day surfing the internet. According to ICT Watch, currently there are 180 million mobile phone users in Indonesia out of around 220 million Indonesians. Instagram users in Indonesia reached 22 million active users according to a press release received by CNN Indonesia [1].. Internet media has become one of the means of product promotion that has very good prospects at this time, where through the internet media sellers can reach consumers widely. Even now the internet has entered into various corners of the country, people who live far from cities can take advantage of this internet facility.

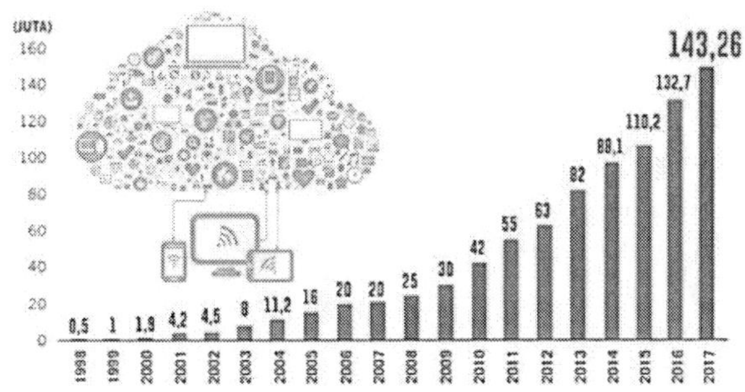

Figure 1. Growth of Internet Users in Indonesia

Based on Figure 1 that the data of internet users in previous years, there was a continued growth. In 2010 internet users in Indonesia were only 42 million users, compared to 2017 users, there was an increase of 3x within 7 years. This research was conducted in the city of Pematangsiantar which aims to recommend the right social media for online businesses, considering that Pematangsiantar City is one of the cities where consumers market their business products through social media. Besides the advantages gained in running an online business is the ease of finding information about products, prices, product selection or availability, pleasure, encouragement in the heart and customer service. Surely this has become positive for online businesses that have long been in the business. What if there are novice consumers who want to do business online but feel confused about choosing the right social media to use. Based on these hypotheses, researchers are interested in providing solutions to these problems. Many branches of computer science can solve complex problems both for predicting, estimating, ranking, classifying, associating and others. The branches of computer science are Artificial Intelligence such as datamining [2], [3], Decision Support Systems (SPK) [4]–[7], expert systems [8], Artificial Neural Networks (ANN) [9]–[11], fuzzy logic [12] and others.

Based on these problems, researchers used a decision support system with PROMETHEE II algorithm in solving these problems. Related research was carried out [1]. This study discusses the analysis of one social media on the formation of brand attachments. Research to be conducted is different from previous studies. The difference is in the settlement method used, the social media assessment criteria and the selection of more than one alternative for social media. It is expected that the results of the research can provide social media recommendations for online businesses, especially the pematangsiantar city community.

2. Methodology

2.1 Decision Support System

Decision Support System is a computer-based information system that approaches to produce various alternative decisions to help certain parties in dealing with problems [13], [14].

2.2 Social media

Social media is an online media, with its users can easily participate, share and create content including blogs, social networks, wikis, forums and the virtual world. Social media has the following characteristics:

a) The message delivered is not just for one person but can be many people for example messages via SMS or the internet;
b) The message is free, without having to go through a Gatekeeper;
c) The message conveyed tends to be faster than other media;
d) Recipient of the message that determines the interaction time [1].

2.3 PROMETHEE II (Preference Ranking Organization Methods for Enrichment Evaluation) Methods

PROMETHEE II is one of several methods for determining the order or priority in multicriteria analysis, which offers a flexible and simple way for users (decision makers) to analyze multi-criteria problems. This method is able to accommodate the selection criteria that are quantitative and qualitative. The main problem is simplicity, clarity and stability [14].

2.4 Research Method

This study uses a Decision Support System (SPK) and Preference Ranking Organization Methods for Enrichment Evaluation (PROMETHEE II) conducted in Pematangsiantar City. The data collection process was carried out by interviewing and giving questionnaires to 300 respondents who conducted online business in the city of Pematangsiantar by random sampling. The assessment criteria used to choose the right social media for online business are security (C1), application features (C2), community (C3), ease of access (C4) and response speed (C5). Alternative social media used in the study are Facebook (A1), Instagram (A2), Line (A3) and WhatsApp (A4).

3. Results and Discussion

3.1 Alternative Determination Stage

In determining social media alternatives, the author uses 4 alternatives as shown in the following table:

Table 1. Alternatives

Social Media	Information
Facebook	A1
Instagram	A2
Line	A3
WhatsApp	A4

3.2 Criteria Determination Stage

The criteria used in the selection of the best social media for online businesses as shown in the following table:

Table 2. Social Media Selection Criteria

Criteria	Information	Alias
Criteria 1	Security	C1
Criteria 2	Application Features	C2
Criteria 3	Community	C3
Criteria 4	Ease of Access	C4
Criteria 5	Response Speed	C5

3.3 Application of the PROMETHEE Method II

The following research data are used to carry out calculations using the PROMETHEE II method. Previously the data had been recapitulated using Microsoft Excel software. The following research data will be processed using PROMETHEE II:

Table 3. Match Ratings of Each Alternative on Each Criteria

Alias	Min/ Max	Criteria			
		A1	A2	A3	A4
C1	Max	0.9	0.7	0.5	0.9
C2	Max	0.9	0.9	0.2	0.7
C3	Max	0.7	0.7	0.9	0.7
C4	Max	0.9	0.7	0.7	0.9
C5	Max	0.7	0.9	0.5	0.7

After determining the match rating for each alternative on each criterion, calculate the preference value. The following are the completion steps

a. Step I: Calculate the Preference Value

The existing criteria will then be calculated as a value and preference index as follows:

Usual Criteria

1. Security Criteria Value (C1)

C1(A1,A2) : d = C1(A1) – C1(A2)
d = 0.9 – 0.7
d = 0,2
d > 0 *then* H|d| = 1

C1(A2,A1) : d = C1(A2) – C1(A1)
d = 0.7 – 0.9
d = -0,2
d > 0 *then* H|d| = 0

C1(A1,A3) : d = C1(A1) – C1(A3)
d = 0.9 – 0.5
d = 0,4
d > 0 *then* H|d| = 1

C1(A3,A1) : d = C1(A3) – C1(A1)
d = 0.5 - 0.9
d = -0.4
d < 0 *then* H|d| = 0

C1(A1,A4) : d = C1(A1) – C1(A4)
d = 0.9 – 0.9
d = 0
d = 0 *then* H|d| = 0

C1(A4,A1) : d = C1(A4) – C1(A1)
d = 0.9 - 0.9
d = 0
d > 0 *then* H|d| = 0

C1(A2,A3) : d = C1(A2) – C1(A3)
d = 0.7 – 0.5
d = 0.2
d > 0 *then* H|d| = 1

C1(A3,A2) : d = C1(A3) – C1(A2)
d = 0.5 - 0.7
d = -0.2
d < 0 *then* H|d| = 0

C1(A2,A4) : d = C1(A2) – C1(A4)
d = 0.7 – 0.9
d = -0.2
d < 0 *then* H|d| = 0

C1(A4,A2) : d = C1(A4) – C1(A2)
d = 0.9 - 0.7
d = 0.2
d > 0 *then* H|d| = 1

C1(A3,A4) : d = C1(A3) – C1(A4)
d = 0.5 – 0.9
d = -0,4
d < 0 *then* H|d| = 0

C1(A4,A3) : d = C1(A4) – C1(A3)
d = 0.9 – 0.5
d = 0,4
d > 0 *then* H|d| = 1

2. Application Features Criteria Value (C2)

C2(A1,A2) : d = C2(A1) – C2(A2)
d = 0.9 – 0.9
d = 0
d = 0 *then* H|d| = 0

C2(A2,A1) : d = C2(A2) – C2(A1)
d = 0,9 – 0.9
d = 0
d = 0 *then* H|d| = 0

C2(A1,A3) : d = C2(A1) – C2(A3)
d = 0.9 – 0.2
d = 0,7
d > 0 *then* H|d| = 1

C2(A3,A1) : d = C2(A3) – C2(A1)
d = 0.2 - 0.9
d = -0,7
d < 0 *then* H|d| = 0

C2(A1,A4) : d = C2(A1) – C2(A4)
d = 0.9 – 0.7
d = 0,2
d > 0 *then* H|d| = 1

C2(A4,A1) : d = C2(A4) – C2(A1)
d = 0.7 – 0.9
d = -0,2
d < 0 *then* H|d| = 0

C2(A2,A3) : d = C2(A2) – C2(A3)
d = 0.9 – 0.2
d = 0,7
d > 0 *then* H|d| = 1

C2(A3,A2) : d = C2(A3) – C2(A2)
d = 0.2 - 0.9
d = -0,7
d < 0 *then* H|d| = 0

C2(A2,A4) : d = C2(A2) – C2(A4)
d = 0.9 – 0.7
d = 0,2
d > 0 *then* H|d| = 1

C2(A4,A2) : d = C2(A4) – C2(A2)
d = 0.7 – 0.9
d = -0,2
d < 0 *then* H|d| = 0

C2(A3,A4) : d = C2(A3) – C2(A4)
d = 0.2 - 0.7
d = -0,5
d < 0 *then* H|d| = 0

C2(A4,A3) : d = C2(A4) – C2(A3)
d = 0.7 – 0.2
d = 0.5
d > 0 *then* H|d| = 1

3. Community Criteria Value (C3)

C3(A1,A2) : d = C3(A1) – C3(A2)
d = 0.7 – 0.7
d = 0
d = 0 *then* H|d| = 0

C3(A2,A1) : d = C3(A2) – C3(A1)
d = 0.7 – 0.7
d = 0
d = 0 *then* H|d| = 0

C3(A1,A3) : d = C3(A1) – C3(A3)
d = 0.7 – 0.9
d = -0.2
d < 0 *then* H|d| = 0

C3(A3,A1) : d = C3(A3) – C3(A1)
d = 0.9 - 0.7
d = 0.2
d > 0 *then* H|d| = 1

C3(A1,A4) : d = C3(A1) – C3(A4)
d = 0.7 – 0.7
d = 0
d = 0 *then* H|d| = 0

C3(A4,A1) : d = C3(A4) – C3(A1)
d = 0.7 – 0.7
d = 0
d = 0 *then* H|d| = 0

C3(A2,A3) : d = C3(A2) – C3(A3)
d = 0.7 – 0.9
d = -0.2
d < 0 *then* H|d| = 0

C3(A3,A2) : d = C3(A3) – C3(A2)
d = 0.9 - 0.7
d = 0.2
d > 0 *then* H|d| = 1

C3(A2,A4) : d = C3(A2) – C3(A4)
d = 0.7 – 0.7
d = 0
d = 0 *then* H|d| = 0

C3(A4,A2) : d = C3(A4) – C3(A2)
d = 0.7 – 0.7
d = 0
d = 0 *then* H|d| = 0

C3(A3,A4) : d = C3(A3) – C3(A4)
d = 0.9 - 0.7
d = 0.2
d > 0 *then* H|d| = 1

C3(A4,A3) : d = C3(A4) – C3(A3)
d = 0.7 – 0.9
d = -0.2
d < 0 *then* H|d| = 0

4. Ease of Access Criteria Value (C4)

C4(A1,A2) : d = C4(A1) – C4(A2)
d = 0.9 - 0.7
d = 0.2
d > 0 *then* H|d| = 1

C4(A2,A1) : d = C4(A2) – C4(A1)
d = 0.7 – 0.9
d = -0.2
d < 0 *then* H|d| = 0

C4(A1,A3) : d = C4(A1) – C4(A3)
d = 0.9 - 0.7
d = 0.2
d > 0 *then* H|d| = 1

C4(A3,A1) : d = C4(A3) – C4(A1)
d = 0.7 – 0.9
d = -0.2
d < 0 *then* H|d| = 0

C4(A1,A4) : d = C4(A1) – C4(A4)
d = 0.9 – 0.9
d = 0
d = 0 *then* H|d| = 0

C4(A4,A1) : d = C4(A4) – C4(A1)
d = 0.9 – 0.9
d = 0
d = 0 *then* H|d| = 0

C4(A2,A3) : d = C4(A2) – C4(A3)
d = 0.7 – 0.7
d = 0
d = 0 *then* H|d| = 0

C4(A3,A2) : d = C4(A3) – C4(A2)
d = 0.7 – 0.7
d = 0
d = 0 *then* H|d| = 0

C4(A2,A4) : d = C4(A2) – C4(A4)
d = 0.7 – 0.9
d = -0.2
d < 0 *then* H|d| = 0

C4(A4,A2) : d = C4(A4) – C4(A2)
d = 0.9 - 0.7
d = 0.2
d > 0 *then* H|d| = 1

C4(A3,A4) : d = C4(A3) – C4(A4)
d = 0.7 – 0.9
d = -0.2
d < 0 *then* H|d| = 0

C4(A4,A3) : d = C4(A4) – C4(A3)
d = 0.9 - 0.7
d = 0.2
d > 0 *then* H|d| = 1

5. Response Speed Criteria Value (C5)

C5(A1,A2) : d = C4(A1) – C4(A2)
d = 0.7 – 0.9
d = -0.2
d < 0 *then* H|d| = 0

C5(A2,A1) : d = C4(A2) - C4(A1)

d = 0.9 - 0.7
d = 0.2
d > 0 *then* H|d| = 1

C5(A1,A3) : d = C4(A1) – C4(A3)
d = 0.7 - 0.5
d = 0.2
d > 0 *then* H|d| = 1

C5(A3,A1) : d = C4(A3) – C4(A1)
d = 0.5 – 0.7
d = -0.2
d < 0 *then* H|d| = 0

C5(A1,A4) : d = C4(A1) – C4(A4)
d = 0.7 – 0.7
d = 0
d = 0 *then* H|d| = 0

C5(A4,A1) : d = C4(A4) – C4(A1)
d = 0.7 – 0.7
d = 0
d = 0 *then* H|d| = 0

C5(A2,A3) : d = C4(A2) – C4(A3)
d = 0.9 – 0.5
d = 0.4
d > 0 *then* H|d| = 1

C5(A3,A2) : d = C4(A3) – C4(A2)
d = 0.5 – 0.9
d = -0.4
d < 0 *then* H|d| = 0

C5(A2,A4) : d = C4(A2) – C4(A4)
d = 0.9 - 0.7
d = 0.2
d > 0 *then* H|d| = 1

C5(A4,A2) : d = C4(A4) – C4(A2)
d = 0.7 – 0.9
d = -0.2
d < 0 *then* H|d| = 0

C5(A3,A4) : d = C4(A3) – C4(A4)
d = 0.5 – 0.7
d = -0.2
d < 0 *then* H|d| = 0

C5(A4,A3) : d = C4(A4) – C4(A3)
d = 0.7 - 0.5
d = 0.2
d > 0 *then* H|d| = 1

b. Step II: Calculate the Preferences Index

Alternative partner (A1,A2)
(A1,A2) = (1 + 0 + 0 + 1 + 0) = 2/5 = 0.4
Alternative partner (A2,A1)
(A2,A1) = (0 + 0 + 0 + 0 + 1) = 1/5 = 0.2
Alternative partner (A1,A3)
(A1,A3) = (1 + 1 + 0 + 1 + 1) = 4/5 = 0.8
Alternative partner (A3,A1)
(A3,A1) = (0 + 0 + 1 + 0 + 0) = 1/5 = 0.2
Alternative partner (A1,A4)
(A1,A4) = (0 + 1 + 0 + 0 + 0) = 1/5 = 0,2
Alternative partner (A4,A1)
(A4,A1) = (0 + 0 + 0 + 0 + 0) = 0/5 = 0
Alternative partner (A2,A3)
(A2,A3) = (1 + 1 + 0 + 0 + 1) = 2/5 =0.4
Alternative partner (A3,A2)
(A3,A2) = (0 + 0 + 1 + 0 + 0) = 1/5 =0.2
Alternative partner (A2,A4)
(A2,A4) = (0 + 1 + 0 + 0 + 1) = 2/5 = 0.4
Alternative partner (A4,A2)
(A4,A2) = (1 + 0 + 0 + 1 + 0) = 2/5 = 0.4
Alternative partner (A3,A4)
(A3,A4) = (0 + 0 + 1 + 0 + 0) = 1/5 = 0.2
Alternative partner (A4,A3)
(A4,A3) = (1 + 1 + 0 + 1 + 1) = 4/5 = 0.8

Following are the complete results of the Preference Index Value as shown in the following table:

Table 4. Preference Index Value

	A1	A2	A3	A4
A1	-	0.4	0.8	0.2
A2	0.2	-	0.4	0.4
A3	0.2	0.2	-	0.2
A4	0	0.4	0.8	-

c. Step III: Calculate PROMETHEE II

1.Calculate Leaving Flow

Then obtained:

A1 = 1 / (5-1) (0.4 + 0.8 + 0,2) = 0.35
A2 = 1 / (5-1) (0.2 + 0.4 + 0.4) = 0.25
A3 = 1 / (5-1) (0.2 + 0.2 + 0.2) = 0.15
A4 = 1 / (5-1) (0 + 0.4 + 0.8) = 0,3

2. Calculate Entering Flow

Then obtained:

A1 = 1 / (5-1) (0 + 0.2 + 0.2) = 0.1
A2 = 1 / (5-1) (0.4 + 0.2 + 0.4) = 0.25
A3 = 1 / (5-1) (0.8 + 0.4 + 0.8) = 0.5
A4 = 1 / (5-1) (0,2 + 0.4 + 0.2) = 0.2

Following are the complete results of Leaving Flow and Entering Flow calculations as shown in the following table:

Table 5. Leaving Flow and Entering Flow calculations

	Leaving Flow	Entering Flow
Facebook	0,35	0.1
Instagram	0.25	0.25
Line	0.15	0.5
Whatsapp	0.3	0.2

d. Step IV: Calculate Ranking

Was obtained:

A = 0.35 – 0.1 = 0.25
B = 0.25 – 0.25 = 0
C = 0.15 – 0.5 = -0.35
D = 0.3 – 0.2 = 0.1

Table 6. Rangking *PROMETHEE II*

	Net Flow	Ranking
Facebook	0.25	1
Instagram	0	3
Line	-0.35	4
Whatsapp	0.1	2

Based on net flow from table 5 above, then you can get a ranking of each social media alternative for online business. The alternative with the largest net flow is the alternative with the top ranking. If the alternative is minus, it means the value of entering flow is greater than leaving flow, this means that the comparison of some of the criteria is not better. As for the social media alternative for Facebook's highest online business (A1) with a net flow of 0.25.

4. Conclusion

Based on these results it can be concluded that the PROMETHEE II method can be applied in selecting the right social media for online businesses with the highest ranking results Facebook (A1) with net flow 0.25 and followed by whatsapp (A4) with net flow = 0.1.

References

[1] I. G. N. A. Lesmana, "Analisis Pengaruh Media Sosial Twitter terhadap Pembentukan Brand Attachment (Studi : PT . XL AXIATA)," Universitas Indonesia, 2012.

[2] W. Katrina, H. J. Damanik, F. Parhusip, D. Hartama, A. P. Windarto, and A. Wanto, "C.45 Classification Rules Model for Determining Students Level of Understanding of the Subject," *J. Phys. Conf. Ser.*, vol. 1255, no. 012005, pp. 1–7, 2019, doi: 10.1088/1742-6596/1255/1/012005.

[3] A. P. Windarto *et al.*, "Analysis of the K-Means Algorithm on Clean Water Customers Based on the Province," *J. Phys. Conf. Ser.*, vol. 1255, no. 1, 2019, doi: 10.1088/1742-6596/1255/1/012001.

[4] S. R. Ningsih, R. Wulansari, D. Hartama, A. P. Windarto, and A. Wanto, "Analysis of PROMETHEE II Method on Selection of Lecturer Community Service Grant Proposals," *J. Phys. Conf. Ser.*, vol. 1255, no. 1, pp. 1–7, 2019, doi: 10.1088/1742-6596/1255/1/012004.

[5] T. Imandasari, M. G. Sadewo, A. P. Windarto, A. Wanto, H. O. Lingga Wijaya, and R. Kurniawan, "Analysis of the Selection Factor of Online Transportation in the VIKOR Method in Pematangsiantar City," *J. Phys. Conf. Ser.*, vol. 1255, no. 012008, pp. 1–7, 2019, doi: 10.1088/1742-6596/1255/1/012008.

[6] K. Fatmawati *et al.*, "Analysis of Promothee II Method in the Selection of the Best Formula for Infants under Three Years," *J. Phys. Conf. Ser.*, vol. 1255, no. 1, 2019, doi: 10.1088/1742-6596/1255/1/012009.

[7] D. R. Sari, N. Rofiqo, D. Hartama, A. P. Windarto, and A. Wanto, "Analysis of the Factors Causing Lazy Students to Study Using the ELECTRE II Algorithm," *J. Phys. Conf. Ser.*, vol. 1255, no. 1, 2019, doi: 10.1088/1742-6596/1255/1/012007.

[8] Y. Malhotra, "Expert systems for knowledge management: crossing the chasm between information processing and sense making," *Expert Syst. Appl.*, vol. 20, no. 1, pp. 7–16, 2001, doi: 10.1016/S0957-4174(00)00045-2.

[9] Sumijan, A. P. Windarto, A. Muhammad, and Budiharjo, "Implementation of Neural Networks in Predicting the Understanding Level of Students Subject," *Int. J. Softw. Eng. Its Appl.*, vol. 10, no. 10, pp. 189–204, 2016.

[10] Budiharjo, T. Soemartono, A. P. Windarto, and T. Herawan, "Predicting tuition fee payment problem using backpropagation neural network model," *Int. J. Adv. Sci. Technol.*, vol. 120, pp. 85–96, 2018, doi: 10.14257/ijast.2018.120.07.

[11] Budiharjo, T. Soemartono, A. P. Windarto, and T. Herawan, "Predicting School Participation in Indonesia using Back-Propagation Algorithm Model," *Int. J. Control Autom.*, vol. 11, no. 11, pp. 57–68, 2018.

[12] A. Mulyanto and Abdul Haris, "Penerapan Metode Fuzzy Tsukamoto Untuk Menentukan Jumlah Jam Overtime Pada Produksi Barang di PT Asahi Best Base Indonesia (ABBI) Bekasi," *J. Inform. SIMANTIK Vol.1,* vol. 1, no. 1, pp. 1–11, 2016.

[13] D. N. Batubara, D. R. Sitorus P, and A. P. Windarto, "Penerapan Metode PROMETHEE II Pada Pemilihan Situs Travel Berdasarkan Konsumen," *J. Sisfokom (Sistem Inf. dan Komputer)*, vol. 8, no. 1, pp. 46–52, 2019, doi: 10.32736/sisfokom.v8i1.598.

[14] K. F. Irnanda, F. N. Arifah, M. R. Raharjo, A. Arifin, and A. P. Windarto, "The selection of Calcium Milk Products that are appropriate for advanced age using PROMETHEE II Algorithm," *J. Phys. Conf. Ser.*, vol. 1381, no. 1, 2019, doi: 10.1088/1742-6596/1381/1/012070.

AUTHOR INDEX